Tropical Zooplankton

Developments in Hydrobiology 23

Series editor
H. J. Dumont

Tropical Zooplankton

Edited by
H. J. Dumont and J. G. Tundisi

Reprinted from Hydrobiologia, vol. 113 (1984)

1984 **DR W. JUNK PUBLISHERS**
A MEMBER OF THE KLUWER ACADEMIC PUBLISHERS GROUP
THE HAGUE / BOSTON / LANCASTER

Distributors

For the United States and Canada: Kluwer Boston, Inc., 190 Old Derby Street, Hingham, MA 02043, USA
for the UK and Ireland: Kluwer Academic Publishers, MTP Press Limited, Falcon House, Queen Square, Lancaster LA1 1RN, England
for all other countries: Kluwer Academic Publishers Group, Distribution Center, P.O. Box 322, 3300 AH Dordrecht, The Netherlands

Library of Congress Cataloging in Publication Data

Main entry under title:

Tropical zooplankton.

(Developments in hydrobiology ; 23)
1. Freshwater zooplankton--Tropics--Congresses.
I. Dumont, H. J. (Henry J.) II. Tundisi, J. G.
III. Hydrobiologia. IV. Series.
QL143.T67 1984 592 84-826

ISBN 978-90-481-8522-1

Cover design: Max Velthuijs

Top left and right, two views of the participants to the symposium in the meeting room; bottom, closing ceremony.

Preface

Our knowledge of the limnology of the waters situated, roughly, between the tropics of cancer and of capricorn, has depended for a long time on the expedition-approach, and therefore developed in a rather irregular, haphazard way, with the personal incentive of a small number of individuals as the main driving force.

Things slowly started to change in the 1950s, and at an accelerating rate in the 1960s and 1970s. The IBP, and later the SCOPE and MAB programs, whatever their shortcomings are or may have been, promoted in-depth research of a small number of tropical lakes. For one thing, they showed the need for the creation of *in situ* limnological research institutes.

When, in the 1970s, limnological research facilities or their nuclei began to appear in the tropical zones of all continents, an interesting phenomenon occurred: while most of the young native limnologists had received their training in advanced centres or courses held in the temperate (and developed) climatic zones, quite a few of their former supervisors or their associates became interested in warm lakes and rivers as well, using the new or newly expanded local institutes. We are, today, still in this phase and it is, apparently, expanding even further. Although not all experiments of this kind lead to happy marriages, a few were quite successful, and several papers contained in the present volume are hoped to reflect this.

Because tropical limnology has now become as fashionable as tropical tourism, there has been a boom in the number of manuscripts offered for publication. The number of cryptic local journals that accommodate some of the weaker of these has increased too, and consequently, the need for a quality control of some kind has become acute.

Not only should one stimulate research, training and motivation (quite a few newly trained limnologists are drained away into bureaucracy on return to their home countries), but also the most immediate needs and priorities should be defined, and duplication of efforts prevented. To achieve this, some amount of organisation and coordination on a world scale is necessary. As a minimum requirement, information should circulate: the different workers and groups active in tropical limnology should be aware of each other's activities.

When at the SIL-conference in Kyoto in 1980 a special session was devoted to tropical limnology, both its success and its heterogeneity were striking. It was evident to all who were present that more and diversified efforts were needed to promote this relatively new field. It was thus decided to greatly emphasize tropical limnology at the SIL-conference in Lyon in 1983, and also to stimulate smaller meetings on selected topics, preferably to be held in a tropical country.

When the editors of the present volume met in July 1981 in the USSR during a SCOPE-meeting on problems of shallow lakes, the idea of having a symposium on tropical zooplankton was born.

Limnology in South America, and especially in Brazil, is developing rapidly, and therefore the latter country seemed extremely appropriate to host this meeting. It was agreed upon that the term zooplankton should be broadly taken, to accommodate not only the limnetic community, but also the littoral, and groups that may occur as planktonts only sporadically, or in the larval stage only. Contributors were invited to focus on taxonomy, ecology, distribution, and managemental aspects of the zooplankton community.

About 60 scientists replied to our first circular, and the final attendance was in that order, with approximately 35 oral and poster presentations of original research work, time for formal and informal discussions, and half a day of workshop meetings in small discussion groups. While the meetings, by their location, certainly highlighted South American limnology, the four tropical continents were all represented.

A midweek excursion took the participants to the reservoir of Barra Bonita in the centre of Sao Paulo state, and a small group stayed for a week in the magnificent environment of the Pantanal lakes of Matto Grosso, a truly unforgettable experience.

Although Dr W. Junk Publishers agreed to publish the proceedings of our symposium in Hydrobiologia/ Developments in Hydrobiology, the whole undertaking would not have been possible without the financial assistance of CAPES (Coordonacao de Ajerjeicamento do Pessoal de Nivel Superior), FAPESP (Fundacao de Amparo a Pesquisa do Estado de Sao Paulo), and CNPq (Conselho Nacional de Desenvolvimento Cientifico e Tecnologico), who provided general support and travel expenses.

Facilities provided by the University of Sao Carlos should also be acknowledged. Brazilian limnology first developed in the Amazon region under the auspices of the Max Planck Gesellschaft. In southern Brazil, limnology has always been connected with sanitary engineering and water quality control. However, in the last decade, a considerable effort was made at the Federal University of Sao Carlos to organize limnological research with an aim to creating a centre of excellence in this field, and to create possibilities for training people at an advanced level. For these reasons, Sao Carlos seemed the most appropriate site for the symposium to take place. The local organising committee did an outstanding job in organising the sessions, and Mrs N. Wendell Magalhaes was a perfect guide in the post-symposium excursion.

Finally, we thank all participants in the group discussions for permitting us to use the results of their deliberations in the epilogue to this volume.

Caraguatatuba, December 1982 Henri J. Dumont & Jose G. Tundisi

Contents

Structure and dynamics of zooplankton communities, Alligator Rivers Region, N.T., Australia

R. D. Tait[1], R. J. Shiel[2] & W. Koste[3]
[1] *Pancontinental Mining Ltd., Jabiluka Division, Environmental Laboratory, Northern Territory, Australia*
Present address: Esso Australia Ltd., P.O. Box 4047, Sydney 2001, Australia
[2] *Department of Biology, University of Waterloo, Waterloo, Ontario, N2L 3G1, Canada*
Present address: Department of Botany, University of Adelaide, 5001 South Australia
[3] *Ludwig Brill Strasse 5, Quakenbrück D-4570, Federal Republic of Germany*

Keywords: tropical zooplankton, Rotifera, Cladocera, Copepoda, tropical Australia

Abstract

Biological monitoring in the Alligator Rivers region, northern Australia, provides baseline ecological information to assess the impact of uranium mining and milling and settlement in the area. Spatial and temporal variations 1978–1980 in zooplankton communities of the Magela Creek, a tributary of the East Alligator River, are described. Extremely diverse plankton assemblages occur late in the wet season (Dec.–May), with up to 80 taxa of rotifers and microcrustacea in some billabongs (= ox-bows), while there is a decrease in diversity but increase in population density as the dry season progresses. Natural fluctuations in water quality may be extreme, and limiting to plankters common elsewhere in the tropics. The plankton is composed largely of littoral or epiphytic taxa, with endemic species in all groups.

Introduction

A continuing water quality and biological monitoring programme in the Alligator Rivers region of northern Australia is providing basic ecological information against which to assess the impact of mining and milling of uranium in the catchment. Most of the limited data on tropical freshwater zooplankton from the Australian continent have derived from this area (see review by Dussart *et al,* 1984).

This report is drawn from a two year study of the species composition and population dynamics of plankton and littoral microfauna and billabongs (= ox-bows, cut off meanders or pools isolated from a river) on the Magela Creek, particularly the microcrustacea (Tait, 1982). A preliminary report documented extreme natural perturbations in the zooplankton (Tait, 1981), and seasonal extremes in rotifer communities were described by Koste (1981) and Koste & Shiel (1982). There is no comparable published information on floodplain plankton communities for the continent, and only a single study of littoral microcrustacea in a billabong (Shiel, 1976).

To extend these reports, we summarise seasonal fluctuations in water quality in the study area, provide a checklist of all taxa (excluding Protozoa) recorded in the zooplankton, and give examples of spatial and seasonal variation in community composition. Data on the littoral microfauna will be published elsewhere (Shiel & Tait, in prep.).

The study area

Magela Creek (Fig. 1) is an ephemeral tributary of the East Alligator River 250 km east of Darwin, N.T. The region receives 97% of the annual rainfall ($\bar{X}$ 1350 mm) Oct.–May, during which the floodplain may be submerged to a depth of 2–3 m. Between May and December, the floodplain dries out

Hydrobiologia 113, 1–13 (1984).

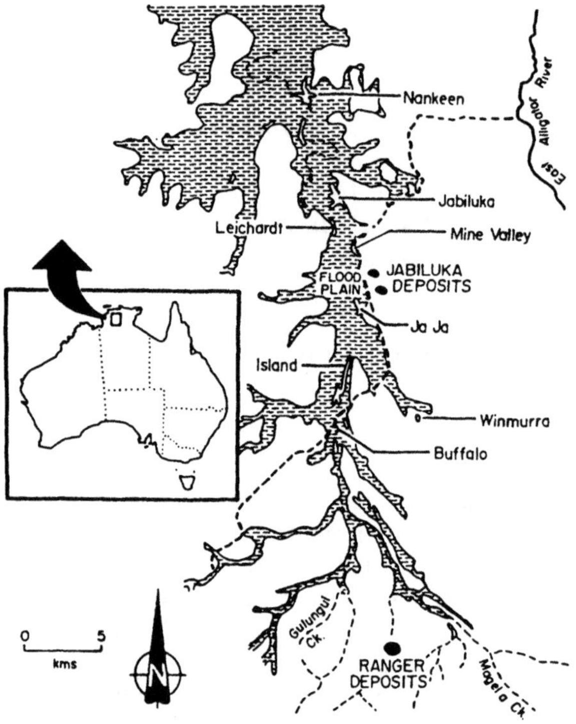

Fig. 1. Billabongs sampled 1978–80, Magela Creek, N.T., showing proximity to uranium mining sites.

progressively, concentrating aquatic life in the remaining billabongs.

Eight billabongs on a 30-km section of the floodplain downstream of two uranium ore deposits, Ranger and Jabiluka, were sampled monthly Nov. 1978–Oct. 1980. Three habitat categories are recognized (cf. Hart & McGregor, 1980), i.e. channel billabongs (Buffalo, Island), in the main creek channel, which have sandy substrates and clear water to 6 m deep. Floodplain billabongs (Nankeen, Jabiluka, Mine Valley, Leichhardt and Ja Ja), 3-5 m deep, with muddy bottoms, are subject to high turbidities over the dry season when used by buffalo and water birds (cf. also Recher & Holmes, 1982). A small and shallow backflow billabong (Winmurra), on a tributary catchment, is filled initially from its own stream, and later by backflow of water, sediments and organic detritus from Magela Creek in high flows. Some 200–400 km^2 of floodplain may be submerged in the wet season (Hart & McGregor, 1980), and all the billabongs are inundated. Physical differences between the habitats have important implications for water quality and maintenance of the biotic community during the dry season.

Water quality of the Magela Creek billabongs

Seasonal fluctuations in water levels approximate those shown in Fig. 2a. Only Winmurra dried out during the study period (late 1979). Water temperatures (e.g. Fig. 2b) generally were within the range 22 °C (July) to 38 °C (November) reported from upstream billabongs (Marchant, 1980). Temperature profiles from all billabongs indicated thermal stratification at some time during the year. Daily stratification was particularly pronounced Nov–March, when ambient air temperatures were highest. Similar temperature regimes were recorded in upstream billabongs by Walker & Tyler (1979).

Surface and bottom dissolved oxygen concentrations (DO) (Fig. 2c) showed no clear trends in any billabong. Surface DO ranged from 0.6 mg l^{-1} to 9.5 mg l^{-1}. Of the billabong series, only Leichhardt had significant oxygen depletion in bottom waters for much of the year (to 0.10 mg l^{-1} in July).

All billabongs had lowest conductivity in the wet season (cf. Fig. 2d) increasing to a maximum in December. Conductivity ranges were greater in the floodplain than the channel billabongs, e.g. Mine Valley reached 798 μS cm^{-1} in the 1979 dry season, a result of substrate differences, more pronounced concentration effects, and groundwater ingress.

pH in all billabongs (e.g. Fig. 2e) was less than 7.1, commonly 5.0-5.6 in the wet season and 6.2-6.9 at the beginning of the dry season. One acid flush in Ja Ja (pH 3.3-3.7, Jan. 1980) resulted when local runoff after rain dissolved a sulphate crust around the billabong margin (Brown *et al*, 1983).

Oxidation reduction potential (e.g. Fig. 2f) generally was within the range +0.3 – +0.5 mV.

Turbidities (e.g. Fig. 3a) were lowest in the wet season (<2 NTU) and increased during the dry season (280 NTU recorded in Nankeen in Dec.), with higher turbidities from fine clay resuspension in the floodplain billabongs. Channel billabongs, with sandy sediments, had less abiogenic turbidity, but had phytoplankton blooms during the dry season.

Total organic carbon (cf. Fig. 3b) increased over the dry season, with higher levels in the backflow and floodplain billabongs than in the channel billabongs.

Ionic concentrations showed sodium/chloride dominance, with increases in magnesium/sulphate in the dry season.

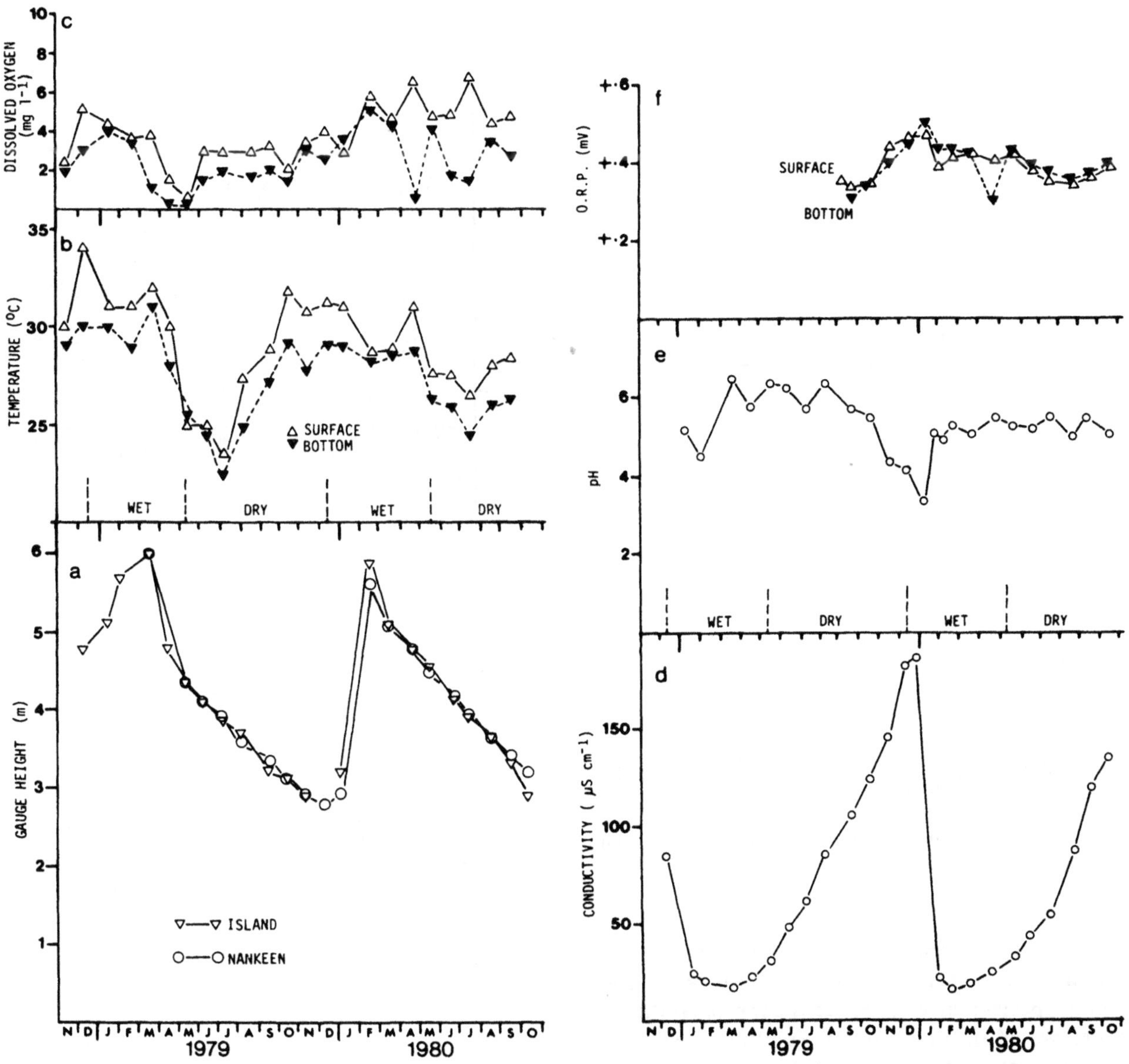

Fig. 2. Examples of seasonal variation 1978–80 in (a) water level (Island, Nankeen) and for Ja Ja billabong: (b) water temperature, (c) dissolved oxygen, (d) conductivity, (e) pH, (f) oxidation-reduction potential.

Nutrients (N and P) (Fig. 4a–e) were variable between habitats, but in general were lowest in the wet season, highest in the dry. The highest total-N recorded was 24.8 mg l^{-1} (Mine Valley) and the highest total-P was 2.5 mg l^{-1} (Winmurra prior to drying out in Nov. 1979). The relationship between nutrients and phytoplankton in the Magela Creek billabongs is considered further by Burgman & Tait (in prep.). Nutrient levels generally are indicative of mesotrophy, with seasonal eutrophy.

The effects on the zooplankton of these seasonal perturbations in water quality, which may be extreme and abrupt (cf. Tait, 1981), are considered in the discussion.

Zooplankton of the Magela Creek billabongs

All taxa recorded in plankton tows from Magela Creek billabongs are listed in Table 1 (227 species of

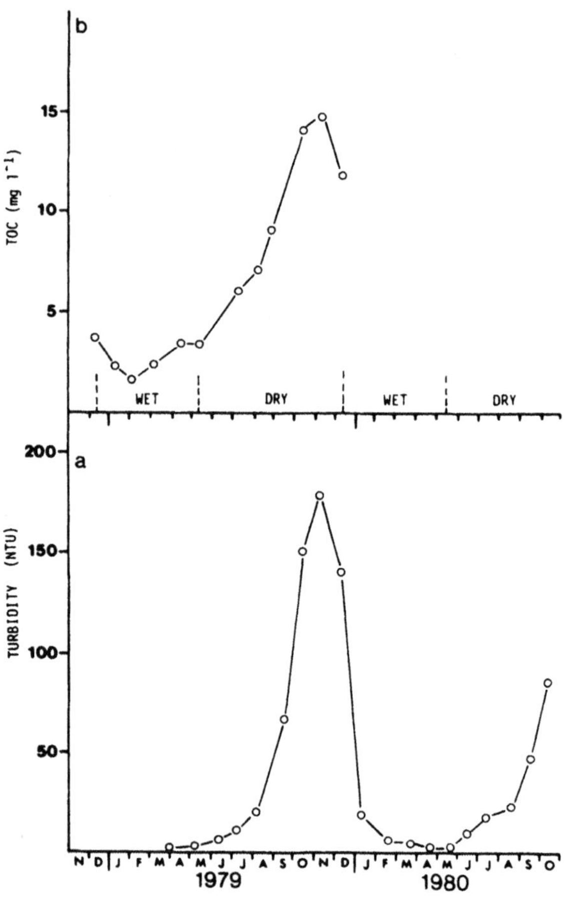

Fig. 3. Seasonal variation (Ja Ja) in (a) turbidity, and (b) total organic carbon.

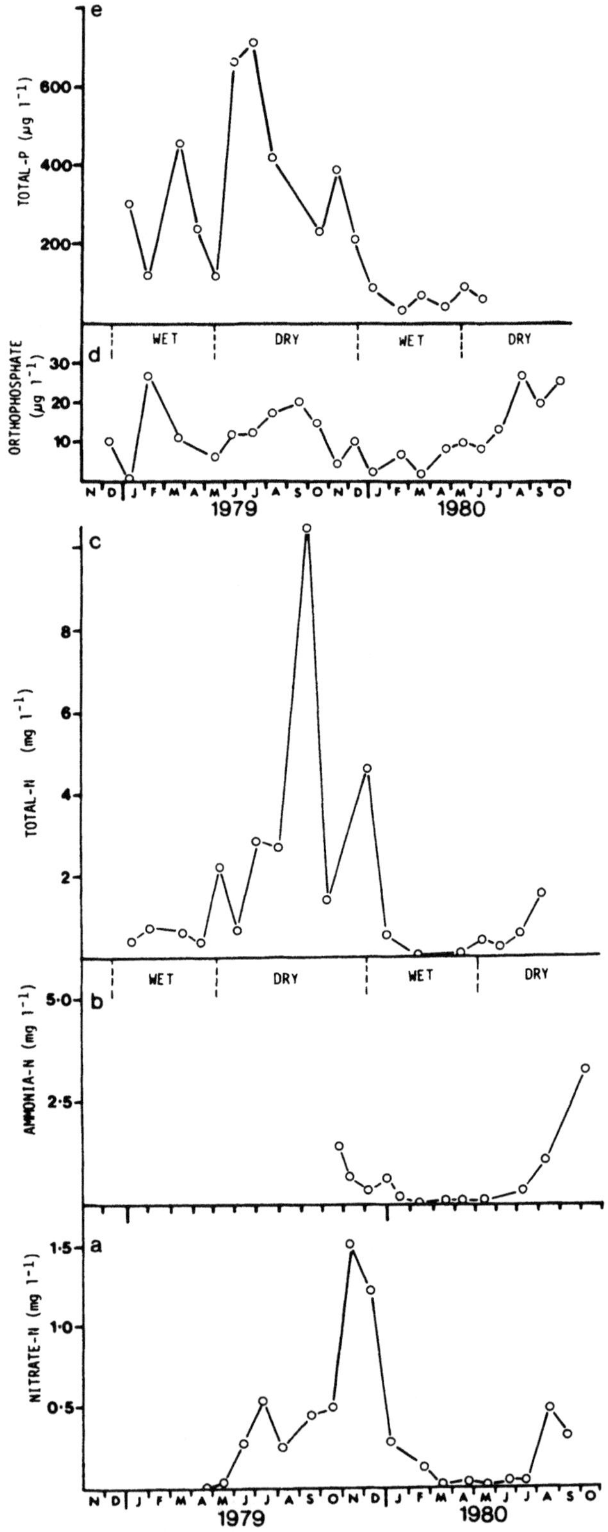

Fig. 4. Seasonal variation (Ja Ja) in (a) Nitrate-N, (b) Ammonia-N, (c) total-N, (d) Orthophosphate, and (e) total-P.

Rotifera, 14 of Copepoda, 35 of Cladocera and 5 of Ostracoda). A large proportion of littoral or epiphytic species is evident, reflecting the instability of 'open water' in these habitats resulting from level fluctuations, and the strong influence of marginal emergent and submergent macrophytes on plankton community composition.

Features of the species assemblages present in tropical Australia are summarized by Dussart *et al* (1984), who also review the literature. Additional comments here derive from the 1978–80 study (Tait, 1982), which was primarily on the microcrustacea.

Rotifera

Most of the rotifers listed in Table 1 were rare in the plankton, occurring on single occasions as one

Table 1. Systematic list of Rotifera and Microcrustacea recorded from the Alligator Rivers region, N.T.: (* = recorded only from region, E = endemic).

ROTIFERA
Bdelloida
Fam. Habrotrochidae
1. **Habrotrocha angusticollis* (Murray) 1905

Fam. Philodinidae
2. *Dissotrocha aculeata* (Ehrenberg) 1832
3. *D. macrostyla* (Ehrenberg) 1838

Monogononta
Fam. Epiphanidae
4. *Epiphanes clavulata* (Ehrenberg) 1832

Fam. Brachionidae
5. *Platyias quadricornis* (Ehrenberg) 1832
6. *Brachionus angularis* Gosse 1851
7. *B. bidentata* Anderson 1889
8. *B. bidentata testudinarius* Jakubski 1912
9. *B. budapestinensis* (Daday) 1885
10. **B. caudatus personatus* Ahlstrom 1940
11. *B. dichotomus* Shephard 1911 (E)
12. *B. dichotomus reductus* Koste & Shiel 1980 (E)
13. *B. falcatus* Zacharias 1898
14. **B. falcatus reductus* Koste & Shiel 1982 (E)
15. *B. lyratus* Shephard 1911 (E)
16. *B. patulus* Muller 1786
17. *B. quadridentatus* Hermann 1783
18. *B. quadridentatus melheni* Barrois & Daday 1894
19. *B. quadridentatus* undescribed dwarf form (E)
20. **B. urceolaris sericus* (Rousselet) 1907
21. **B. urceolaris sessilis* (Varga) 1951
22. *Keratella cochlearis* (Gosse) 1851
23. *K. javana* Hauer 1937
24. *K. lenzi* Hauer 1953
25. *K. procurva* (Thorpe) 1891
26. *K. tropica* (Apstein) 1907
27. *Anuraeopsis coelata* De Beauchamp 1932
29. *A. navicula* Rousselet 1910

Fam. Euchlanidae
30. **Beauchampiella eudactylota* (Gosse) 1886
31. *Euchlanis dilatata* Ehrenberg 1832
32. *E. incisa* Carlin 1939
33. *E. meneta* Myers 1930
34. *E. oropha* Gosse 1887
35. *E. parva* Rousselet 1892
36. *E. triquetra* Ehrenberg 1838
37. **Dipleuchlanis propatula* (Gosse) 1886
38. **D. propatula macrodactyla* (Hauer) 1965

Fam. Mytilinidae
39. **M. bisulcata* Lucks 1912
40. **Mytilina crassipes* (Lucks) 1912
41. **M. macracantha* (Gosse) 1886
42. *M. ventralis* (Ehrenberg) 1832
43. *M. ventralis brevispina* (Ehrenberg) 1832
44. **M. ventralis longidactyla* Wulfert 1965
45. *Lophocharis salpina* (Ehrenberg) 1834

Fam. Trichotridae
46. *Trichotria tetractis* (Ehrenberg) 1830
47. *T. truncata* (Whitelegge) 1889
48. *T. tetractis similis* Stenroos 1898
49. **Macrochaetus collinsi* (Gosse) 1867
50. **M. danneeli* Koste & Shiel 1982
51. *M. subquadratus* (Perty) 1850

Fam. Colurellidae
52. *Colurella obtusa* (Gosse) 1886
53. *C. uncinata* (Muller) 1773
54. *Squatinella mutica* (Ehrenberg) 1832
55. *Lepadella acuminata* (Ehrenberg) 1834
56. *L. apsicora* Myers 1934
57. **L. apsida* Harring 1916
58. **L. benjamini* Harring 1916
59. *L. ehrenbergi* (Perty) 1850
60. **L. elliptica* Wulfert 1939
61. *L. latusinus* (Hilgendorf) 1899
62. **L. lindaui* Koste 1981
63. **L. minorui* Koste 1981
64. **L. monodactyla* Berzins 1960
65. *L. ovalis* (Muller) 1786
66. *L. patella* (Muller) 1773
67. *L. quinquecostata* (Lucks) 1912
68. *L. rhomboides* (Gosse) 1886
69. **L. triptera* (Ehrenberg) 1830
70. **L. williamsi* Koste & Shiel 1980
71. *Heterolepadella ehrenbergi* (Perty) 1950
72. **H. heterodactyla* Fadeew 1925

Fam. Lecanidae
73. *Lecane acronycha* Harring & Myers 1926
74. **L. aculeata* (Jakubski) 1912
75. *L. batillifer* (Murray) 1913
76. *L. bulla* (Gosse) 1851
77. *L. closterocerca* (Schmarda) 1895
78. *L. crenata* (Harring) 1913
79. **L. crepida* Harring 1914
80. **L. curvicornis* (Murray) 1913
81. **L. curvicornis nitida* (Murray) 1913
82. **L. decipiens* (Murray) 1913
83. **L. doryssa* Harring 1914
84. *L. flexilis* (Gosse) 1886
85. **L. furcata* (Murray) 1913
86. **L. grandis* (Murray) 1913
87. *L. hamata* (Stokes) 1896
88. *L. hornemanni* (Ehrenberg) 1834
89. **L. inopinata* Harring & Myers 1926
90. **L. leontina* (Turner) 1892
91. **L. levistyla* (Olafson) 1917
92. **L. ludwigi* (Eckstein) 1893
93. *L. luna* (Muller) 1776
94. *L. lunaris* (Ehrenberg) 1832
95. *L. lunaris crenata* (Harring) 1923
96. **L. monostyla* (Daday) 1897
97. **L. nana* (Murray) 1913
98. **L. nodosa* Hauer 1937
99. **L. obtusa* (Murray) 1913

Table 1. (Continued).

100. •*L. papuana* (Murray) 1913
101. •*L. pertica* Harring & Myers 1926
102. •*L. pusilla* Harring 1914
103. •*L. pyriformis* (Daday) 1905
104. *L. quadridentata* (Ehrenberg) 1832
105. •*L. scutata* (Harring & Myers) 1926
106. •*L. signifera* (Jennings) 1896
107. *L. signifera ploenensis* (Voigt) 1902
108. *L. stenroosi* (Meissner) 1908
109. *L. stichea* Harring 1913
110. •*L. tenuiseta* Harring 1914
111. •*L. tethis* (Harring & Myers) 1926
112. •*L. unguitata* (Fadeew) 1925
113. •*L. ungulata* (Gosse) 1887
114. *L. ungulata australiensis* Koste 1979

Fam. Proalidae
115. *Proales decipiens* (Ehrenberg) 1831
116. •*P. fallaciosa* Wulfert 1937

Fam. Notommatidae
117. *Tetrasiphon hydrocora* Ehrenberg 1840
118. *Scaridium longicaudum* (Muller) 1786
119. •*Monommata actices* Myers 1930
120. •*M. grandis* Tessin 1890
121. *M. longiseta* (Muller) 1786
122. •*M. maculata* Harring & Myers 1924
123. •*Taphrocampa annulosa* Gosse 1851
124. *T. selenura* Gosse 1887
125. •*Eosphora thoides* Wulfert 1935
126. •*Resticula melandocus* (Gosse) 1887
127. *Notommata collaris* Ehrenberg 1832
128. *N. copeus* Ehrenberg 1834
129. *N. pachyura* (Gosse) 1886
130. •*N.* cf. *triangularis* (Kirkman) 1906
131. *N. tripus* Ehrenberg 1838
132. •*Cephalodella apocolea* Myers 1924
133. *C. forficula* (Ehrenberg) 1832
134. *C. gibba* (Ehrenberg) 1832
135. •*C. intuta* Myers 1924
136. •*C. misgurnis* Wulfert 1937
137. *C. mucronata* Myers 1924
138. •*C. panarista* Myers 1924
139. •*C. sterea* (Gosse) 1887
140. •*C. tantilloides* Hauer 1935
141. *C. tinca* Wulfert 1937

Fam. Trichocercidae
142. •*Trichocerca bicristata* (Gosse) 1887
143. •*T. birostris* (Minkiwicz) 1900
144. *T. brachyura* (Gosse) 1851
145. *T. capucina* (Wierzejski & Zacharias) 1893
146. *T. cavia* (Gosse) 1889
147. •*T. collaris* (Rousselet) 1896
148. •*T. chattoni* (De Beauchamp) 1907
149. *T. elongata* (Gosse) 1886
150. *T. flagellata* Hauer 1937
151. •*T. fusiformis* Levander 1894
152. *T. insignis* (Herrick) 1885
153. •*T. insulana* Hauer 1937
154. •*T. intermedia* (Stenroos) 1898
155. •*T. jenningsi* Voigt 1957
156. *T. longiseta* (Schrank) 1802
157. *T. mus* Hauer 1938
158. •*T. myersi* (Hauer) 1931
159. *T. porcellus* (Gosse) 1886
160. *T. pusilla* (Jennings) 1903
161. *T. rattus* (Muller) 1776
162. *T. rattus carinata* (Ehrenberg) 1830
163. *T. similis* (Wierzejski) 1893
164. *T. stylata* (Gosse) 1851
165. *T. tenuior* (Gosse) 1886
166. *T. tigris* (Muller) 1786
167. *T. weberi* (Jennings) 1903

Fam. Gastropodidae
168. *Gastropus minor* (Rousselet) 1892
169. •*G. stylifer* Imhof 1891
170. *Ascomorpha ecaudis* (Perty) 1850
171. *A. ovalis* (Carlin) 1943
172. *A. saltans* Bartsch 1870

Fam. Synchaetidae
173. *Synchaeta longipes* (Gosse) 1887
174. *S. pectinata* Ehrenberg 1832
175. •*Polyarthra minor* Voigt 1904
176. *P. vulgaris* Carlin 1943

Fam. Asplanchnidae
177. *Asplanchnopus multiceps* (Schrank) 1793
178. *Asplanchna brightwelli* Gosse 1850
179. *A. sieboldi* (Leydig) 1854

Fam. Dicranophoridae
180. •*Dicranophorus claviger australiensis* Koste & Shiel 1980
181. •*D. epicharis* Harring & Myers 1928
182. *D. forcipatus* (Muller) 1786
183. •*D. grandis* (Ehrenberg) 1832
184. •*D. halbachi* Koste 1981
185. •*D. hercules* Wisniewski 1932
186. *Encentrum* spec.
187. •*Albertia naidis* Bousfield 1886
188. •*Balatro calvus* Calparide 1867

Fam. Testudinellidae
189. •*Testudinella amphora* Hauer 1938
190. *T. emarginula* (Stenroos) 1898
191. •*T. greeni* Koste 1981
192. •*T. incisa* (Ternetz) 1892
193. •*T. insinuata* Hauer 1938
194. •*T. parva* (Ternetz) 1892
195. *T. patina* (Hermann) 1783
196. •*T. semiparva* Hauer 1938
197. •*T. tridentata* Smirnov 1931
198. •*T. walkeri* Koste & Shiel 1980
199. *Pompholyx complanata* Gosse 1851

Fam. Flosculariidae
200. *Beauchampia crucigera* (Dutrochet) 1812
201. *Limnias ceratophylli* Schrank 1803

Table 1. (Continued).

202. *L. melicerta* Wiesse 1848
203. *Floscularia janus* (Hudson) 1881
204. *F. melicerta* (Ehrenberg) 1832
205. *F. ringens* (Linnaeus) 1758
206. **Octotrocha speciosa* Thorpe 1893
207. **Ptygura brachiata* (Hudson) 1886
208. *P. crystallina* (Ehrenberg) 1834
209. *P. melicerta* (Ehrenberg) 1832
210. *P. pilula* (Cubitt) 1872
211. *P. tacita* Edmondson 1940
212. **Sinantherina spinosa* (Thorpe) 1893
213. *Lacinularia flosculosa* (Muller) 1758

Fam. Conochilidae
214. *Conochilus dossuarius* (Hudson) 1885
215. *C. unicornis* Rousselet 1892

Fam. Hexarthridae
216. *Hexarthra intermedia* (Wisniewski) 1929
217. *H. mira* (Hudson) 1871

Fam. Filiniidae
218. *Filinia australiensis* Koste 1980
219. *F. longiseta* (Ehrenberg) 1834
220. *F. longiseta limnetica* (Zacharias) 1893
221. *F. opoliensis* (Zacharias) 1898
222. *F. passa* (Muller) 1786
223. *F. pejleri* Hutchinson 1964

Fam. Trochosphaeridae
224. *Trochosphaera equatorialis* Semper 1872

Fam. Collothecidae
225. *Collotheca campanulata* (Dobie) 1849
226. *C. trilobata* (Collins) 1872

Fam. Atrochidae
227. *Cupelopagis vorax* (Leidy) 1857

CLADOCERA
Fam. Sididae
228. *Diaphanosoma excisum* Sars 1885
229. *D. sarsi* Richard 1894
230. *D. unguiculatum* Gurney 1927 (E)

Fam. Chydoridae
231. *Pleuroxus aduncus* (Jurine 1820)
232. *Alonella exigua* (Lilljeborg 1853)
233. *Chydorus sphaericus* s.l. Muller 1785 including *faviformis* type.
234. *C. eurynotus* Sars 1901
235. *Ephemeroporus barroisi* (Richard 1894)
236. *Dunhevedia crassa* King 1853
237. *Dadaya macrops* (Daday 1898)
238. *Alona costata* Sars 1862
239. *A. davidi* Richard 1895
240. *A. guttata* Sars 1862
241. *Kurzia longirostris* (Daday 1898)
242. *Camptocercus australis* Sars 1896 (E)
243. *Graptoleberis testudinaria* Sars 1901
244. *Leydigia leydigi* (Schodler 1863)
245. *Biapertura affinis* (Leydig 1860)
246. *B.* cf. *duodonata* (Henry 1922) (E)
247. *B. karua* (King 1852)
248. *Euryalona orientalis* (Daday 1898)
249. *Australochydorus aporus* Smirnov & Timms 1983 (E)

Fam. Macrothricidae
250. *Ilyocryptus brevidentatus* Ekman 1905
251. *I. sordidus* Lieven 1848
252. *Macrothrix spinosa* King 1853
253. *M. breviseta* Smirnov 1976 (E)
254. *Echinisca* spec.

Fam. Moinidae
255. *Moina micrura* Kurz 1874
256. *M. australiensis* Sars 1896 (E)

Fam. Bosminidae
257. *Bosminopsis dietersi* Richard 1895
258. *Bosmina meridionalis* Sars 1903

Daphniidae
259. *Simocephalus acutirostratus* (King 1853)
260. *S. latirostris* Stingelin 1906
261. *Ceriodaphnia dubia* Richard 1895
262. *C. cornuta* Sars 1886

OSTRACODA
263. n. gen., n. spec. A
264. n. gen., n. spec. B
265. n. gen., n. spec. C
266. n. gen., n. spec. D
267. *Limnocythere* spec.

COPEPODA
Calanoida
Fam. Diaptomidae
268. *Eodiaptomus lumholtzi* (Sars 1889)

Fam. Centropagidae
269. *Calamoecia ultima* (Brehm 1960)

Cyclopoida
Fam. Cyclopidae
270. *Microcyclops dengizicus* Lepeshkin 1900
271. *M. varicans* Sars 1860
272. *Microcyclops* spec.
273. *Ectocyclops medius* Kiefer 1930
274. *Mesocyclops thermocyclopoides australiensis* Kiefer 1981
275. *M. decipiens* Kiefer 1929
276. *M. hyalinus* (Rehberg 1880)
277. *M. notius* Kiefer 1981
278. *M.* cf. *annae* Kiefer 1930
279. *Tropocyclops* spec.
280. *Eucyclops agilis* (Koch 1839)

Harpacticoida
Fam. Canthocamptidae
281. cf. *Canthocamptus* spec.

or two specimens. Abundant species were *A. brightwelli, B. falcatus, F. opoliensis, K. lenzi* and *K. tropica,* with densities up to 70 000 l^{-1} recorded from some billabongs, among the highest population densities of Rotifera yet recorded. Large populations of other species were restricted to single billabongs, commonly under extreme conditions when community diversity was depressed. *B. urceolaris sericus*, for example, known from strongly acidified (sulphate) lakes (ph 2.8–4.0, cf. Koste, 1978) reached dense populations only in Mine Valley. In general, the rotifer assemblages were acidophilic; species common elsewhere in the tropics, particularly brachionids, were sparse or absent. Aberrant or dwarf forms of several *Brachionus* taxa recorded (cf. Shiel, 1983) were possibly a response to the extreme biotopes. Contrasts between the rotifer community species composition and diversity in the wet and the dry season are described by Koste (1981) and Koste & Shiel (1982).

Copepoda

The Cyclopoida of northern Australia are poorly known. Cyclopoids were represented by at least 11 species (Table 1), only two of which *(M. notius* and *M. varicans)* were abundant. A large carnivorous species recorded from Nankeen and Winmurra in July 1979 (Fig. 5) resembles very closely *Mesocyclops annae* Kiefer, previously recorded only from Madagascar (Kiefer, 1930, 1981). Several other cyclopoids, including *Tropocyclops* sp., are unde-

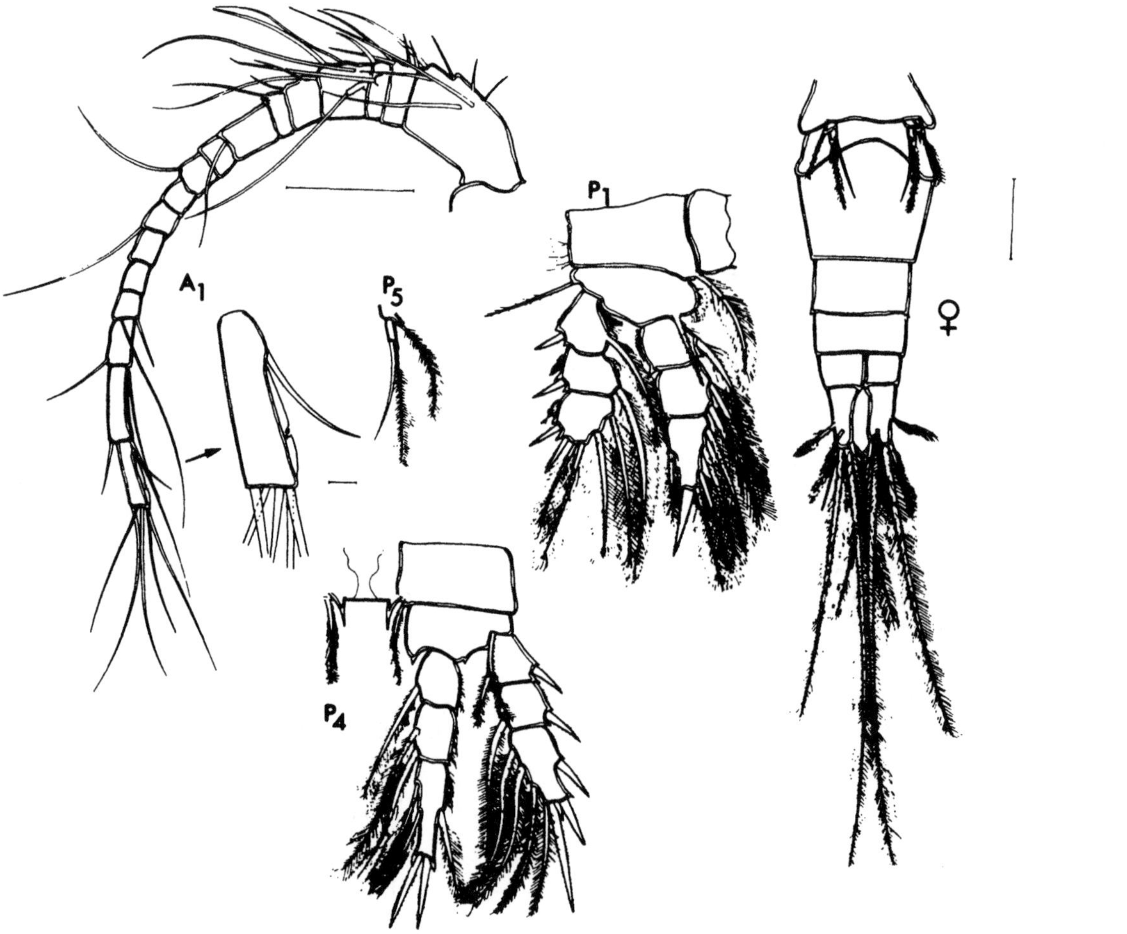

Fig. 5. Mesocyclops cf. *annae* Kiefer, Winmurra billabong, 13 June 1979. Scalars: A_1, P_1, P_4, P_5, urosome = 100 μm; terminal segment A_1 = 10 μm.

scribed (D. W. Morton, Monash University, pers. commun.).

Calanoids were represented by only two species, *E. lumholtzi* (dry season) and *C. ultima* (perennial). These two species may have been competing for the open-water niche in the billabongs; one species always dominated in co-occurring populations. Compared to other tropical regions, e.g. South America, where calanoids have radiated (cf. Dussart *et al* (1984), or to southern Australia, with a high degree of endemism (cf. Bayly, 1964), the group is depauperate in the Magela zooplankton.

Cladocera

Most of the 35 taxa recorded were littoral incursions; only seven species were abundant in the plankton, and not all were found in all billabongs, i.e. *D. excisum* (maxima at the end of the dry season), *C. sphaericus* s.l. (wet season), *M. spinosa* (late dry), *B. deitersi, S. acutirostratus, G. testudinaria* and *C. cornuta*. Of these, only *B. deitersi* appears restricted to tropical Australia.

Notably, the genus *Daphnia* was not recorded. Temperature and food particle size may be limiting factors; the genus is noted to be rare in other tropical studies (cf. Fernando, 1980).

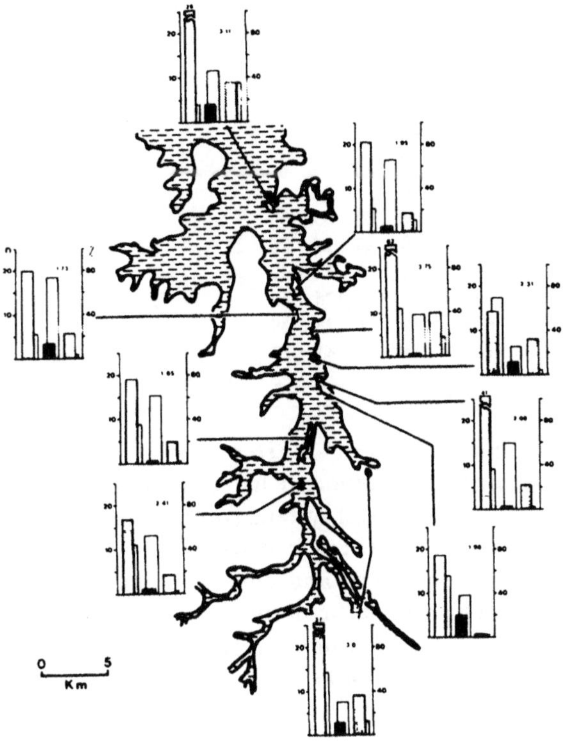

Fig. 6. Comparative plankton communities in ten Magela billabongs June 13–14 1979, showing number of species of Rotifera ▥, Copepoda ■ and Cladocera ⊟ (left axis), per cent composition of each group (right axis, rear bars) and Shannon-Wiener diversity (H′).

Spatial variation in zooplankton communities

To provide information on the momentary zooplankton species composition, 30 m tows were taken from an air boat in each of the 8 study habitats and two additional sites (Ja Ja Extension and an unnamed billabong north of Ja Ja) on June 13–14 1979. This was the beginning of the dry season, and floodwaters were receding, but all billabongs except Winmurra were still submerged.

In view of the apparent continuous connection between the habitats, the disparity in species composition summarised in Fig. 6 is remarkable. From 22 taxa (Buffalo) to 80 (Mine Valley) were present, with rotifers the most abundant in terms of species (62 in Mine Valley, including 16 *Lecane*). Only in three of the sites did rotifers comprise more than 50% of the community [Winmurra, Ja Ja, unnamed, with dominants *F. opoliensis* (39%), *P. vulgaris* (44%) and *P. vulgaris* (34%) respectively]. All other sites were numerically dominated by calanoid nauplii and copepodites. Only in Nankeen, concurrent with a bloom of filamentous green algae, were Cladocera abundant (36%, principally *C. sphaericus* s.l.).

Of the 160 species identified from the ten samples (127 Rotifera, 24 Cladocera, 6 Copepoda and 3 Ostracoda), only ten occurred in more than five sites: *K. tropica, K. lenzi, B. quadridentatus mehleni, P. vulgaris, L. bulla, T. pusilla, T. similis, F. longiseta, F. opoliensis, M. varicans* and *C. sphaericus* s.l. The remaining taxa occurred in less than 50% of the sites, of these 73 (46%) were recorded from single collections. Table 2 shows low per cent similarity between the June samples as measured by the Sorenson Index. Community species diversity, as measured by the Shannon-Wiener Index, is included in Fig. 6. The most diverse communities were Mine Valley, Nankeen and Winmurra, i.e. with a higher proportion of littoral incursion species.

Table 2. Per cent similarity of zooplankton communities, Magela Creek billabongs, June 13–14, 1979 (Sorenson Index).

Billabong										
1. Leichhardt's										
2. Jabiluka	46									
3. Mine Valley	17	18								
4. Ja Ja	48	31	26							
5. Winmurra	36	23	35	34						
6. Buffalo	20	25	27	26	27					
7. Island	22	12	30	32	26	27				
8. Nankeen	32	17	25	49	33	27	23			
9. Ja Ja Ext.	44	31	28	48	36	26	44	31		
10. unnamed	23	24	41	27	35	30	24	23	24	
cf.	1	2	3	4	5	6	7	8	9	10

Similar species-rich, heterogeneous plankton communities are reported from billabongs of the Murray-Darling river system, southeastern Australia. Spatial differences in distribution and abundance of submerged and emergent macrophytes, water chemistry (particularly nutrient sources) and flow contribute to produce disparate plankton assemblages (Shiel, 1981).

Temporal variation in zooplankton communities

Seasonal variations are described only for Ja Ja billabong. Similar trends occurred in all sites, although the species dominants or synchrony of events were not necessarily the same.

Zooplankton abundance (Fig. 7a) is expressed as total numbers of zooplankters per sample and not per standard volume. Schindler samples collected too fewer animals. Zooplankton increased from July-August until the start of the wet season, when the billabongs were flushed by floodwaters from the upper Magela Creek catchment. The drastic decline in Ja Ja zooplankton in January was documented by Tait (1981), but no clear causal relationship was found. Recent work by Bishop *et al* (1982) suggests that naturally occurring soluble toxic compounds produced by fringing vegetation (e.g. freshwater mangrove, *Barringtonia acutangula)* are implicated. Aborigines use such plants as ichthyocides, and it is possible that leaching of compounds such as saponins into the billabongs is responsibe for both fish kills and decimation of the zooplankton.

Brown *et al* (1983) concluded that aluminium and low pH were the cause of the large fish kill in Ja Ja during January 1980, and could also have

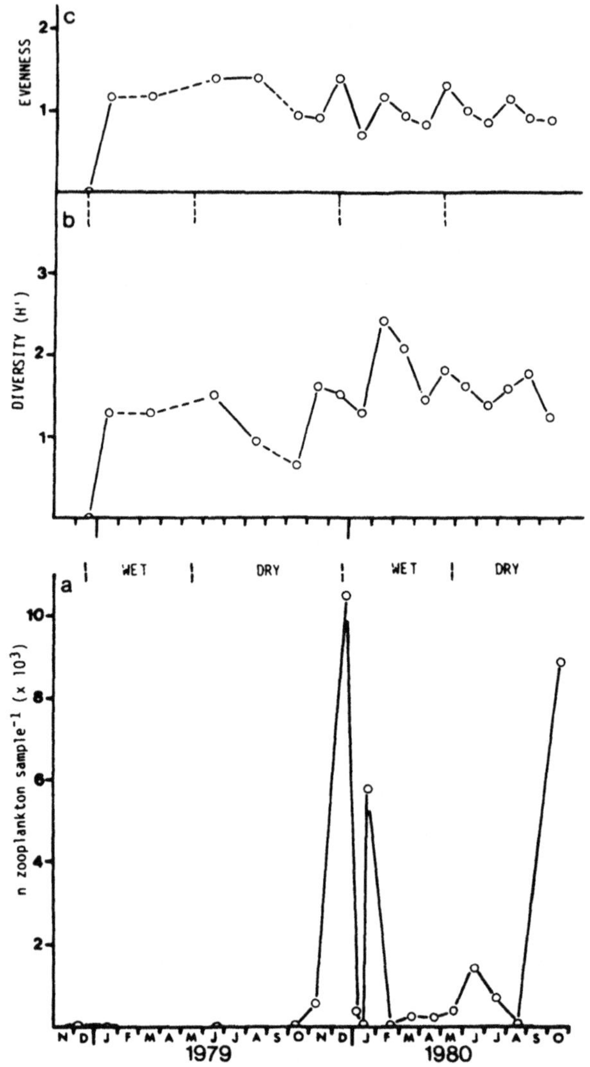

Fig. 7. Changes in zooplankton numbers per sample (a), species diversity (b) and evenness (c) in Ja Ja billabong.

been the cause of the reduction in zooplankton numbers. Further work clearly is necessary.

Species diversity (Fig. 7b) increased in the wet season and decreased as the dry season progressed. Microcrustacean diversity was slightly higher in Ja Ja than the other floodplain billabongs, and the community was more even (Fig. 7c) throughout the study.

Seasonality of the most abundant species in Ja Ja is shown in Fig. 8 (species code as in Table 1). *M. notius* was perennial, forming large populations lasting three or four months before declining. *E. lumholtzi* occurred between June and September after flow had ceased and before turbidity markedly increased. *C. ultima* was collected in both wet and

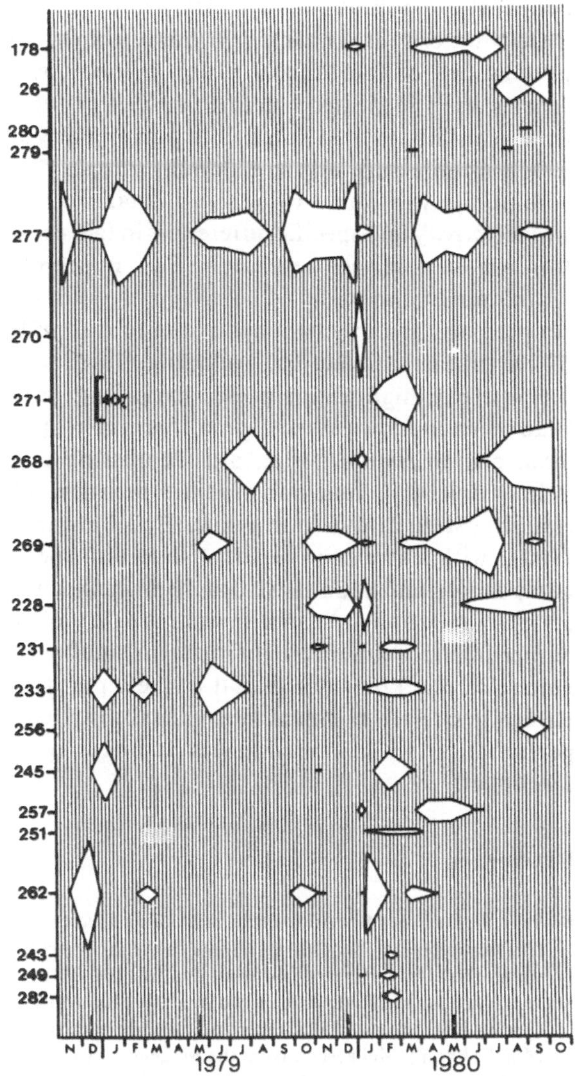

Fig. 8. Seasonal occurrence and relative abundance of the dominant zooplankton in Ja Ja billabong.

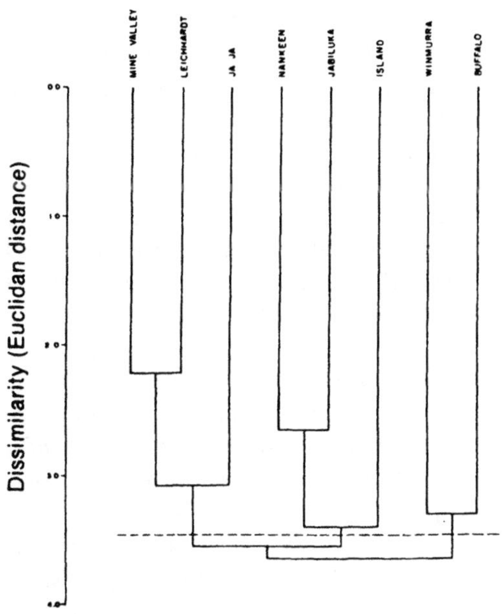

Fig. 9. Dendrogram resulting from cluster analysis of plankton composition data showing taxonomic resemblance of the billabong communities.

dry seasons. *C. cornuta* peaked at the end of the dry season. Several littoral taxa were relatively abundant in the high flows of the 1980 wet season *(P. aduncus, C. sphaericus* s.l., *B. affinis* and *I. sordidus).*

Taxonomic resemblance was used to group the billabongs by clustering composite zooplankton species lists. The resulting dendrogram (Fig. 9) identifies three groups. The first, Mine Valley, Leichhardt and Ja Ja, are all floodplain billabongs located close together and furthest upstream on the floodplain proper. A mixed assemblage of limnetic and littoral taxa resulted from variations in flow conditions. The second group, Nankeen, Jabiluka and Island have in common large limnetic areas, and tended to be dominated by true limnetic species, with very rapid compositional changes. Winmurra and Buffalo, the smallest and furthest upstream, also had great oscillation in species diversity, with most species tending to be present for less than four months. These two billabongs also were more extreme habitats than the others: Winmurra dried out in the 1979 dry season, and was recolonized (probably by resting eggs in the bottom mud, cf. el Moghraby 1977) largely by littoral Cladocera. High flows in Buffalo during the wet season precluded establishment of a zooplankton population.

Zooplankton and environmental interactions

Water quality deteriorated as concentration due to evapotranspiration occurred over the dry season. Many correlations between water quality parameters and zooplankton species and abundance were detected, however there are inherent difficulties in accepting such correlations outright (cf. Dumont,

1977). Ranges of conductivities between wet and dry seasons, for example, are all within the recorded ranges (Shiel, 1981) for the majority of species in the billabongs.

Flow velocities in the wet were too high to allow zooplankton populations to remain in the billabongs (cf. Welcomme, 1979), unlike the slow flows and minimal perturbation reported from a billabong in southeastern Australia (Shiel, 1976). Dilution also accounted for the great reduction in abundance.

Variations in water temperature and dissolved oxygen concentrations had little apparent impact on the zooplankton. Similarly, no direct influence of conductivity changes should be concluded, despite apparent correlations between some species and conductivity. pH, however, as reported elsewhere (e.g. Sprules, 1975), markedly affected the community composition. A more simplified plankton community occurred in each billabong as the dry season progressed and the pH fell, e.g. mean momentary species compositions of 2.0 and 4.4 microcrustacea per billabong were recorded in December 1978 and 1979 respectively, vs. 3.7 and 5.0 in June 1979 and 1980. The simplification of rotifer communities in response to pH changes over the dry season was described by Koste & Shiel (1982), and an extreme effect of acidification in Ja Ja was described by Tait (1981).

Food was probably not limiting to the Magela zooplankton community. Spearman Rank correlation coefficients for total zooplankton numbers per sample with total organic carbon (measure of organic detritus) and chlorophyll-a for each billabong were positive. Grazing pressure on the phytoplankton and detritus was probably never high enough to reduce the available food. Examination of gut contents of several species of dominant plankters at different times of the year showed more particulate organic matter (probably the remains of decayed vascular hydrophytes) than phytoplankton cells, especially during the dry season. This correlates with the greatly increased amount of detrital food available at this time of the year. The community appears to be detritus based, depending on decaying vascular hydrophytes as the main source of food. The majority of plankters are herbivorous or detritivorous; few carnivores were recorded (e.g. *A. brightwelli, Mesocyclops* spp.).

The effects of invertebrate and fish predation on zooplankton populations are little known in Australian waters. In a concurrent study on the fish populations of the Magela Creek billabongs, 38 species were identified, more than half of which were planktivorous at some time during the year. Predation pressure on the billabong zooplankton community was found to be substantial. Data on size selectivity and specific differences in fish predation will be published separately (Tait, in prep.).

In summary, the reduced number of true limnetic zooplankton in the billabongs is due to a complex of environmental factors, including flow, water quality (e.g. high temperatures excluding larger Cladocera and low pH excluding common tropical taxa, e.g. *Brachionus* spp., and depressing community diversity), food and predator pressure. The resulting state of dynamic balance (cf. Fernando, 1980), with extremely rapid changes in species composition and abundance, results from the interaction of the zooplankton with water quality, algae, higher plants, zooplanktonic predators and herbivores and planktivorous fish. It is likely that these naturally stressed communities will exhibit greater perturbation with increasing development in the catchment.

References

Bayly, I. A. E., 1964. A revision of the Australian species of the freshwater genera *Boeckella* and *Hemiboeckella* (Copepoda: Calanoida). Aust. J. mar. Freshwat. Res. 15: 180–238.

Bishop, K. A., L. M. Baker & B. N. Noller, 1982. Naturally-occurring Ichthyocides and a report on Owenia vernicosa F. Muell. (Family Meliaceae), from the Magela Creek system, Northern Territory. Search 13: 150–153.

Brown, T. E., A. W. Morley, N. T. Sanderson & R. D. Tait, 1983. Report of a large fish kill resulting from natural acid water conditions in Australia. J. Fish. Biol. 22: 335–350.

Dumont, H. J., 1977. Biotic factors in the population dynamics of rotifers. Ergebn. Limnol. 8: 98–122.

Dussart, B. H., C. H. Fernando, T. Matsumura-Tundisi & R. J. Shiel, 1984. A review of systematics, distribution and ecology of tropical freshwater zooplankton. Hydrobiologia, this volume.

el Moghraby, A. I., 1977. A study on diapause of zooplankton in a tropical river, the Blue Nile. Freshwat. Biol. 7: 207–217.

Fernando, C. H., 1980. The freshwater zooplankton of Sri Lanka with a discussion of tropical freshwater zooplankton composition. Int. Revue ges. Hydrobiol. 65: 85–125.

Hart, B. T. & R. J. McGregor, 1980. Limnological survey of eight billabongs in the Magela Creek catchment, Northern Territory. Aust. J. mar. Freshwat. Res. 31: 611–626.

Kiefer, F., 1930. Zur Kenntnis der freilebenden Copepoden Madagaskars. Zool. Anz. 87: 42–46.

Kiefer, F., 1981. Beitrag zur Kenntnis von Morphologie, Taxonomie und geographischer Verbreitung von *Mesocyclops leuckarti* auctorum. Arch. Hydrobiol., Suppl. 62: 148-190.

Koste, W., 1978. Rotatoria, die Rädertiere Mitteleuropas (Uberordnung Monogononta. Revision after M. Voigt 1956/57, 2. Borntraeger, Stuttgart.

Koste, W., 1981. Zur Morphologie, Systematik und Ökologie von neuen monogononten Rädertiere aus dem Uberschwemmungsgebiet des Magela Creek in der Alligator River Region Australiens, N.T., 1. Osnabrücker naturw. Mitt. 8: 97-126.

Koste, W. & R. J. Shiel, 1982. Morphology, systematics and ecology of new monogonont Rotifera from the Alligator River region. N.T., 2. Trans. r. Soc. S. Aust. 107: 109-121.

Marchant, R., 1980. The macroinvertebrates of Magela Creek, Northern Territory. Off. Supervising Scientist Res. Rep. 1, Aust. Govt. Publ. Service, Canberra, 40 pp.

Recher, H. F. & R. T. Holmes, 1982. The foraging behaviour of herons and egrets on the Magela Creek floodplain, Northern Territory. Off. Supervising Scientist, Tech. Mem. 4, Aust. Govt. Publ. Service, Canberra, 16 pp.

Shiel, R. J., 1976. Associations of Entomostraca with weedbeds in a billabong of the Goulburn River, Victoria. Aust. J. mar. Freshwat. Res. 27: 533-549.

Shiel, R. J., 1981. Plankton of the Murray-Darling River system. Ph.D. Thesis, Univ. Adelaide. 286 pp.

Shiel, R. J., 1983. The genus *Brachionus* (Rotifera: Brachionidae) in Australia, with a description of a new species. Proc. r. Soc. Vict. 95: 33-37.

Sprules, W. G., 1975. Factors affecting the structure of limnetic crustacean zooplankton communities in Central Ontario lakes. Verh. int. Ver. Limnol. 19: 635-643.

Tait, R. D., 1981. Natural seasonal changes in zooplankton population in a billabong of the Magela Creek, Northern Territory, Australia, prior to mining in the area. In M. Sudzuki (ed.), Some Approaches to Saprobiological Problems. Sanseido, Tokyo: 29-35.

Tait, R. D., 1982. The ecology of the plankton of selected Magela catchment billabongs, Northern Territory. M.Sc. Thesis, Macquarie Univ. 256 pp.

Walker, T. D. & P. A. Tyler, 1979. A limnological survey of the Magela Creek, Alligator Rivers region, Northern Territory. An interim report to the Supervising Scientist, covering the period April 1978 to December 1978. Univ. Tasmania. 184 pp.

Welcomme, R. L., 1979. Fisheries Ecology of Floodplain Rivers. Longman, N.Y., 317 pp.

Feeding in adult females of *Argyrodiaptomus furcatus* (Sars, 1901), Copepoda-Calanoida, of Lobo Reservoir (Broa), São Carlos, São Paulo, Brazil

L. H. S. Tavares & T. Matsumura-Tundisi
Universidade Federal de São Carlos, Departamento de Ciências Biológicas, Via Washington Luiz, Km 235, Caixa Postal: 676, 13.560 São Carlos, SP, Brazil

Keywords: tropical zooplankton, Copepoda, feeding, experiments, filtration rate, assimilation rate, ^{14}C, algal culture.

Abstract

The aim of this work was to study the feeding process of *Argyrodiaptomus furcatus* (Copepoda-Calanoida) in the Lobo Reservoir (São Carlos, SP, Brazil). Non-ovigerous adult females and the ^{14}C technique were used to measure filtration and assimilation rates. The diet contained the following phytoplankton species: *Chlamydomonas sp., Ankistrodesmus gracilis, Melosira italica, Scenedesmus quadricauda* and *Chlorella zoofingensis*.

The experiments were carried out using unialgal and mixed cultures during 2-, 4- and 6-h periods. The results of the filtration and assimilation rates were compared.

The data obtained by statistical tests showed the highest assimilation rate in *Argyrodiaptomus furcatus* fed *Chlamydomonas* sp. in both culture types. However, *Chlorella zoofingensis* and *Scenedesmus quadricauda* were the most filtered species in unialgal and mixed cultures, respectively. A higher filtration rate was observed for the 2-h period than for the 4- and 6-h periods.

Culture agent was also important. Higher assimilation and filtration rates were obtained during the log phase of *Chlamydomonas* sp. growth than during the stationary phase.

Introduction

The study of herbivorous plankton feeding is important since they are the main organisms to serve as a link between producers and other consumers. For this reason, a knowledge of their feeding habits and the quantity of food they require provides a good estimate of the production process, even considering a restricted group such as freshwater Calanoida, whose feeding behavior is variable among species and different stages of development within the same species. However, food preference will depend on phytoplanktonic community which varies in lakes showing different conditions.

Feeding studies in copepods can be made through the morphology of filtering appendages, examination of the digestive tract content of preserved animals or a feeding test of living animals upon a single or mixed type of phytoplankton labeled with ^{32}P or ^{14}C.

Although there is much information on alimentary diets in marine copepods (Marshall & Orr, 1955, 1956; Mullin, 1963; Conover, 1966), freshwater studies are mainly concentrated on Cladocera species as shown by Ryther (1954), MacMahon & Righer (1965), Burns (1968), Ward *et al.* (1970), Arnold (1971), Narita & Ward (1972); Lampert (1974, 1977), Geller & Muller (1981) and others; litte is known about copepods, especially Calanoida.

Certain zooplanktonic organisms are highly selective for size and type of phytoplanktonic food. The community of filter-feeders is usually divided, for selectivity, into: microfiltrators, including most rotifers, small cladocerans and larval stages of co-

Hydrobiologia 113, 15–23 (1984).

pepods, and macrofiltrators comprising mainly Calanoida, Cladocera and Rotifera of larger size. In the microfiltrators food, there is a predominance of particles measuring up to 5 microns, mostly composed of bacteria and fine detritus, while macrofiltrators present larger particles, mainly nannoplanktonic algae (Kajak *et al.*, 1972).

Gliwicz (1977), emphasizes the importance of two phytoplanktonic fractions which influence zooplanktonic community in different ways: nannophytoplankton, a fraction smaller than 50 μm which is well consumed by filter feeders and microphytoplankton, a fraction larger than 50 μm which is not available as food but somehow influences the filtration process of animals.

The purpose of this work was to study the feeding process in order to calculate filtration and assimilation rates of non-ovigerous adult females of *Argyrodiaptomus furcatus*, the only species of Calanoida (Copepoda) of the Lobo Reservoir (Broa) (Matsumura-Tundisi & Tundisi, 1976). In terms of biomass, *Argyrodiaptomus furcatus* is important for the reservoir zooplanktonic community. Consequently, it plays an influential role in the trophic web, constituting an important link in the food chain.

Material and methods

Field proceedings

Zooplankton samples were collected in the shallow portion of the Lobo Reservoir (Broa) between the Itaqueri River and the Lobo Stream (Fig. 1). A

Fig. 1. Map of the Lobo Reservoir with localization of the sampling station.

plankton net with 68 μm-mesh aperture was used and the material was placed in a container in order to be worked in the laboratory.

The experiments were carried out in the following months: March, April, May, July, December, 1981 and January, 1982.

Laboratory proceedings

a) *Animal preparation for feeding.* Zooplankton was placed in an incubator chamber under controlled temperature (20 °C), remaining in the dark for acclimation. It was then isolated from the rest and placed in another container filled with previously filtered and sterilized water from the reservoir; after this, it was left with no food. This zooplankton 'washing' process was repeated many times.

Non-ovigerous adult females of *Argyrodiaptomus furcatus* were isolated and left in the dark and under controlled temperature for a 41-h period.

b) *Phytoplankton culture.* The algal species used as food were those most abundant in the Lobo Reservoir (Tundisi & Hino, 1981; Watanabe, 1981). Besides abundance, size, amplitude and form were also considered. The following species were studied: *Melosira italica, Ankistrodesmus gracilis, Chlamydomonas* sp., *Chlorella zoofingensis* and *Scenedesmus quadricauda*.

Two types of culture were utilized in this work: unialgal and mixed, the latter obtained by mixing the five algal species already mentioned; moreover, a culture was always kept with ^{14}C; all cultures were at exponential phase of growth (log phase) with an age of eight days.

c) *Method and techniques utilized.* The carbon-14 method was used to measure the filtration and assimilation rates of *A. furcatus*, after Marshall & Orr (1955) and Edmondson & Winberg (1971).

For the unialgal and mixed culture experiments three sets of 26 flasks [1] were mounted, each of them corresponding to a different schedule: 2, 4 and 6 h. There were ten females in each experimental flask.

To verify the influence of culture age on the filtration and assimilation rates only a *Chlamydomonas* culture was utilized and one set of 5 flasks was mounted for a 2-h experiment. A growth curve was plotted, points were selected at different curve phases (4, 8, 11, 15, 77 and 116 days) and experiments were carried out.

The experimental flasks were submitted to a rotatory movement for food homogenization.

d) *Sample activity determination.* In the feeding experiments, copepods and fecal pellets were isolated and the rest filtered with 0.25 μm pore size Millipore filters (HAWP). Afterwards, the organisms were placed in an excavated slide and washed three times with previously filtered and sterilized water from the reservoir, the same being done for the fecal pellets. The copepods were manually macerated, while the fecal pellets and the filtered content were dissolved in scintillator liquid.

The activity of the samples (copepods, fecal pellets and filtered content) was determined by placing them in 20 ml vials^{-1} containing 10 ml of Bray-scintillator liquid, a combination described by Bray (1960). Afterwards, they were counted for 10 min in a LS 100 Beckman-scintillator liquid, an American product with an efficiency of 87%.

e) *Calculation of filtration and assimilation rates.* According to Edmondson & Winberg (1971), filtration rate refers to the number of cells ingested by an animal per time unit in a determined volume. Its usage does not imply that the water volume flown through filtering appendages is known or that all particles captured by filter-feeders are consumed. Other terms used as synonymous of filtration rate are 'filtration capacity', 'grazing rate', 'filtered rate' and even 'feeding rate'.

Filtration rate is calculated from the equation given by Gauld (1951):

$$F = v \frac{\log Co - \log Ct}{t \log e}$$

where: F = filtration rate
Co = radioactive initial concentration
Ct = radioactive algal concentration at time t
v = water volume:animal
t = period of experiment
e = neperian algarism.

Data obtained were expressed as ind h^{-1}.

Assimilation rate defined as absorption of material in the intestine (Edmondson & Winberg, 1971), was calculated through a method slightly modified by Marshall & Orr (1955). It can be calculated by the formula:

$$\text{Assimilation } \% = \frac{B + E}{B + E + F}$$

where: B = animal radioactivity.
E = radioactivity of eggs produced during the experiment.*
F = radioactivity of fecal pellets produced during the experiment.

All data obtained were expressed in terms of percentage.

Statistical analysis

To analyze the variation of filtration and assimilation rates, a univaried variance analysis was applied to both types of culture. For the experiments, two factors were considered (Spiegel, 1978). One factor corresponded to phytoplankton species and another to the duration of each experiment.

For contrast of means the test of Tukey (Cochran & Cox, 1957) was applied in order to complement the univaried variance analysis applied to the result of the utilized treatments. Overall mean values of filtration and assimilation rates were estimated for each phytoplankton species and exposure time.

Results

Laboratory experiments using ^{14}C for calculating assimilation and filtration rates

The present study utilizes the values of filtration and assimilation rates to evaluate feeding consumption of *Argyrodiaptomus furcatus* on five phytoplankton species during different experimental periods.

Table 1 shows the number of cells present in 7.5 ml aliquots used for each experiment with unialgal and mixed cultures.

Despite identical conditions of light intensity, temperature, age and medium of culture, a variation was observed in cell concentration in two lots: A (sampling carried out from 4/27/81 to 5/31/81) and B (sampling carried out from 3/16/81 until

* In this work, E was equal to zero, since non-ovigerous females were considered.

Table 1. Cell numbers of each phytoplankton species labeled with ^{14}C, obtained from 7.5 ml aliquots utilized for (two) lots A and B in experiments with unialgal and mixed cultures.

Lots	Date of experiment	Labeled species	Number of cells: 7.5 ml	
			Mixed	Unialgal
A	16/03	*Melosira italica*	11.625	46.500
	23/03	*Chlorella zoofingensis*	47.625	190.500
	30/03	*Ankistrodesmus gracilis*	16.013	64.050
	06/04	*Chlamydomonas* sp	18.375	73.500
	13/04	*Scenedesmus quadricauda*	17.663	70.650
B	27/04	*Ankistrodesmus gracilis*	18.863	75.450
	04/05	*Chlorella zoofingensis*	82.950	331.800
	18/05	*Chlamydomonas* sp.	15.375	75.338
	25/05	*Scenedesmus quadricauda*	22.163	88.650
	31/05	*Melosira italica*	15.000	60.000

4/13/81). In general, the cell concentrations for all phytoplankton species in lot B were higher than in lot A either in unialgal or in mixed culture. *Chlorella zoofingensis* showed the highest cell concentration and *Melosira italica* the lowest.

Figure 2 shows the variation in overall means of assimilation rates of *A. furcatus* for unialgal and mixed cultures at different periods of experiments. It can be observed that in unialgal culture experiments *Scenedesmus quadricauda* and *Chlorella zoofingensis* were the least assimilated species even considering overall means. In mixed culture, *Scenedesmus quadricauda* was highly assimilated (80%) at the 6-h experiment; this unexpected abrupt increase may be related to errors of experimental handling. In unialgal culture, *Chlorella zoofingensis* was the least assimilated species ranging from 67% to 58%. In general, *Chlamydomonas* sp. was the most assimilated species in both types of culture (unialgal and mixed).

For calculating the assimilation rate, the number of eggs produced was not computed in the formula since only non-ovigerous females were used. Feces were collected and their radioactivity measured, which varied, for the five phytoplankton species, from 4% to 40% in unialgal culture and from 6% to 38% in mixed culture.

Figure 3 shows the overall means of filtration rates of *A. furcatus* on five phytoplankton species in unialgal and mixed cultures. In the 2-h experiments, a high filtration rate was observed for the different species in the following order: *Chlo-*

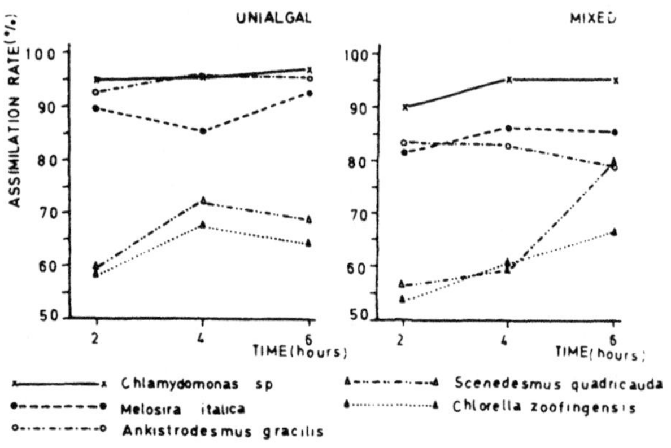

Fig. 2. Variations of the assimilation rate mean values of *Argyrodiaptomus furcatus* for five phytoplankton species in unialgal and mixed cultures in function of time (hours).

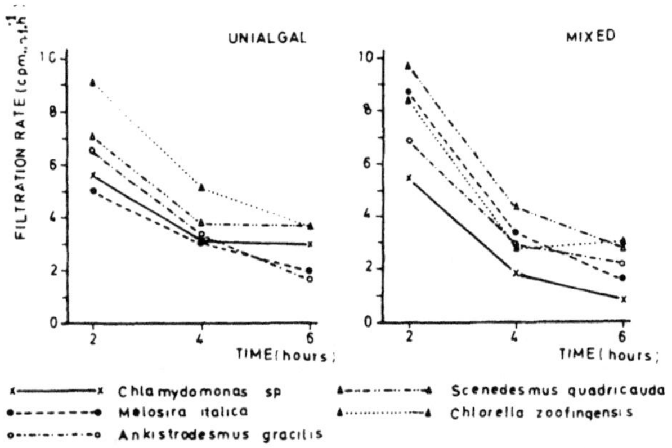

Fig. 3. Variation of the filtration rate mean values of *Argyrodiaptomus furcatus* for five phytoplankton species in unialgal and mixed cultures in function of time (hours).

rella zoofingensis, Scenedesmus quadricauda, Ankistrodesmus gracilis, Chlamydomonas sp. and *Melosira italica*. The same order was established in the 4-h experiment. In contrast, the 6-h experiment showed a more irregular order: *Chlorella zoofingensis, Scenedesmus quadricauda, Chlamydomonas* sp., *Melosira italica* and *Ankistrodesmus gracilis*. In mixed culture, it could also be observed that the 2-h experiment provided the highest filtration rate of *A. furcatus* in the following sequence: *Scenedesmus quadricauda, M. italica, Chlorella zoofingensis, A. gracilis* and *Chlamydomonas* sp. During the last periods this sequence was alterated. In the 4-h experiment, *Chlorella zoofingensis* was less filtered than *A. gracilis* and in the 6-h experiment, it was the most filtered species. Moreover it was observed that *Chlamydomonas* sp. was always the least filtered species.

Variations between lots A and B for assimilation and filtration rates in both cultures can be related to the individuals under study whose behavior may intrinsically vary within the species.

Statistical analysis of the filtration and assimilation rates in unialgal and mixed cultures

The application of statistical tests to the experimental data is intended to give a better assessment of the degree of significance of the results with respect to filtration and assimilation rates for both types of culture considering the following variables: species interactions, preferential species and preferential time.

Table 2 summarizes the results obtained through univaried variance analysis.

The second test considered was the Test of Tukey which intends to establish, through pooling and recombination, the significance order of the data obtained from filtration and assimilation rates in both cultures, considering the different species and periods of experiments.

Either species or periods of experiments were ranked in a decreasing order of mean values (Table 3). Thus, the results do not show significative differences in relation to filtration rates in both cultures during the 4- and 6-h periods (T_4 and T_6). On the other hand the 2-h period showed the highest filtration rate.

Considering the unialgal cultures, *Chlorella zoofingensis* and *Scenedesmus quadricauda* were more filtered than the other algae considered. In mixed culture, *Scenedesmus quadricauda, M. italica* and *C. zoofingensis* were the most filtered species and *A. gracilis* and *Chlorella* sp. the least. *M. italica, C. zoofingensis* and *A. gracilis* were placed intermediately. *Chlorella* sp. was the least filtered species in both cultures.

Table 2. A univaried variance analysis of filtration and assimilation rates of *A. furcatus* in mixed and unialgal cultures.

Type	Variation	Filtration	Assimilation
Unialgal	Species interaction	Not significant (*A. furcatus* has similarly consumed the five phytoplankton species)	Not significant (*A. furcatus* has similarly consumed the five phytoplankton species)
	Preferential species (algae)	Significant (The copepod shows preference according to the species)	Significant (The copepod shows preference according to the species)
	Preferential time (hours)	Significant (The *A. furcatus* filtration rate varies in function of time).	Not significant (The *A. furcatus* assimilation rate varies in function of time).
Mixed	Species interaction	Not significant (*A. furcatus* has similarly consumed the five phytoplankton species).	Not significant (*A. furcatus* has similarly consumed the five phytoplankton species).
	Preferential species (algae)	Significant (The copepod shows food preference according to the species).	Significant (The copepod shows food preference according to the species).
	Preferential time (hours)	Significant (The *A. furcatus* filtration rate varies in function of time).	Significant (The *A. furcatus* assimilation rate varies in function of time).

Table 3. Grouping of filtration and assimilation rates values (in decreasing order) obtained through the test of Tukey in experimental periods of 2, 4 and 6 hours and for different species of algae.

Type	Factor	Filtration	Assimilation
Unialgal	Time[a] (hours)	T_2 T_4 T_6	T_6 T_4 T_2
	Species[b] (algae)	A_5 A_4 A_1 A_3 A_2	A_1 A_3 A_2 A_4 A_5
Mixed	Time[a] (hours)	T_2 T_4 T_6	T_6 T_4 T_2
	Species[b] (algae)	A_4 A_2 A_5 A_3 A_1	A_1 A_2 A_3 A_4 A_5

[a] $T_2 = 2$ hours; $T_4 = 4$ hours; $T_6 = 6$ hours.

[b] A_1 = *Chlamydomonas* sp.; A_2 = *Melosira italica*; A_3 = *Ankistrodesmus gracilis*; A_4 = *Scenedesmus quadricauda*; A_5 = *Chlorella zoofingensis*.

In relation to assimilation rate, there were no significative differences among the three experimental periods, the same not happening to the mixed culture.

With respect to species of algae, *Chlamydomonas* sp., *M. italica* and *A. gracilis* were the most assimilated species while *C. zoofingensis* and *Scenedesmus quadricauda* were the least assimilated species in both cultures. However, in mixed culture, *M. italica* and *A. gracilis* were considered as of medium assimilation.

Age culture influence on assimilation and filtration rates

It is known that culture age is a significant factor in food behavior of aquatic herbivores, since the food value of cells in culture may vary with age and nutrients present in the medium of culture. Therefore, algae collected at different phases of their growth curve can influence animal feeding in different ways.

Some experiments were carried out with a *Chlamydomonas* sp. culture chosen for its highest assimilation by *Argyrodiaptomus furcatus*.

Figure 4 provides the growth curve of *Chlamydomonas* sp. with a sharp log phase. On the eleventh day the growth curve reaches the stationary phase, after which there are fluctuations in populational density that last for approximately 180 days.

The results of filtration and assimilation rates of *Argyrodiaptomus furcatus* for *Chlamydomonas* sp. culture on the 4th, 8th, 15th, 77th, and 116th day are shown in Table 4. The highest filtration rate was 10.8 ind. h^{-1} obtained on the 8th day, and the lowest rate was 2.3 ind. h^{-1} reached on the 77th day. Considering assimilation rates, the highest values, 96.5 and 94%, were obtained in the log phase of the growth curve on the 4th and 8th days, respectively,

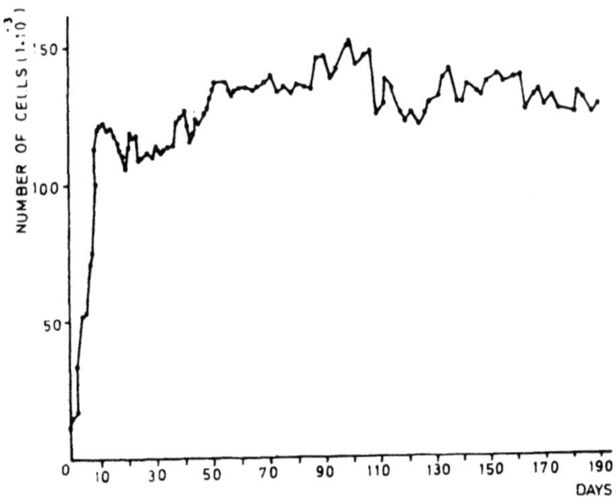

Fig. 4. Growth curve of *Chlamydomonas* sp.

Table 4. Experimental results showing the variation of filtration and assimilation rates of *A. furcatus* for *Chlamydomonas* sp., culture, in function of culture age.

Culture age (days)	Filtration rate (cpm ml h^{-1})	Assimilation rate (%)
4	7.3	96.5
8	10.8	94.0
11	5.2	88.0
15	5.3	93.5
77	2.3	83.3
116	5.4	79.6

while the lowest rate, 79.6% was reached on the 116th day. Neither filtration nor assimilation rates decreased significantly as the *Chlamydomonas* sp. culture aged.

Discussion

Information on feeding of fresh-water copepods, particularly Calanoida, is scarce. Recent studies have indicated nannophytoplankton as the main source of energy for these organisms considering the fact that bacteria and detritus may function as an additional source and sometimes as an alternative one (Gliwicz, 1969; McQueen, 1970).

Marshall & Orr (1955), Porter (1973), Wilson (1973) and Campbell & Haase (1971) have observed that copepods have a food preference, selecting certain phytoplankton species depending on their form and size while rejecting others that are poorly digested. The feeding behavior of copepods is influenced by many factors. The use of unialgal or mixed cultures for *Argyrodiaptomus furcatus* has been significant since food preferences for types of algae among organisms vary from one culture to another. In unialgal cultures, *Argyrodiaptomus furcatus* filtered more *Chlorella zoofingensis* at the highest filtration rate. However, it was poorly absorbed, and at the lowest assimilation rates. Its low assimilation may be related to the production of some toxic substances of inhibitory effect. Pratt & Fong (1944) have shown the same fact for *Chlorella vulgaris*, which produces a substance inhibitory of its own growth called chlorelline. This substance may create unfavorable conditions to the animal, altering its metabolism. Ryther (1954) has also verified that *Chlorella* can inhibit feeding in *Daphnia* when added in sufficiently high concentrations.

In mixed culture, *A. furcatus* shows a higher filtration rate for *Scenedesmus quadricauda* than for *Chlorella zoofingensis*, although the latter species was present in a higher concentration. However it had a higher assimilation rate for *Chlamydomonas* sp. in spite of these low filtration rates.

A high filtration rate for *Chlorella* has also been observed by Kibby (1971), working on *Diaptomus gracilis* but its highest assimilation rate was obtained with *Ankistrodesmus* sp. It seems that *Chlamydomonas, Ankistrodesmus, Scenedesmus,* among others are the most consumed species by herbivorous zooplankton (McQueen, 1970; Arnold, 1971; Narita & Ward, 1972).

The filtration rate could be also influenced by cell length and to some extent this affected *Melosira italica* and *A. gracilis* filtration by *A. furcatus.* Another situation which affects filtration rate is that cells may pass intact through the digestive tract of copepods.

The differences observed in filtration and assimilation rates by *A. furcatus* for different algae, may be associated with differences in nutrient content or with the cell morphology. That is the case of certain algae which have a thick wall, and are an inferior source of food when compared with flagellates with no carapace. They can also present a reduced food value when untolerable or even toxic to animals used in laboratory experiments.

The food concentration and its relation to assimilation percentage under laboratory conditions has been frequently discussed. Marshall & Orr (1955) have verified that the assimilation rate was independent of the available quantity of food. On the other hand, Gauld (1951) found that *Calanus* had a filtration rate for *Chlamydomonas* independent of cell concentration, while Mullin (1963) observed a dependent relation between *Calanus* 'grazing' rate and cell concentration.

High filtration and assimilation rates for most organisms generally occur during log phase of culture growths. Marshall & Orr (1955) have found, when working on *Lauderia borealis*, a higher assimilation rate at log phase than at senescent phase.

For many zooplanktonic organisms, culture age seems to have greater influence on filtration than on assimilation. In the case of *A. furcatus*, the filtration rate was higher at *Chlamydomonas* sp. log phase (between the 4th and 8th day), but the assimilation rate showed practically no variation between log and stationary phases. Mullin (1963) showed that *Calanus hyperboreus*, fed *Thalassiosira fluviatilis*, showed a drecreasing filtration rate with culture aging; in contrast, Conover (1966), working on the same *Calanus* species found no significant difference between log and stationary phases of *Thalassiosira fluviatilis* for assimilation rate. In addition, Mac Mahon & Rigler (1965) observed that in *Daphnia* species fed on *Chlorella vulgaris*, senescent cells cells could cause a decrease in filtration rate.

The duration of experiments has considerable influence on the filtration rate but not on the assimilation rate. Haney (1971) found that for *Ceriodaphnia, Daphnia rosea*, and *D. galeata*, the ideal period to feed on *Rhodotorula* sp. ranges from 30 s to 5 min. Then, a change in food uptake occurred. As it was observed for the five phytoplankton species given as food, the filtration rate of *A. furcatus* was higher for the 2-h experiment. Mullin (1963) observed that copepods feed irregularly, showing high rates during the first hours of feeding due to the fast to which they had been submitted. After a period of 41 h with almost no food, *A. furcatus* started filtrating quickly, decreasing in intensity posteriorly. This may be a random process considerably influenced by the size of cells. This is the case of *Chlamydomonas* sp., which was the least filtered species when compared with *C. zoofingensis* and *Scenedesmus quadricauda*, although its medium size is intermediate between both species. Also, Narita & Ward (1972), working on *Daphnia similis* considering three different *Scenedesmus* cultures, have verified that ingestion within the first 15 min was much higher than after 30 min or 1 h of experiment.

Zooplanktonic herbivores can select food particles according to size. Quantitative and qualitative features have been providing important knowledge of zooplankton feeding behavior (Infante, 1978). According to Marshall (1973) and Monakov (1972), the filtering appendages largely influence particle size selection, mainly in herbivorous filter-feeders. Furthermore, Burns (1968), working on Cladocera, related particle size to body size.

The Lobo Reservoir represents a eutrophicated region with an abundant aquatic vegetation where *A. furcatus* reproduction takes place. In this region, Watanabe (1981) has found a nannophytoplankton density smaller than that of a region near the Lobo Reservoir dam and this fact can be attributed to nannophytoplankton consumption by *A. furcatus*. Rocha (1978) found in this region an inverse relation between nannophytoplankton biomass (measured by chlorophyll *a*) and *A. furcatus* population density.

In spite of evidence of nannophytoplankton utilization by *A. furcatus*, these organisms probably use other sources of food when nannophytoplankton population decreases in the reservoir. In this case, bacteria may function as an additional source of food for three main reasons: first, they occur in nature within a size range from 0.2 tot 2 μm (Geller & Muller, 1981); second, this region, where the copepods were collected for this study, shows high concentrations of bacteria when compared with the other regions of the reservoir (Godinho, 1976), and finally, this region is considered mesotrophic tending to eutrophic (Tundisi *et al.*, 1972a, b; Matsumura-Tundisi & Tundisi, 1976).

There has been much controversy with respect to bacteria as a source of food for marine and freshwater copepods. Geller & Muller (1981) point out that in oligotrophic lakes, zooplanktonic organisms present a low efficiency on bacteria, being in many cases exclusively herbivores. In mesotrophic lakes bacteria can be considered an important additional food source, however.

References

Arnold, D. E., 1971. Ingestion, assimilation, survival and reproduction by *Daphnia pulex* fed seven species of blue-green algae. Limnol. Oceanogr. 16: 906–920.

Bray, G. A., 1960. A simple efficient liquid scintillation method for counting aqueous solution in a liquid scintillation counter. Analyt. Biochem. 1: 279–285.

Burns, C. W., 1968. The relationship between body size of filterfeeding Cladocera and the maximum size of particle ingested. Limnol. Oceanogr. 13: 675–678.

Campbell, J. M. & B. L. Haase, 1981. Availability of suitable phytoplankton food for zooplankton in an ice-covered lake. Hydrobiologia 79: 113–119.

Cochran, W. G. & G. M. Cox, 1957. Experimental Design. 2nd Edn. John Wiley & Sons, Inc. Lond., 611 pp.

Conover, R. J., 1966. Factors affecting the assimilation of organic matter by zooplankton and the question of superfluous feeding. Limnol. Oceanogr. 11: 346–354.

Edmondson, W. T. & G. G. Winberg, 1971. A Manual of Methods for Assessment of Secondary Productivity in Fresh Waters. I.B.P. Handbook 17, Blackwell Scientific Publications, Oxford, 358 pp.

Gauld, D. T., 1951. The grazing rate of planktonic copepods. J. mar. Biol. Assoc. U.K. 29: 695–706.

Geller, W. & H. Muller, 1981. The filtration apparatus of Cladocera: Filter mesh-sizes and their implications food selectivity. Oecologia 49: 316–321.

Gliwicz, Z. M., 1969. Studies on the feeding of pelagic zooplankton in lakes with varying trophy. Ekol. Pol. 17: 1–46.

Gliwicz, Z. M., 1977. Food size selection and seasonal succession of filter feeding zooplankton in a eutrophic lake. Ekol. Pol. 25: 179–225.

Godinho, M. J. L., 1976. Ciclo sazonal 'Standing-Stock' e Distribução do Bacterioplancton em Ecossistemas Lacustre Artificial (Represa do Lobo, São Carlos). Dissertação de Mestrado, Univ. S. Paulo, 209 pp.

Haney, J. F., 1971. An in situ method for the measurement of zooplankton grazing rates. Limnol. Oceanogr. 16: 970-977.

Infante, A., 1978. Natural food of herbivorous zooplankton of Lake Valencia (Venezuela). Arch. Hydrobiol. 52: 347-358.

Kajak, Z., A. Hillbricht-Ilkowska & E. Pieczynska, 1972. The production processes in several Polish lakes, In Proceedings of the IBP-UNESCO Symposium on Productivity Problems of Freshwaters. Kazimiers Dolnym Poland: 129-147.

Kibby, H. V., 1971. Energetics and population dynamics of *Diaptomus gracilis*. Ecol. Monogr. 41: 311-327.

Lampert, W., 1974. A method for determining food by zooplankton. Limnol. Oceanogr. 19: 995-998.

Lampert, W., 1977. Studies on the balance of *Daphnia pulex* as related to environmental condition, 1. Methodological problems of the use of ^{14}C for the measurement of Carbon assimilation. Arch. Hydrobiol. 48: 287-309.

Marshall, S. M., 1973. Respiration and Feeding in copepods. Adv. mar. Biol. 11: 57-120.

Marshall, S. M. & A. P. Orr, 1955. On the biology of *Calanus finmarchicus*, 8. Food uptake, assimilation and excretion in adult and stage V *Calanus*. J. mar. Biol. Ass. U.K. 34: 495-529.

Marshall, S. M. & A. P. Orr, 1956. Experimental feeding of the copepod *Calanus finmarchicus* (Gunner) on phytoplankton cultures labelled with radioactive carbon (^{14}C). Mar. Biol. Oceanogr.: 110-114.

Matsumura-Tundisi, T. & J. G. Tundisi, 1976. Plankton studies in a lacustrine environment, 1. Preliminary data on zooplankton ecology of Broa Reservoir. Oecologia 25: 265-270.

Mc Mahon, J. W. & F. H. Rigler, 1965. Feeding rate of *Daphnia magna* Straus in different foods labeled with radioactive phophorus. Limnol. Oceanogr. 10: 105-114.

Mac Queen, D. J., 1970. Grazing rates and food selection in *Diaptomus oregonensis* (Copepoda) from Marion Lake, British Columbia. J. Fish. Res. Bd Can. 27: 13-20.

Monakov, A. V., 1972. Review of studies on feeding of aquatic invertebrates conducted at the Institute of Biology of Inland Waters, Academy of Science, USSR. J. Fish Res. Bd Can. 29: 363-383.

Mullin, M. M., 1963. Some factors affecting the feeding of Marine copepods of the genus *Calanus*. Limnol. Oceanogr. 8: 239-250.

Narita, T. & F. J. Ward, 1972. Ingestion and incorporation by *Daphnia similis* from carbon-14 labeled *Scenedesmus* cultures. Jap. J. Limnol. 33: 36-43.

Porter, K. J., 1973. Selective grazing and differential digestion of algae by zooplankton. Nature 244: 179-180.

Pratt, R. & J. Fong, 1944. Chlorelline, an antibacterial substance from *Chlorella*. Science 99: 351-352.

Rocha, O., 1978. Flutuação Sazonal da População de *Diaptomus furcatus*, Sars (Copepoda, Calanoida), na Represa do Lobo ('Broa'), São Carlos, SP. Diss. Mestrado, Univ. S. Paulo, 147 pp.

Ryther, J. H., 1954. Inhibitory effects of Phytoplankton upon the feeding of *Daphnia magna* with reference to growth, reproduction and survival. Ecology 35: 522-533.

Spiegel, M. R., 1978. Probabilidade e Estatística. Coleção Schaum. Editora Mc Graw-Hill do Brasil, Ltda., 518 pp.

Tundisi, J. G., T. M. Tundisi, G. Strixino, M. A. Martins, E. Mattos & E. P. dos Santos, 1972a. Estudos ecológicos na Represa do Broa, 1. (Progress Rep.). Univ. fedl. S. Carlos, SP., 21 pp.

Tundisi, J. G., T. M. Tundisi, G. Strixino, M. A. Martins, E. Mattos & E. P. dos Santos, 1972b. Estudos ecológicos na Repressa do Broa, 2. (Progress Rep.). Univ. fedl. S. Carlos, SP., 61 pp.

Tundisi, J. G. & K. Hino, 1981. List of species and growth seasons of phytoplankton from Lobo (Broa) Reservoir. Revta bras. Biol. 41: 63-68.

Ward, F. J., B. Wong & G. G. C. Robinson, 1970. A liquid scintillation procedure for determining the effect of size on self-absorption of ^{14}C in *Daphnia pulex*. Limnol. Oceanogr. 15: 648-651.

Watanabe, T., 1981. Flutuação sazonal e distribuição espacial do nano e microfitoplancton na Represa do Lobo ('Broa'), São Carlos, SP. Diss. Mestrado. Univ. fedl. S. Carlos, SP., 160 pp.

Wilson, D. S., 1973. Food size selection among copepods. Ecology 54: 909-914.

Some Crustacea Copepoda from Venezuela

B. H. Dussart
Station Biologique de l'Université de Paris VI, F 24620 Les Eyzies, France

Keywords: tropical zooplankton, Crustacea, Copepoda, Venezuela, biogeography, ecology

Abstract

The study of 38 samples of aquatic fauna from Venezuela increased the number of known species here from 28 to 66. Fifteen new species for science are described. A list of Copepoda known from the other regions of South America is presented. From this list, it can be said that only 50% of the inland water Copepoda living actually in Venezuela are known.

In spite of the diversity of aquatic environments in Venezuela, their crustacean fauna is not well known. Only the lake of Valencia has been regularly prospected (Pearse, 1921; Kiefer, 1954, 1956; de Infante *et al*, 1979). The Orinoco delta and the surroundings of Lake Maracaibo were studied by Gessner (Kiefer, 1956) and Deevey (Bowman, 1973). Zoppi & Michelangeli were interested in the Caracas region and state of Guarico (region of Calabozo). In the Caribbean sea, waters of some isles were prospected (Kiefer, 1933; Bowman, 1979; Dussart, 1982a).

This region is biogeographically interesting because it is at the cross-section of the Caribbean, Andes, Amazonia and Guyana Massifs. Thus, I was pleased to have the opportunity to collect samples there, some of which have been rarely prospected in Venezuela.

Stations studied (Fig. 1)

a. Collection B. Dussart

1. Lake of Valencia (littoral, South), 12.10.1981
   Mesocyclops meridianus (Kiefer, 1926)
   Thermocyclops decipiens (Kiefer, 1929)
2. Lake of Valencia (other station in the surroundings), 12.10.1981
   Microcyclops anceps (Richard, 1897)
   Thermocyclops decipiens
3. Lake of Valencia (littoral, South, with *Typha* sp.) 12.10.1981
   Microcyclops anceps
   Mesocyclops meridianus
   Thermocyclops decipiens
4. Small pool near and in communication with lake of Valencia, 12.10.1981, numerous larvae of Culicids
   Mesocyclops anceps
   Mesocyclops meridianus
5. Lake of Valencia (littoral, South, with *Eichhornia*), 12.10.1981
   Notodiaptomus deeveyorus nov. nom.
   Microcyclops anceps
6. Rio near Magdalena (Carabobo), among periphyton (biotecton), 12.10.1981
   Mesocyclops aspericornis (Daday, 1906)
7. Zuata reservoir near Cagua (Aragua), 12.10.1981
   Prionodiaptomus colombiensis (Thiébaud, 1912)
   Microcyclops anceps
   Mesocyclops meridianus

Hydrobiologia 113, 25–67 (1984).

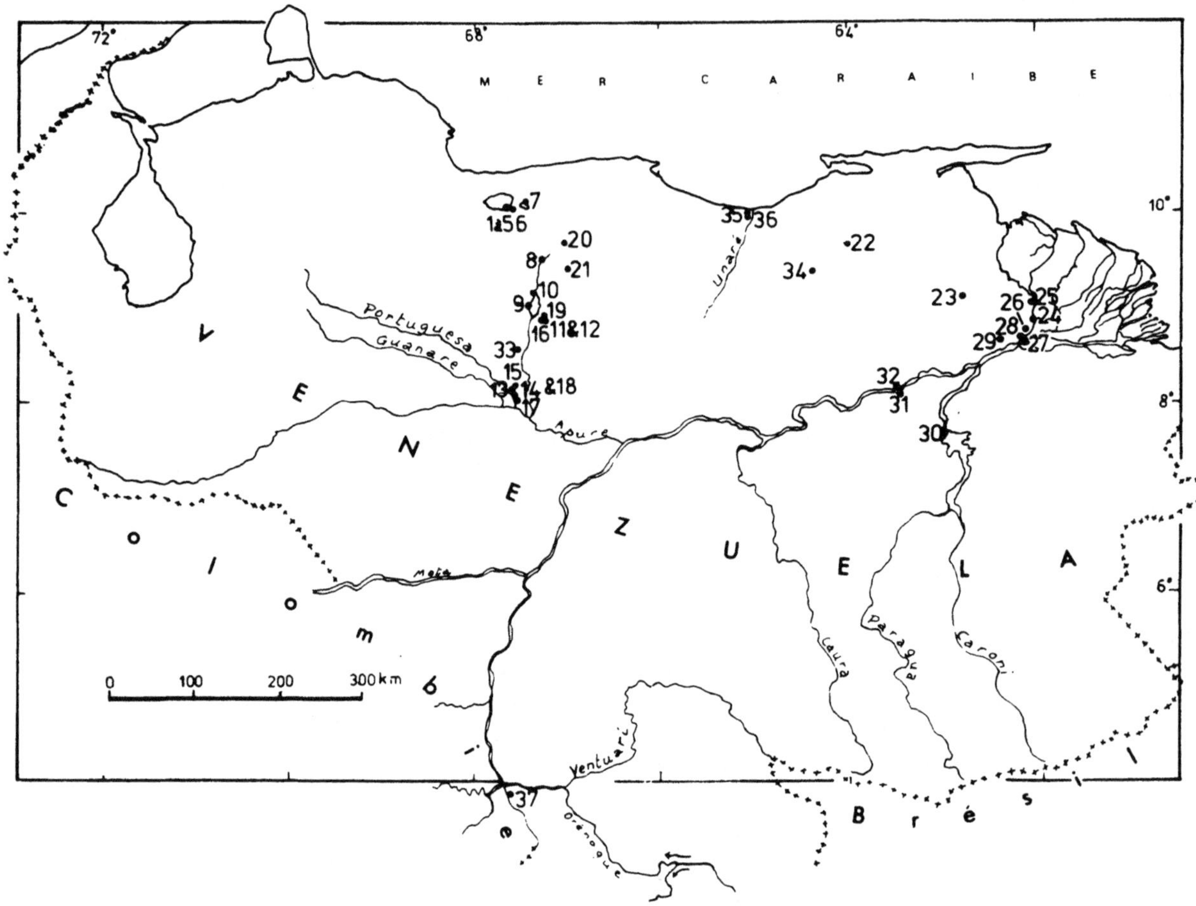

Fig. 1. Situation of sampling stations

Thermocyclops decipiens

8. Swamp near Dos Caminos (near Ortiz) (Guarico), 15.10.1981
   Acanthocyclops sp. ? juv.
   Mesocyclops longisetus (Thiébaud, 1914)
   Mesocyclops meridianus
9. Guarico reservoir near Calabozo (Guarico) at front of the pump plant, 15.10.1981
   Notodiaptomus cearensis (Wright, 1936)
   Mesocyclops meridianus
   Thermocyclops decipiens
10. Guarico reservoir 'up stream', 15.10.1981
    Microcyclops anceps
    Mesocyclops meridianus
11. Peat bog with Morichal, farm near Calabozo, 16.10.1981
    Macrocyclops sp. juv.
    Ectocyclops herbsti nov. nom.
    Microcyclops varicans (Sars, 1863)
12. *Drinking tank at the farm near Calabozo, 16.10.1981*
    Metacyclops tredecimus (Lowndes, 1934)
    Mesocyclops longisetus
13. Caño Falcon, Rio Portuguesa near San Fernando de Apure (Guarico), 16.10.1981
    Notodiaptomus cearensis
    Macrocyclops albidus principalis Herbst, 1963
    Microcyclops anceps
    Thermocyclops decipiens
    Paramphiascella aquaedulcis nov. spec.
    Canthocamptus cf. *microstaphylinus* Wolf, 1905
    Attheyella orinocoensis nov. spec.
    Elaphoïdella bidens (Schmeil, 1894)
14. Rio Portuguesa at Camaguan (Guarico), 16.10.1981
    Notodiaptomus sp. juv.
    Macrocyclops albidus albidus (Jurine, 1820)

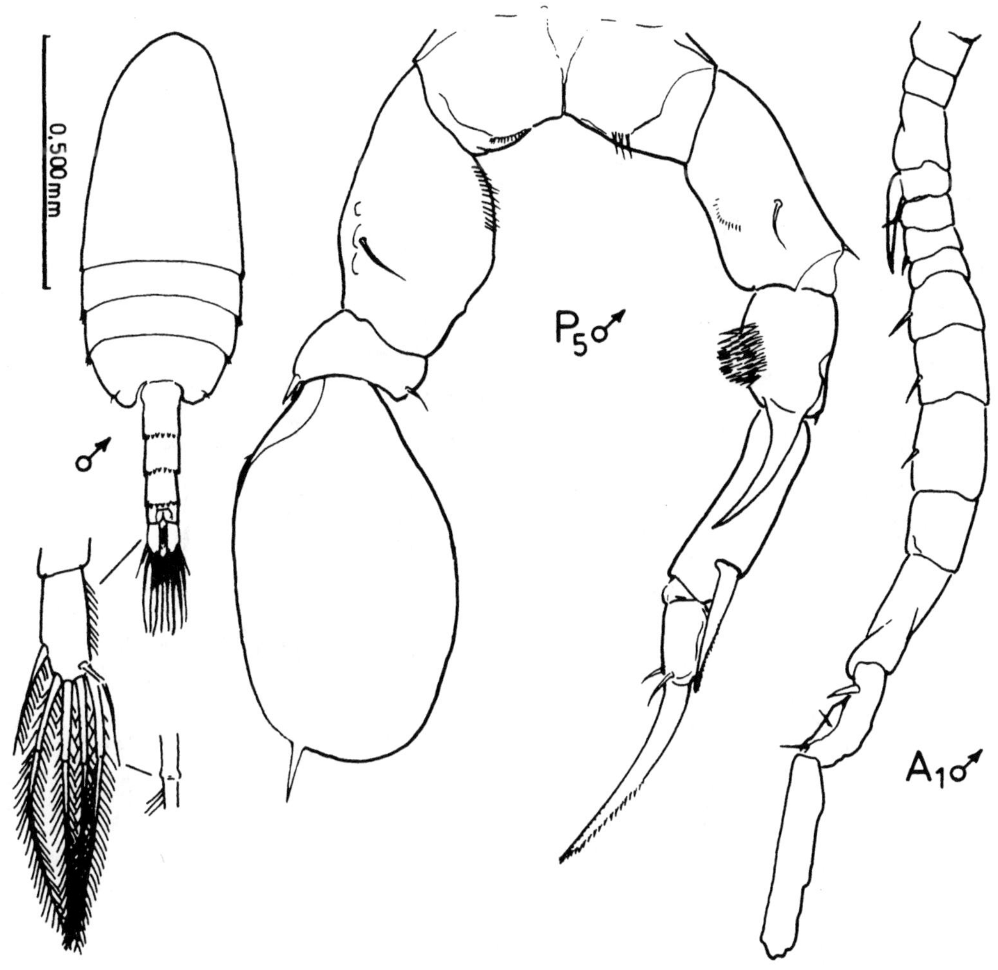

Fig. 2. Pseudodiaptomus gracilis, (orig.)

Microcyclops anceps
Thermocyclops decipiens

15. 'Estero'[1] de Camaguan (Guarico), near the road n° 2, 16.10.1981
Prionodiaptomus colombiensis
Notodiaptomus sp.
Microcyclops varicans
Mesocyclops meridianus
Thermocyclops minutus (Lowndes, 1934)

16. Limnocrene spring at El Carmen (Hato Becerra) near Calabozo, 16.10.1981
Ectocyclops cf. *bromelicola* juv.
Elaphoïdella bispina nov. spec.
Forficatocaris forficata crenensis nov. ssp.

[1] Local name for the back waters left after floods of the rivers in the plains (llanos) and return in their bed. An 'estero' is also partially filled with rain waters.

17. Rio Portuguesa at Camaguan, 'gas oil' station (dock), 16.10.1981
Mesocyclops ellipticus Kiefer, 1936

18. Pool with *Azolla* and leaves, under trees near Camaguan, 16.10.1981
Microcyclops diversus (Kiefer, 1935)
Microcyclops finitimus nov. spec.
Neutrocyclops brevifurca (Lowndes, 1934)
Mesocyclops meridianus
Thermocyclops decipiens
Thermocyclops minutus

19. Pond (natural) 'los Patos' near the field biological station of Calabozo, 16.10.1981
Notodiaptomus cearensis
Microcyclops finitimus
Mesocyclops meridianus, juv.

20. Man-made lake at Camatagua (Aragua), 17.10.1981

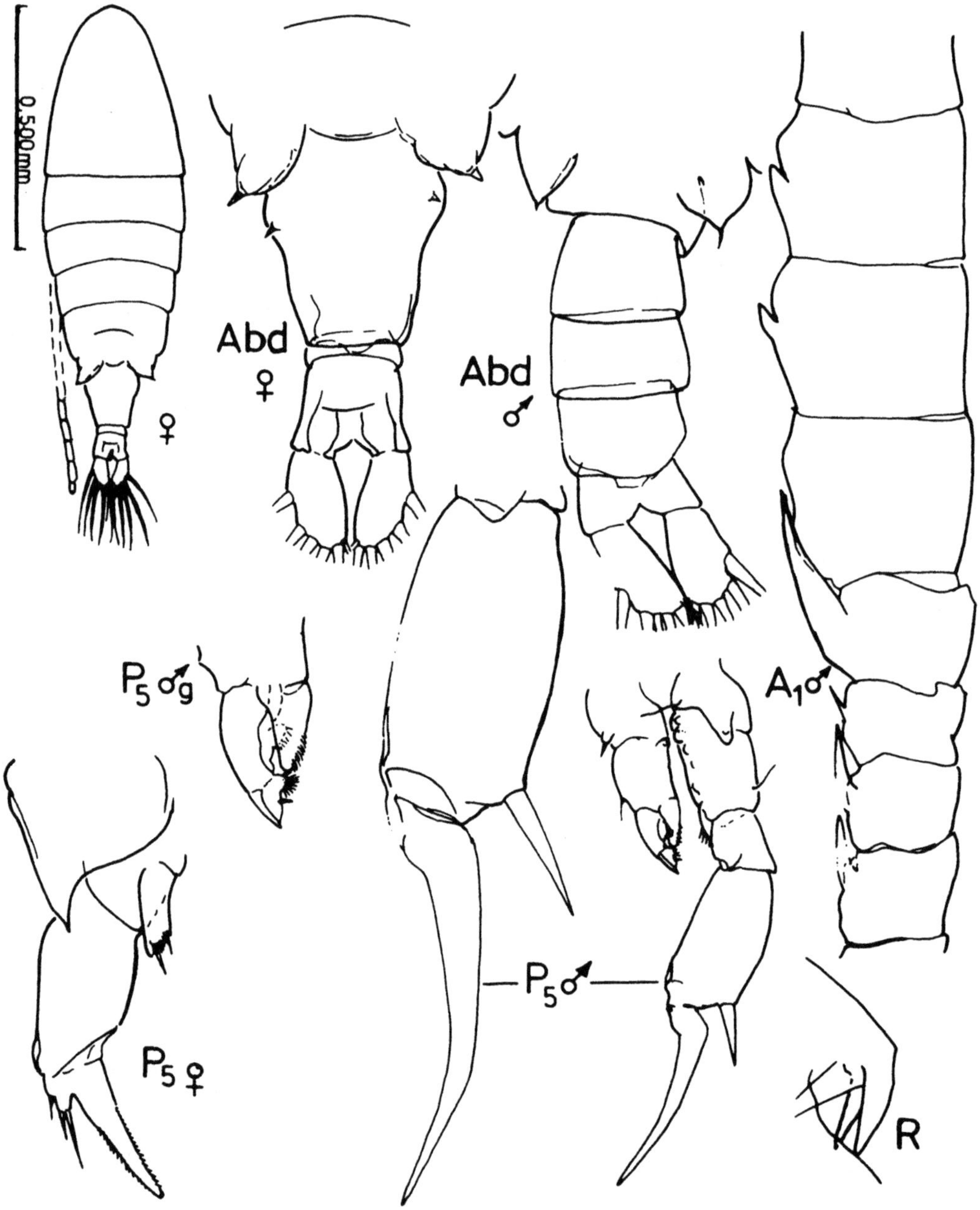

Fig. 3. Notodiaptomus henseni (= *N. venezolanus*) (Paratypes of this 'species') (orig.)

Notodiaptomus cearensis
Macrocyclops albidus albidus
Thermocyclops decipiens

21. Pond (natural) near El Sombrero (Guarico), with important littoral zone of macrophytes, 17.10.1981
Notodiaptomus cearensis
Mesocyclops meridianus
Thermocyclops decipiens
Thermocyclops minutus

22. Pond between Barcelona and Maturin, near Urica (road n° 13) (Anzoategui), 23.10.1981
Notodiaptomus cearensis
Ectocyclops herbsti
Microcyclops finitimus
Microcyclops dubitabilis (Kiefer, 1934)

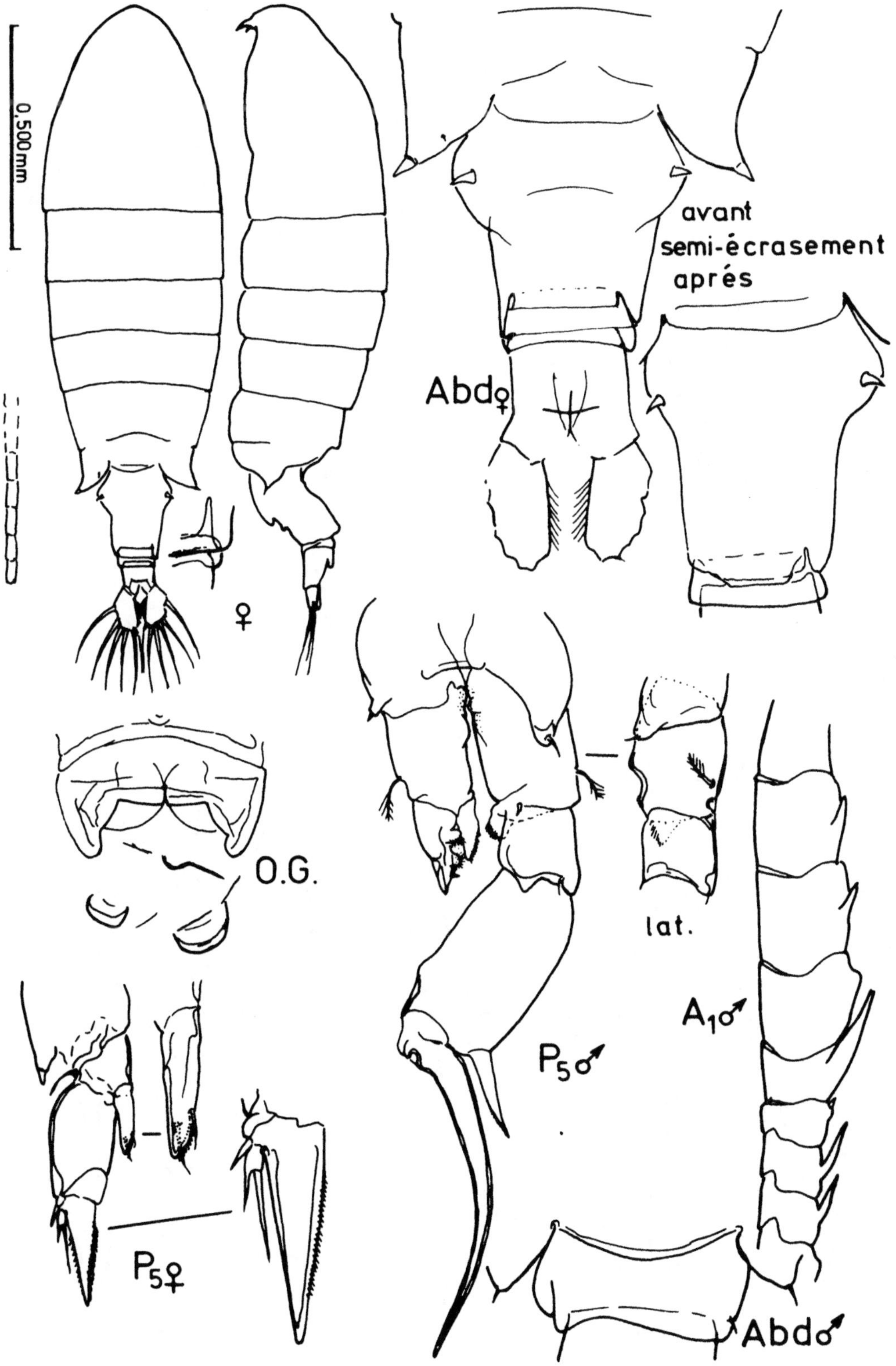

Fig. 4. Notodiaptomus deeveyorus nov. nom. (orig.)

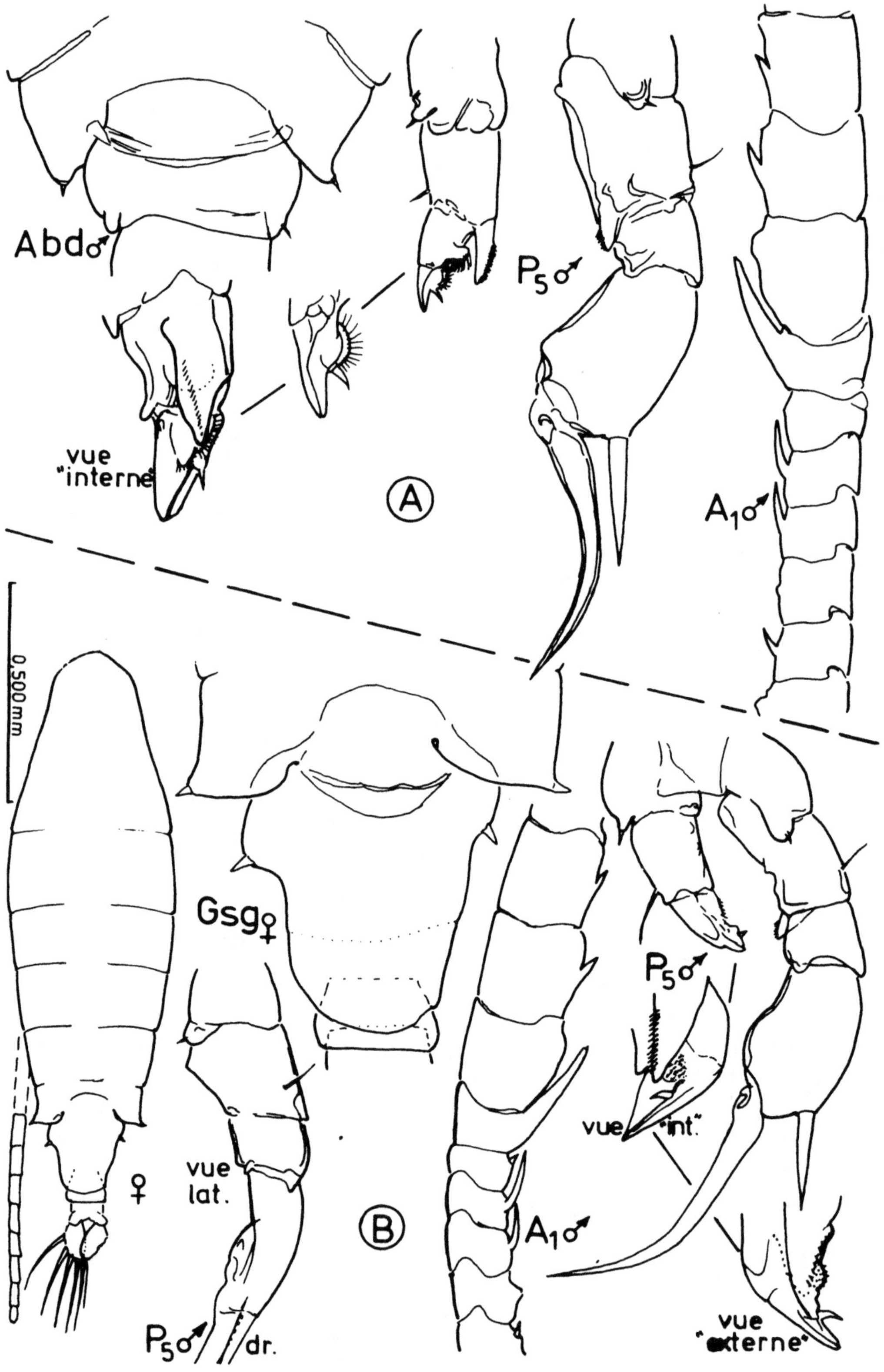

Fig. 5. A. *Notodiaptomus amazonicus;* B. *Notodiaptomus nordestinus* (orig.)

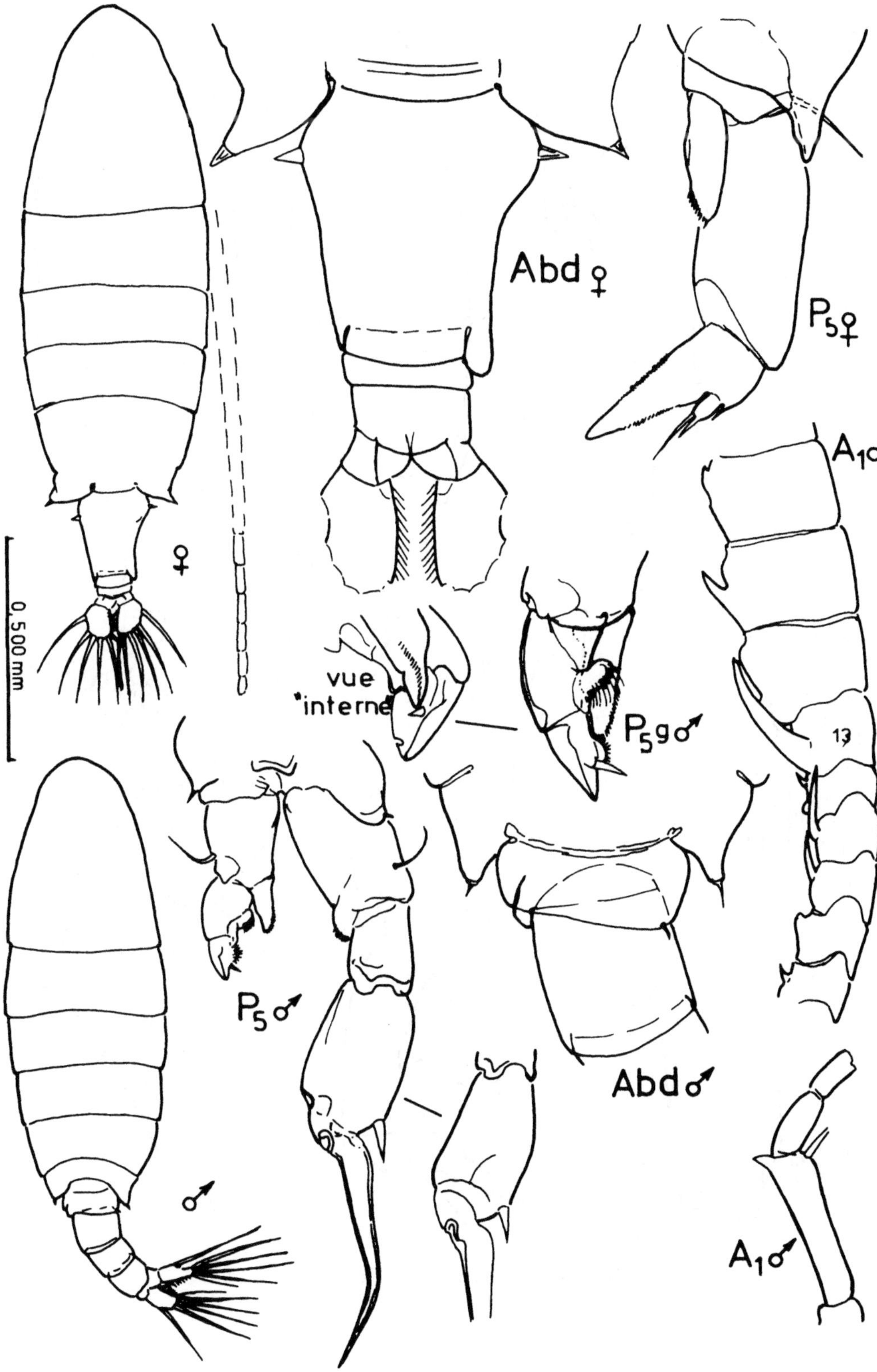

Fig. 6. Notodiaptomus cearensis (orig.)

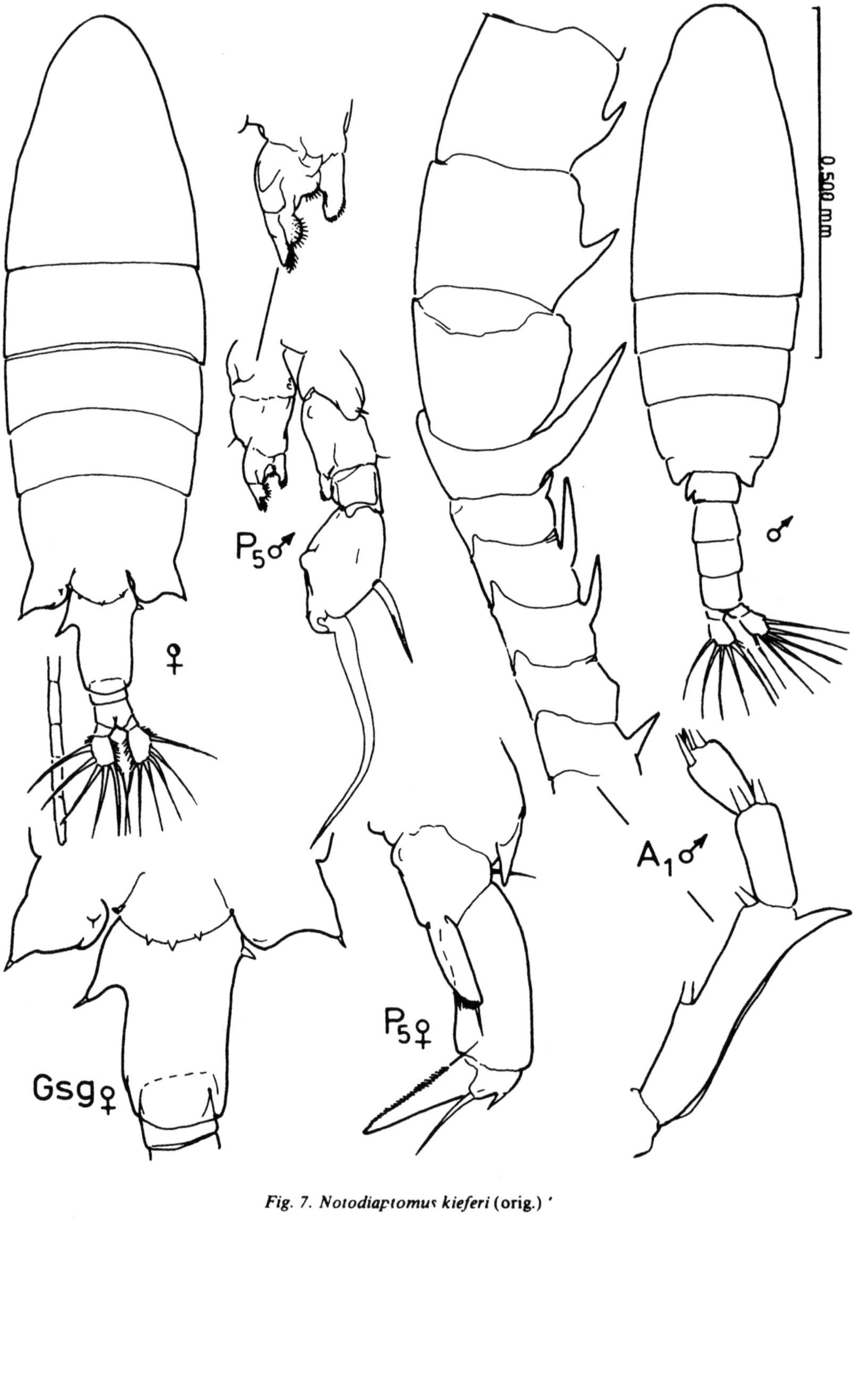

Fig. 7. Notodiaptomus kieferi (orig.)

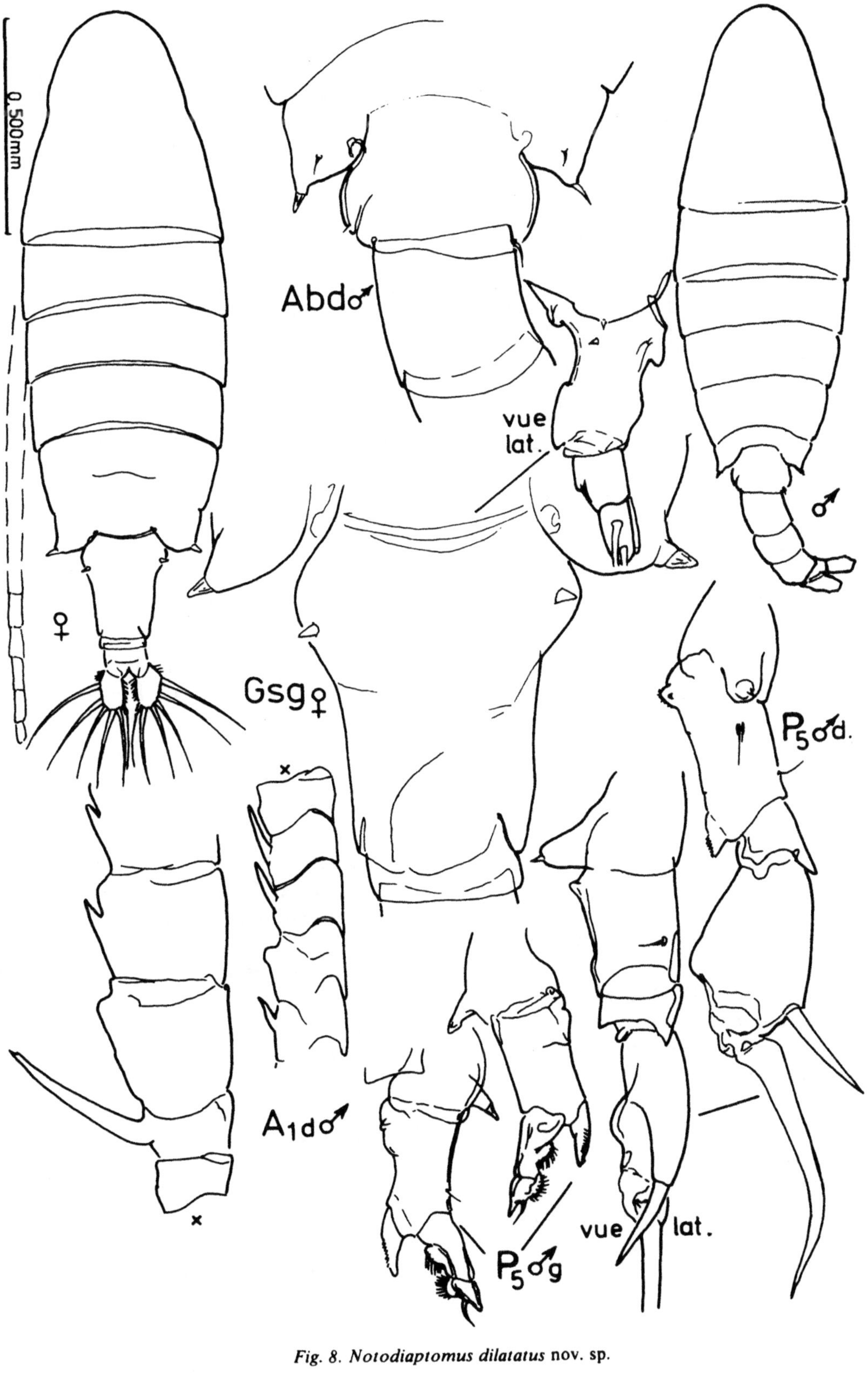

Fig. 8. Notodiaptomus dilatatus nov. sp.

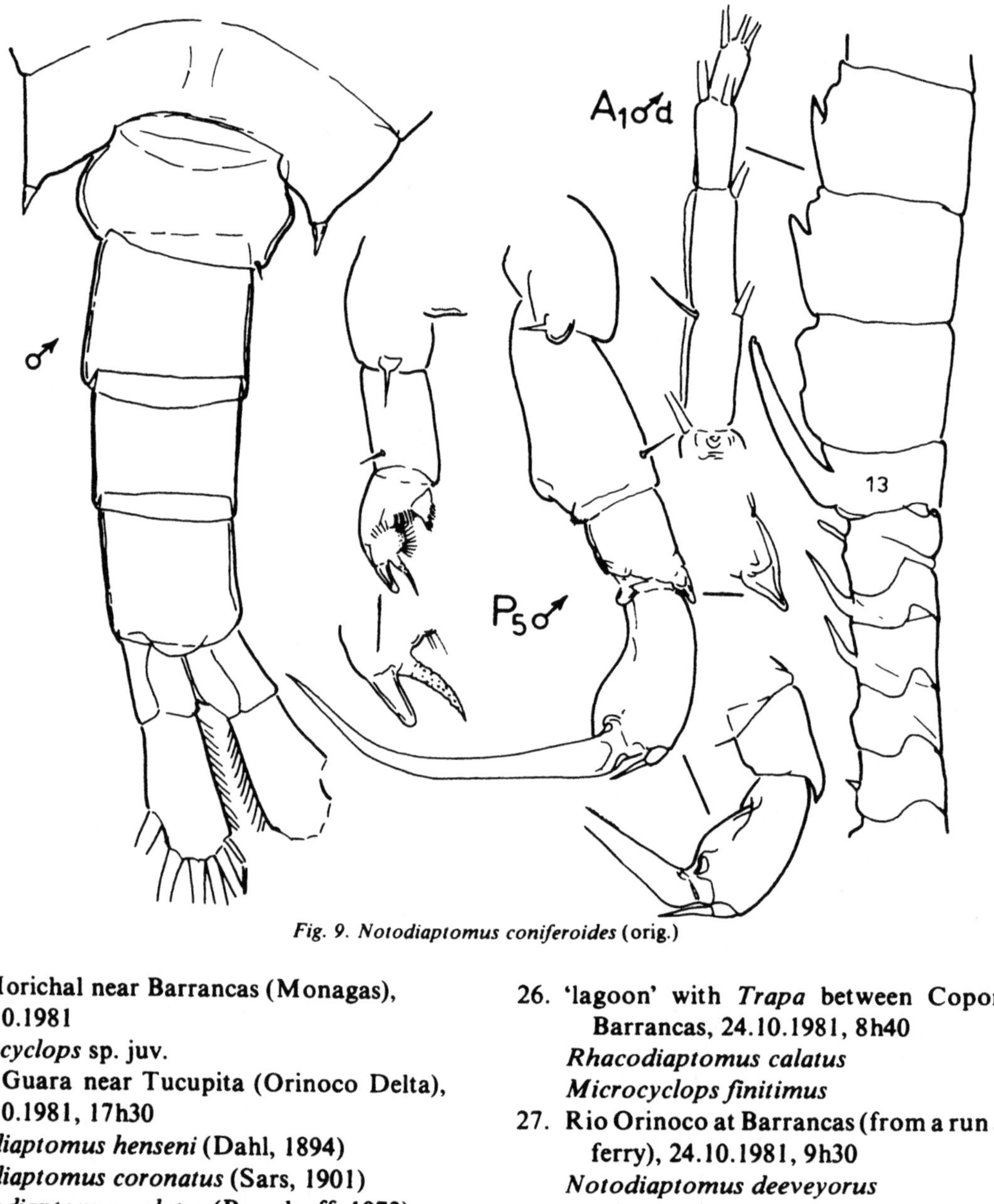

Fig. 9. Notodiaptomus coniferoides (orig.)

23. Rio Morichal near Barrancas (Monagas), 23.10.1981
    Microcyclops sp. juv.
24. Caño Guara near Tucupita (Orinoco Delta), 23.10.1981, 17h30
    Notodiaptomus henseni (Dahl, 1894)
    Notodiaptomus coronatus (Sars, 1901)
    Rhacodiaptomus calatus (Brandorff, 1973)
    'Diaptomus' negrensis (Andrade & Brandorff, 1975)
    Oithona sp.
    Macrocyclops albidus albidus
    Mesocyclops meridianus
25. Caño Manamo near Tucupita, 24.10.1981, 8h
    Notodiaptomus deeveyorus
    Notodiaptomus cearensis
    Notodiaptomus henseni
    Ectocyclops herbsti
    Mesocyclops meridianus
26. 'lagoon' with *Trapa* between Coporito and Barrancas, 24.10.1981, 8h40
    Rhacodiaptomus calatus
    Microcyclops finitimus
27. Rio Orinoco at Barrancas (from a run aground ferry), 24.10.1981, 9h30
    Notodiaptomus deeveyorus
    Notodiaptomus amazonicus (Wright, 1935)
    Notodiaptomus cearensis
    Notodiaptomus coniferoides (Wright, 1927)
    Rhacodiaptomus calatus
    'Diaptomus' negrensis
    Dactylodiaptomus pearsei (Wright, 1927)
    Oithona amazonica (Burckhardt, 1913)
    Macrocyclops albidus juv.
    Paracyclops pilosus nov. spec.
    Microcyclops anceps
    Mesocyclops ellipticus

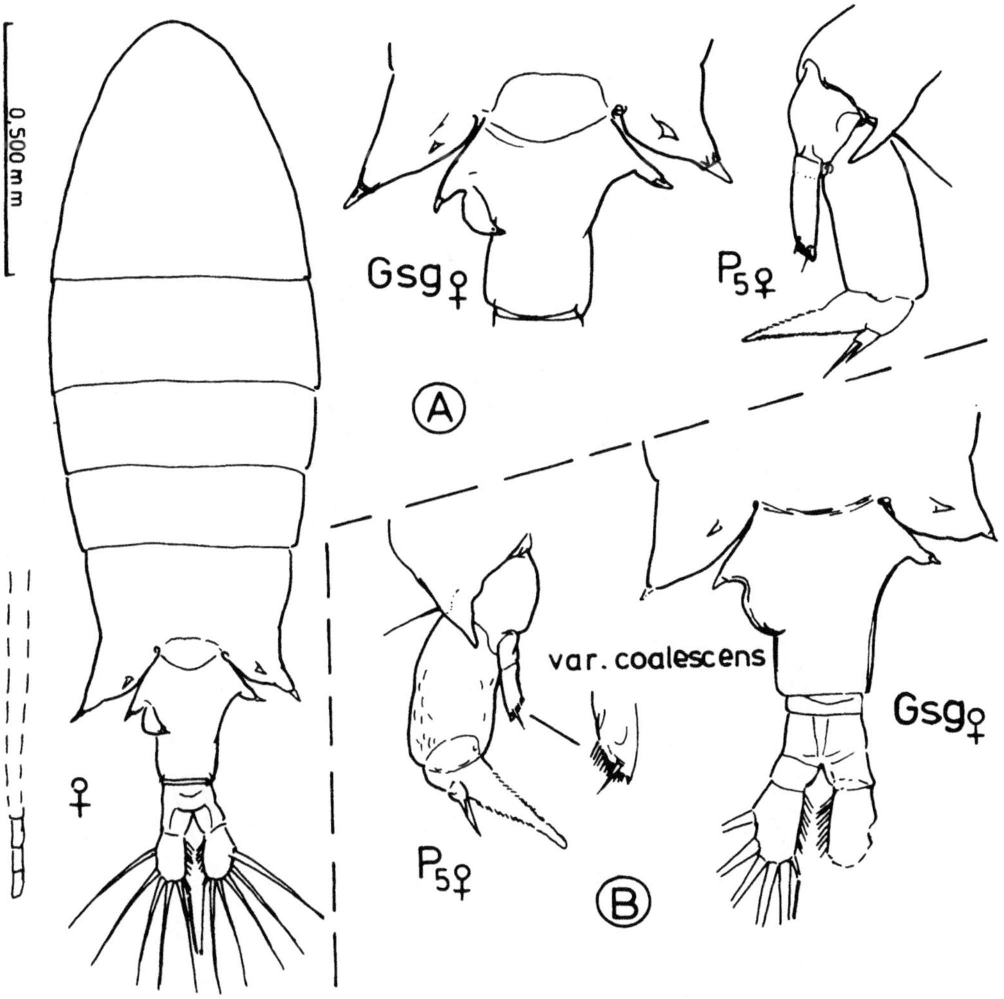

Fig. 10. A. *Rhacodiaptomus calatus;* B. *Rhacodiaptomus calatus coalescens* nov. sspec. (orig.)

28. Caño (pond) lateral and near Orinoco at Barrancas (south from the town), 24.10.1981, 10h
    Pseudodiaptomus gracilis (Dahl, 1894)
    Notodiaptomus deeveyorus
    Notodiaptomus coniferoides
    Rhacodiaptomus calatus
    'Diaptomus' negrensis
    Oithona amazonica
    Macrocyclops sp., juv.
    Ectocyclops compactus (Sars, 1909)
    Microcyclops finitimus
    Thermocyclops minutus
    Attheyella (Chappuisiella) orinocoensis nov. spec.
29. Rio Guarguapo with Morichal near Barrancas, 24.10.1981, 11h30
    Microcyclops finitimus
    Metacyclops curtispinosus nov. spec.
30. Guri, man-made lake near the dam on Caroni River, 24.10.1981, 16h
    Notodiaptomus deeveyorus
    Notodiaptomus cearensis
    Notodiaptomus kieferi Brandorff, 1972
    Mesocyclops sp. juv.
    Thermocyclops minutus
31. Rio Orinoco, right side, at Ciudad Bolivar, 25.10.1981
    Notodiaptomus deeveyorus
    Notodiaptomus amazonicus
    Notodiaptomus dilatatus nov. spec.
    Notodiaptomus kieferi
    Rhacodiaptomus calatus
    Dactylodiaptomus pearsei

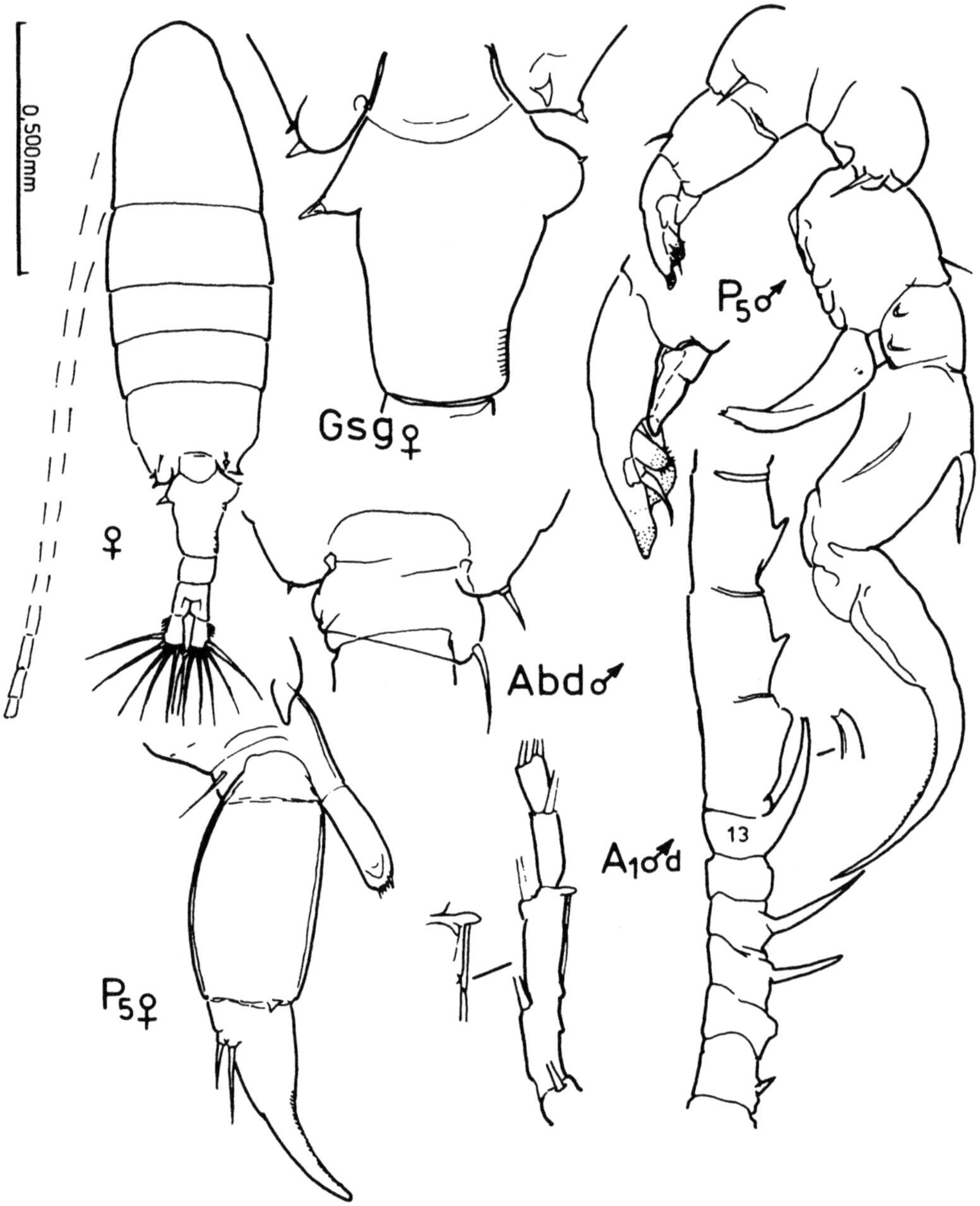

Fig. 11. 'Diaptomus' negrensis (orig.)

'Diaptomus' negrensis
Oithona amazonica
Macrocyclops albidus
Paracyclops pilosus nov. spec.
Microcyclops anceps
Metacyclops subaequalis nov. spec.
Allocyclops neotropicalis nov. spec.
Mesocyclops meridianus
Mesocyclops ellipticus
Themocyclops decipiens

32. Rio Orinoco, left side at Soledad, 25.10.1981, 10h
Notodiaptomus cearensis
'Diaptomus' negrensis
Paracyclops pilosus
Mesocyclops meridianus
Thermocyclops decipiens
33. Shady pool, km 245, road n° 2, between Camaguan and Calabozo, 16.10.1982
Prionodiaptomus colombiensis

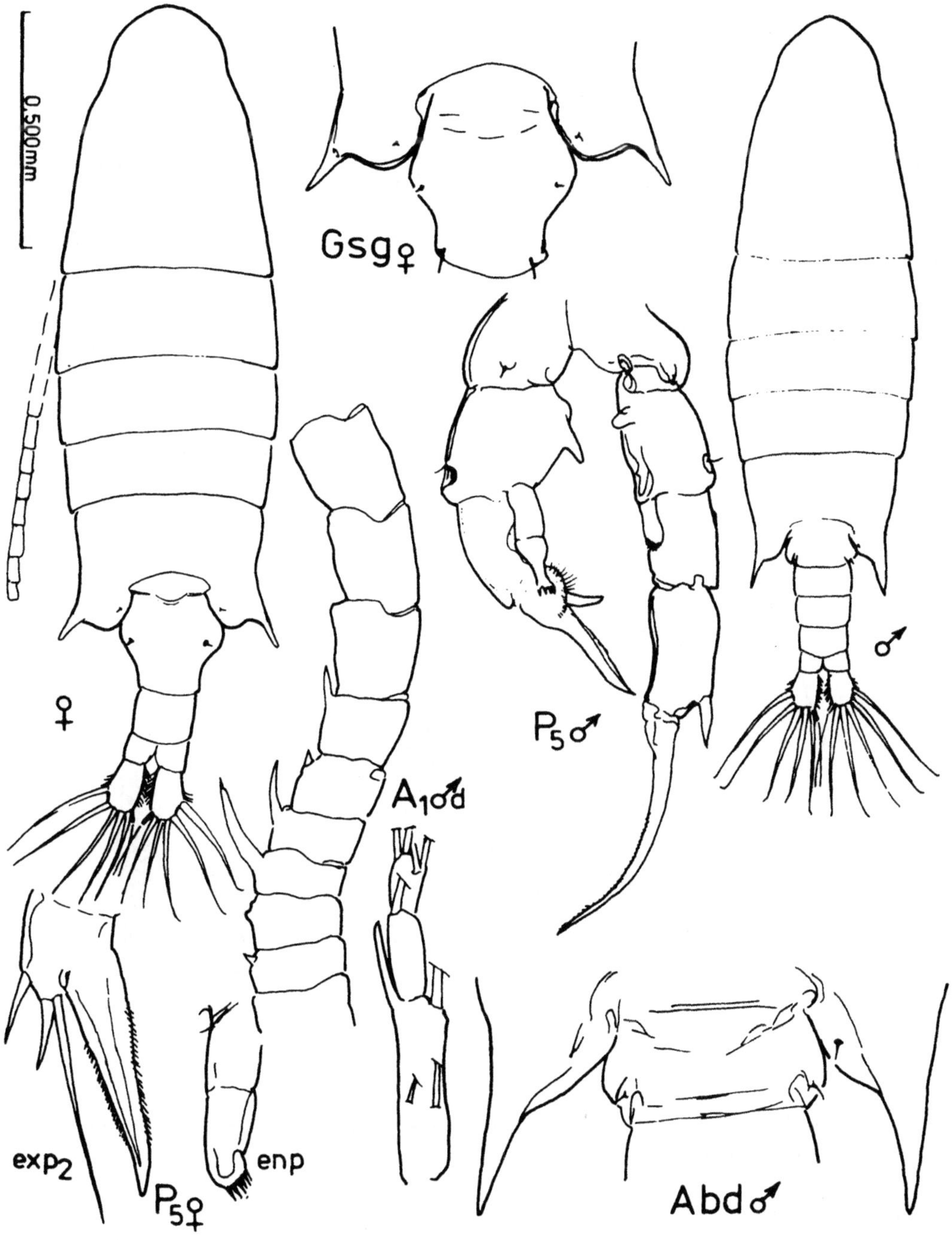

Fig. 12. Dactylodiaptomus pearsei (orig.)

34. 'Estero' between Cantaura and Anaco (cross of Buena Vista), 25.10.1981
Microcyclops cf. *varicans*
Mesocyclops meridianus
Mesocyclops ellipticus

b. Collection G. Pereira

35. Charca 1, near Unaré river at Clarines, 13.4.1981
Prionodiaptomus colombiensis
Microcyclops anceps var. *minor* nov. var.

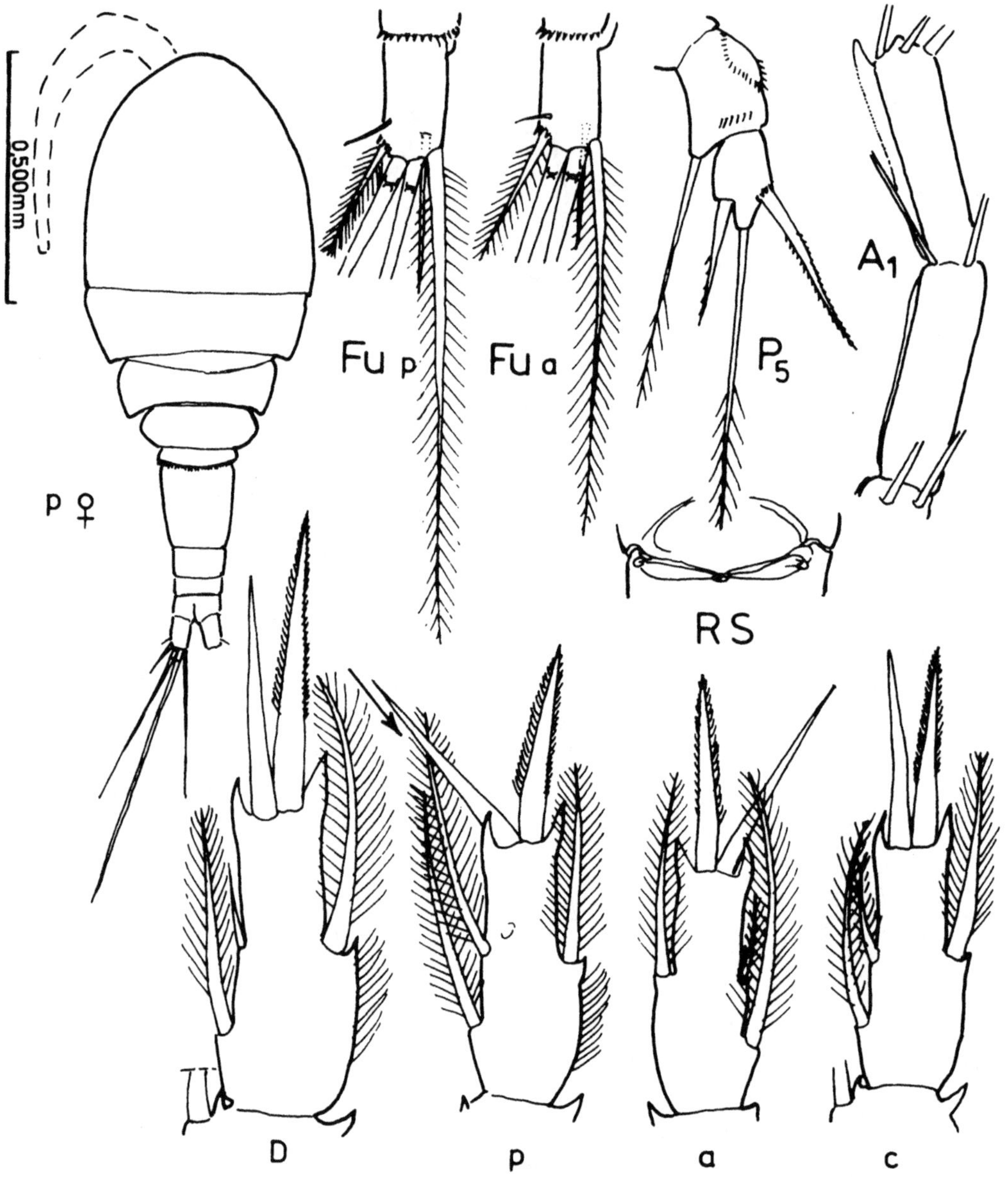

Fig. 13. Macrocyclops albidus. D = individual of Dordogne (France); a = individual of Rio Portuguesa; c = individual of Camaguan; p = individual of Caño Falcon (rio Portuguesa) *(M. a. principalis)* (orig.)

Mesocyclops aspericornis
Thermocyclops decipiens

36. Charca 2, near Unaré River, at Clarines, 13.4.1981
Notodiaptomus cearensis
Thermocyclops decipiens

c. Collection E. Zoppi de Roa

37. Rio Atabapo, 1, 24.2.1974
Notodiaptomus deeveyorus
Notodiaptomus coniferoides
Notodiaptomus kieferi
Notodiaptomus dilatatus nov. spec.
Rhacodiaptomus calatus coalescens nov. spec.
Mesocyclops meridianus

38. Laguna Mucubaji (Andes), 5.10.1980 (altitude 3 600 m)
Eucyclops pseudoensifer nov. spec.
Metacyclops leptopus mucubajiensis Kiefer, 1956

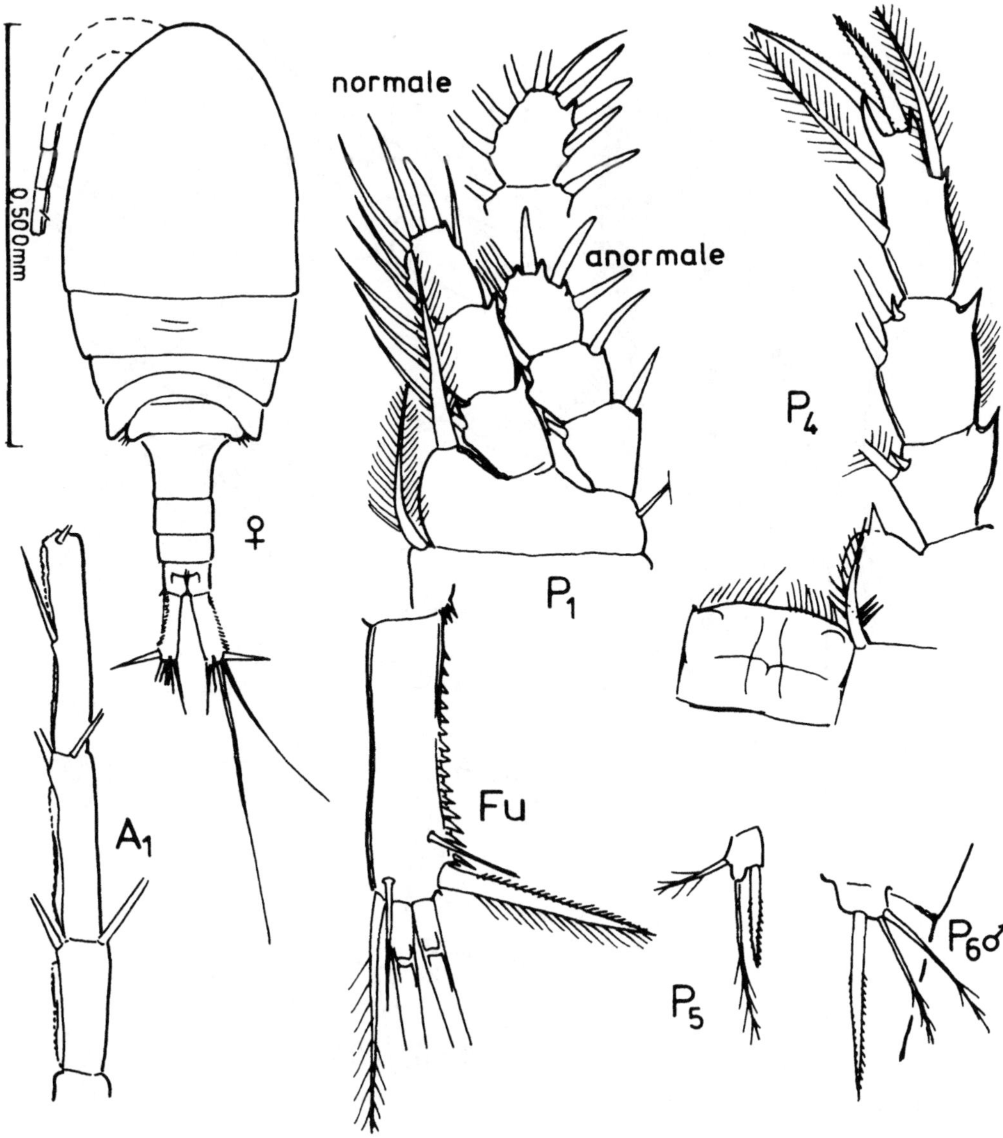

Fig. 14. Eucyclops pseudoensifer nov. sp. (orig.)

List of species observed

Order CALANOIDA
  Family Pseudodiaptomidae
    Pseudodiaptomus gracilis (Dahl, 1894)
  Family Diaptomidae
    Prionodiaptomus colombiensis (Thiébaud, 1912)
    Notodiaptomus henseni (= venezolanus) (Dahl, 1894)
    N. deeveyorus nov. nom.
    N. amazonicus (Wright, 1935)
    N. coniferoides (Wright, 1927)
    N. cearensis (Wright, 1936)
    N. coronatus (Sars, 1901)
    N. kieferi Brandorff, 1972
    N. dilatatus nov. spec.
    Rhacodiaptomus calatus Brandorff, 1973
    R. calatus coalescens nov. sspec.
    'Diaptomus' negrensis Andrade et Brandorff, 1975
    Dactylodiaptomus pearsei (Wright, 1927)

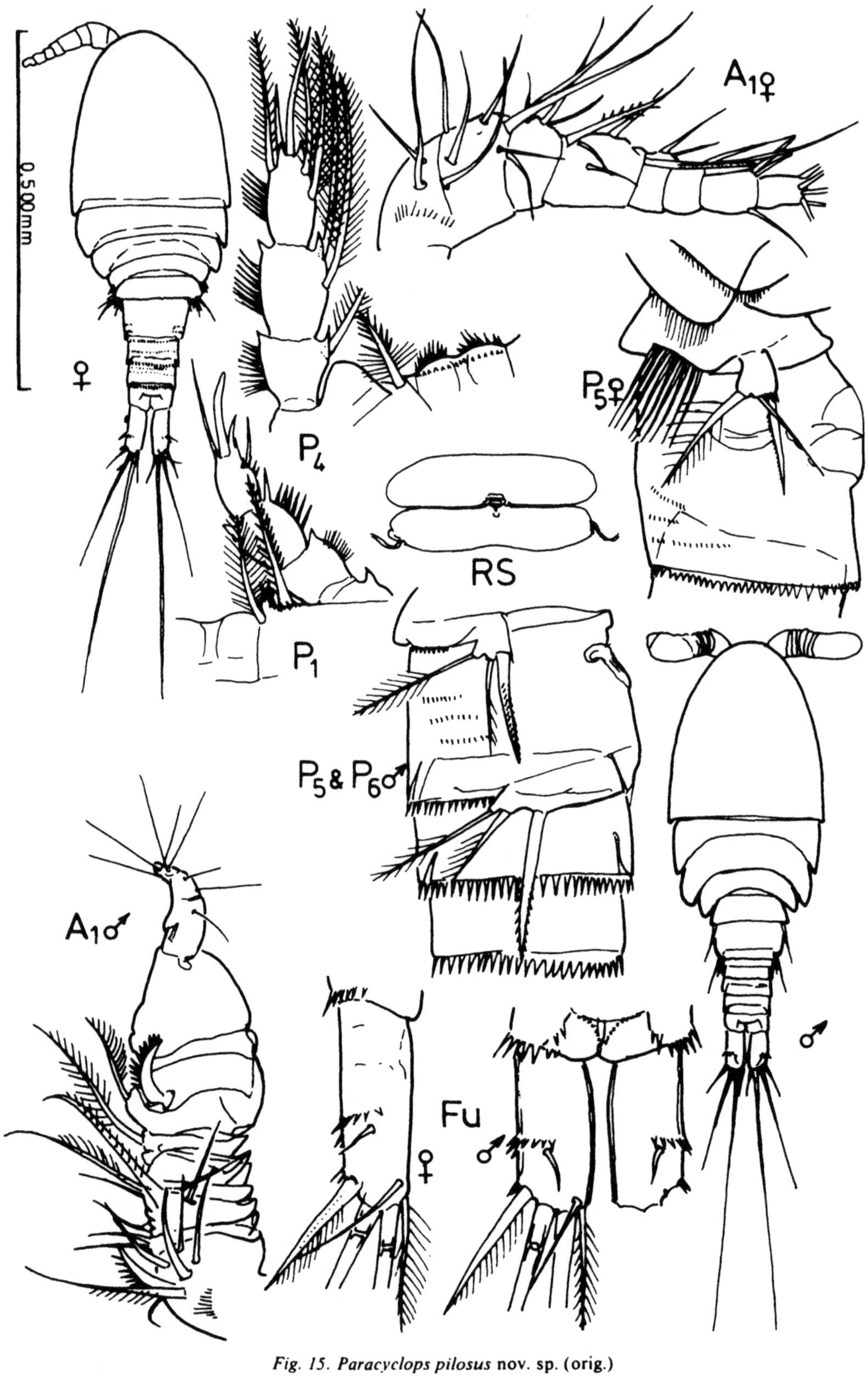

Fig. 15. Paracyclops pilosus nov. sp. (orig.)

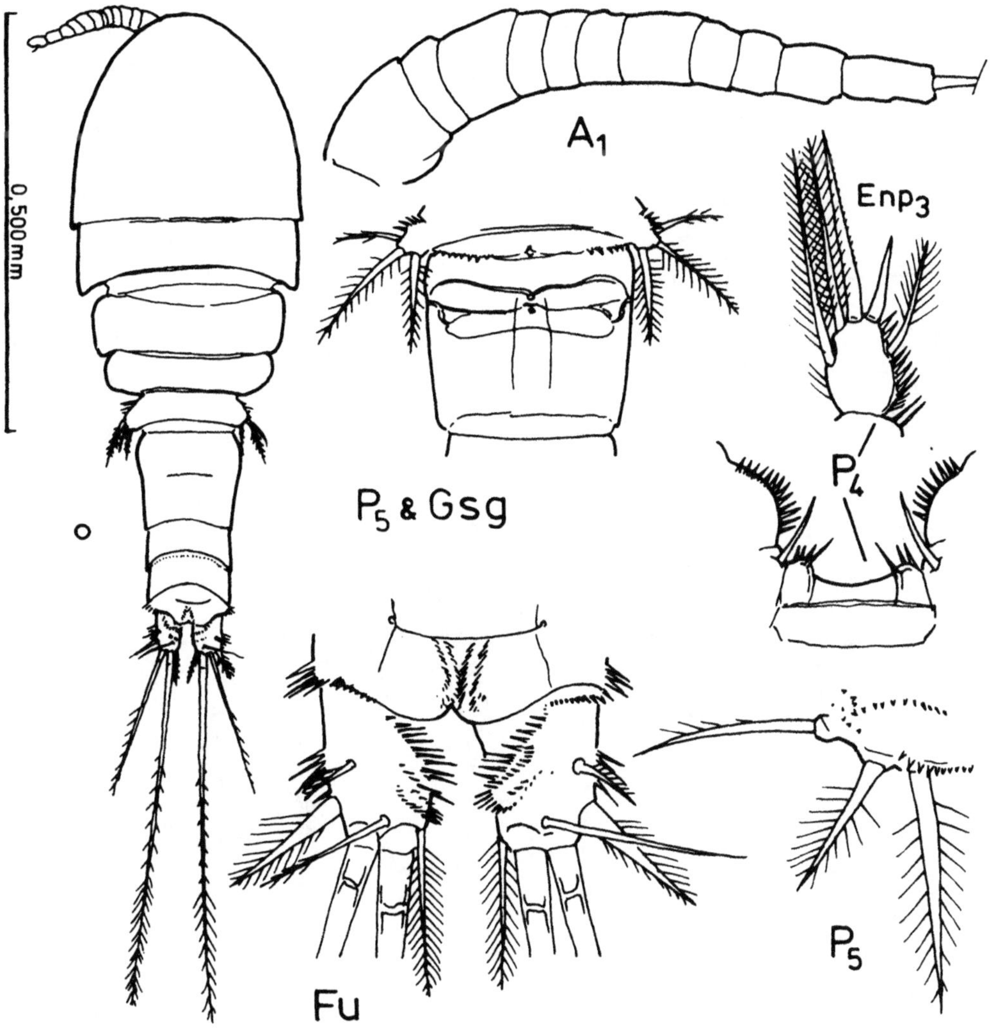

Fig. 16. Ectocyclops herbsti nov. sp. (orig.)

Order CYCLOPOIDA
- Family Oithonidae
  - *Oithona* sp.
  - *Oithona amazonica* Burckhardt, 1913
- Family Cyclopidae
  - *Macrocyclops albidus* (Jurine, 1820)
  - *Macrocyclops albidus principalis* Herbst, 1963
  - *Eucyclops pseudoensifer* nov. spec.
  - *Ectocyclops compactus* (Sars, 1909)
  - *E. herbsti* nov. nom.
  - *E.* cf. *bromelicola* Kiefer, 1935
  - *Paracyclops pilosus* nov. spec.
  - *Microcyclops anceps* (Richard, 1887)
  - *M. anceps* var. *minor* nov. var.
  - *M. diversus* (Kiefer, 1935)
  - *M. dubitabilis* (Kiefer, 1934).
  - *M. finitimus* nov. spec.
  - *M. varicans* (Sars, 1863)
  - *Metacyclops tredecimus* (Lowndes, 1934)
  - *Metacyclops curtispinosus* nov. spec.
  - *M. subaequalis* nov. spec.
  - *M. leptopus mucubajiensis* Kiefer, 1956
  - *Neutrocyclops brevifurca* (Lowndes, 1934)
  - *Allocyclops neotropicalis* nov. spec.
  - *Mesocyclops aspericornis* (Daday, 1906)
  - *M. meridianus* (Kiefer, 1926)
  - *M. longisetus* (Thiébaud, 1914)
  - *M. ellipticus* Kiefer, 1936
  - *Thermocyclops decipiens* (Kiefer, 1929)
  - *T. minutus* (Lowndes, 1934)

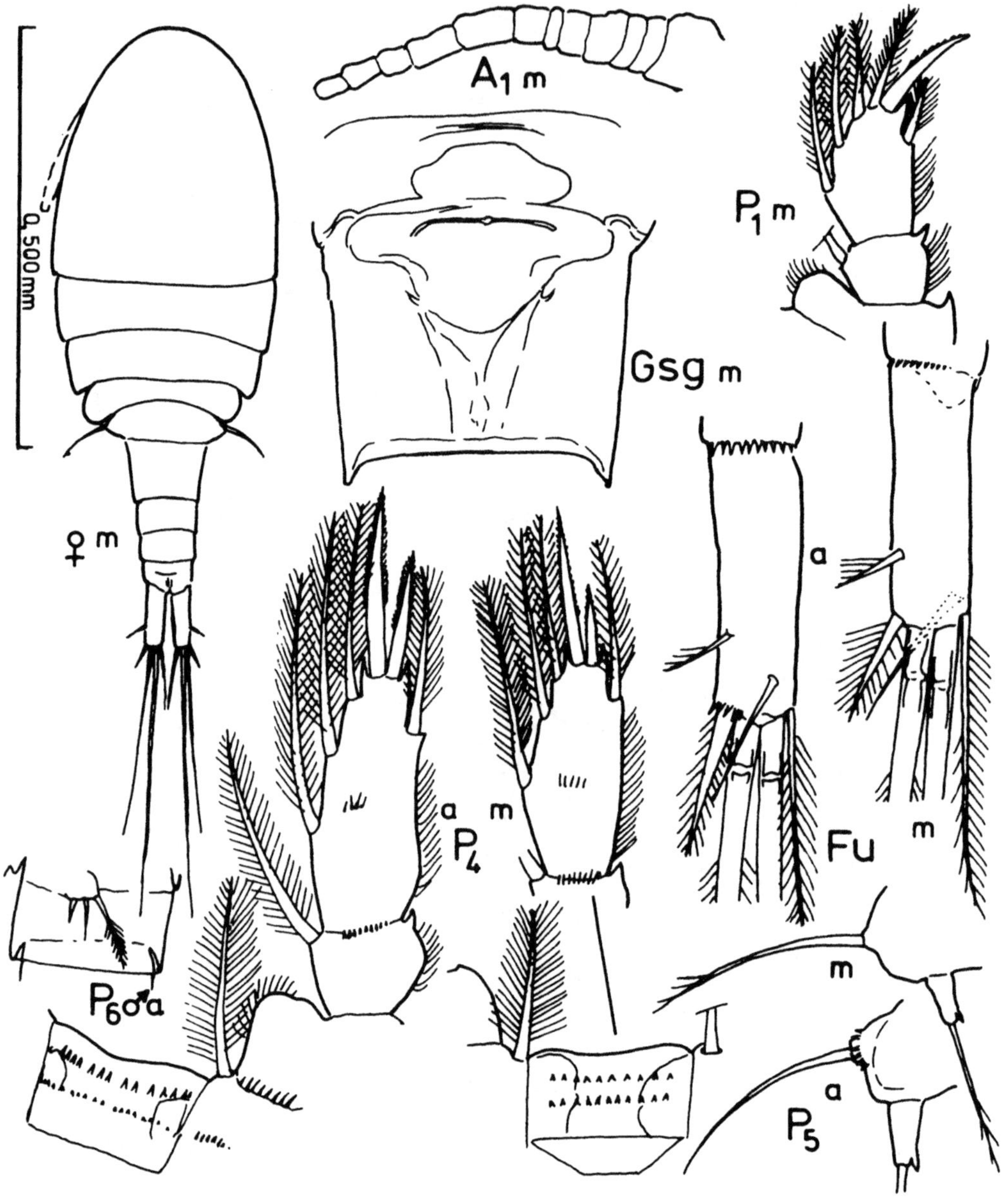

Fig. 17. Microcyclops anceps. a = *M. anceps* m = var. *minor* nov. var. (orig.)

Order HARPACTICOIDA

Family Diosaccidae

Paramphiascella aquaedulcis nov. spec.

Family Canthocamptidae

Canthocamptus cf. *microstaphylinus* Wolf, 1905

Attheyella (Chappuisiella) orinocoensis nov. spec.

Elaphoïdella bidens (Schmeil, 1894)

Elaphoïdella bispina nov. spec.

Family Parastenocaridae

Forficatocaris forficata crenensis nov. ssp.

Notes on some species[2]

[2] Types of some species and subspecies are placed in the author's collection, Station Biologique de l'Université Pierre et Marie Curie, at F.-24620 Les Eyzies and will be stocked thereafter in the collection of Museum National d'Histoire Naturelle in Paris.

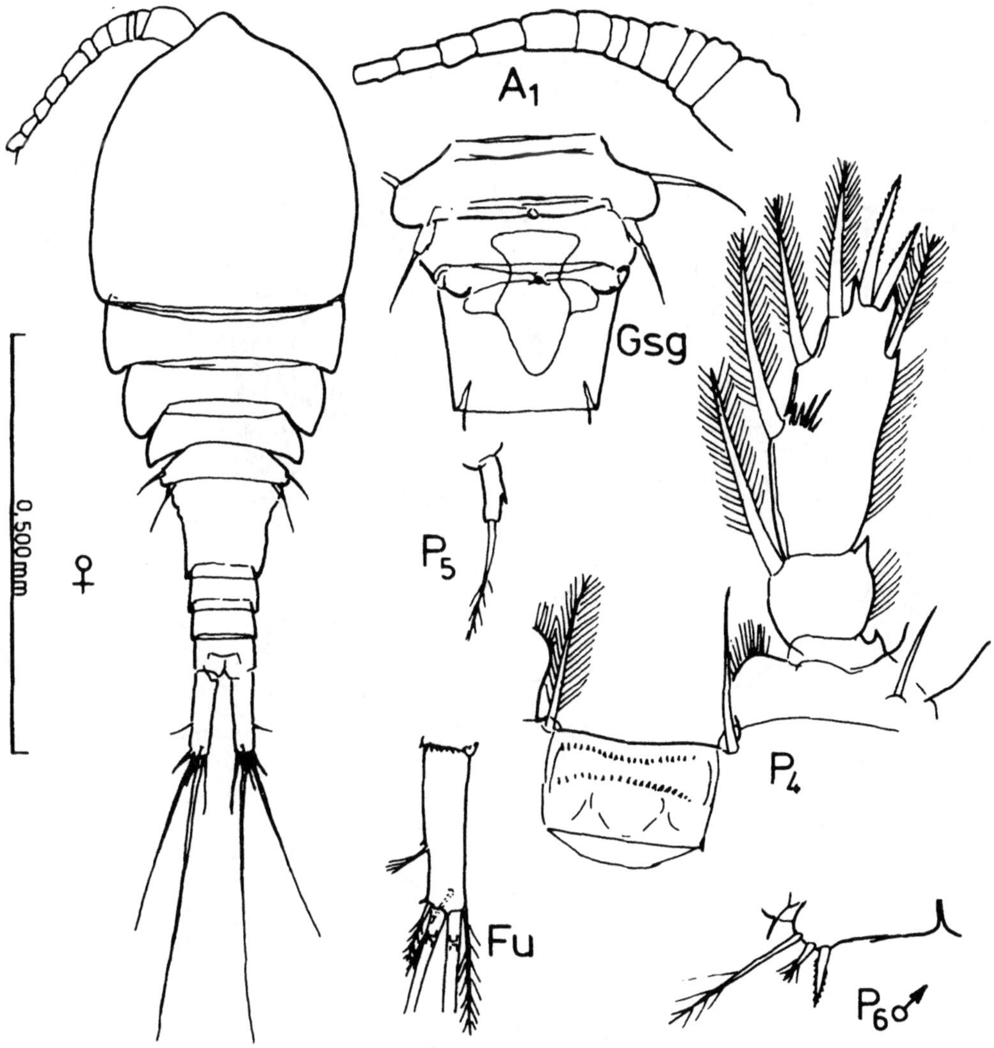

Fig. 18. Microcyclops diversus (orig.)

Pseudodiaptomus gracilis (Fig. 2)

Until now, no *Pseudodiaptomus* was found in Venezuela. Thus, the presence of *P. gracilis* near Tucupita is not noteworthy. This species is common in freshwaters such as the lower Amazone, rio Tocantin and lake Arary. Wright (1936) believed that it was endemic to the Amazonian region.

Only several males of the species were found; they were small and easy to recognize.

Notodiaptomus henseni (Fig. 3)

Contrary to what Wright stated (1935a), the 'description' of Dahl (1894) of *N. henseni* is precise enough, although reduced to some drawings and measurements. Wright's drawings, on the other hand, are insufficient.

The configuration of the right P5 of male is difficult to present in only one drawing because the knobs on exopodite 2 are only obvious in an oblique view.

The internal sides of basopodites right and left are granular as is the knob near the proximal inner angle of basopodite of right P5.

N. henseni presents no other noticeable peculiarity at the right antennule of male. Of the spiniform processes of the joints 8, 10, 11, 13, 15 and 16, those

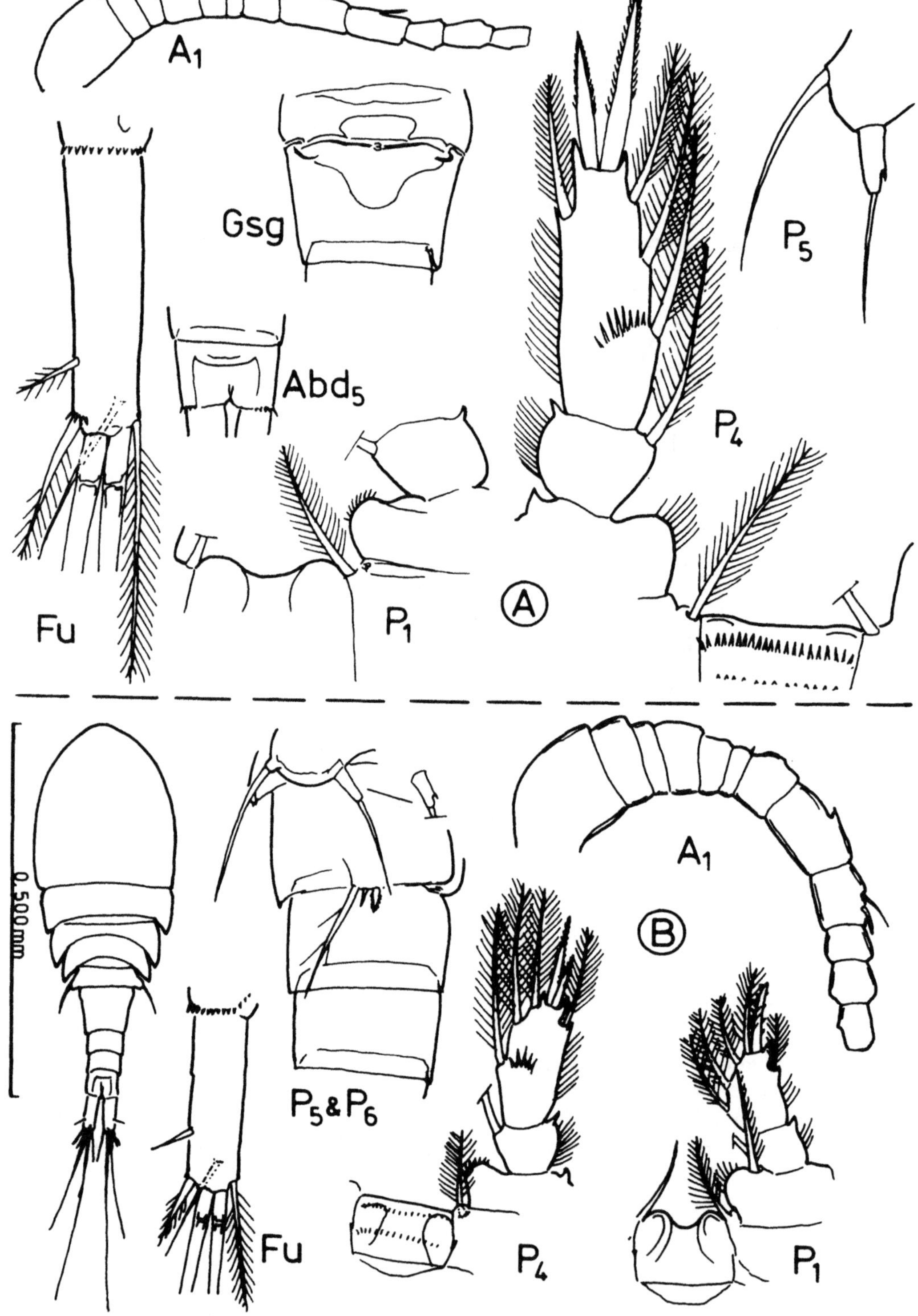

Fig. 19. A. *Microcyclops finitimus* nov. sp.; B. *Microcyclops cf. varicans* (orig.)

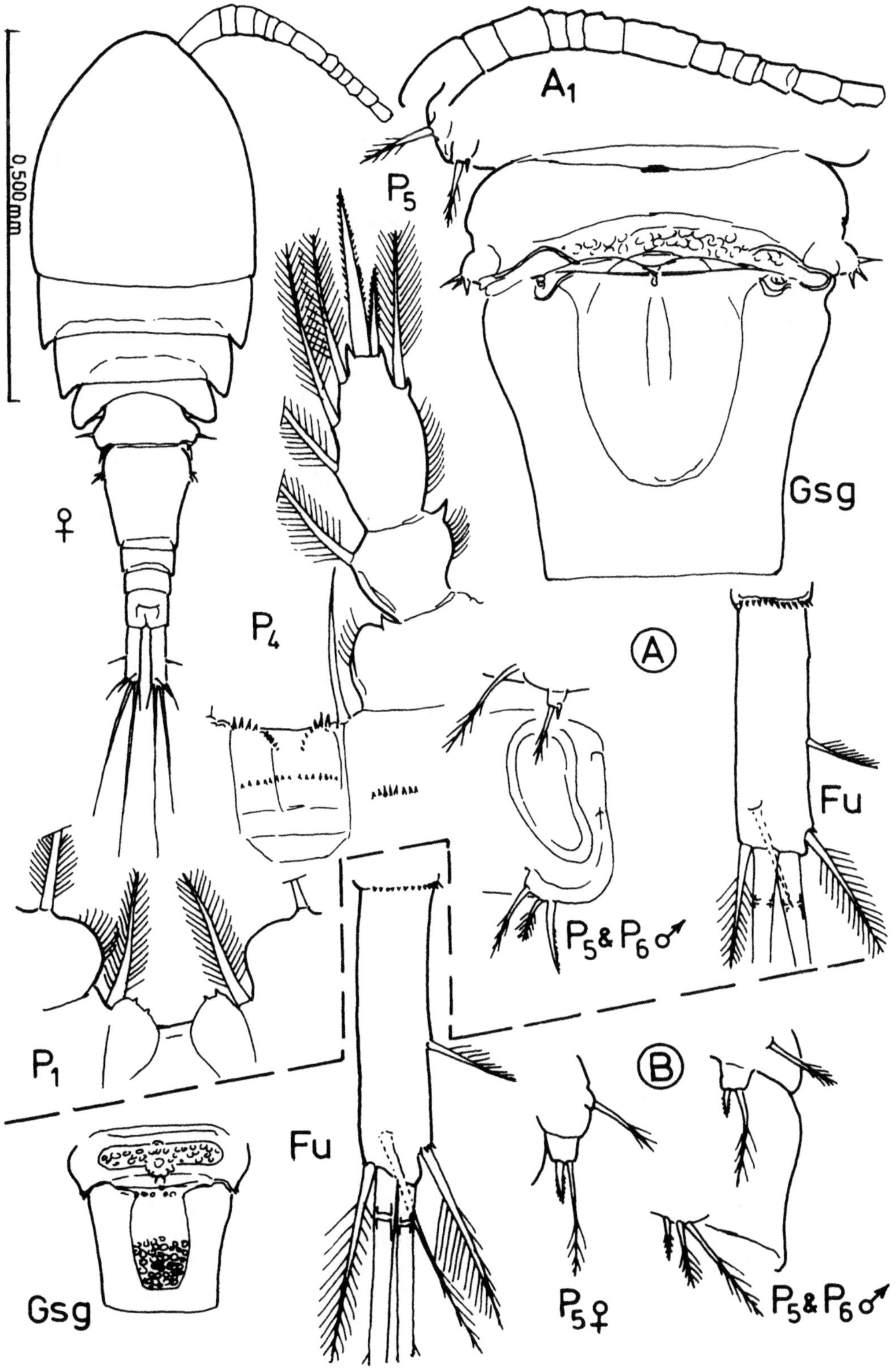

Fig. 20. A. *Metacyclops tredecimus;* B. *Metacyclops leptopus mucubajiensis* (orig.)

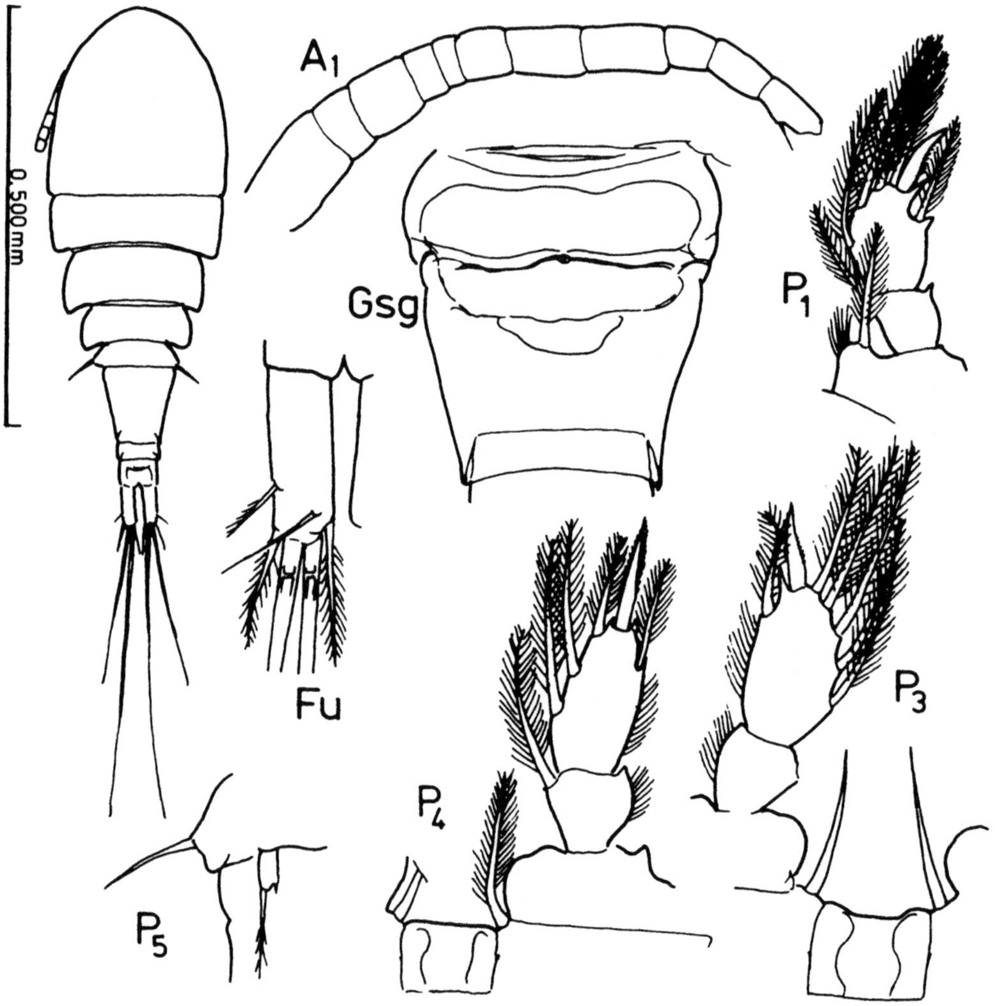

Fig. 21. Metacyclops curtispinosus nov. sp.

of 8, 15 and 16 are the smallest.

With the granulation of the internal side of basopodite of left P5, *N. henseni* makes transition with the genus *Argyrodiaptomus.* However, the species of this genus have an exopodite different at left and right P5.

I have studied paratypes of *N. venezolanus*[3] and I did not find any noticeable difference between this *N. venezolanus* and the *D. henseni* of Dahl. In comparing the illustrations of the first of these species by Kiefer (1954), I have come to the conclusion that these two species are synonymous.

[3] I thank T. Bowman from the Smithsonian Institution who kindly offered me this possibility.

Notodiaptomus deeveyorus nov. nom. (Fig. 4)

Bowman (1973) described a subspecies of *N. venezolanus* and named it *N. venezolanus deeveyorum.* He did not refer to the work of Dahl (1894) on *N. henseni.*

The differences between this last species and the subspecies *deeveyorum* are such that it is possible and necessary to elevate the subspecies to species. I suggest the name *N. deeveyorus.* Bowman (1973) was already of this opinion when he wrote: 'this decision could be altered when the now largely unknown diaptomid fauna of Northern Venezuela becomes adequately studied'.

N. deeveyorus is very close to that of *N. nordestinus* Wright, 1935. By comparison with some speci-

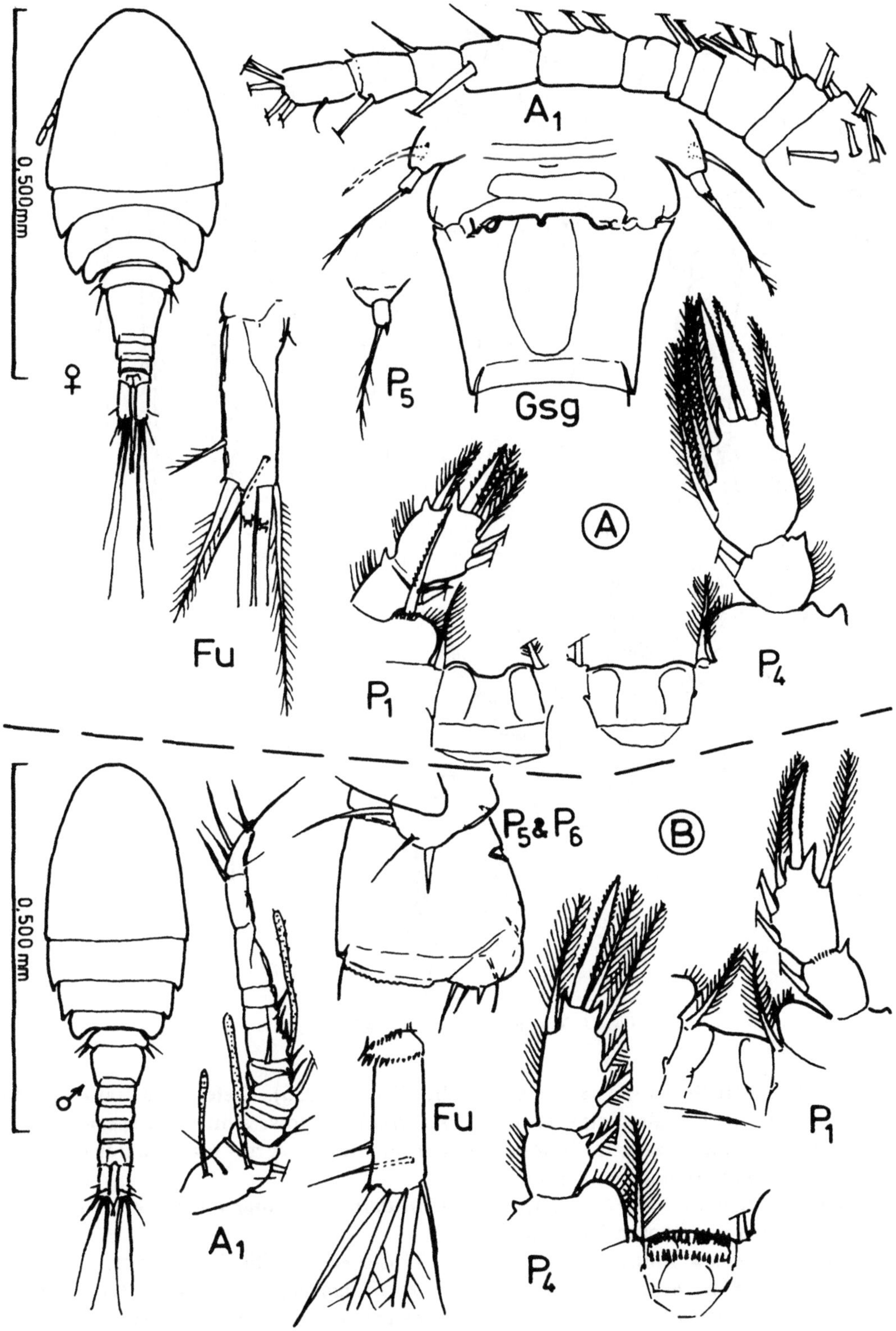

Fig. 22. A. *Metacyclops subaequalis* nov. sp.; B. *Allocyclops neotropicalis* nov. sp. (Mâle)

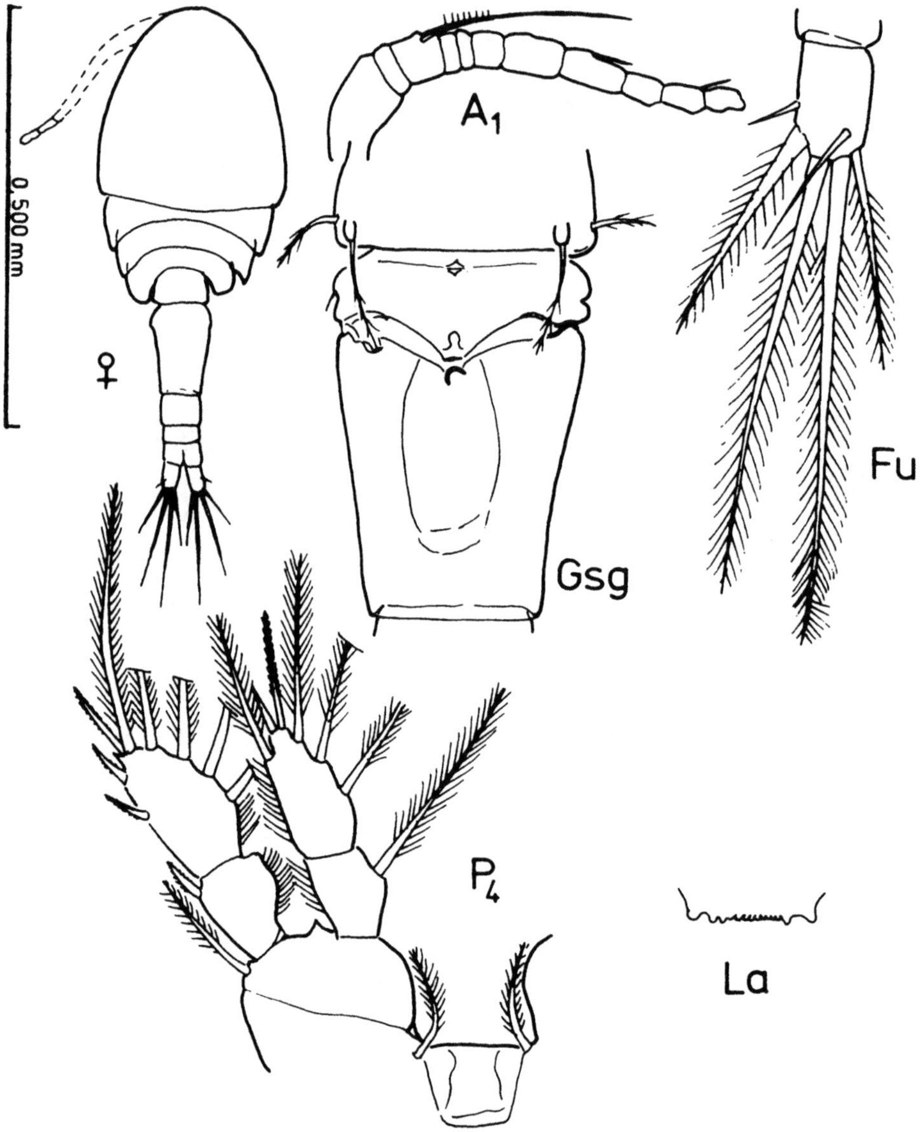

Fig. 23. Neutrocyclops brevifurca (orig.)

mens collected by S. Wright in Brazil and named *N. nordestinus,* I recognized some analogies between females and differences between males: the basopodite of left P5 of *N. deeveyorus* male has a 'ponctuated' hyaline lamella; that of right P5, at the proximal inner corner is a ponctuated knob; the exopodite 2 of right P5 is twice as long as it is wide. At the right antennule, the thirteenth joint has a spiniform process which is particularly strong.

Thus, *N. nordestinus* and *N. deeveyorus* are species which must be distinguished in spite of their resemblance and the succinct description of the first by Wright (1935). Kiefer's illustration (1936) of *N. nordestinus* shows a different species related to *N. deeveyorus* but surely not this, nor *N. nordestinus.*

Notodiaptomus amazonicus (Fig. 5)

This species, difficult to separate from some other *Notodiaptomus* existing in the Orinoco-Amazone region, is easy to characterize by the right P5 of the male: the last joint is strong and the lateral spine (aculeus) is long and almost half the terminal claw which is regularly curved, relatively short and

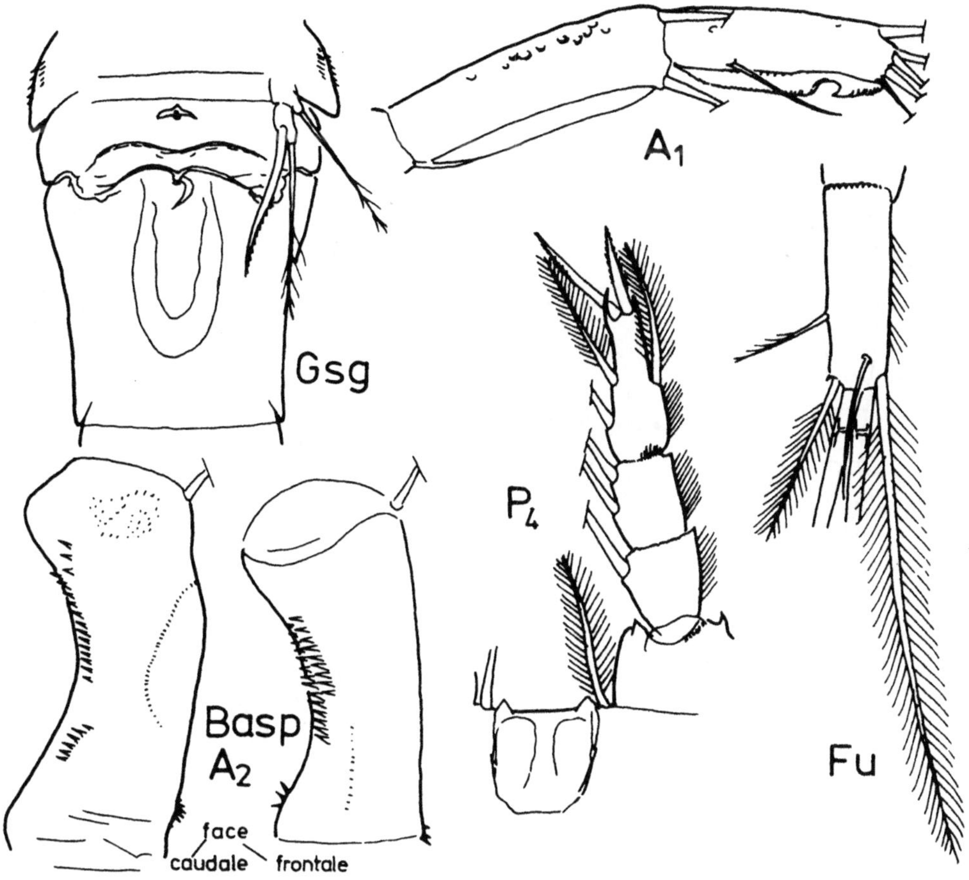

Fig. 24. Mesocyclops aspericornis (orig.)

significantly sclerified.

The length of the male corresponds to that of *N. deeveyorus* but the eight joint of its right A1 has a spiniform process relatively long and strong and that of the sixteenth joint is reduced.

Notodiaptomus cearensis (Fig. 6)

Wright (1936) already noted some variations in *N. cearensis* particularly in the 20th joint of antennule which may have a distal 'spur'.

I also noted some differences in the joints 10 and 11 which have spiniform processes subparalleles or not and in the joint 13 where the corresponding process may be more or less curved towards the antennule. The processes sometimes show with the joint an angle of 30 to 45 °.

N. cearensis also has a lamella on the posterior face of exopodite 2 of the right P5 male, lamella not described by Wright (1936) or Bowman (1973).

Notodiaptomus kieferi (Fig. 7)

This species is characterized as follows.

Male: Abdomen with the four first segments subequal, the second, the longest; Fu dissymetric, smooth. A1 with a spur at the antepenultimate joint not long enough to reach the middle of the next joint; the joint 8 has a well-developed spiniform process and also at the joints 10, 11, 15 and 16; at the joint 13, it is strong and reaches the end of joint 14.

The right P5 has a coxopodite with a well-developed lobe and an exopodite 1 characteristic of the genus. Exopodite 2 presents a lateral internal development which is similar in *N. falcifer*. The lateral spine (aculeus) is long, slender and fixed at the middle of the segment. The terminal claw is very long and slender.

The left P5 has a short bristle near the top, weakly feathered and going beyond the last finger-like

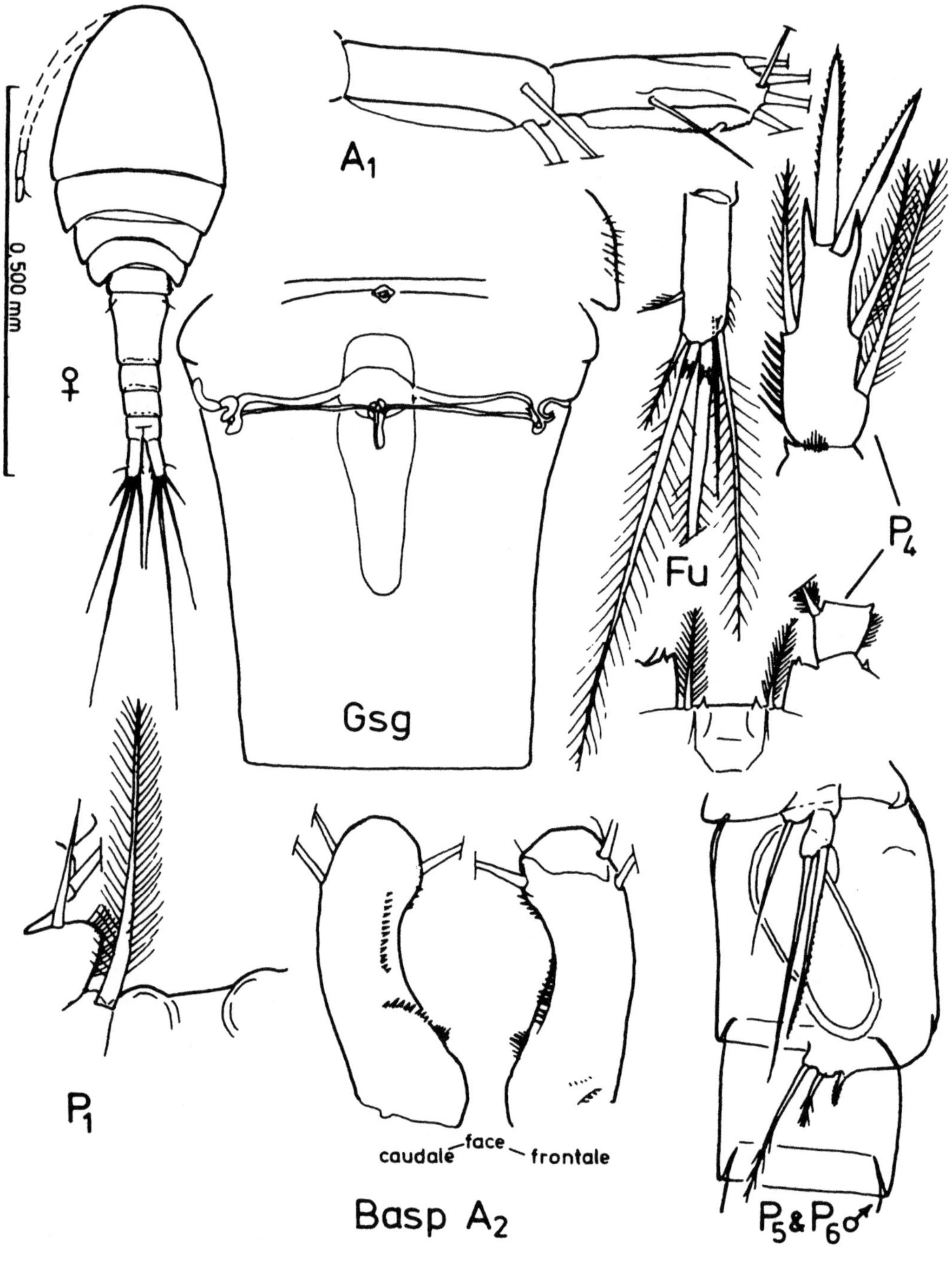

Fig. 25. Mesocyclops ellipticus (orig.)

process.

Length: 0.90 mm (0.83 to 1.03)

Female: 4th and 5th thoracic segments fused. Wings bilobates and slightly dissymetric, each armed by a strong enough spine. Abdomen with 4 distinct segments. Genital segment long and assymetric with the left side proximally expanded in a fingerlike process orientated posteriorly and armed by a spine; the corresponding spine at right side is still more proximal.

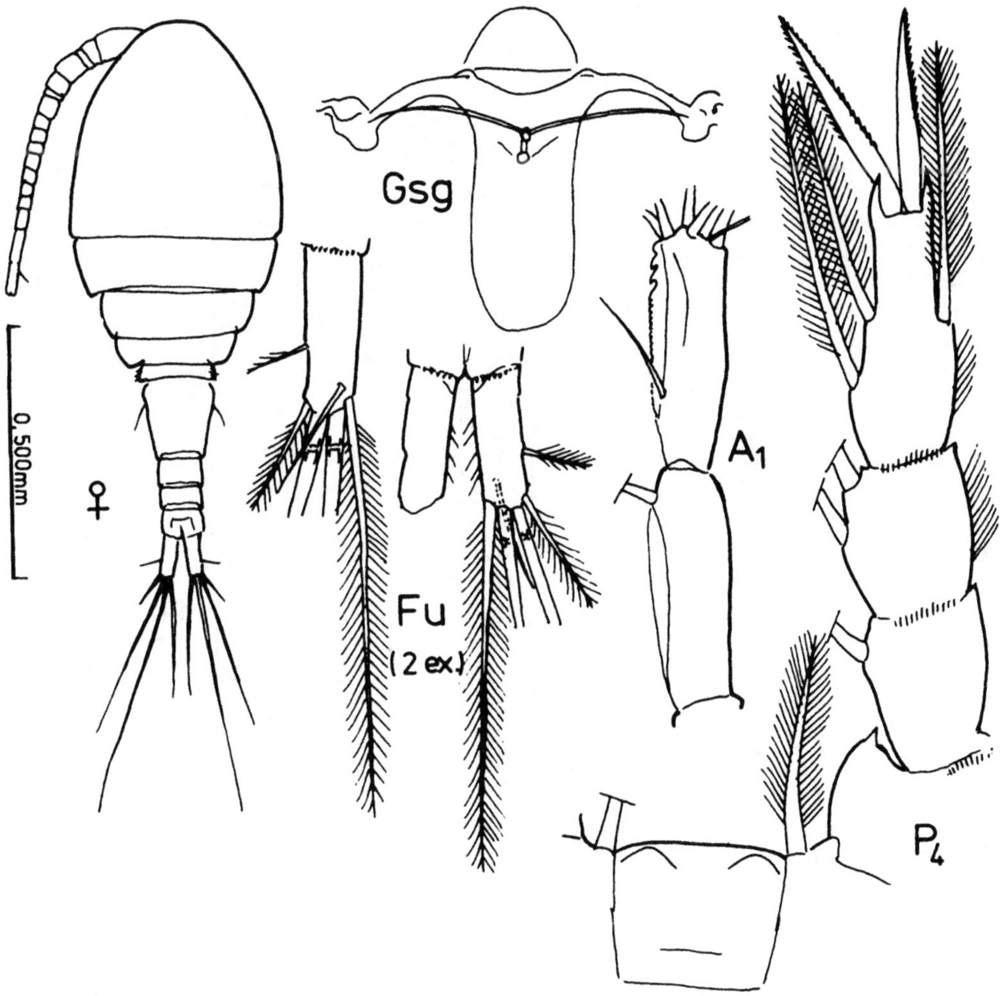

Fig. 26. Mesocyclops meridianus (orig.)

Antennules very long, even exceeding the furcal setae. The coxopodite of P5 has a lobe with a strong spiniform process. The exopodite 3 of this P5 is well distinguishable; its endopodite is unisegmented and has at the tip a long spine, a brush and a subterminal spine less conspicuous.

Length: 1.08 mm (until at least 1.10).

This small species, very characteristic, presents some analogy with *Dactylodiaptomus pearsei* for the female (excepted the size) and with *N. falcifer* for the male.

I found it in the Caroni River – man-made lake of Guri – and it might be found in Guyana Massive. Elsewhere, it was described from Amazonia (Brandorff, 1973) near Manaus.

Notodiaptomus dilatatus nov. sp. (Fig. 8)

Holotype: a male taken in Orinoco River at Ciudad Bolivar (coll. B. Dussart, 25.10.1981, prep. B. Dussart).

Allotype: a female taken at the same station (prep. B. Dussart).

Similar to *N. amazonicus* this species is characterized by several features.

Description. Male: 5th thoracic segment distinct from the fourth and terminating in acutely produced corners pointing posteriorly, with spines, the left one hyaline and stout, the right fine and slender. Moreover, two small spines are placed on dorsal face of these wings.

First abdominal segment with a slender spine

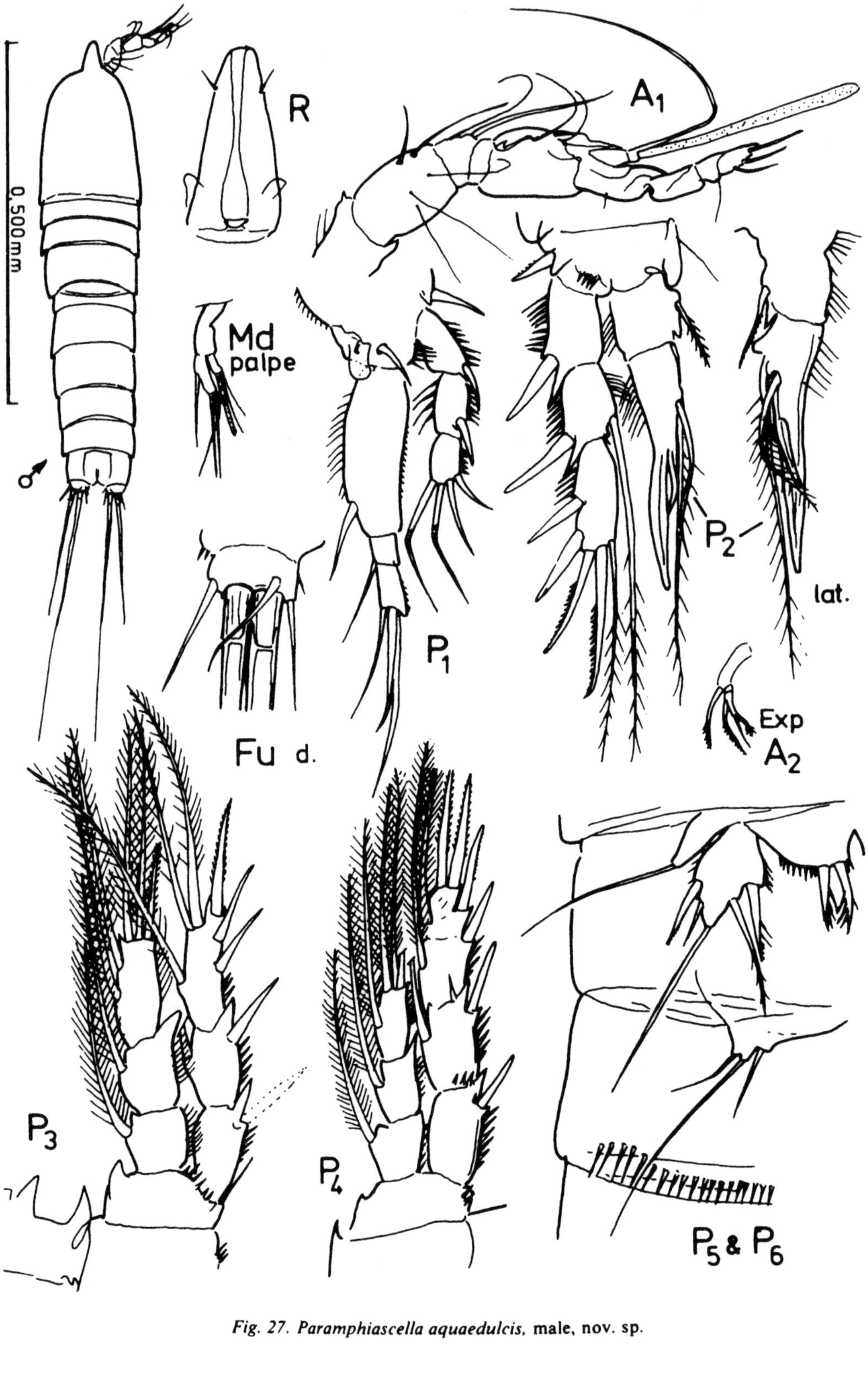

Fig. 27. Paramphiascella aquaedulcis, male, nov. sp.

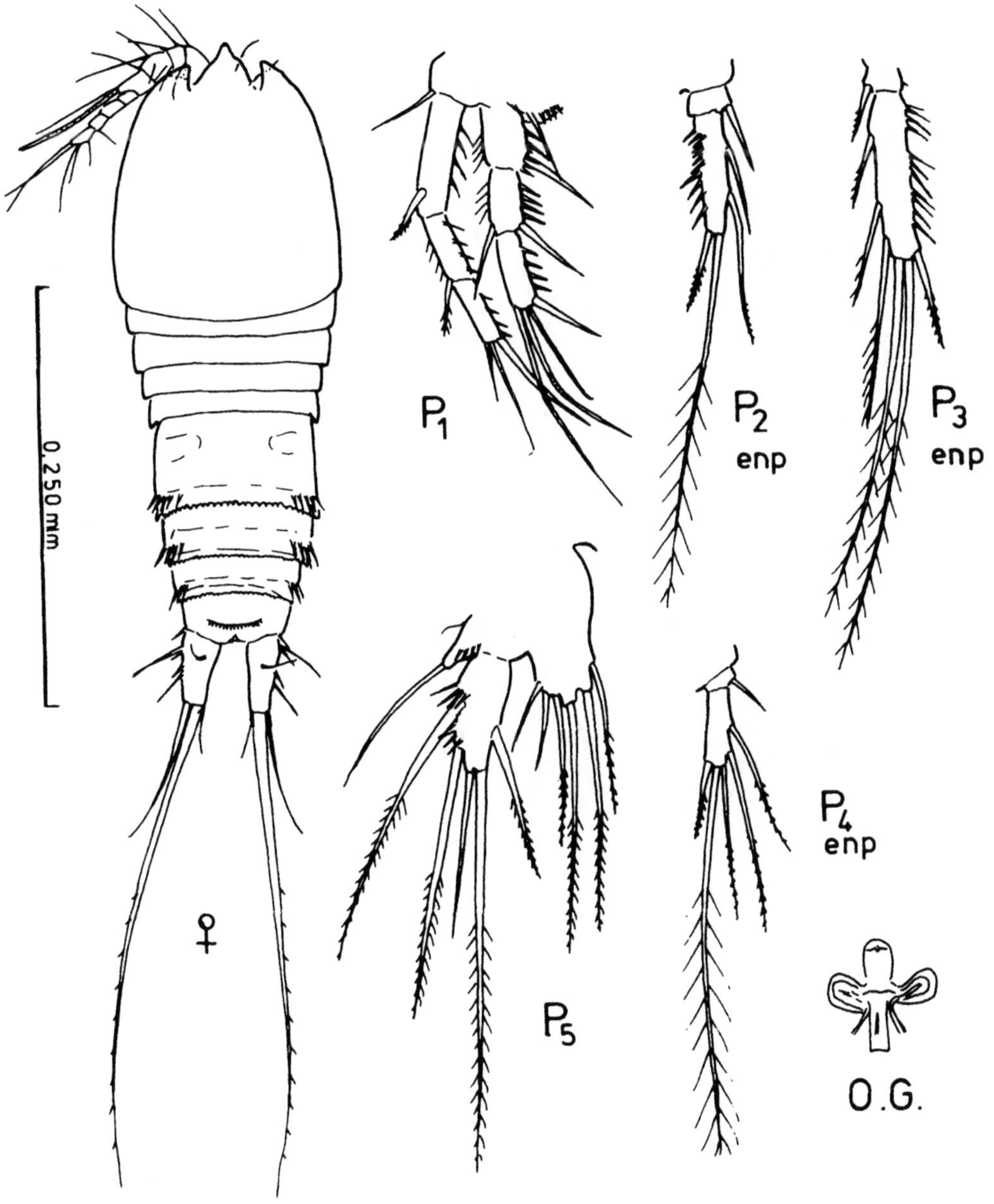

Fig. 28. Attheyella (Chappuisiella) orinocoensis nov. sp.

posteriorly on each side, the right one long and curved. Right antennule with a spiniform process at the 13th joint particularly long and the spine of 8th joint is relatively long. The antepenultimate joint has no hook but a conspicuous hyaline lamella.

The right P5 has a basopodite with a verrucose knob at proximal inner corner; its posterior face is expanded in its first third; its exopodite 1 is similar to that in *N. amazonicus*. Exopodite 2 is expanded interiorly and forms a knob well visible laterally; lateral spine of exopodite 2 relatively long and almost straight; terminal claw strong, short and curved.

The left P5 is special by the proximal inner corner of basopodite slightly expanded and bringing a small button. The spine of coxopodite is conical and attached to a tronconical expansion of the joint.

At the end of the leg, the subterminal internal setae is long, curved and smooth.

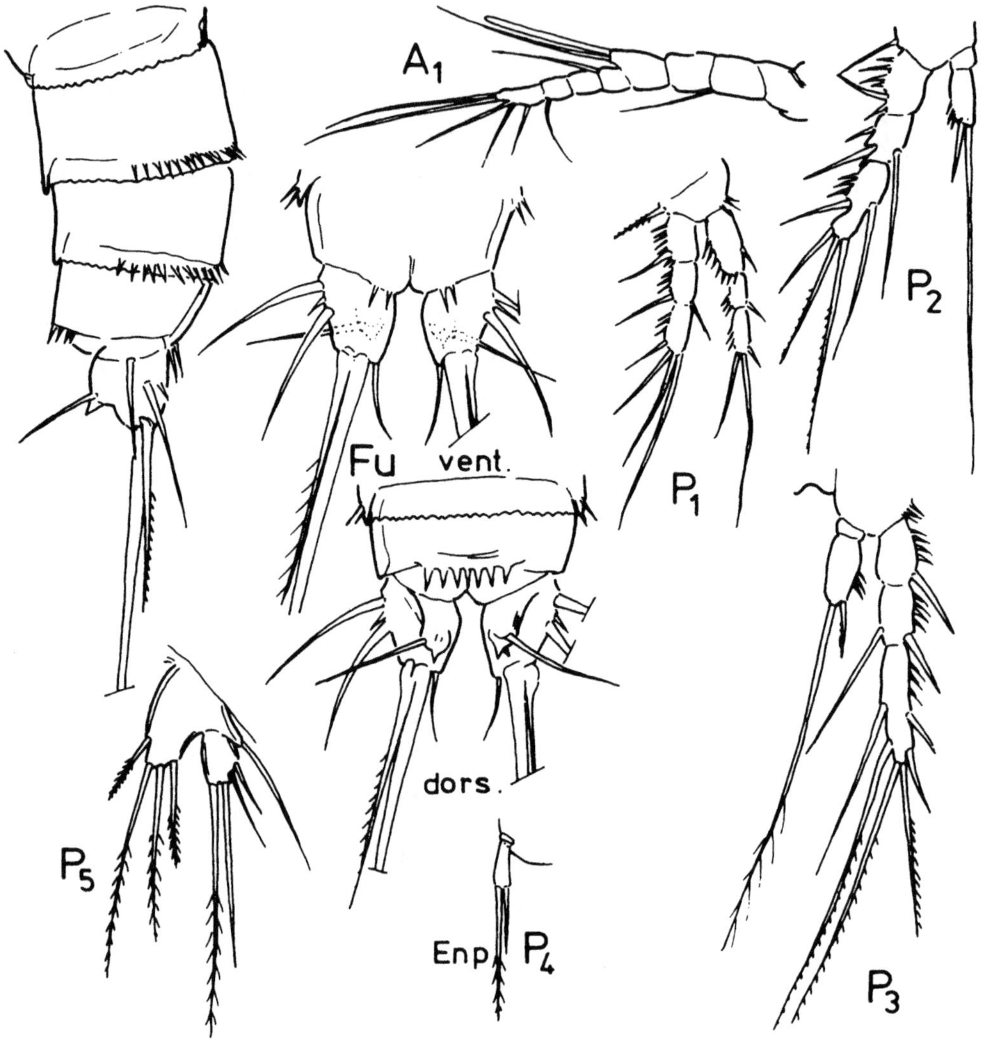

Fig. 29. Elaphoïdella bispina nov. sp.

Length: about 1.5 mm.

Female: Only slightly larger as the male. Wings of Th5 dissymetric, orientated outwards, each terminating with a strong and short spine. Genital segment relatively wide, expanded anteriorly (more at left) and deepened as a saddle dorsally. Posterior corners overlapping the next segment, particularly at right. Second abdominal segment short. Fu ciliated externally and internally. P5 with endopodite relatively elongated, exceeding the 3 quarters of exopodite 1. Spines of exopodites 2 and 3 strong.

Found in Orinoco River, this species might be found along the river in the calm zones.

Notodiaptomus coniferoides (Fig. 9)

Largely distributed in South America, this species was simultaneously described by Wright (1927) and Pesta (1928) who gave it the name *Diaptomus lobifer*. Brehm (1957, 1958) observed it in Argentina. Noone noted the particularity of this species having a spine verrucose at the end of left P5 of male, near the finger-like terminal process. At the right P5, the external angle of exopodite 1 is very sharp and hyaline; the exopodite 3 is conspicuous. The 13th joint of right antennule of male has a spiniform process elongated and curved once or twice; the antepenultimate joint has a hyaline lamel-

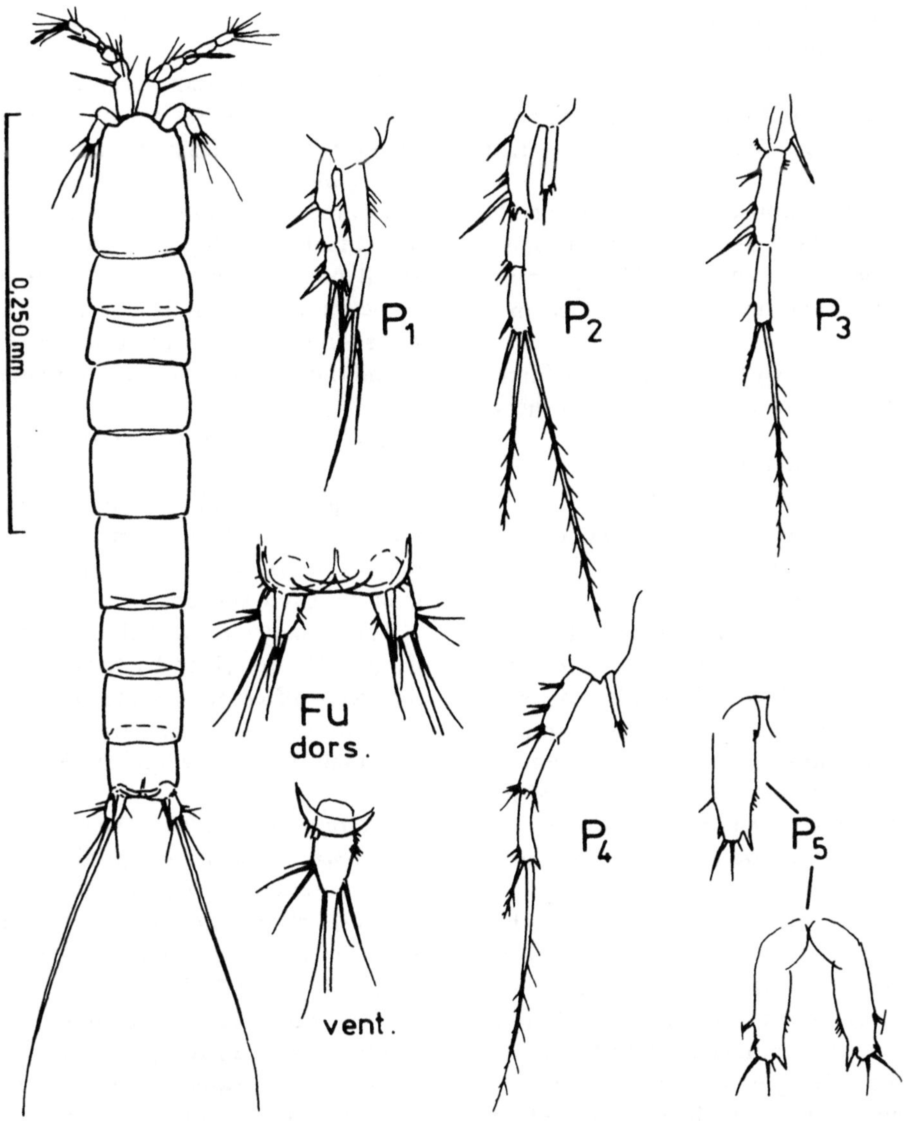

Fig. 30. Forficatocaris forficata crenensis nov. subspec.

la, narrow but conspicuous.

This species was found in the whole Orinoco basin.

Rhacodiaptomus calatus (Fig. 10)

Only females were found. These females had two features: some presented the left expansion of the genital segment as an acute semicircular tongue curved towards the back as in the type described by Brandorff (1973). Others had this tongue fused with this back. P5 were decorated with wrinkles on exopodite 1. This variety was relatively small: up to 0.70 mm only. I suggest calling it *R. calatus coalescens* because of the solder of the semicircular lobe with the genital segment.

'Diaptomus' negrensis (Fig. 11)

Andrade & Brandorff (1975) describe this species found in lateral lakes to Rio Negro near Manaus (Brazil) and compare it to *'Diaptomus' alter* Herbst, 1960. This last species is in fact the *D. proximus* described by Kiefer (1936). Herbst (1960) compares

'D.' alter to *Prionodiaptomus colombiensis* described by Thiébaud (1914) and named *D. marshi* by Juday (1914).

It is difficult to put all these species in only one genus close to *Arctodiaptomus* as Light (1939) tried to do.

Because we do not know this group sufficiently well in South America, I agree with the author's decision to put them in a 'supergenus' *'Diaptomus'* which will be with time revised and divided into several new genera.

I noted in *'D.' negrensis* the lateral right furcal seta which is twice as wide at the base as the others.

Dactylodiaptomus pearsei (Fig. 12)

After the discovery of this species by Wright (1927), it seems that it was difficult to find new specimens, males or females.

In the samples collected in the Orinoco basin, I found males corresponding to those described by Wright. But the females linked, if they had defined affinities with the males present, had no common characteristics with those described by Wright. In these conditions, I do not know what kind of females must be referred to the described males.

I consider the females found here as the good ones. Their description is:

Body torpedo-like. 5th thoracic segment not defined from the preceding one and with lateral corners expanded in sharp wings orientated towards the tail. Only the left wing has a sort of spine at the end. Moreover every internal lobe has a tiny posterior spinule. The genital segment is nearly symmetric, expanded in the middle and armed on each side with a small spine. Second abdominal segment well-developed, as long as wide. Fu ciliated on each side. Furcal setae curved at the end. Antennules short, extending only until the posterior edge of thorax. P5 elongated: endopodite as the two-third of exopoditel, more or less two-segmented, with a shew brush of spinules near the tip. Exopodite 3 with a seta longer as exopodite 2.

Only when it is possible to observe several (monospecific) populations of *'D.' pearsei* will we be sure that these females correspond to the males observed and described by Wright. Recently Brandorff (1982) arrived at the same conclusions.

Macrocyclops albidus (Fig. 13)

This cosmopolitan species exists in Venezuela under two subspecies, the nominal one, *Macrocyclops albidus albidus* and the subspecies *M. albidus principalis.*

I noted in the nominal form a variability in the second seta of endopodite 3 of P4. Sometimes it is a small one, unciliated as in European form and sometimes a 'normally developed seta', although not so long as the first internal lateral seta. I never found the form *oligolasius* so frequent in Africa.

Eucyclops pseudoensifer n. sp. (Fig. 14)

In laguna Mucubaji, some specimens of *Eucyclops* were present with some *Metacyclops leptopus mucubajiensis,* a form described by Kiefer (1956) from the same station.

These *Eucyclops* were almost similar to *E. ensifer* from Chili, described by Kiefer (1936). However, they were different by some characters.

Holotype: a female from Laguna Mucubaji. 5.10.1980. Coll. E. Zoppi de Roa (prep. B. Dussart).

Allotype: a male from the same station.

Female: P4 with endopodite 3 relatively long, more than twice as long as wide; inner apical spine relatively short, only a little longer as the joint. Setae 'normal'. P5 with inner spine 'normal'.

Length: about 0.77 mm.

Male: a little smaller (0.76 mm), the male has a P6 formed with a strong and long spine (55 μm) and two setae of the same length. Fu without longitudinal spines (serra) but with some transversal spines near the marginal external seta.

This species and *E. ensifer* are similar in their precoxal lamella and their coxopodite of P4. I therefore suggest naming this population of laguna Mucubaji *E. pseudoensifer.* This species is to be related to *E. leptacanthus* described by Kiefer (1956) and to *E. bondi* from Haïti (Kiefer, 1934).

I noted moreover an anomaly in exopodite 3 of P1 of *E. pseudoensifer:* in one specimen (Paratype) the internal edge was armed with 3 instead of 5 setae (Fig. 14).

Paracyclops pilosus n. sp. (Fig. 15)

Kiefer (1957) described *P. andinus* with the peculiarity of having an antennule with 8 joints, Fu

2.4 times as long as wide, a P5 with median seta as long as the inner one, a P4 with endopodite 3 about 1.65 as long as wide and armed with 2 spines, the inner twice to 2.5 as long as the outer.

In Orinoco River lives another *Paracyclops*, related to *P. andinus*.

Holotype: a female found in Orinoco River at Barrancas. 24 Oct. 1981 (coll. B. Dussart) (prep. B. Dussart).

Allotype: a male from the same station, same date (prep. B. Dussart).

Female: Fu long, 2.9 times as long as wide. Inner terminal seta as long as the outer. Dorsal seta of the same length, smooth. Thoracic segments 3, 4 and 5 with hairs at the edge. Abdomen with cuticule finely striated and so the Fu ventrally. P1 with precoxal lamella smooth. That of P4 with spinules. Spine formula of swimming legs: 3.4.4.3.

Antennule with 8 joints, the two first partially confluent. Receptaculum seminis symetric, genital opening particular.

Length: ca. 0.595 mm.

Male: of the same size, the geniculate antennules are compact. P6 with inner spine 1.5 times as long as the second abdominal segment. The medial seta is much shorter and the external seta is of the same length of that corresponding of P5.

This species was found in the littoral zone of flowing waters of Orinoco at Barrancas and at Ciudad Bolivar.

Ectocyclops herbsti nov. nom. (Fig. 16)

Herbst (1959) has described an *Ectocyclops* under the name *E. hirsutus* which has some characteristics different from those of the species described by Kiefer (1930) from animals collected in Madagascar (see Dussart, 1982).

The essential characteristics of this species are:

Antennule with 11 joints as in *E. phaleratus*. Fu short, 1.2 to 1.8 times as long as wide only; P5 with setae (and spine) as 100, 65 and 74, the internal very ciliated. P6 constituted by three appendages more or less of the same size.

The armature of furcal rami is very particular and different from *E. hirsutus:* the proximal row of spinules is formed with long ones. The dorsal furcal seta is long, the inner apical seta is longer than the outer ones, this last strong and very ciliated.

This species was first found in a peat bog with *Sphagnum* in East Brazil, at Cananea near São Paulo and was thus considered a synonym to *E. hirsutus*, but I also found it in a swamp near Calabozo supplied by a 'black water' stream and in a 'Morichal' environment and I consider *E. hirsutus* as a different species.

I suggest naming it *Ectocyclops herbsti* nov. nom. in remembrance of the first describer. It is related to *E. bromelicola* found in Brazil by F. Kiefer (1935).

Microcyclops anceps var. *minor* (Fig. 17)

The different descriptions of *M. anceps* in South America show that furcal setae and also endopodite P4 are very variable in size.

Near Unaré, I found individuals with terminal inner seta long, about 2.5 times the outer, relatively stout. Endopodite 2 P4 is a little more twice as long as wide and has two terminal spines, the inner twice the outer and 0.75 times the joint.

The length of observed animals was about 0.73 mm only. It is why I propose to name this population (at least) variety *minor*, to differenciate it from the 'normal' ones which measure 0.8 to 0.9 mm in general.

Moreover, the seta of Th5 has no spinules near its base on the segment.

Microcyclops diversus (Fig. 18)

Very near to *M. varicans*, this species is differentiated particularly by a group of microcharacters. In Venezuela, the animals were small relatively (0.88 mm) in comparison with those studied by Kiefer (1935) from Uruguay. Moreover, their furcal rami were longer (4.5 times as long as wide) but for the other characteristics, the animals were similar to the type (particularly the receptaculum seminis and the P4).

Microcyclops finitimus nov. sp. (Fig. 19A)

In *Microcyclops* in general, the basopodite of P1 has a spine at inner corner. However, Lindberg (1957) recalls that *M. anceps* and *M. crassipes* are exceptions to this rule.

In a pool near Camaguan, I found a female closely resembling *M. anceps* but with a P5, the lateral spine of which is fixed at the beginning of the second half of the joint and not at its tip. So, this P5 is quite close to that of *M. diversus*. The furcal rami are longer as those of *M. anceps*, the inner dorsal setae also longer (compared with the outer).

A1 with 12 joints and endopodite of P4 a little more elongated as in *M. anceps*.

Because of the relationship of this species with *M. anceps* I suggest calling it *M. finitimus*.

Holotype: one female from a pool in the shade of trees near Camaguan along the road. 16.10.1981 (coll. B. Dussart) (prep. B. Dussart).

The main measurements of the individuals studied are:

Length (without furcal setae): 0.89 mm
Ratio L:1 Fu = 4.10
T_i:T_e = 1.7
Ratio L:1 Enp2P4 = 2.6
Inner spine of Enp2P4:L Enp2P4 = 0.75

Microcyclops cf. *varicans* (Fig. 19B)

In an 'Estero' near the Buena Vista Cross, I found a curious individual of *M. varicans*, a male, except for the antennule which was constituted as in a female. The P6 was similar to that of *M. diversus* and the endopodite of P4 was like in *M. longiramus*.

Because I was observing only one individual of this curious species, I named it cf. *varicans* waiting for more samples to indicate what it is.

Metacyclops tredecimus (Fig. 20A)

The description of Lowndes (1934) is sufficient. However, I noted in the populations of Venezuela that the antennules are shorter and do not reach the Th2. Moreover, the precoxal lamella are armed differently in P1 and P4.

Male was unknown. I give some drawings, particularly of P6. Lindberg (1961) considers *M. tredecimus* as a good species. I agree. It was discovered in Paraguay and I found it in a trough of a farm near Calabozo (in the Llanos).

Metacyclops leptopus mucubajiensis (Fig. 20B)

Found for the first time by Kiefer (1956) in the same station, the individuals observed were quite similar to those of the type. Kiefer *(loc. cit.)* put it close to *M. leptopus venezolanus*, which seems to me a *M. mendocinus*. The findings of Lindberg (1961), the suggestions of Löffler (1963) and my observations (Dussart, 1979) show the necessity to describe again what could be called two good species: *M. mendocinus* and *M. leptopus*. In any case, the fact that the first of these species was found in a pelagic zone disproves the assertion of Löffler (1963).

Metacyclops curtispinosus nov. sp. (Fig. 21)

This small species could be named *M. subdolus*, particularly as it is characterized by furcal setae (Lindberg, 1961).

Holotype: one female found in the rio Guarguapo near Barrancas; 24.10.81. (coll. B. Dussart) (prep. D. Dussart).

Female: Antennules short, with 11 joints; they do not attain the posterior border of cephalothorax. Swimming legs tri-segmented; spine formula of exopodites: 3.4.4.3.

P1 with a long spine slender at the inner edge of the basopodite. P4 with endopodite 2 less than twice as long as wide. Terminal seta short (0.8 times the length of the joint).

Furcal rami with a lateral outer seta at 1/3 of the border. The dorsal seta is almost as long as the outer and this last, shorter as the inner.

The P5 has the particularity of having only one joint with a terminal seta relatively long and a spinule at the outer angle.

Length: 0.6 mm.

The environment of this species is special enough to suggest that it would be interesting to collect more fauna in the water, aquatic plants, sand banks, etc. which constitute this 'milieu' protected by Morichal.

Metacyclops subaequalis nov. sp. (Fig. 22A)

In Orinoco River, lives one of the 13 species of *Metacyclops* known in South America (cf. addendum).

It is small and its swimming legs have spines in spatula.

Holotype: a female observed at Ciudad Bolivar in Orinoco river, 25.10.81 (coll. B. Dussart) (prep. B. Dussart).

Length: 0.56 mm.

Body suboval, like an *Eucyclops cf. serrulatus*, but smaller. Th5 with the spine fixed subdorsally (representing the first joint of P5). Abdomen with posterior edge of segments smooth. Fu 3 times as long as wide, with an outer lateral spine fixed at the end of the three quarters of the segment. The apical medial setae are long and feathered but heteronomous. The inner seta is 1.5 times the outer. The dorsal seta is a bit shorter than the outer and is not as long as the furcal rami.

Antennule with 11 joints. P1 with a long spine inserted at the inner corner of basopodite; apical spine of endopodite stout and almost straight. P4 with endopodite 2 short, armed with two terminal spines almost equal (whence the name of the species); precoxal lamella with symetric knobs smooth and small.

Male: unknown.

This species is related to *M. brauni* Herbst (1962) but P4 is different as are the length and the number of joints of antennules.

Allocyclops neotropicalis nov. sp. (Fig. 22B)

Until now, the genus *Allocyclops* was only known in Africa and Europe. If they are subterranean species in Europe, in Africa they live in temporary waters more or less phreatic.

It was surprising to find an *Allocyclops* in Orinoco River in the littoral vegetation.

Holotype: one male from Orinoco river, 25.10.-1981 at Ciudad Bolivar (coll. B. Dussart) (prep. B. Dussart).

Description: Anterior part of the body ovoïdal; Th4 wrapping, with round corners. First abdominal segment well-developed, a little more wide than long. All abdominal segments with a posterior border serrate. Anal operculum simply convex. Furcal rami with spinules at the base all around; lateral seta inserted at the end of the second third of the segment. Apical outer seta stout at the base, with spinules inserted at the junction with the furcal rami. Apical inner seta as long as the furcal rami and slightly shorter than the outer. Dorsal seta relatively short.

Antennule geniculate with three important aesthetascs, two at the first joint and one at the ninth. P1 with rami two-jointed and with endopodite armed with an outer seta, a stout apical spine, slightly curved and three setae; coxal setae well developed. P4 with precoxal lamella cambered and decorated with two rows of spinules; endopodite 2 1.7 times longer than wide, with three inner setae, one subapical seta, a spine aside of almost the same length and an outer seta. P5 formed by a spine, a seta and the normal lateral and long seta which is a vestige of the first joint. P6 constituted by a short stout spine and two setae.

Length: 0.59 mm.

Because of the lack of possibility to describe the female, I consider as provisional the taxonomic place of this male and therefore suggest naming it *neotropicalis*, to remember the neotropical region where it was found.

Neutrocyclops brevifurca (Fig. 23)

This curious species, well described by Lowndes (1934) and Kiefer (1956) presents apical furcal setae feathered helicoïdally at the end. It is recognizable by several characters, particularly the relative length of the furcal setae and furcal rami. The antennule has 11 joints, not 10.

Biogeographically, it is a South American species common in Venezuela and present at least from Brazil to Colombia.

Mesocyclops aspericornis (Fig. 24)

This species was recently redescribed by Kiefer (1981). It was known in Asia and from the Pacific isles (Mariannes, Marshall, Hawaï, Tahiti) and its presence in Africa is controverted. Its presence in South America increases its extension in the intertropical regions of biosphere. The same may be true of *M. thermocyclopoïdes* which is actually chiefly known from East and Far East.

Mesocyclops ellipticus (Fig. 25)

This species is easy to recognize by its receptaculum seminis, which is similar to that of *M. tenuisaccus* or *M. annulatus*.

Found in Brazil (Bom Conselho) for the first time, it was observed in Venezuela by Kiefer (1956). Herbst (1962) found it in Amazonia.

However, comparison of the various descriptions shows significant differences and this could be explained by the existence of two or several species called *M. ellipticus*. Only a study of an adequate number of samples would permit a conclusive identification. The individuals observed in Rio Portuguesa and in Orinoco were already different from those collected in the Estero de Buena Vista. Unfortunately, I did not obtain a sufficient number of individuals from this last station to describe this 'species' sufficiently.

Mesocyclops meridianus (Fig. 26)

After a very short description of *M. meridianus* from Paraguay (Kiefer, 1926), Kiefer (1933) described a *M. brasilianus* from the Amazon region near Manaus.

By comparing drawings, it seems that *M. meridianus* from Paraguay is very similar to *M. ellipticus* found in Pernambouc (east Brazil). Likewise, very few differences distinguish *M. meridianus* and the Brazilian *M. brasilianus*. One of these species is frequent in Venezuela and there, some '*M. meridianus*' have a receptaculum seminis very close to that of *M. brasilianus*. Moreover, the furcal rami may have some internal hairs. Kiefer (in litt.) does not possess more types of these 'species' but he thinks that their variability is sufficient to explain the confusion between these last two species. Harding (1955) and Löffler (1963) also consider *M. brasilianus* and *M. meridianus* as synonyms. I agree. In both species, size, position of egg-sacs, and number of eggs per sac (20–25) are similar.

The P6 of the male has a particularly long medial seta, half the external one. The inner spine is always stout and short. The Kiefer's (1956) illustrations are an extreme case. Further studies will determine the precise number and characteristics of species present in South America under the collective name: *M. meridianus*.

Paramphiascella aquaedulcis nov. sp. (Fig. 27)

Quite new was the finding between Camaguan and San Fernando de Apure, in freshwaters, of a representative of the family Diosaccidae, genus *Paramphiascella*.

The station is characterized by the presence of rio Portuguesa seasonally flooding the plain along the road where only a male was collected.

Holotype: a male collected in Rio Portuguesa (Caño Falcon) near Camaguan, 16.10.1981 (coll. B. Dussart) (prep. B. Dussart).

Body almost cylindrical. Rostrum projecting. Last abdominal segment rather elongated. Furcal rami very short. Two furcal setae well-developed; inner and outer setae subequal. Antennule geniculate with a long aesthetasc exceeding the last joint. Antenna with last joint armed with three appendages. Mandibule with a palpe well-developed. P1 with endopodite 1 longer than the exopodite and with a subrectangular expansion. P2 with the last joint of endopodite modified and armed with two unequal spiniform processes. P3 and P4 with exopodite and endopodite three-jointed. Setae-spinal formula (without the outer marginal spines of exopodites):

	Exopodite	*Endopodite*
P1	0 · 0 · 121	1 · 1 · 120
P2	0 · 1 · 023	1 transf.
P3	0 · 1 · 123	1 · 1 · 221
P4	0 · 1 · 222	1 · 1 · 121

Exopodite P5 armed with 5 setae, two of which are spiniform; basoendopodite with 2 spines (more or less feathered) and several outer spinules. P6 constituted by 3 setaes more or less spiniform, the middle one the longest.

Length: 0.63 mm

Female: unknown.

Since this species was found only one time, in one place and only in one individual, it is not yet possible to compare it with other populations coming from similar environments. I suggest provisionally naming it *Paramphiascella aquae dulcis* nov. sp. considering that it is very close to *P. xiphophora* Lang, 1965.

Canthocamptus cf. *microstaphylinus*

Between the water plants growing in Caño Falcon, I obtained another Harpacticoida easy to recognize by its form and the armature of its last abdominal segment and also by the P4 and P5. The nearest known species is *Canthocamptus microstaphylinus*.

However, the individual was damaged; the difficulty in observing all details and the fact that it is the first time that such a species is found in South America leaves doubt about its determination. New samplings are needed to confirm this observation.

Attheyella (Chappuisiella) orinocoensis nov. sp. (Fig. 28)

Holotype: a female from Barrancas in a caño near Orinoco river, 24.10.1981 (coll. B. Dussart) (prep. B. Dussart)

This species is related to *A. (Ch.) fuhrmanni* and *A. (Ch.) camposi*. But it has a sharp rostrum, the furcal rami are elongated, hairless at the inner edge, the genital aperture has a different armature and the endopodite of P2 has only one terminal seta (instead of two).

The female is small (about 0.41 mm). The last abdominal segment has only three lateral spines at posterior edge. The other abdominal segments have this edge denticulated and are armed with, the genital segment 5 long spines laterally, the two other some ventral and lateral spines, these last the shortest.

The formula of setae and spines of the four first swimming legs is:

	Exopodite	*Endopodite*
P1	0 · 1 · 121	1 · 1 · 120
P2	0 · 1 · 123	1 · 211
P3	0 · 1 · 223	1 · 321
P4	0 · 1 · 233	1 · 221

P1 and P5 are similar to the corresponding legs in *A. fuhrmanni*.

Male: unknown.

Elaphoïdella bispina nov. sp. (Fig. 29)

In a water trickle between stones and gravels on a path near a farm at Calabozo, I collected in a few minutes three species of Copepods: an *Ectocyclops* too young to be denominated with certainty, a Parastenocarid (see further) and an *Elaphoïdella* related to *E. surinamensis* (Delachaux, 1924).

Holotype: a female from this station in Venezuela, 16.10.1981 (B. Dussart coll. and prep.);

Description: Posterior to the genital segment, the two other abdominal segments have ventral spinules and a dorsal serrate edge. The anal operculum is armed with 7 stout spines. The furcal rami are subconical and present a ridge terminated by a strong hook. P1 biramous with exopodite and endopodite 3-segmented. Endopodite P2 to P4 2 segmented and with 2 very unequal setae. P5 with 4 setae at basoendopodite and at exopodite. Length of this female: 0.42 mm.

I do not know of any other *Elaphoïdella* with similar characteristics, especially such simple endopodites.

Male: unknown, but it could be easy to collect it by regularly sampling this 'spring'.

Forficatocaris forficata crenensis nov. subsp. (Fig. 30)

In samples collected in the littoral region of Amazon river near Santarem, Noodt (1963) described a *Parastenocaris* (sensu lato) which belonged to a new group. Jakobi (1969) designated this group as *Forficatocaris* nov. gen.

In the spring described above, I found a female of such a species but with some small differences.

These differences are:

Size relatively larger: 0.43 mm (instead of 0.35 mm for the type). Anal segment with apical spines slender, straight. No other ornementation at this segment; anal operculum smooth. Antennules with 6 joints. P5 slightly different from the denominate type, especially by some hairs on the inner side.

I did not find any male and as I find these above differences significant, I suggest designating it as a new subspecies *crenensis* of *F. forficata*.

Considering the distance between these two findings (Santarem and Calabozo), it is probable that new samplings in the phreatic environment of Orinoco and the Amazona river will be rich in other surprises.

Summary

Venezuela, with 912 050 km^2, has a special biogeographical interest:

- In the South, it is part of Amazon basin thanks to the Casiquiare which connects the Orinoco and Rio Negro.
- In the East and already from the Caroni River, it is a part of the Guyan Massif, old and long isolated from the rest of South America.
- In the West, it is related to the Andes and part of Central America and all mountains of the Cordillera in the South.
- In the North, it is the route of colonization of the Caribbean isles and the contact point between marine, brackish and freshwater environment.

Finally, the center of the country has a uniqueness of its own, with the Orinoco Basin and where marine species may have the opportunity to live in freshwater and to be adapted to this new environment.

So, it is 'normal' to find in Venezuela a diversified although insufficiently known fauna. 368 species are actually described in freshwaters in South America (cf. addendum) and 66 (18%) in Venezuela; 19 other species may be present. On these 66 species, 47 were found during this survey. It would be desirable to study the upper region of Orinoco, and Atabapo, Ventuari and Caura rivers.

Some synonymies were detected:

'Notodiaptomus venezolanus' is a synonym of *N. henseni* described by Dahl (1894). *'D.' alter* Herbst (1960) is in fact *'D.' proximus* Kiefer, 1936.

I also described what I consider to be the true female of *Dactylodiaptomus pearsei*. It is always difficult to decide whether males and females are of the same species in plurispecific communities.

Moreover, it is rare to find so many species of Copepods in the same sample (here up to 16!).

Finally, South American Copepoda are often known only partially and the drawings of some authors are sketchy. If you except those of Mrázek (1901) and Sars (1901), only the male or female is described. Sometimes, however, the drawings are sufficiently precise to clarify the taxonomical position of a species. For instance, *Ectocyclops* described by Herbst (1960) is not *E. hirsutus* but a new species I found again.

Biogeographically, the study of samples collected in 1981 is very interesting because it extends the distribution of some Brazilian species: *N. coronatus, R. calatus, D. pearsei* for example or *M. tredecimus* known only in Paraguay and so on. Moreover, the genus *Allocyclops* is new for South America whereas *Pseudodiaptomus* and *Paramphiascella* are new for Venezuela.

Ecologically too, this collection is informative. For the first time, a member of the Pseudodiaptomidae has been collected in Venezuela. Several marine or brackish water species were found in freshwater: two species of *Oithona* are living in Orinoco waters as far as Ciudad Bolivar. A harpacticoid of the marine genus *Paramphiascella* is also living in freshwater.

Thus, this region, rich in primitive forms, can be considered as a center of expansion of many freshwater species which could start the conquest of salt waters.

The complex Amazona-Orinoco shows faunistic homogeneity. Only a more in-depth knowledge of the fauna of Guyana would enable a better understanding of the South American fauna distribution and allow conclusions on their ecological requirements and their role in the ecosystems.

Resumen

El estudio de 38 muestras de fauna acuatica recogidas y/o examinadas por el autor en Venezuela permitió reencontrar casi todas las especies determinadas por Kiefer (1956), Chappuis (1956), Infante *et al* (1979), Montiel y Zoppi (1979).

Las especies identificadas pasan de 28 a 66, cuyas 15 son nuevas para la ciencia y descritas.

Si se compara con las listas de Copepodos conocidos en las regiones vecinas de America del Sur, es posible concluir que Venezuela aun tiene otras tantas especies que descubrir, particularmente Calanoidos y Harpacticoidos. La descripción y ilustración de varias especies son completadas.

Acknowledgements

I thank Pedro Roa, Director of the Tropical Zoology Institute of the Central University of Caracas for his help and use of the facilities, and Dr Evelyn Zoppi de Roa who helped me in the field and in the laboratory.

Addendum

Copepods actually known in inland waters of South America and their relative (x) or probable (?) distribution

	Regions (see Fig. 31) 1	2	3	4	5	6	7	8	9
Order CALANOIDA									
Acartiidae									
Acartia tonsa		x		x					
Centropagidae									
Parabroteas sarsi									x
Boeckella longicornis								x	

Fig. 31. Biogeographic zones in South America. Regions: 1 - magdalenian; 2 - orinoco-venezuelian; 3 - Guyanas; 4 - amazonian; 5 - east brazilian; 6 - andine; 7 - paranean; 8 - chilian; 9 - patagonian.

Addendum continued.

	Regions (see Fig. 31) 1	2	3	4	5	6	7	8	9
B. bergi							x		
B. b. conesae							x		
B. b. cornuta							x		
B. b. serrifera							x		
B. gracilipes						x			x
B. michaelseni									x
B. gracilis							x		x
B. occidentalis						x			
B. pooponensis						x			x
B. meteoris									x
B. m. dentifera								x	
B. schwabei						x			
B. birabeni (= rahmi?)							x		
B. titicacae						x			
B. bilobata							x		
B. kinzeli						x			
Pseudoboeckella braziliensis									x
P. brevicaudata									x

Addendum continued.

	Regions (see Fig. 31) 1	2	3	4	5	6	7	8	9
P. poppei									x
P. dubia									x
P. entzei									x
P. longicauda									x
P. silvestrii									x
P. vallentini									x
P. erubescens								x	
P. gibbosa								x	
P. thomseni							x		
P. calcaris						x			
P. palustris						x			
P. peruviensis						x			
Pseudodiaptomidae									
Pseudodiaptomus acutus				x	x				
P. a. leptopus						x			
P. richardi				x	x		x		
P. r. inaequalis							x		
P. gracilis		x		x	x				
P. marshi					x				
P. wrighti					x				
Diaptomidae									
Prionodiaptomus colombiensis	x	x							
Argyrodiaptomus granulosus							x		
A. bergi						x	x		?
A. furcatus					x		x		
A. f. macrochaetus							x		
A. aculeatus (= furcatus?)					x		x		
A. (spiniger) incerta sedis (cf. *Notod.*)							x		
A. denticulatus				x			x		
A. azevedoi		?		x	x				
A. argentinus							x		
A. neglectus					x				
Calodiaptomus merrillae		?		x	x				
C. perelegans		?		x					
Notodiaptomus gibber					x		x		
N. deitersi					x		x		
N. henseni (= venezolanus)		x	x	x					
N. conifer		x		?	x		x		
N. coronatus		x		x	x				
N. anisitsi (= inflexus?)					x		x		
N. incompositus					x		x		x
N. lobifer							x		
N. coniferoides		x		x			x		
N. santaremensis		?		x					
N. transitans							x		
N. inflatus		?		x					
N. carteri							x		
N. amazonicus		x	x	x		x	x		
N. a. occidentalis						x			
N. iheringi					x				
N. nordestinus				x	x				
N. cearensis		x			x				

Addendum continued.

	Regions (see Fig. 31)								
	1	2	3	4	5	6	7	8	9
N. (corderoi) (?)					x				
N. isabelae					x				
N. dahli		?		x	x				
N. jatobensis				x	x				
N. maracaibensis		x							
N. deeveyorus		x		?	?				
N. (anceps) (?)							x		
N. (bidigitatus) (?)							x		
N. kieferi		x		x					
N. orellanai							x		
N. dilatatus		x							
Rhacodiaptomus calamensis		?		x					
R. flexipes		?		x					
R. insolitus				x	x				
R. retroflexus		?		x					
R. calatus		x		x					
R. c. coalescens		x							
Tumeodiaptomus vivianae								x	
T. diabolicus								x	
Dactylodiaptomus pearsei		x		x					
Odontodiaptomus thomseni							x		
O. michaelseni		?					x		
O. paulistanus		?			x				
Idiodiaptomus gracilipes					x				
Aspinus acicularis		?		x					
'Diaptomus' proximus						x			
D. carinifera							x		
D. echinatus							x		
D. falcifer							x		
D. inexpectatus							x		
D. linus		?		x					
D. meridionalis							x		
D. mildredae							x		
D. negrensis		x		x					
D. ohlei				x					
D. santafesinus							x		
D. silvaticus		?		x					
D. s. infrequens				x					
Order CYCLOPOIDA									
Oithonidae									
Oithona plumifera	x				x	x			
O. p. atlantica		?				x			x
O. similis		?				x			
O. vivida		?							
O. brevicornis		?				?			
O. hebes						x			
O. amazonica		x							
O. a. continentalis				x					
O. ovalis				x					
O. gessneri		x							

Addendum continued.

	Regions (see Fig. 31)								
	1	2	3	4	5	6	7	8	9
O. attenuata	x					?			
O. simplex					x				
O. sp.		x							
O. neotropica		x							
Dioithona minuta				x	x	x			
Cyclopinidae									
Cyclopinopsis brasiliensis					x				
Procyclopina polyarthra					x				
Cyclopidae									
(Halicyclopinae)									
Halicyclops crassicornis					x				
H. venezuelaensis	x								
Neocyclops medius					x				
(Eucyclopinae)									
Macrocyclops albidus		x				x	x	x	
M. a. principalis		x		x					
?M. fuscus							x		
M. ater				x					
Eucyclops serrulatus	x	x		x	x	x	x	x	x
E. s. chilensis								x	
E. delachauxi						x			
E. silvestrii		x					x		
E. neumani				x	x	x	x	x	
E. n. titicacae						x			
E. prionophorus							x		
E. ensifer					x	x		x	x
E. neotropicus				x					
E. leptacanthus		x							
E. alticola						x			
E. de = macedoi						x			
E. solitarius					x				
E. breviramatus						x			
E. siolii				x					
E. pseudoensifer		x							
Paracyclops fimbriatus		x			x	x	x	x	
P. f. chiltoni						x		x	
P. finitimus						x	x	x	
P. poppei?							x		
P. andinus						x	x	x	
P. pilosus		x							
Ectocyclops rubescens				x		x			
E. compactus		x							
E. phaleratus	?	?	?		?		?		
E. bromelicola		x			x				
E. strenskei				x	x				
E. herbsti		x			x				
E. hirsutus						x			
Tropocyclops prasinus		x					x	x	
T. p. peruviana						x			
T. p. meridionalis					x	x			x

Addendum continued.

	Regions (see Fig. 31) 1	2	3	4	5	6	7	8	9
T. schubarti					x				
T. s. dispar				x					
(Cyclopinae)									
Megacyclops viridis		x						x	x
Acanthocyclops robustus						x	x		
A. vernalis						?			
A. v. michaelseni							x	x	x
A. v. skottsbergi						x		x	
Diacyclops uruguayensis							x		
Microcyclops varicans		x			x	x	x		x
M. v. subaequalis		x				x			
M. anceps	x	x			x	x	x	x	x
M. a. minor		x							
M. a. pauxensis				x					
M. furcatus							x		
M. dubitabilis		x							
M. diversus		x					x		
M. alius						x	x		
M. paludicola					x				
M. finitimus		x							
M. elongatus						x			
Metacyclops gracilis							x		
M. mendocinus	x					x	x	x	
M. m. venezolanus		x							
M. leptopus						x			
M. l. mucubajiensis		x							
M. laticornis							x		
M. tredecimus		x					x		
M. grandis							x		
M. hartmanni	?								
M. brauni				x					
M. curtispinosus		x							
M. subaequalis		x							
M. distans		x							
Neutrocyclops brevifurca		x			x				
Allocyclops neotropicalis		x							
Mesocyclops cf. *leuckarti*				?			?		
M. edax	?	x							
M. annulatus					x	x	x		x
M. a. diversus				x					
M. aspericornis		x							
M. longisetus	x	x			x		x	x	
M. l. araucanus								x	
M. meridianus (= brasilianus)		x				x	x		
M. minutus					x		x		
M. ellipticus		x		x					
M. nicaraguensis						x			
Apocyclops procerus					x	x			
Thermocyclops tenuis					x		x		
T. brehmi							x		
T. decipiens		x							
T. minutus		x			x		x		
T. inversus					x				

Addendum continued.

	Regions (see Fig. 31) 1	2	3	4	5	6	7	8	9
Muscocyclops operculatus			x		x				
Menzeliella staheli			x						
Order HARPACTICOIDA									
Phyllognathopodidae									
Allophyllognathopus brasiliensis				x					
Phyllognathopus viguieri	x	x							
Tachidiidae									
Euterpina acutifrons				x	x				
Harpacticidae									
Harpacticus dubitabilis						x			
Tigriopus angulatus								x	
Thalestridae									
Parathalestris ganio								x	
Diosaccidae									
Paramphiascella aquae dulcis		x							
Schizopera vicina						x			
Ameiridae									
Stygonitocrella montana							x		
Canthocamptidae									
Canthocamptus staphylinus							x		
C. microstaphylinus		x							
Attheyella (? Atth.) brasiliensis						x			
A. farelloensis						x			
A. santaremensis						x			
A. septemarticulata						x			
Attheyella (Chappuisiella) crenulata								x	x
A. fuhrmanni	x	x					x		x
A. huaronensis						x			
A. guyanensis			x						
A. oculta								x	
A. subdola							x		
A. palustris								x	
A. pichilafquensis						x			
A. quillehuensis								x	
A. levigata								x	
A. bullata								x	
A. camposi								x	
A. chilensis								x	
A. laciniata								x	
A. vivianii								x	
A. orinocoensis		x							
A. koepkei						x			
A. brasiliana						x			
A. salviniae						x			
Attheyella (?Canthosella) bromelicola						x			

Addendum continued.

	Regions (see Fig. 31) 1	2	3	4	5	6	7	8	9
A. goeldii						x			
A. montana						x			
Attheyella (Delachauxiella) lanata								x	x
A. trigonura									x
A. aculeata							x		
A. horvathi							x		
A. dadayi							x		
A. hannae								x	
A. incae								x	
A. schindleri						x			
A. biarticulata								x	
A. ciliata								x	
A. nuda								x	
A. ornata								x	
A. serrata								x	
A. wieseri								x	
A. freyi						x			
A. arequipensis						?			
A. clavigera						x			
A. ensifer						x			
A. ferox						x			
A. inconstans						x			
A. i. egena						x			
A. insignis						x			
A. lanceolata						x			
A. maxima						x			
A. peruana						x			
Attheyella (? Elaph.) sphagnobiotica						x			
Attheyella (? Neocamptus) cordillierica						x			
Attheyella (? Neocamptus) nebulosa						x			
'Attheyella' gessneri			x						
Maraenobiotus fontinalis						x			
M. fontinaloides						x			
Moraria neotropica								x	
M. kummeroworum								x	
Löfflerella chilensis								x	
L. rouchi								x	
L. trisaetosa								x	
Epactophanes richardi								x	
Elaphoïdella bidens		x							
E. laciniata				x					
E. surinamensis			x						
E. pectinata		x							
E. humboldi						x			
E. paraplesia				x					
E. siolii				x					
E. bispina		x							
E. armata						x			
E. grandidieri						x			
E. schubarti						x			
Pseudocamptus peruanus						?			
Antarctobiotus koenigi									x
A. rapoporti									x

Addendum continued.

	Regions (see Fig. 31) 1	2	3	4	5	6	7	8	9
A. bahamondei								x	
Spelaeocamptus neotropicus						x			
Parastenocaridae									
Parastenocaris[4] *brasilibathynellae*					x				
P. hurdi					x				
P. chelifer			x						
P. clandestina				x					
P. digitata				x					
P. icoaraci				x					
P. jakobi				x					
P. paraensis				x					
P. p. bulbifera				x					
P. amazonicus						x			
P. arequipensis						x			
P. hexacantha	x					x			
P. panamericana						x			
P. staheli	x					x		x	
P. surinamensis	x					x		x	
P. paraguayensis							x		
P. p. phylloides							x		
P. remanei							x		
P. santaremensis				x					
P. sioli				x					
P. tapajoensis				x					
P. jujuyensis							x		
P. argentina							x		
P. sierrae							x		
P. cordobaensis							x		
P. ciliata							x		
P. rhizophora							x		
P. hecate							x		
P. persephone							x		
P. pluto							x		
P. oncophora							x		
P. sanctiludovici							x		
P. membranacea							x		
P. dactyloides				x					
P. drepanoephora				x					
P. bolbodes				x					
P. columbiensis		x							
P. divae					x				
P. kubitskii		x							
P. röttgeri		x							
P. tagae					x				
P. ignotus[5]							x		
Forficatocaris amazonensis				x					
F. tetracantha						x			

[4] Jakobi, in 1972, proposed divisions of the genus. Except for *Forficatocaris*, a more complete knowledge of this group seems necessary before creation of new genera.

Addendum continued.

	Regions (see Fig. 31) 1	2	3	4	5	6	7	8	9
F. forficata				x					
F. f. crenensis		x							
F. guarani					x				
F. noodti					x				
F. claudii					x				
F. lilianae					x				
F. evelinae					x				
F. jakobii					x				
F. affinis[5]							x		
Cletodidae									
Cletocamptus deitersi							x		x
C. d. ecuadorianus						x			
C. bicolor						x			

Total: 16 families, 61 genera, 368 species or subspecies.

References

Andrade, A. R. & G.-O. Brandorff, 1975. Uma nova espécie de Diaptomidae (Crustacea, Copepoda) '*Diaptomus' negrensis* das aguas pretas perto de Manaus. Acta Amazonica 5 (1): 97–103.

Bowman, T. E., 1973. Two freshwater copepods from Barrancas, Venezuela: *Notodiaptomus cearensis* (Wright) and *N. venezolanus deeveyorum*, new subspecies (Calanoida: Diaptomidae). Proc. biol. Soc. Wash. 86: 193–202.

Brandorff, G.-O., 1973. Neue freilebende Calanoide Copepoden (Crustacea) aus dem Amazonasgebiet. Amazoniana 4 (2): 205–218.

Brehm, V., 1957. Sobre los copepodos hallados por el Profesor Biraben en la Argentina. 4a comun. Neotropica 3 (11): 53–64.

Brehm, V., 1958. Bemerkungen zu einigen Kopepoden Südamerikas. Sitz. Ber. Akad. Wiss. Wien 1, 167, 3 (4): 139–171.

Dahl, F., 1894. Die Copepodenfauna des unteren Amazonas. Ber. Naturf. Ges. Freiburg b. 8: 10–23.

de Infante, A., W. Riehl & J. F. Saunders, 1979. Los Copépodos del Lago de Valencia, Venezuela. Acta Cient. Venezolana 30: 224–233.

Delachaux, Th., 1924. Zur Kenntnis der Copepodenfauna von Surinam. Zool. Anz. 59, 1,2: 1–16.

Dussart, B., 1979. Algunos Copépodos de America del Sur. Publ. Oc. 30. Mus. nac. Hist. nat., Santiago de Chile, 13 pp.

Dussart, B. H., 1981. Sobre algunos Copépodos de America del Sur, 2. Comun. cient. CECOAL, Argentina, 12: 6 pp.

Dussart, B. H., 1982. Copépodes des Antilles françaises. Rev. Hydrobiol trop. 5, 4: 313–324.

Dussart, B. H., 1982. Faune de Madagascar, 58. Crustacés Copépodes des eaux intérieurs. ORSTOM/CNRS, Paris, 146 pp.

[5] Species in course of description (Dussart, 1983).

Dussart, B. H., 1983. Sobre algunos Copépodos de America del Sur, 3 (sous presse).

Harding, J. P., 1955. The Percy Sladen expedition to lake Titicaca in 1937 - Crustacea, Copepoda. Trans. linn. Soc. Lond. 1 (3): 216–247.

Herbst, H. V., 1960. Copepoden (Crustacea, Entomostraka) aus Nicaragua und Südperu. Gewäss. Abwäss. 27: 27–54.

Jakobi, H., 1972. Trends (Enp P4 ♂) innerhalb der Parastenocarididen (Copepoda Harpacticoidea). Crustaceana 22 (2): 127–146.

Kiefer, F., 1926. Beiträge zur Copepodenkunde (IV). Zool. Anz. 69, 1,2: 21–26.

Kiefer, F., 1933. Süsswassercopepoden aus Brasilien. Zool. Anz. 105, 1,2: 38–43.

Kiefer, F., 1935. Neue Süsswassercyclopiden (Crustacea, Copepoda) aus Uruguay. Zool. Anz. 109, 7,8: 181–188.

Kiefer, F., 1936. Über die Systematik der südamerikanischen Diaptomiden (Crustacea, Copepoda). Zool. Anz. 116, 7,8: 194–200.

Kiefer, F., 1956. Freilebende Ruderfusskrebse (Crustacea, Copepoda), 1. Calanoida und Cyclopoida. Ergebn. deutsch. limnol. Venezuela-Exped. 1952 1: 233–268.

Kiefer, F., 1957. Freilebende Ruderfusskrebse (Crustacea, Copepoda) des Titicacasees. Veröff. Zool. Staatssamml. München 4: 125–150.

Kiefer, F., 1981. Beitrag zur Kenntnis von Morphologie, Taxonomie und geographischer Verbreitung von *Mesocyclops leuckarti* auctorum. Arch. Hydrobiol., Suppl. 62 (1); 148–190.

Lang, K., 1965. Copepoda Harpacticoidea from the Californian Pacific Coast. K. Svenska vetensk. Handl. 10 (2): 560 pp.

Light, S. F., 1939. On american subgenera of *Diaptomus* (Copepoda, Calanoida). Trans. am. microsc. Soc. 58 (4): 473–484.

Lindberg, K., 1961. Remarques sur le genre *Metacyclops* (Kiefer, 1927) et description d'un *Metacyclops* nouveau du Portugal. K. fysiogr. Sällsk. Lund För. 31, 14: 133–145.

Löffler, H., 1963. Zur Ostracoden- und Copepodenfauna Ekuadors. Arch. Hydrobiol. 59 (2): 196–234.

Löffler, H., 1981. Copepoda. In S. H. Hurlbert, G. Rodriguez & N. D. Dos Santos, Aquatic Biota of Tropical South America, 1. Arthropoda: 14–19.

Lowndes, A. G., 1934. Results of expedition to Brazil and Paraguay in 1926–27. Copepoda. I. Linn. Soc. Lond. Zool. 39: 83–131.

Noodt, W., 1963. Subterrane Crustaceen der zentralen Neotropis. Zur Frage mariner Relikte des Rio Paraguay-Parana-Amazonas-Systems. Zool. Anz. 171: 114–147.

Noodt, W., 1965. Crustacea subterranea aus Argentinien. Beitr. neotrop. Fauna: 84–129.

Pearse, A. S., 1921. Crustacea from Lake Valencia, Venezuela. Proc. U.S. nac. Mus. 59: 459–462.

Pesta, O., 1928. Ein Beitrag zur Kenntnis der Copepodenfauna von Argentinien. Zool. Anz. 73, 3,4: 67–80.

Wright, S., 1927. A revision of the south american species of *Diaptomus*. Trans am. microsc. Soc. 46, 2: 73–121.

Wright, S., 1935. Three new species of *Diaptomus* from Northeast Brazil. Ann. Acad. brasil. Cienc. 7: 213–233.

Wright, S., 1936. Preliminary report on six species of *Diaptomus* from Brazil. Ann. Acad. brasil. Cienc. 8 (2): 79–85.

Some remarks on the latitudinal distribution of Cladocera on the Indian subcontinent

C. H. Fernando & A. Kanduru
Department of Biology, University of Waterloo, Waterloo, Ontario, Canada N2L 3G1

Keywords: tropical zooplankton, Cladocera, latitudinal distribution, Indian subcontinent

Abstract

One hundred and thirty species of Cladocera have been recorded from the Indian subcontinent which extends from 6° N to 37° N latitude and covers an area of 4.5 million km^2. The equatorial region has few *Daphnia* species, all belonging to the sub-genus *Ctenodaphnia* while the more northern parts have more species of *Daphnia* including *Daphnia* s. str. The limnetic Cladocera lacks the carnivorous Polyphemidae and Leptodoridae at lower latitudes (equatorial). The common limnetic species of the equatorial region are eurytopic and extend throughout the subcontinent.

Introduction

The Indian sub-continent (Bangladesh, India, Nepal, Pakistan and Sri Lanka) spans the latitudinal range from 6° N (Sri Lanka) to 37° N (Kashmir, India). The land area it occupies is about 4.5 million km^2. Although this region is part of what is referred to as the humid tropics, some areas are dry to almost arid. However most of the region is well supplied with both lentic and lotic habitats. Natural lakes though, are notable for their scarcity, only a few occurring in the northernmost region and Nepal. However, man-made lakes (reservoirs) are quite numerous (Fernando, in press). Until very recently the zooplankton of this region, except perhaps Sri Lanka, was very poorly known. Studies made at the University of Waterloo have added considerably to this meagre knowledge in the Cladocera. The data we have collected in Waterloo together with the previous data are used to evaluate the role of latitude on the distribution of Cladocera. This exercise was undertaken to test the hypothesis that Cladocera, especially limnetic species, are less diverse in the equatorial than the temperate region (Fernando, 1980a, 1980b). The present study is restricted to the Cladocera. Green (1972) has shown latitudinal variation in associations of Rotifera and Fernando & Ponyi (1981) have provided some evidence that fewer Cyclopoida are found in tropical than in temperate freshwaters.

The present study shows that there are very few Cladocera restricted to the equatorial (>12° N) latitudes, while far more are restricted to the sub-tropical to temperate region (>24° N). The intermediate zone (12–24° N), which we will designate as tropical for purposes of the present study has a species association consisting of those found in the sub-tropical temperate region and widely distributed across the whole latitudinal range.

The most interesting findings are: (1) There are far fewer species of *Daphnia* in the equatorial zone as compared to more northern regions and there is a complete absence of Polyphemidae and Leptodoridae south of 22° N. These groups of Cladocera provide a high proportion of limnetic species. The absence of these two families and the paucity of species of the genus *Daphnia* means that in the tropics there are relatively few limnetic Cladocera; (2) The limnetic region in the equatorial and tropical zones has mainly eurytopic Cladocera species as

Hydrobiologia 113, 69–76 (1984).

previously shown for Sri Lanka by Fernando (1980a); (3) The limnetic Cladocera of the equatorial and tropical zones are species with a wide latitudinal range generally. This range extends in the Indian subcontinent to regions which are almost temperate. Some species are very rare and their occurrence or non-occurrence cannot be used as evidence in this study of latitudinal distribution. Some genera e.g. *Pleuroxus* are definitely Northern in distribution. *Eurycercus* sp. was found as far South as Jabalpur. This temperate genus was carried by a river originating in cold regions of the Himalayas (Adholia, 1979).

Materials and methods

Over 300 samples from India were studied and their Cladocera identified to species. These samples came from 19 different areas which span the latitudinal range of India (Fig. 1).

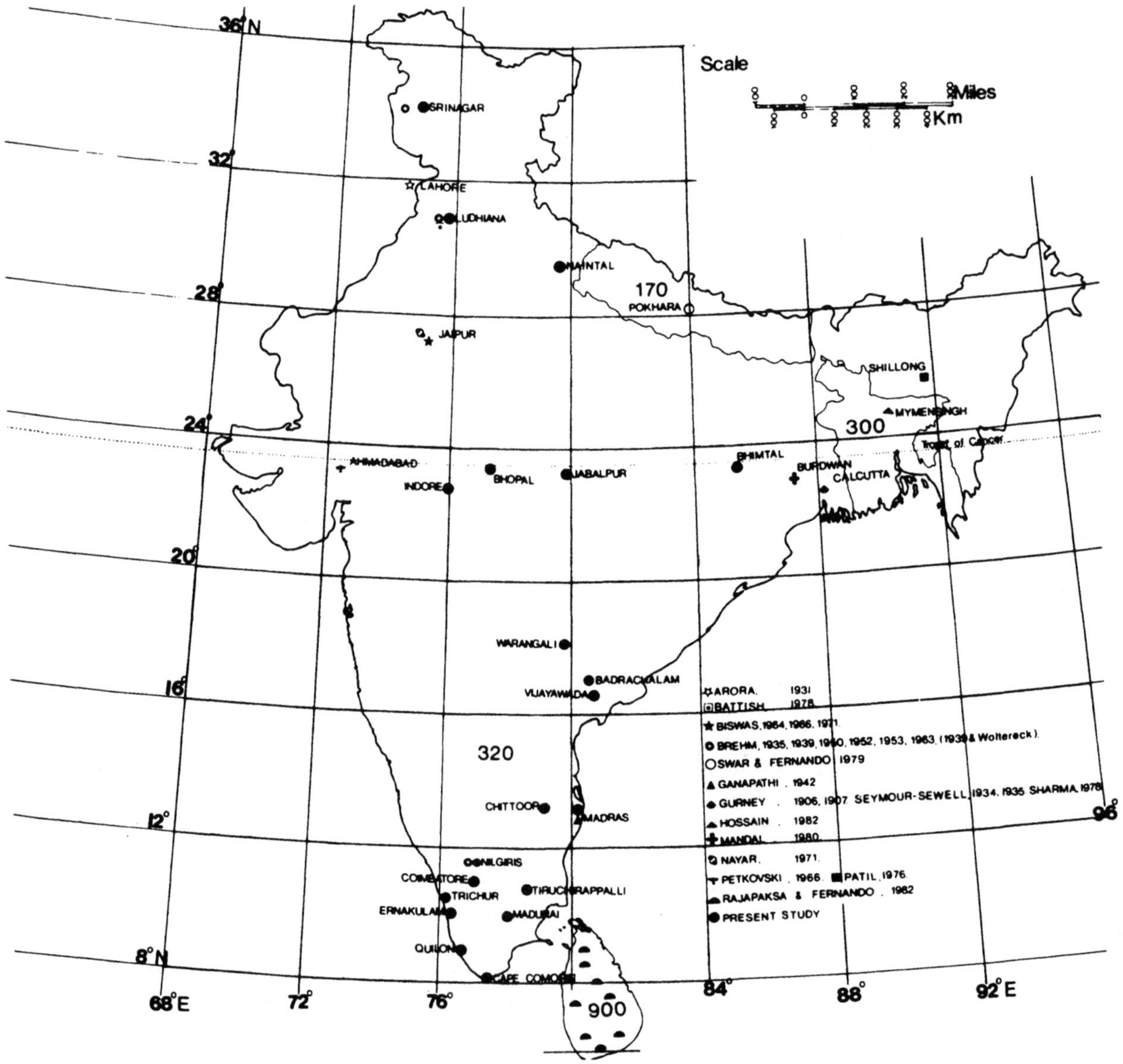

Fig. 1. Areas where zooplankton samples were collected for study or reported in the literature for Cladocera. The 28 'sites' cover a much wider area than shown. The number of samples from each country are given (except Pakistan).

Data from material studied in the laboratory of the senior author since 1968 from published and unpublished work was used. This includes the work of Hossain (1982), Bangladesh; Kanduru (1982) India; Rajapaksa & Fernando (1982) and unpublished data, Sri Lanka; and Swar & Fernando (1979), Nepal.

Other published data was used too, including systematics papers or theses. There are about 23 such works available since 1860. We have used most of these except when only a few species were recorded. These include: Arora (1931), incidentally the only paper on Pakistan Cladocera, Battish (1978), Biswas (1964, 1966, 1971), Brehm (1936, 1950, 1952, 1963), Brehm & Woltereck (1939), Gurney (1906, 1907), Nayar (1971), Patil (1976), Petkovski (1966), Seymour-Sewell (1934). A few studies on biology and distribution also were used for species lists. These were in regions where reords were few, had extensive lists of species (relatively) or were recent and had identifications to the species level. These were Ganapati (1943), Mandel (1980) and Seymour-Sewell (1935). For Sri Lanka, Bangladesh and Nepal, where our material was much more extensive than that of previous workers we omitted previous records. The number of samples examined by previous workers is difficult to estimate. However the figure of >1 200 for India seems reasonable since Battish (1978) examined 800 samples. The number of samples studied in the Department of Biology, University of Waterloo are given in Fig. 1. A list of species of Cladocera recorded in the Indian subcontinent is given in Table 1. The actual areas of sampling cover a much wider geographic area than indicated in the map. For purposes of analysing species associations the area under study was divided as follows and are:

1. 4–8°N Sri Lanka
   8–12°N Nilgiris, Coimbatore, Tiruchchirappalli, Trichur, Ernakulam, Madurai, Quilon and Cape Comorin

   These two zones will be referred to as *equatorial*.
2. 12–20°N Warangali, Badrachalam, Wijewada, Chitoor, and Madras. Little data was available from this zone.
   20–24°N Bhimtal, Burdwan, Jabalpur, Bhopal, Indore and Calcutta

   These two zones will be referred to as *tropical*.
3. 24–28°N Jaipur, Shillong, Mymensingh
   28–32°N Lahore, Ludhiana, Naintal, Pokhara
   32–38°N Srinagar.

   These three zones will be referred to as *sub-tropical-temperate*.

The species lists in the literature were checked for inaccuracies and synonyms. e.g. *Daphnia carinata* has been recorded many times from India and Sri Lanka. However Hebert (1977) has shown that this species is restricted to the Australasian region. The Indian region records refer to *D. cephalata*. There are, as expected, quite a few synonyms in the literature.

The species lists from each of the 28 'sites' (see Fig. 1), were scrutinized for obvious Northern (temperate) and Southern (equatorial) species. The complete lists of Cladocera of Europe (Hrbàcek *et al*, 1978) and Sri Lanka (Rajapaksa & Fernando, 1982) were used for comparison of species associations using Sorensen's index. The northernmost Indian site Srinagar has been very poorly sampled and could not be used as a reliable region for a temperate species list. Cluster analysis was used separately for each of the six families of Cladocera. The latitudinal distribution of limnetic species was scrutinized using all the data available.

Results and discussion

A total of 87 species were identified in the 320 samples examined from India. Taking all published records, 130 species are known to occur on the sub-continent. Michael & Sharma (in press) in a monograph on Indian Cladocera, list and describe 89 species. As a comparison Europe has 150 species (Hrbàcek *et al*, 1978). The Australian region has about 125 recorded species (Smirnov & Timms, 1983). The species list for the Indian subcontinent can therefore be considered reasonably complete.

The latitudinal distribution shows the following features on scrutiny:

There is an exclusively northern group of species not reported south of 22° N. (Occurrences in 1 locality only are omitted.) This list consists of *Leptodora kindtii*, *Polyphemus pediculus*, *Sida crystallina*, *Diaphanosoma brachyurum*, *Daphniopsis*, *Daphnia* s. str. (9 species), *Eurycercus* sp., *Pleuroxus laevis*, *Pleuroxus trigonellus*, *Alonella exigua*, *A. nana*, *Alona quadrangularis*, *Rhynchotalona falcata*, and *Moina rectirostris*.

Only *Grimaldina brazzai* can be considered a strictly equatorial species occurring below 12° N.

Table 1. List of Cladocera recorded in the Indian sub-continent. Some synonyms are retained because they very probably refer to species not in the list.

Leptodoridae
1. Leptodora kindti

Polyphemidae
2. Polyphemus pediculus

Sididae
3. Diaphanosoma aspinosum
4. Diaphanosoma brachyurum
5. Diaphanosoma dubia
6. Diaphanosoma excisum
7. Diaphanosoma modigliani
8. Diaphanosoma paucispinosum
9. Diaphanosoma sarsi
10. Diaphanosoma senegal
11. Latonopsis australis
12. Latonopsis occidentalis
13. Pseudosida bidentata
14. Sida crystallina

Daphniidae
15. Ceriodaphnia cornuta
16. Ceriodaphnia dubia
17. Ceriodaphnia laticaudata
18. Ceriodaphnia pulchella
19. Ceriodaphnia quadrangula
20. Ceriodaphnia reticulata

(Ctenodaphnia)
21. Daphnia cephalata
22. Daphnia lumholtzi
23. Daphnia magna
24. Daphnia similis
25. Daphnia psittacea

(Daphnia)
26. Daphnia fusca
27. Daphnia hyalina
28. Daphnia longispina
29. Daphnia obtusa
30. Daphnia pamirensis
31. Daphnia pulex
32. Daphnia rosea

(Daphniopsis)
33. Daphnia tibetana

34. Scapholeberis aurita
35. Scapholeberis kingi
36. Simocephalus acutirostratus
37. Simocephalus australiensis
38. Simocephalus exspinosus
39. Simocephalus latirostris
40. Simocephalus serrulatus
41. Simocephalus vetulus

Moinidae
42. Moina banffyi
43. Moina belli
44. Moina brachiata
45. Moina flagellata
46. Moina macrocopa
47. Moina micrura
48. Moina rectirostris
49. Moina reticulata
50. Moina weismanni
51. Moinadaphnia mcleayi

Bosminidae
52. Bosmina coregoni
53. Bosmina longirostris
54. Bosminopsis dietersi

Macrothricidae
55. Echinisca capensis
56. Echinisca odiosa
57. Echinisca rosea
58. Echinisca shadini
59. Echinisca triserialis
60. Grimaldina brazzai
61. Guernella raphaelis
62. Ilyocryptus spinifer
63. Macrothrix goeldi
64. Macrothryx hirsuticornis
65. Macrothrix laticornis
66. Mecrothrix rectirostris
67. Macrothrix spinosa
68. Macrothrix tobaensis
69. Streblocercus serricaudatus

Chydoridae
70. Acroperus harpae
71. Alona anodonta
72. Alona bukobensis
73. Alona costata
74. Alona davidi
75. Alona glabra
76. Alona guttata
77. Alona harpularia
78. Alona holdeni
79. Alona ladacensis
80. Alona monacantha
81. Alona pulchella
82. Alona quadrangularis
83. Alona rectangula
84. Alonella excisa
85. Alonella exigua
86. Alonella nana
87. Biapertura affinis
88. Biapertura intermedia
89. Biapertura karua
90. Biapertura verrucosa
91. Camptocercus australis
92. Camptocercus rectirostris
93. Chydorus barroisi (Ephemeroporus)
94. Chydorus ceylonicus
95. Chydorus ciliatus
96. Chydorus eurynotus
97. Chydorus faviformis
98. Chydorus gibbus
99. Chydorus invaginatus
100. Chydorus kallipygos
101. Chydorus parvus
102. Chydorus reticulatus
103. Chydorus sphaericus
104. Chydorus ventricosus
105. Dadaya macrops
106. Dunhevedia crassa
107. Dunhevedia serrata
108. Euryalona occidentalis
109. Euryalona orientalis
110. Eurycercus sp.
111. Kurzia longirostris
112. Kurzia latissima
113. Graptoleberis testudinaria
114. Leydigia acanthocercoides
115. Leydigia australis
116. Leydigia macrodonta
117. Indialona annandelei
118. Indialona ganapati
119. Indialona globulosa
120. Indialona macronyx
121. Oxyurella sinhalensis
122. Oxyurella tenuicornis
123. Pleuroxus aduncus
124. Pleuroxus denticulatus
125. Pleuroxus laevis
126. Pleuroxus parmirensis
127. Pleuroxus similis
128. Pleuroxus trigonellus
129. Pseudochydorus globosus
130. Rhynchotalona falcata

Many species are widely distributed throughout the sub-continent. These include: *Pseudosida bidentata, Daphnia similis, D. lumholtzi, Scapholeberis kingi, Alonella excisa, Chydorus eurynotus, C. ventricosus, Graptoleberis testudinaria, Alona pulchella, A. guttata, A. rectangula* and *Biapertura verrucosa.*

A group of species occurred at all latitudes except the Srinagar area (32–20° N). These are *Diaphanosoma excisum, D. sarsi, Echinicia triserialis, Kur-*

zia longirostris, Alona davidi, Biapertura karua, B. affinis, and *Leydigia acanthocercoides.* A head shield of *Eurycercus* sp. was found in Jabalpur. This temperate genus was also recorded in Nepal (Dumont & Van de Velde, 1977), and Kashmir. It is likely that this southern record is of a specimen brought down by a river arising in the cold Himalayan region.

A number of rare species were found. Some of them – *Echinisca shadini, Chydorus ceylonicus* and *Indialona macronyx* – have not been recorded outside Sri Lanka on the Indian sub-continent. They have been recorded elsewhere except *Ch. ceylonicus.* Sri Lanka lacks the genera, *Camptocercus* and *Acroperus,* found in equatorial India.

For the Sorensen index all species of Cladocera were used at intervals of 4 degrees of latitude (except the Warangali zone where few data were available). The results are shown in Fig. 2. If a comparison is made with Europe then all the zones except Sri Lanka had Mymensingh are grouped relatively closely. Mymesingh and Sri Lanka are not very close together either. The low values for this index is a reflection of the use of all the European species for comparison with a relatively poorly known fauna for each area (except Sri Lanka) and the use of a much wider latitudinal range in Europe. When the Sri Lanka zone is compared with other zones. Srinagar stands out rather distinctly while Mymensingh, Madurai and Jabalpur are rather close. The

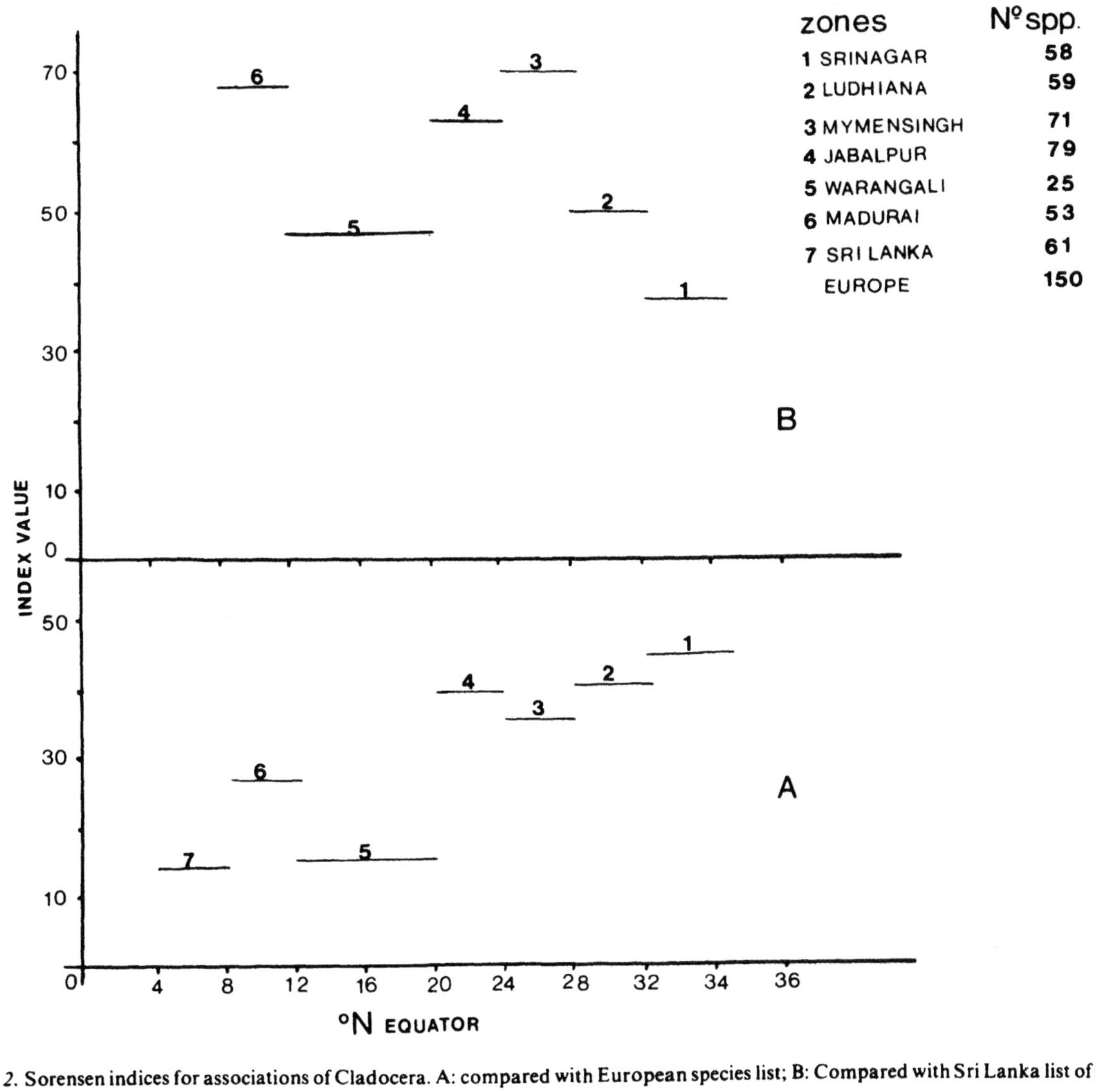

Fig. 2. Sorensen indices for associations of Cladocera. A: compared with European species list; B: Compared with Sri Lanka list of species.

higher index is due to the use of a well studied but relatively small range of species in Sri Lanka. However, the present data cannot be considered really adequate for this type of index.

The cluster analysis did not give any clearcut results. The data available for many families is meagre. These are the Leptodoridae, Polyphemidae, Sididae, Moinidae, and Bosminidae. The widely distributed Chydoridae show hardly any clustering. The Daphniidae on the other hand show a distinct separation of Srinagar and Sri Lanka. If the *Ctenodaphnia* are separated from *Daphnia* (s. str.) and *Daphniopsis*, the separation would be even greater. The cluster maps do separate the temperate (Srinagar), Jabalpur and Mymesingh (sub-tropical) and the equatorial (Sri Lanka) regions generally. More data is required from across the whole latitudinal range for the cluster analysis to give

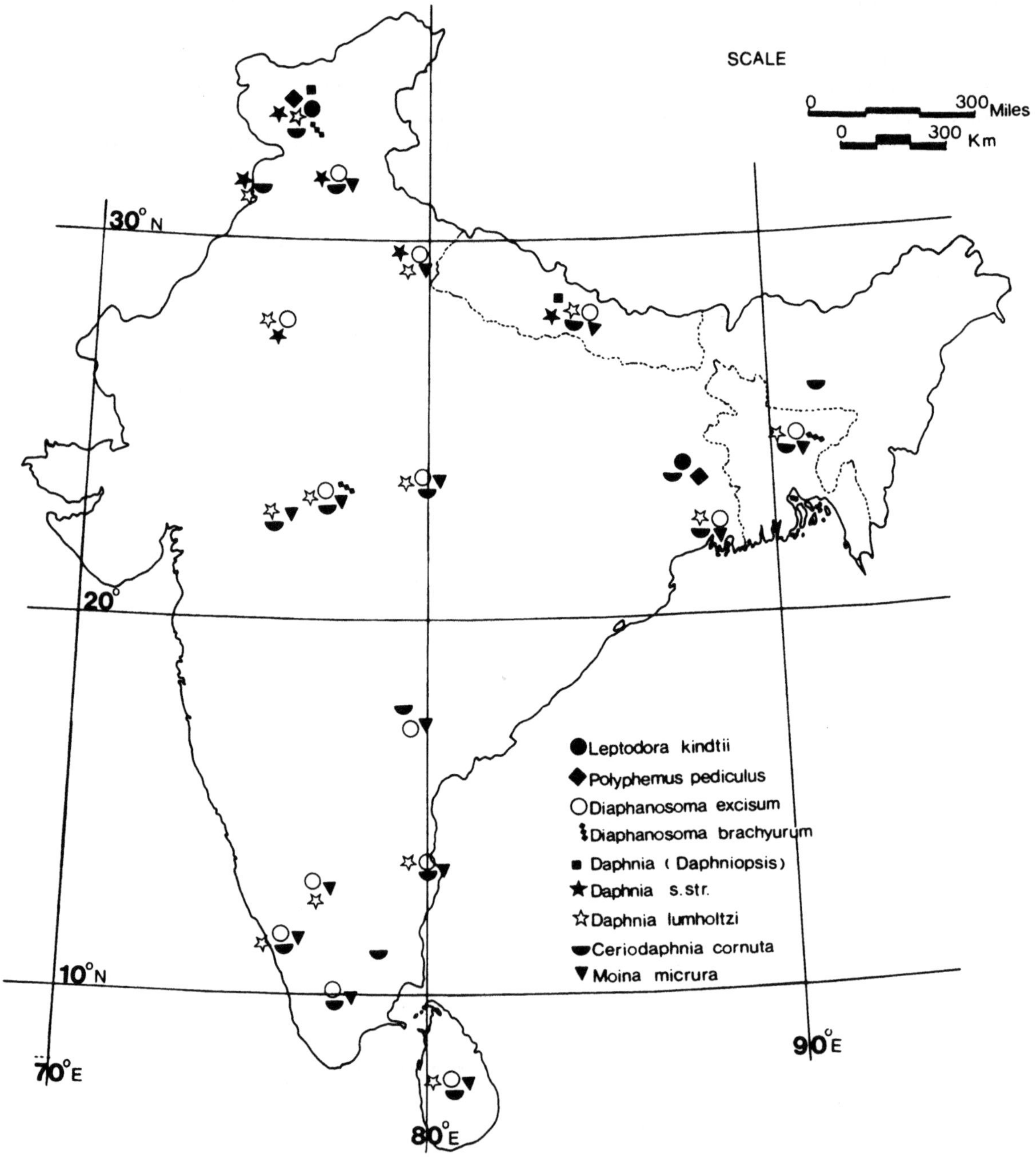

Fig. 3. The distribution of some limnetic species including all the common species found <24° N latitude.

more meaningful results than those which can be obtained by simple inspection of the species lists for each locality.

The distribution of some limnetic species is shown in Fig. 3. In all, 7 *Daphnia* spp. (S. Str.), 5 *Daphnia (Ctenodaphnia)* and 1 *Daphniopsis* species were recorded. All the *Daphnia* spp. and *Daphniopsis* were restricted to >24° N. On the other hand only *Ctenodaphnia* was recorded below 24° N. *Daphnia similis* was widespread but occurred in astatic waters in the lower latitudes. The limnetic *Daphnia* species consist of *Daphnia lumholtzi*, a small species of *Ctenodaphnia* and all the *Daphnia* (S. Str.) species and *Daphniopsis* found in more northern latitudes. The genus *Diaphanosoma* is represented by *D. brachyurum* which is a northern species while *D. excisum* and *D. sarsi* occur across the whole latitudinal range except Srinagar and Lahore as do the other common tropical species. The limnetic species *Ceriodaphnia cornuta* and *Moina micrura* occur throughout the latitudinal range of the sub-continent. *Bosminopsis deitersi* has a more southern distribution from Sri Lanka to Mymensingh and *Bosmina longirostris* has a more northern distribution from Srinagar to Calcutta and the hilly areas, the Nilgiris and Coimbatore. *Polyphemus pediculus* and *Leptodora kindtii* are restricted in their distribution to >24° N.

It is quite evident that the equatorial region has relatively few limnetic Cladocera. This has been shown before for the South East Asian region by (Fernando, 1980a) who also pointed out the reduction of *Daphnia* (S. l) as one goes towards the equator (Fernando, 1980c). Also all the common limnetic species found in the equatorial zone extend their range to sub-tropical to temperate regions of the sub-continent. This suggests that the limnetic Cladocera are of the temperate origin. These limnetic species not only have a wide latitudinal range but they are also eurytopic. Fernando (1980a) has shown that these limnetic species occur commonly in the whole range of tropical freshwater habitats. In temperate regions with a high diversity of limnetic species they are eliminated, e.g. *Diaphanosoma excisum* and *Ceriodaphnia cornuta* in Europe and North America; or are restricted to non-limnetic situations mainly, e.g. *Moina micrura*. There are no strictly limnetic Cladocera in the equatorial and tropical parts of the Indian subcontinent.

The present study is limited by the small number of samples available from India and Pakistan considering their vast areas. However the general picture of latitudinal species association is not likely to be changed much by more data.

There are many rare species among those found or listed in the literature. Their distribution cannot be discussed from the latitudinal standpoint.

Latitude is certainly not the only factor determining Cladocera species distribution. The availability of a wide range of habitats, river drainage, altitude, trophic status of freshwaters and human influence on freshwater are other factors influencing distribution.

Summary

Based on the study of 320 samples from India and more extensive samples from Sri Lanka, Bangladesh and Nepal and from lists of species in the literature, an analysis of the role of latitude on Cladoceran distribution was attempted for the Indian sub-continent.

Very few species are limited to the equatorial region (<12° N). Far more species are, sub-tropical-temperate, occurring only >24° N. Most species span a wide latitudinal range and do not belong to equatorial, tropical or temperate groups. Sorensen's index and cluster analysis separates some most northern species from the most southern species groupings but the most interesting results are that *Daphnia* (S. Str.) is absent in the tropical and equatorial region while this sub-genus is well represented in latitudes >23° N. The limnetic Cladocera of tropical and equatorial regions are all the same species that occur at latitudes >24° N. However many limnetic species found at these higher latitudes do not occur in the tropics. There appear to be no truly limnetic Cladocera in the tropical and equatorial regions. Welch (1952) rightly points out that the tropical origin of freshwater zooplankton is not substantiated. It appears that the common limnetic zooplanktonic Cladocera are more northern in origin but tolerant of the higher temperatures in tropical and equatorial regions. These species are also eurytropic in the lower latitudes.

Acknowledgements

We wish to thank Mr. M. A. Hossain and Mrs. R. Rajapaksa for allowing us to use their unpublished data. Funds for the study were provided by a UNESCO Fellowship to one of us (A.K.) and a grant from the National Science and Engineering Council of Canada. (NSERC) No. A 3478 to C.H.F. Many scientists from India, Bangladesh, Nepal and Sri Lanka have helped with sampels of zooplankton.

References

Adholia, U. N., 1979. Zooplankton of the River Betwa. Geo-Eco-Trop. 3: 267-271.

Arora, G. L., 1931. Fauna of Lahore 2. Entomostraca (Water-fleas) of Lahore. Bull. Dept. Zool. Punjabi Univ. 1: 62-100.

Battish, S. K., 1978. A survey of the free-living entomostracean (Crustacean) fauna of Punjab with special reference to recent ostracoda. Ph.D. Thesis, Punjabi Univ., 274 pp.

Biswas, S., 1964. A new species of the cladoceran genus *Latona* Strauss, 1820 (Crustacea; Cladocera: Sididae) from Rajasthan, India. Proc. zool. Soc. Calcutta 17: 149-152.

Biswas, S., 1966. Five species of Daphnidae (Crustacea: Cladocera) from Simla Hills in India with a new record of *Alona costata* Sars, from Kameng Division, Nefa. J. zool. Soc. India 16: 92-98.

Biswas, S., 1971. Fauna of Rajasthan India Part 2. Crustacea, Cladocera. Rec. Zool. Survey, India 63: 95-141.

Brehm, V., 1935. Report on Cladocera, Yale North India Expedition. Mem. Conn. Acad. Arts Sci. 10: 283-293.

Brehm, V., 1950. Contributions to the freshwater fauna of India Parts 1 and 2. Rec. Indian Mus. 48: 1-28.

Brehm, V., 1952. *Diaphanosoma hydrocephalus* nov. Spec. Eine Eigenartige Sididae aus Vorder-Indian. Zool. Anz. 149: 138-140.

Brehm, V., 1953. Indische Diaptomiden, Pseudodiaptomiden und Cladoceran. Ost. zool. Z. 4: 241-345.

Brehm, V., 1963. Einige Bermarkungen Zu Wien Indischen Entomostraken. Int. Revue ges. Hydrobiol. 48: 159-172.

Brehm, V. & R. Woltereck, 1939. Die Daphniden der Yale - Northindia Expedition. Int. Revue ges. Hydrobiol. 39: 1-19.

Dumont, H. J. & I. Van de Velde, 1977. Report on a collection of Cladocera and Copepoda from Nepal. Hydrobiologia 53: 55-65.

Fernando, C. H., 1980a. The freshwater zooplankton of Sri Lanka with a discussion of tropical freshwater zooplankton composition. Int. Revue ges. Hydrobiol. 65: 85-125.

Fernando, C. H., 1980b. The species and size composition of tropical freshwater zooplankton, with special reference to the Oriental region (South East Asia) Int. Revue ges. Hydrobiol. 85: 411-426.

Fernando, C. H., 1980c. The freshwater zooplankton of South East Asia. In J. I. Furtado (ed.), Tropical Ecology and Development. Kuala Lumpur, Malaysia: 845-853.

Fernando, C. H., in press. Reservoir and lake ecosystems in South East Asia (Oriental Region) In F. B. Taub (ed.), Lake and Reservoir Ecosystems. Elsevier, Amsterdam.

Fernando, C. H. & J. E. Ponyi, 1981. The freeliving cyclopoid Copepoda (Crustacea) of Malaysia and Singapore. Hydrobiologia 78: 113-123.

Ganapati, S. V., 1943. An ecological study of a garden pond containing abundant zooplankton. J. indian Acad. Sci. 17: 41-58.

Green, J., 1972. Latitudinal variation in associations of planktonic Rotifera. J. Zool., Lond. 167: 31-39.

Gurney, R., 1906. On some freshwater Entomostraca in the collection of the Indian Museum. J. asiat. Soc. Beng. 11: 273-280.

Gurney, R., 1907. Further notes on Indian freshwater Entomostraca. Rec. indian Mus. 1: 21-23.

Hebert, P. D. N., 1977. A revision of the taxonomy of the genus *Daphnia* (Crustacea: Daphniidae) in South East Australia. Aust. J. Zool. 25: 371-398.

Hossain, M. A., 1982. A taxonomical study of the freshwater zooplankton (Rotifera, Cladocera and Copepoda) of Bangladesh. M.Sc. Thesis, Univ. Waterloo, 284 pp.

Hrbáček, J., V. Kořínek & D. G. Frey, 1978. Limnofauna Europaea. A check list of the animals inhabiting European Inland Waters with accounts of their distribution (except Protozoa). Cladocera: 189-195.

Kanduru, A. I., 1981. The Cladocera (Crustacea: Branchiopoda) of the Indian sub-continent and their latitudinal distribution. M.Sc. Thesis, Univ. Waterloo, 313 pp.

Mandel, B. K., 1980. Limnological studies of a freshwater fish pond at Burdwan, West Bengal, India. Jap. J. Limnol. 41: 10-18.

Michael, R. G., B. K. Sharma (in press). A monograph on the Cladocera of India. Fauna India Ser., Calcutta.

Nayer C. G. K., 1971. Cladocera of Rajasthan. Hydrobiologia 17: 509-519.

Patil, S. G., 1976. Freshwater Cladocera (Arthropoda: Crustacea) from Northeast India. Curr. Sci. 45: 312-313.

Petkovski, T. K., 1966. Eine neue gattung aus dem Western Indiens, *Indialona ganapati*. N. gen. et N. sp. Fragm. balanc. 5: 159-163.

Rajapaksa, R. & C. H. Fernando, 1982. The Cladocera of Sri Lanka with remarks on some species. Hydrobiologia 94: 49-69.

Seymour-Sewell, R. B. S., 1934. A study of the fauna of the salt lakes, Calcutta. Rec. indian Mus. 36: 45-121.

Seymour-Sewell, R. B. S., 1935. Studies on the bionomics of freshwaters of India 2. On the fauna of the tank in the Indian Museum compound and the seasonal changes observed. Int. Revue ges. Hydrobiol. 31: 203-238.

Sharma, B. K., 1978. A note on freshwater Cladocerans from West Bengal. Bangladesh J. Sci. 6: 149-151.

Smirnov, N. N. & B. V. Timms, 1983. A revision of the Australian Cladocera (Crustacea). Rec. aust. Mus. Suppl. 1.

Swar, D. B. & C. H. Fernando, 1979. Cladocera from Pokhara Valley, Nepal with notes on distribution. Hydrobiologia 66: 113-128.

Welch, P. S., 1952. Limnology. McGraw-Hill, New York, 538 pp.

A review of systematics, distribution and ecology of tropical freshwater zooplankton

B. H. Dussart[1], C. H. Fernando[2], T. Matsumura-Tundisi[3] & R. J. Shiel[2]
[1] *Station Biologique, Les Eyzies, France*
[2] *Department of Biology, University of Waterloo, Ontario, Canada*
[3] *Departemento de Ciencias Biologicas, Universidade Federal de São Carlos, Brazil*

Keywords: tropical zooplankton, systematics, distribution, ecology review

Abstract

Studies on tropical freshwater zooplankton which commenced in the mid-19th century have been intensified during the past twenty years or so. The whole region, barring a few areas, has been investigated, including very recently tropical Australia. The widely scattered literature is briefly summarized with comments. Some general distributional patterns are emerging. There is a dearth of ecological studies especially on seasonality and production. Systematics lags behind the North Temperate zone in most tropical regions.

Introduction

Studies on all aspects of tropical zooplankton have lagged far behind, especially in the North Temperate zone. During the last ten years or so, however, the mostly fregmentary data on systematics, distribution and ecology of tropical zooplankton have at times been brought together at least on a national basis in Asia, Africa, and South America. Systematic work has spread to cover most tropical regions, though notable gaps exist in Australia and New Guinea (Australasia) and some other tropical countries: Burma, Laos and Kampuchea (Asia), some regions of tropical Africa and most of the Caribbean. In addition to regional and national studies, world monographs enabling accurate species diagnosis on small or large groups of zooplanktonic organisms are now available for the Rotifera, Cladocera and Cyclopoida. These studies vary from region to region and from group to group. The most unevenly known are the Calanoida which are restricted in their distribution. Minor zooplanktonic groups like Protozoa, Chaoborinae and Ostracoda are also still poorly known systematically.

In this presentation we will deal with the state of our knowledge of systematics, distribution and ecology of zooplankton in tropical, sub-tropical and to a lesser extent, sub-temperate regions. At the present time it is perhaps more satisfactory to deal with systematics and distribution on a regional basis, i.e. Africa, America, and Australia, because most of the studies are regional or national and because each author has special knowledge of one region only.

The geographic areas where studies have been made are shown in Fig. 1. The area covered is considerable in relation to the total area of land between the tropics (i.e. 23°N–23°S). However, many studies have been based on only a few collections or are restricted to one group of zooplankton (e.g. Cyclopoida, Calanoida, Chydoridae).

A problem is the designation of 'tropical'. We have used this to cover the area between the tropics (i.e. 23°N–23°S). However we have reviewed some studies in ecology made outside these areas because of the lack of literature in this field. Strictly speaking altitude must be taken into account when considering 'tropical' climates. At the present time we have not converted altitude to latitude in considering distribution of zooplankton.

In dealing with the four continents with true tropical regions, we have not attempted a strict

Hydrobiologia 113, 77–91 (1984).

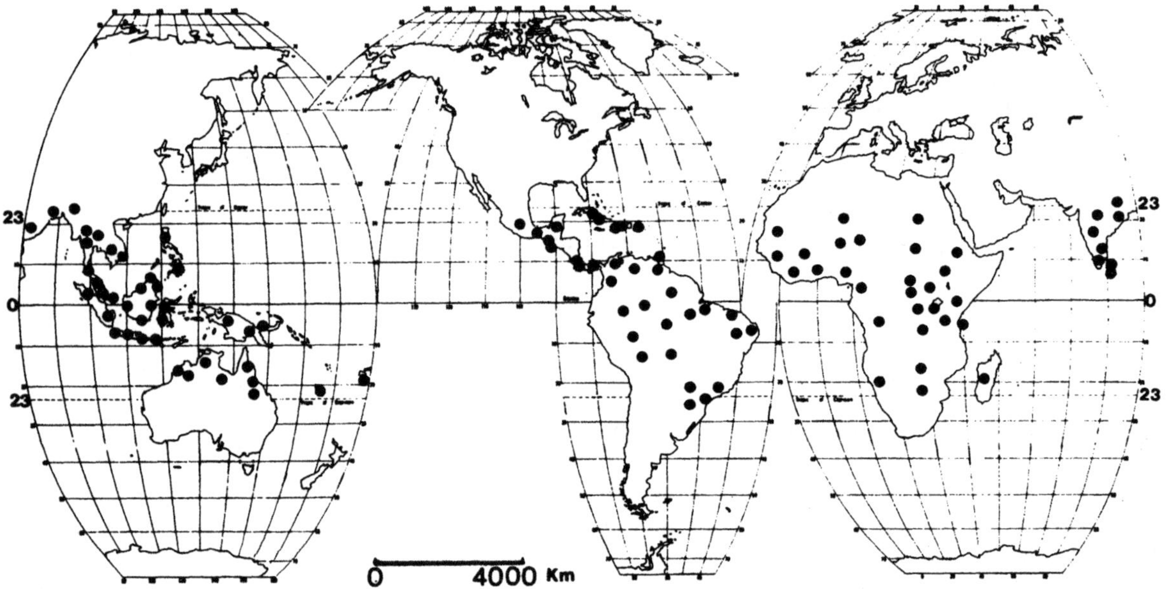

Fig. 1. Distribution of study areas for zooplankton in the lattitudinal zone 23°N-23°S.

uniformity. The state of knowledge is different in each of these areas. Australasia is the least known. There is much work done in Africa but the strictly tropical regions have not been intensively investigated and few broad-based studies exist even at the national level. The South American and Australasian work is relatively recent and is therefore treated slightly more extensively. Some Asian work has been summarized before on a national and areal basis and this area is also the best investigated in systematics, distribution and ecology. Hence the treatment of Asia is more brief.

Africa

Systematics

The study of tropical freshwater zooplankton commenced at the end of the 19th century. The relative proximity to Europe and the mounting of expeditions were the prime reasons for the accumulation of knowledge in the 20th century.

There is scattered literature dealing with the African zooplankton. Much of this work however is in Mediterranean and sub-temperate regions of North Africa and South Africa. If we restrict ourselves to the region 23°N-23°S there is much less concentration of studies. The following are the more extensive, recent or less known studies. Dumont (1979) has lisited much of the African work. The other studies which fall into the categories stated are as follows: Brunelli & Cannicci (1938), Daday (1910), Dumont *et al* (1981), Dussart (1974, 1977, 1980, 1982), Einsle (1971), Fryer (1957), Gauthier (1939, 1951), Green (1962a, b, c), Gurney (1928), Harding (1942), Kiefer (1928, 1933, 1952, 1956, 1981, 1982), Kiss (1960), Lindberg (1950a, b, c, 1951a, b), Lowndes (1930, 1936), Marques (1958, 1961, 1963, 1966), Onabamiro (1952, 1957), Rzoska (1968, 1974), Rzoska, Brooke & Prowse (1955), Sars (1909) and Van de Velde (1978). Proszynska (1967) has provided a bibliography of the crustacean zooplankton of African inland waters. Moreover, Dussart (1982b) has described the Copepod fauna of Madagascar.

Specifically dealing with Rotifers are the papers of Beauchamp (1955), Green (1960a), Pourriot (1968, 1980) and Ridder (1981).

There are relatively few studies covering even one order of zooplankton organisms comprehensively for any country. Only a few studies can be considered in the category: e.g. Dumont (1979), Gauthier (1951) and Jeje (1982).

Some elements of the plankton of tropical freshwater are noteworthy such as the medusa stage of *Limnocnida* (Green, 1980b; Goy, 1980).

For the lesser known protozoans we must await the continuation of the studies of Dragesco (1972a, b, c, 1973) to discover the diversity of the African forms. Dragesco (1980) has already summarized the African work.

Zoogeography

Of the 82 species of Calanoida reported from Africa more than a quarter have been observed only once or in the same group of habitats. This is the same for some of the 191 species and sub-species of Cyclopoida. Certain genera are represented by only one species while others have more, e.g. *Tropodiaptomus*, widely distributed genus in Asia, Australia and Africa, has 32 different species in Africa.

If our knowledge of systematics is very fragmentary then our knowledge of distribution and ecology is even more fragmentary.

From North to South, Africa can be divided into zones. Of the 191 Cyclopoida species for example, 33 were found in North Africa and 97 south of the Tropic of Cancer. Lake Tanganyika alone has 23 of these species, all of which are endemics.

It should be noted that there is a large difference between species populations of North Africa, Egypt and the Mediterranean on the one hand and the forms of Gondwana origin. These latter forms are found in South Africa, Australia, Chile and Argentina. Between the Gondwana and Eurasian faunas is a diversified intertropical fauna, part of which is limited to water masses of ancient origin like Lake Tanganyika.

Therefore there are three directions of entry of fauna into Africa: (1) The Eurasiatic forms which reached Africa but do not reach the Sahara; (2) the Mediterranean forms which are relicts of the ancient Tethys sea which reach the Soudanian zone and ((3) the western and central intertropical forms and eastern forms which evolved along the rift valley that extends from Ethiopia to Malawi and the meridional forms which extend from the Kalahari desert to the East and north-east to Mozambique.

A special place has to be given to Madagascar: on 85 species known of Copepoda, 33 seem to be endemic to this isle-continent, particularly Cyclopoïda and Harpacticoïda (Parastenocaridae).

Americas

Systematics and distribution

Studies of zooplankton systematics and their geographic distribution in tropical America are relatively few. In Brazil, from 0° to 23° South latitude, the Copepoda have been studied by Poppe (1891), Dahl (1894), Sars (1901), Wright (1927, 1935, 1936, 1937, 1938a, b), Kiefer (1933, 1935, 1936), Lowndes (1934), Kleerekoper (1944), Herbst (1959, 1962), Brandorff (1937a, b, 1976, 1978), Cippoli & Carvalho (1973), Andrade & Brandorff (1975), Matsumura-Tundisi & Rocha (in press) and Sendacz & Kubo (1982). The Cladocera, however, are almost unknown and there are not sufficient data to assess the number of species which may occur and their geographic distribution. A small contribution was made by Richard (1897), Herbst (1967) and Brandorff *et al* (1982) to the Cladocera of the Amazon region. Also restricted to the Amazon region are the studies on Rotifera by Ahlstrom (1937), Hauer (1965), Koste (1972) and Brandorff *et al* (1982).

In Venezuela the studies of Pearse (1921), Kiefer (1954, 1956), Bowman (1973), de Infante (1979) on copepods, Hauer (1956) have contributed to the knowledge of the occurrence of zooplankton species in this region.

In Peru, Equador and Colombia many authors have worked with copepods: Thiebaud (1914) and Pearse (1915) in Colombia; Kiefer (1926, 1942, 1957), Delachaux (1928), Harding (1955), Lindberg (1955) and Herbst (1960) in Peru and Löffler (1963) in Equador.

The Central American region (Mexico, Guatemala, Honduras, El Salvador, Panama, Cuba, Haiti, Curacao) has been well studied for the Copepoda by Richard (1895), Marsh (1913, 1919, 1931), Pearse (1911, 1938), Juday (1915), Kiefer (1931, 1933, 1934, 1936, 1938), Ueno (1939), Rioja (1940a, b), Tafall (1941, 1943, 1944), Coker (1943), Brehm (1932, 1939, 1948, 1955), Lindberg (1954), Herbst (1960), Smith & Fernando (1978, 1980) and Fernando & Smith (1982). Rotifers were studied by Ridder (1977) in the Caribbean and the Cladocerans documented by Frey (1982).

Concerning the number of freshwater copepod species, tropical South America has about 35 species of Calanoida belonging to ten genera: *Argyro-*

diaptomus, Calodipatomus, Dactylodiaptomus, Idiodiaptomus, Notodiaptomus, Odontodiaptomus, Prionodiaptomus, Rhacodiaptomus, Aspinus and *Diaptomus*. Most of these genera show a narrow range of distribution. *Prionodiaptomus*, with one species, *P. colombiensis* Thiebaud, has been recorded between the parallel 0° and 15° North latitude (Colombia, Venezuela, Panama, Guatemala, Honduras, El Salvador and Nicaragua). *Dactylodiaptomus* and *Rhacodiaptomus* are restricted between 0° to 10° South latitude. *Calodiaptomus* has a slightly larger range extending to 17° South latitude. *Odontodiaptomus (O. paulistanus)* and *Idiodiaptomus (I. gracilis)* have been recorded only in São Paulo State (23° South latitude). The genus *Argyrodiaptomus* is characteristic of the southern part, however some species such as *A. azevedoi* can be distributed as far north as the Amazon region according to Brandorff (1976). *Notodiaptomus* has a large range of distribution comprising a large number of species (19) and it is the commonest genus from South America. In Central America 18 species have been recorded (Fernando & Smith, 1981) and four genera are known: *Arctodiaptomus, Mastigodiaptomus, Prionodiaptomus* and *Diaptomus*. *Prionodiaptomus colombiensis* Thiebaud, 1914 = *Diaptomus marshi*, Juday, 1914 is the only calanoid species common to both Central and South America.

Among the Cyclopoida, the number of species is much higher than for the Calanoida. About 60 species occur in tropical South America and 30 species in Central America. There are also several species that occur in both regions: e.g. *Eucyclops encifer, E. serrulatus, Paracyclops fimbriatus, Mesocyclops aspericornis, M. longisetus, M. brasilianus, Thermocyclops inversus, T. tenuis, T. crassus, Tropocyclops prasinus, Metacyclops mendocinus.*

Thermocyclops minutus, the commonest cyclopoid from Brazil, does not occur in Central America.

Asia

Survey of major studies on systematics

Asia is well studied as regards zooplankton systematics. An early start and easier accessibility has made this region's zooplankton systematics relatively well known compared to other tropical regions. Studies in Indonesia are based on collections made during the German Sunda Island and the Wallacea Expeditions (Brehm, 1933a, b, c, Hauer, 1938, 1941; Kiefer, 1933; Ruttner, 1943, 1952). Geographically more limited studies have been made in other Asian countries. Sri Lanka is the best known: Daday (1898), Bär (1929), Gurney (1916), Apstein (1907, 1910), and more recent work (Mendis & Fernando, 1962; Fernando, 1974, 1980a, b; Fernando & Rajapaksa, in press; Rajapaksa & Fernando, 1982, in press; Chengalath & Fernando, 1972; Chengalath *et al*, 1973, 1974). Recent studies on Malaysia (Fernando & Ponyi, 1981; Fernando & Zankai, 1981; Idris & Fernando, 1981; Karunakaran & Johnson, 1978 and Lai & Fernando, 1978a, b, 1979, 1980, 1981) has made the knowledge of zooplankton systematics almost complete in this region. Indian work is much more fragmentary and because of the vast areas is restricted to small portions of the country. No monograph is available or the whole country unlike in Sri Lanka, Indonesia (only Java, Bali and Sumatra) and Malaysia. However, the Indian work has been summarised in the paper of Michael (1980). Michael & Sharma (in press) have monographed the Indian Cladocera. A parallel monograph has been completed by Kanduru (1981).

For the Philippines the monograph of Mamaril & Fernando (1979) is a basic guide but this is certainly incomplete as regards species listing. The Philippine Calanoida were monographed by Lai *et al* (1979). The other countries in tropical and subtropical Asia are relatively poorly studied. Recent work in Nepal (Swar & Fernando, 1979; Dumont & Van de Velde, 1977) has provided some data on species in the region. The Thai species have been studied only recently and in a very limited way by Bricker *et al* (1978) and Lai & Fernando (1981). Bangladesh zooplankton has been monographed by Hossain (1982).

For Southern Vietnam the work of Shirota (1968) is a useful zooplankton list but it is based on very limited material. Little is known about the zooplankton of Burma, Laos, Kampuchea and Pakistan.

There are a few monographs dealing with the whole Southeast Asian region or large portions of it. They refer to a single genus (Kiefer, 1981, 1982;

Seymour-Sewell, 1957), zoogeography (Lai & Fernando, 1981) or species composition (Fernando, 1980a, b, c).

A group of zooplankton organisms not usually collected in temperate freshwater although quite common in tropical freshwaters is the Ostracoda. Victor & Fernando (1981) have provided keys to the Oriental genera.

This survey of the literature is confined to major and more recent studies. Since there are many recent compilations of the literature in the references listed, all earlier work has been included in these papers. This makes detailed listing of all the literature superfluous.

Comments on distribution

It is difficult to obtain a general picture of the numbers of zooplankton species and their distribution in tropical Asia because most studies have been based on collections from very limited geographical areas. In general it can be said that the diversity of species is similar to that in other tropical areas studied.

The least known groups are the Protozoa and the rare Coelenterates *Limnocnida* and *Craspedacusta*. Protozoans are of general occurrence and the genera *Arcella, Difflugia* and *Centropyxis* are quite common (Fernando, 1980a).

The Rotifera show in general a typical tropical species composition with many *Brachionus* (Green, 1971) and *Lecane* species. But in Malaysia Fernando & Zankai (1980) found few *Brachionus* species. Acid waters and few permanent lentic habitats combined could account for this.

The Cladocera have the characteristic tropical species composition. *Daphnia* is rare (maximum 3 species in any area), and these are often astatic water species and very rare. The typical limnetic species, *Ceriodaphnia cornuta, Moina micrura* and *Diaphanosoma excisum* also are eurytopic.

The Cyclopoida are not very diverse (Fernando & Ponyi, 1981), but are quite numerous in individual samples. The species *Mesocyclops leuckarti* s.l. (*M. ruttneri, M. thermocyclopoides* and *M. aspericornis*), and *Thermocyclops crassus* s.l. are the commonest limnetic species.

The Calanoida are not very diverse. Lai & Fernando (1980) found that the number of species decreases towards the equator. The acidic waters of Malaysia, Kalimantan and Sumatra seem to limit most genera except *Tropodiaptomus*.

The occurrence of Ostracoda in tropical zooplankton in Asia has been noted by Apstein (1907) and Klie (1935). More recently this phenomenon has been recorded in Africa and especially in Central America (Deevey *et al*, 1980).

Some general remarks on the distribution of zooplankton in South East Asia are given by Fernando (1980a, b and c). There appear to be few limnetic zooplankton species. All the common limnetic species, barring a few rotifers, are also eurytopic and occur in the whole range of habitat types quite commonly.

Australasia

Survey of literature

Northern Australia and islands to the north (including the Indonesian Archipelago east of Wallace's line and Papua New Guinea) are little studied with respect to the zooplankton. A systematic report on the microfauna of New Guinea was given by Daday (1901) and brief references to the zooplankton of the Celebes, Flores and New Caledonia are found in the reports of early expeditions (e.g. Stingelin, 1900, 1915; Heberer, 1923; Brehm, 1933a, b, 1938). Subsequent systematic work referred to collections from tropical regions (e.g. Lindberg, 1954; Russell, 1957, 1961; Brehm, 1959, 1960; McKenzie, 1966). Others described northern ranges of southern Australian taxa (e.g. Bayly, 1961, 1962, 1964; Hebert, 1977; Korovchinsky, 1981), or commented on distributions of northern groups (e.g. Bayly, 1966, Shiel & Koste, 1979; Koste & Shiel, 1980). Recent papers referring to the northern Australasian region include Ball & Glucksman (1978, 1980), Williams (1979), Bayly & Morton (1980), Green (1981), Koste (1981), Tait (1981), Koste & Shiel (1982) and Tait *et al* (1983). Unpublished work containing information on northern Australian zooplankon includes that of Morton (1977), Tait (1982), Berzins (1982), Fernando (1982), Shiel & Koste (in press), and Smirnov & Timms (1983).

Systematics and distribution

Protozoa. There are no published records of protozoans from the region. Genera which are quite common in samples from northern Australia are *Arcella, Difflugia, Euglypha* and *Centropyxis* (Fernando, 1982).

Rotifera. Isolated collections from Fiji, Gilbert Islands, New Hebrides, Queensland and Tonga were reported by Russell (1957, 1961). Brief systematic notes on rotifers of Northern Territory billabongs (= ox-bows) were given by Koste & Shiel (1980), and rotifers of Queensland crater lakes by Green (1981). A study of the monogonont rotifers of the Alligator Rivers region, N.T. (Koste, 1981) is the most comprehensive. Additional records by Koste & Shiel (1982) brought the Rotifera recorded from northern Australia to 224 taxa in 52 genera (see Tait *et al*, 1983), more than from any other tropical region (cf. Hauer, 1933; Chengalath *et al*, 1974, Ridder, 1981). Of these, 90 species (40%) are not recorded from southern Australia, while 80 species (36%) also are found in Indonesia (Hauer, 1937, 1938; Koste & Shiel, 1980; Koste, 1981).

The predominance of lecanids was noted by Russell (1957), and Shiel & Koste (1983) recorded up to 20 species of *Lecane* in single collections from N.T. billabongs, with a total of 40 species in the genus present. Most of these are warm stenotherms, acidophilic, recorded also from Indonesia and tropical South America.

Brachionids were not well represented in contrast to other studies in tropical areas (e.g. Green, 1971; cf. Pejler, 1977), and apparently are limited by the strongly acid conditions (to pH 3.3, Tait, 1981) and the paucity of permanent open waters. A similar lack of dominance of limnetic species of *Brachionus* in Malaysia was attributed to the lack of stable open waters in the region (Fernando & Zankai, 1981). Of ten species of *Brachionus* recorded from northern Australia (cf. 33 taxa known from southern Australia (Shiel, 1983)) only *B. falcatus*, a pantropical thermophile (Koste, 1978) is abundant. Other species are rare or markedly seasonal, including cosmopolitan taxa (*B. angularis, B. budapestinensis*) and endemics (*B. dichotomus, B. lyratus*). Aberrant or dwarf forms recorded from single collections include *B. caudatus personatus, B. dichotomus reductus, B. falcatus reductus, B. urceolaris sericus* (Koste, 1981; Tait, 1981; Koste & Shiel, 1982). These studies did not report *B. calyciflorus* or *B. forficula*, which are common elsewhere (cf. Fernando, 1980b). *Keratella* is similarly poorly represented with only two species common (*K. tropica* and *K. lenzi*) (Tait, 1982).

Seasonal community composition data for the region are sparse. Exceptional diversity in the wet season (Oct.–Apr.) noted by Shiel & Koste (1983) was due in part to flushing of littoral taxa into the temporary limnetic of billabongs, while the dramatic decrease in community diversity over the dry season (May–Sept.) was attributed to extremes of water quality (Tait, 1981; Koste & Shiel, 1982). Highest population densities were recorded in the dry season, with maxima to 70 000 plankters l^{-1} (Tait, 1982).

Cladocera. Most information is in the form of species lists. Records from New Caledonia were provided by Stingelin (1915) and from New Guinea by Brehm (1938, 1959, 1960) and Ball & Glucksman (1978, 1980). Records from tropical northern Australia are few. Sars (1897, 1899) reported collections from northwestern Australia and Queensland, Gurney (1927) from Longreach, Qld, and Williams (1979) from the Kimberly region of W.A. Recent revisions containing information on northern distributions include those of Hebert (1977) on *Daphnia* and Korovchinsky (1981) on *Diaphanosoma*. Ecological information on tropical Australian Cladocera is sparse; the only comprehensive study is that of Tait (1981, 1982).

Some 36 taxa in 23 genera are identified in the early literature, not including synonymies and taxa listed as *incertae sedis* (cf. Smirnov, 1971). A further 38 species and 11 genera were identified by Fernando (1982) in collections from northern Australia. This total compares to 125 species of Cladocera in 40 genera known from the continent (Smirnov & Timms, in press). The list will undoubtedly be expanded, and lack of systematic information remedied, with publication of the Smirnov & Timms' revision (1983).

A marked contrast to the cladoceran fauna of Europe and North America is in the absence from the Australasian region of the Holopedidae, Polyphemidae, Cercopagidae and Leptodoridae. Of the remaining families, the Sididae commonly are represented by *Diaphanosoma excisum* and *D. sarsi*. *D. unguiculatum* was recorded by Gurney (1927) from central Queensland (23°S), apparently the northern limit of its range, and *D. australiensis* was described from Queensland by Korovchinsky (1981), who noted the similarity of the Australian *Diaphanosoma* fauna to that of Asia, with a different species assemblage from that of South America. Two species of *Latonopsis* occur rarely, and *Pseudosida* was recorded for the first time from the continent by Williams (1979).

The Chydoridae are predominantly littoral in habit, and although some 35 species are recorded from tropical Australian waters, only a few are relatively common, e.g. in Tait's (1982) two-year study, only *Chydorus sphaericus s.l.* was facultatively limnetic, with species of *Alona, Biapertura, Graptoleberis* and *Pleuroxus* swept into open water from fringing reedbeds in the wet season, or accompanying algal blooms (Tait, 1981, 1982).

Macrothricidae are rare in the plankton. Most common records are of *Ilyocryptus spinifer, Macrothrix spinosa* and Echinisca spp. Moinids too, are poorly represented, with *M. micrura*, the most widely distributed species of the genus (Goulden, 1968) occasional in northern Australia and *Moinodaphnia macleayi* rare. A single bosminid, *Bosminopsis deitersi* was recorded as common in Tait's study, and *Bosmina meridionalis* was noted as a rare component of N.T. plankton (Fernando, 1982). Finally, Daphniidae are represented by *Ceriodaphnia cornuta*, with several species of *Simocephalus* recorded. *Daphnia* s.l. is extremely rare, with a record of *D. cephalata* from Flores (Heberer, 1923), and a single record of *D. carinata* from New Caledonia (Stingelin, 1915). Hebert's *D. longicephala* from northern Queensland may be conspecific with the New Caledonian form (Hebert, 1977). Hebert also recorded *D. lumholtzi* from Qld. This 'tropical' species is abundant in sub-tropical and temperate Australia (Shiel, 1981).

Ecological information on tropical Australasian Cladocera is sparse. The only long-term study is that of Tait (1981, 1982), who recorded 28 taxa. Only seven of these were abundant in the plankton, *viz. Diaphanosoma excisum, Chydorus sphaericus* s.l., *Macrothrix spinosa, Bosminopsis deitersi, Simocephalus acutirostris, Graptoleberis testudinaria* and *Ceriodaphnia cornuta* (Tait, 1982).

Copepoda. There are few records from Cyclopoida from the region, and the systematics of the genera present is inadequately resolved. Daday (1901) recorded four species (= *Thermocyclops crassus, Eucyclops serrulatus, Ectocyclops phaleratus* and *Paracyclops fimbriatus*) from New Guinea. Bayly & Morton (1980) added *Acanthocyclops* sp., *Tropocyclops* cf. *prasinus* and *Mesocyclops* sp. from the Purari R. catchment, Papua New Guinea. Ten species of cyclopoid were recorded by Tait (1982) from Northern Territory billabongs (including *Mesocyclops notius, Apocyclops dengizicus* and *Microcyclops varicans*). In addition, Fernando (1982) recorded *Mesocyclops aspericornis* and *M. decipiens* from New Caledonia, and *M. pilosus, Tropocyclops* cf. *setifer, Eucyclops euacanthus, Macrocyclops albidus* and *Ectocyclops medius* from Queensland. A species of *Mesocyclops* resembling *M. annae*, previously known only from Madagascar (Kiefer, 1981) recently was recorded from the N.T. (Tait *et al*, 1983).

At least 20 species in seven genera are known, and the total will undoubtedly be expanded with publication of Morton's (1977) revision. The observation that the number of Cyclopoida in tropical freshwaters is markedly lower than at temperate latitudes in other geographical regions (Fernando & Ponyi, 1981) may apply also in Australia (cf. Bayly & Morton, 1978).

The calanoid fauna is better known after the work of Bayly (1961, 1962, 1964, 1966). Endemic species of diaptomids were described from the Celebes by Brehm (1933). *Eodiaptomus lumholtzi* and *Tropodiaptomus australis* are widely distributed across northern Australia and New Guinea (Bayly, 1966; Williams, 1979; Bayly & Morton, 1980a). Brehm's (1959) record of *Tropodiaptomus novaeguinea* is considered inadequately described and figured (Kiefer, 1982), i.e. *incertae sedis*. Two species of centropagid, *Calamoecia ultima* and *C. trifida* are relatively common in tropical areas of Australia and Papua New Guinea, while a third *Boeckella triarticulata* extends into the distribution of *Diaptomus* around 23°S, although there is little overlap possibly as a result of competitive exclusion (Bayly, 1966).

Competition also was suggested in coexisting populations of *E. lumholtzi* and *C. ultima* in N.T. billabongs, with one species always dominant (Tait, 1982). The depauperate calanoid fauna, which may reflect low trophic status of waters in the region (cf. Lai & Fernando, 1980), was noted in Tait's study.

Hitherto there have been no descriptions, which can be regarded as good or even barely adequate by modern standards, of any non-marine harpacticoids from any south-east Asian country or from Australia to the north of lat. 30°S. The reports by Löffler (1973), Bayly & Morton (1978) and Tait (1982), as well as the earlier records collated by Lang (1948), do no more than indicate that there are harpacticoids of various kinds which would repay restudy if topotypes of them can be found. An account of all the '*Canthocamptus*-like' Australian species is almost complete, and hopefully will be followed by similar accounts of other non-marine Australian taxa (R. Hamond, pers. comm.).

Ostracoda. Several species of ostracods were reported by Daday (1901) from New Guinea, McKenzie (1966, 1971, 1980) recorded 16 taxa from northwest Western Australia, two from New Guinea and one from the Moluccas. A species of *Cypretta* was noted by Bayly & Morton (1980), and four taxa, all in undescribed new genera, were recorded by Tait (1982). The ostracod fauna of the region, which apparently has a high degree of endemicity (cf. McKenzie, 1966) is otherwise poorly known.

In summary, most of the information on tropical zooplankton from the Australasian region stems from isolated collections over a wide geographical area, and systematic records are, at best, patchy. The situation will improve with publication of the revisions noted above, and with publication of results of extensive biological monitoring programs conducted in the Alligator Rivers region of the Northern Territory following the discovery and development of uranium deposits in the area. Clearly, from the present records, of the major zooplankton groups, the rotifer and cladoceran fauna are of comparable diversity to other tropical areas, although of different species composition, and the Copepoda are reduced in diversity.

Ecology

There is hardly any quantitative work on production of tropical zooplankton. Burgis (1974) gives a value of 44 mg m^{-2} d^{-1} for the main herbivore, *Thermocyclops hyalinus* (= *crassus*) in Lake Uganda and Lewis (1979) a value of 34.4 μg wet weight l^{-1} d^{-1} as mean herbivore production for the herbivorous zooplankton. Considering the high primary production in these lakes (Lewis, 1974), these values are low.

Lewis (1979), in the most comprehensive study of tropical zooplankton ecology, gives a descriptive account of the plankton followed by an analysis of the whole community, its structure, energy flow and adaptive strategy.

The earliest study of tropical zooplankton ecology was made by Apstein (1907, 1910). He described the seasonal variations in numbers of the zooplankton and attempted to correlate these with temperature and rainfall in two ornamental lakes in Sri Lanka. A similar but more descriptive study was done by Seymour-Sewell (1935) in a pond in Calcutta, India. Ruttner (1952) gave the most extensive descriptive account of tropical zooplankton up to that time and also investigated vertical migration (Ruttner, 1943). All these studies included a detailed diagnosis of Rotifera and crustacean zooplankton. On the other hand the work that followed in Asia till very recently on the biology of zooplankton has little in the way of accurate species diagnosis. Welch (1952) claimed that the composition of tropical zooplankton was the same as that of North Temperate zone. This is manifestly not so (Fernando, 1980a, 1980b).

It is only relatively recently that the study of seasonality initiated by Apstein (1906) was continued. In Africa, seasonality of plankton in the Sokoto River was documented by Holden & Green (1960). Studies on the Nile headwaters are reviewed by Rzoska (1976). Egborge (1976) and Magis (1978) investigated the phenomenon in reservoirs. Lake Chad was investigated by Gras *et al* (1967) and Robinson & Robinson (1976). In India Mandel (1979), Moitra & Mukherjee (1972) and Nasar (1972, 1978) studied changes in species composition seasonally in ponds and fishponds. Swar & Fernando (1980) studied seasonality in a lake in Nepal. A more detailed study was made in a shallow lake, Parakrama Samudra, Sri Lanka by Duncan & Gulati (1981), Duncan (in press) and Fernando & Rajapaksa (in press). They investigated the densities, species composition and their long and short-term changes.

Studies on South and Central American zooplankton ecology were initiated very recently: Matsumura-Tundisi & Tundisi (1976) gave an account of seasonal changes in species composition and Zago (1976) dealt with the seasonal dynamics of Cladocera in a reservoir. Brandorff (1977) has made the most detailed study so far for this region on seasonal dynamics of zooplankton in an Amazonas lake. Robertson (1980) has attempted a similar study, also in an Amazonian lake. Zaret (1980), in a book on predation in zooplankton communities, has drawn mainly on his work on zooplankton ecology in Central America. Deevey & Deevey (1980) studied the structure of zooplankton communities in lakes in Guatemala.

There have been a few studies on individual species of zooplankton in the field. Swar & Fernando (1979) investigated the seasonality and fecundity of *Daphnia lumholtzi* in Nepal and Green (1976) studied the polymorphism caused by predation in the same species in Africa. Zago (1976) studied cyclomorphosis in *Daphnia gessneri* in a reservoir in São Paulo.

Apart from the study of Ruttner (1943) on vertical migration of zooplankton in Asia there have been other studies in Africa by Adeniji (1981), Hart & Allanson (1976) Onabamiro (1952b) and Worthington (1931).

Gras & St. Jean (1976) investigated the horizontal distribution of zooplankton in Lake Chad and Green *et al* (1976) observed changes after overturn in a lake in Java, Indonesia.

Zooplankton ecology in the tropics (and subtropics) remains relatively poorly known in spite of a great increase in studies on zooplankton in the tropics. The present understanding of species composition indicates a lower diversity as the equator is reached (Fernando, 1980a, 1980b). The seasonality is not universally the same, nor is it the same for all groups of zooplankton. The classical bimodal pattern of the North Temperate zone is not evident. Rainfall and temperature influences density and the species composition of reservoirs can change with drastic fluctuations in water levels (Fernando & Rajapaksa, in press).

The meagre data on production indicates a relatively low production of zooplankton. This is so in spite of the universally high phytoplankton production (Lewis, 1974).

Summary

A brief review of the literature on the systematics, distribution and ecology of tropical freshwater zooplankton on a global scale has been attempted.

The present status of our knowledge in systematics and distribution is very uneven. In Asia a fairly comprehensive knowledge of systematics and distributions is available at least in half the countries of the area. In Africa the situation is less satisfactory while in South America and especially Australia most of the work is recent and limited.

We have some general ideas of the distribution of limnetic zooplankton throughout the tropics. However no detailed information is available for any major group as a whole.

As expected it is in ecology that our knowledge is most fragmentary. The study of seasonal variations in zooplankton and production are little known, however a start has been made in research in these fields.

References

Adeniji, H. A., 1981. Circadial vertical migration of zooplankton during stratification and its significance to fish distribution and abundance in Kainji Lake, Nigeria. Verh. int. Ver. Limnol. 21: 1021–1024.

Ahlstrom, E. H., 1937. Plankton Rotatoria from northeast Brazil. Ann. Acad. bras. Cienc. 1: 30–45.

Andrade, E. & G. O. Brandorff, 1975. Una nova especie de *Diaptomus*, (Crustacea, Copepoda) *Diaptomus negrensis* de aguas pretas de Manaus. Acta Amazonica 5: 97–103.

Apstein, C., 1907. Das plankton in Colombo See auf Ceylon. Zool. Jb. Syst. Okol. Geogr. Tiere 25: 201–244.

Apstein, C., 1910. Das plankton des Gregory See auf Ceylon. Zool. Jb. Syst. Okol. Geogr. Tiere 29: 661–680.

Ball, E. E. & J. A. Glucksman, 1978. Limnological studies of Lake Wisdom, a large New Guinea caldera lake with a simple fauna. Freshwat. Biol. 8: 455–468.

Ball, E. E. & J. A. Glucksman, 1980. A limnological survey of Lake Dakataua, a large caldera lake on West New Britain, Papua New Guinea, with comparisons to Lake Wisdom, a younger nearby caldera lake. Freshwat. Biol. 10: 73–84.

Bär, G., 1924. Uber Cladoceren von der Insel Ceylon. Jena Z. Naturw. 60: 80–126.

Bayly, I. A. E., 1961. A revision of the inland water genus *Calamoecia* (Copepoda: Calanoida). Aust. J. mar. Freshwat. Res. 12: 54–91.

Bayly, I. A. E., 1962. Additions to the inland water genus *Calamoecia* (Copepoda: Calanoida). Aust. J. mar. Freshwat. Res. 13: 252–264.

Bayly, I. A. E., 1964. A revision of the Australasian species of the freshwater genera *Boeckella* and *Hemiboeckella* (Copepoda: Calanoida). Aust. J. mar. Freshwat. Res. 15: 180–238.

Bayly, I. A. E., 1966. The Australian species of *Diaptomus* (Copepoda: Calanoida) and their distribution. Aust. J. mar. Freshwat. Res. 17: 123–134.

Bayly, I. A. E. & D. W. Morton, 1978. Aspects of zoogeography of Australian micro-crustaceans. Verh. int. Ver. Limnol. 20: 2537–2540.

Bayly, I. A. E. & D. W. Morton, 1980. A note on zooplankton from four Papua New Guinea lakes (altitudinal range 538–3630 m). In T. Petr (ed.), Aquatic Ecology of the Purari River Catchment, pp. 3–5.

Berzins, B., 1982. Contribution to the knowledge of Rotatoria of Australia. Unpubl. Rep., Lund, 24 pp.

Bowman, T. E., 1973. Two fresh water copepods from Barranca, Venezuela: *Notodiaptomus cearensis* (Wright) and *N. venezolanus deeveyorum*, new subspecies (Calanoida, Diaptomidae). Proc. biol. Soc. Wash. 86: 193–202.

Brandorff, G. O., 1973. Die neotropische Gattung *Rhacodiaptomus* Kiefer (Crustacea, Copepoda). Amazoniana 4: 341–365.

Brandorff, G. O., 1973. Neue Freilebende calanoide Copepoden (Crustacea) aus dem Amazonasgebiet. Amazoniana 4: 205–218.

Brandorff, G. O., 1976. The geographic distribution of the Diaptomidae in South America (Crustacea, Copepoda). Revta bras. Biol. 36: 613–627.

Brandorff, G. O., 1977. Untersuchungen zur Populations-dynamik des Crustaceenplanktons im Lago Castanho (Amazonas, Brazilien). Diss., Christian-Albrechts Univ., Kiel, 108 pp.

Brandorff, G. O., 1978. Ein neuer Diaptomidae (Crustacea, Copepoda) '*Diaptomus*' *ohlei* aus dem Brasilianischen Amazonas Gebiet. Arch. Hydrobiol. 82: 295–299.

Brandorff, G. O., W. Koste & N. N. Smirnov, 1982. The composition and structure of rotiferan and crustacean communities of the lower Rio Nhamunda, Amazonas, Brazil. Stud. neotrop. Faun. Envir. 17: 69–121.

Brehm, V., 1932. Notizen zur Susswasserfauna Guatemalas und Mexico. Zool. Anz.: 63–60.

Brehm, V., 1933. Die Cladoceren der Deutschen Limnologischen Sunda-Expedition. Arch. Hydrobiol., Suppl. 11: 631–771.

Brehm, V., 1933b. Mitteilungen von der Wallacea Expedition Woltereck. 5. Phyllopoden. Zool. Anz. 104: 31–40.

Brehm, V., 1933c. Mitteilungen von der Wallacea Expedition Woltereck. 7. Neue und wenig bekannten Entomostraken. Zool. Anz. 104: 130–142.

Brehm, V., 1938. Die Cladoceren die Wallacea Expedition. Int. Revue ges. Hydrobiol. 38: 99–124.

Brehm, V., 1939. La fauna microscopia del Lago Peten, Guatemala. Esc. nac. Cienc. Biol. An. 1: 173–204.

Brehm, V., 1948. Datos para la fauna de agua dulce de Cuba. Publ. Inst. Biol. apl. 95: 104–106.

Brehm, V., 1955. Mexikanische Entomostraken. Ost. zool. Z. 6: 412–420.

Brehm, V., 1959. Cladocera and Calanoide Copepoden aus Neu Guinea. Nova Guinea 10: 1–10.

Brehm, V., 1960. Scientific results of the Netherlands New Guinea Expedition 1959. 1. Entomostraken aus Zentral New Guinea. Nova Guinea 5: 83–88.

Bricker, K. S., L. Wongrat & J. E. Gannon, 1978. Composition and distribution of crustacean plankton in twelve water bodies of Thailand. Kaesetsart Univ. Fish. Res. Bull. 10: 1–14.

Brunelli, G. & G. Cannicci, 1938. Le caratteristiche biologiche del lago Tana. Atti Reale Accad. Lincei 6, Rc 27, 7: 319–323.

Burgis, M. D., 1974. Revised estimates for the biomass and production of zooplankton in Lake George, Uganda. Freshwat. Biol. 4: 535–541.

Changalath, R. & C. H. Fernando, 1973. Rotifera from Sri Lanka (Ceylon), 1. The Genus *Lecane* with descriptions of two new species. Bull. Fish. Res. Stat. Sri Lanka 24: 13–17.

Changalath, R., C. H. Fernando & W. Koste, 1973. Rotifera from Sri Lanka (Ceylon), 2. Further studies on the Eurotatoria including new records. Bull. Fish. Res. Stat. Sri Lanka 24: 29–62.

Changalath, R., C. H. Fernando & W. Koste, 1975. The Rotifera of Sri Lanka, 3. Description of new species with new records, a list of species in Sri Lanka and discussion on distribution. Bull. Fish. Res. Stat. Sri Lanka 25: 83–96.

Cippolli, M. N. & M. A. J. Carvalho, 1973. Levantamento de Calanoida e Cyclopoida (Copepoda, Crustacea) das aguas da regiao de Guama Capim e Tocantins com nota sobre a fauna acompannante. Pap. avulsos Zool. S. P. 7: 95–110.

Coker, R. E., 1943. *Mesocyclops edax* (S. A. Forbes), *M. leuckarti* (Claus) and related species in America, J. Elisha Mitchell Sci. Soc. 59: 181–200.

Daday, E., 1910. Untersuchungen über die Susswasser-Mikrofauna Deutsch-Ostafrikas. Zoologica 23, 59: 1–314.

Dahl, F., 1894. Die Copepodenfauna des unteren Amazonas. Ber. naturf. Ges. Freiburg 8: 10–23.

Deevey, E. S., G. B. Deevey & M. Brenner, 1980. Structure of zooplankton communities in the Peten District, Guatemala. In: W. C. Kerfoot (ed.), The Evolution and Ecology of Zooplankton Communities. University Press of New England, Hanover: 669–678.

De Infante, A., 1980. Les rotifers del Lago de Valencia. Acta cient. Venezolana 31: 40–47.

DeInfante, A., W. Riehl & J. F. Saunders, 1979. Los copepodos del Lago de Valencia Venezuela. Acta cient. Venezolana 30: 224–233.

Delachaux, T., 1928. Faune invertébrée d'eau douce des hauts plateaux du Pérou. Bull. Soc. neuchatel. Sci. nat. 52: 45–77.

De Ridder, M., 1977. The rotatoria of the Caribbean Region. In: Studies of the fauna of Curacao and other Caribbean Islands 171: 72–130.

De Ridder, M., 1981. Rotifera: Hydrobiological survey of the Lake Bangweulu Luapula River basin Scientific Results. Ed. J. J. Symoens, Brussels. 190 pp.

Dragesco, J., 1972a. Ciliés libres de la cuvette tchadienne. Ann. Fac. Sci. Cameroun 11: 71–91.

Dragesco, J., 1972b. Ciliés libres de l'Ouganda. Ann. Fac. Sci. Cameroun 9: 87–126.

Dragesco, J., 1973. Quelques données écologiques sur les Ciliés libres de l'Afrique. Am. Zool. 13: 231–232.

Dragesco, J., 1980. Les Protozoaires. In Flore et faune aquatiques de l'Afrique Sahelo-Soudanienne. O.R.S.T.O.M., Paris: 153–192.

Dumont, H., 1979. Limnologie van Sahara en Sahel. Rijksuniv. Gent, 557 pp.

Dumont, H. J., J. Pensaert & I. Van de Velde, 1981. The crustacean zooplankton of Mali (West Africa). Hydrobiologia 80: 161–187.

Dumont, H. J. & I. Van de Velde, 1977. Report on a collection of Cladocera and Copepoda from Nepal. Hydrobiologia 53: 55–65.

Duncan, A., in press. The composition, density and distribution of the zooplankton of Parakrama Samudra during March and April 1980. Hydrobiologia.

Duncan, A. & R. D. Gulati, 1981. Parakrama Samudra (Sri Lanka) Project, a study of a tropical lake ecosystem. Composition, density and distribution of the zooplankton in 1979. Verh. int. Ver. Limnol. 21: 1007-1014.

Dussart, B. H., 1974. Contribution a l'étude des Copépodes des eaux douces d'Ethiopie. Bull. I.F.A.N. 36A, 1: 92-116.

Dussart, B. H., 1977. Contribution a l'étude des eaux douces du Ruanda. Bull. I.F.A.N. 39A, 4: 821-840.

Dussart, B. H., 1980. Copépodes. In la flore et la faune aquatiques de l'Afrique sahelo-soudanienne, I. O.R.S.T.O.M. 44: 333-356.

Dussart, B. H., 1981. Copépodes du Bassin du Niger. Bull. I.F.A.N., 3A, 1-2.

Dussart, B. H., 1982b. Crustacés Copépodes des eaux intérieures. In Faune de Madagascar 58: 146 pp.

Dussart, B. H. & R. Gras, 1966. Faune planctonique du lac Tchad, 1. Crustacés Copépodes. Cah. O.R.S.T.O.M., Sér. Oceanogr. 4: 3: 77-91.

Egborge, A. B. M., 1981. The composition, seasonal variation and distribution of zooplankton in Lake Asejire, Nigeria. Revue zool. afr. 95: 137-180.

Einsle, U., 1971. Copépodes libres. Explor. hydrobiol. bassin du lac Bangweulu et du Luapula 13, 1: 74 pp.

Fernando, C. H., 1974. A guide to the freshwater fauna of Sri Lanka (Ceylon), Suppl. 4. Bull. Fish. Res. Stat. Sri Lanka (Ceylon) 25: 27-81.

Fernando, C. H., 1980a. The freshwater zooplankton of Sri Lanka with a discussion of tropical freshwater zooplankton. Int. Revue ges. Hydrobiol. 65: 85-125.

Fernando, C. H., 1980b. The species and size composition of tropical freshwater zooplankton with special reference to the Oriental Region (South East Asia). Int. Revue ges. Hydrobiol. 65: 411-426.

Fernando, C. H., 1981. The freshwater invertebrate fauna of Sri Lanka. Colombo Centennary Vol. Spolia Zeylan 35: 15-42.

Fernando, C. H., 1982. An interim report on the freshwater zooplankton of tropical and subtropical Australasia. Unpublished MS, Univ. Adelaide, 46 pp.

Fernando, C. H. & J. E. Ponyi, 1981. The freeliving freshwater cyclopoid Copepoda (Crustacea) of Malaysia and Singapore. Hydrobiologia 78: 113-123.

Fernando, C. H. & N. P. Zankai, 1981. The Rotifera of Malaysia and Singapore with remarks on some species. Hydrobiologia 78: 205-219.

Fernando, C. H. & K. E. Smith, 1982. Copepoda. In S. H. Hurlbert (ed.), Aquatic Biota of Mexico, Central America and the West Indies. Publ. San Diego State Univ., Calif.: 192-195.

Fernando, C. & R. Rajapaksa, in press. Some remarks on long-term and seasonal changes in the zooplankton of Parakrama Samudra. Hydrobiologia.

Frey, D. G., 1982. Cladocera. In S. H. Hurbert (ed.), Aquatic Biota of Mexico, Central America and the West Indies. Publ. San Diego State Univ., Calif.

Fryer, G., 1957. Free-living freshwater Crustacea from lake Nyasa and adjoining waters. Arch. Hydrobiol. 53, 1: 62-86.

Gauthier, H., 1939. Contribution à l'étude de la faune dulcaquicole de la région du Tchad et particulièrement des Branchiopodes et des Ostracodes. Bull. I.F.A.N. 1: 110-244.

Gauthier, H., 1951. Contribution a l'étude de la faune des eaux douces du Sénégal (Entomostracés). Imp. Minerva, Alger, 169 pp.

Gessner, F., 1955. Die limnologischen Verhaltnisse in den Seen und Flussen von Venezuela. Verh. int. Ver. Limnol. 12: 284-295.

Gessner, F., 1956. Das Plankton des Lago Maracaibo. Ergebn. dt. Limnol. Venezuela-Exped. 1: 67-92.

Goulden, C., 1968. The systematics and evolution of the Moinidae. Trans. am. phil. Soc. 58: 1-101.

Goy, J., 1980. Cnidaires. In Flore et faune aquatiques de l'Afrique Sahelo-soudanienne. O.R.S.T.O.M., Paris: 192-198.

Gras, R. & L. Saint-Jean, 1976. Etude de la Répartition spatiale du zooplancton dans le Lac Chad: Variation de la dispersion en fonction des modalités d'échantillonnage et de conditions hydrodynamiques du Milieu. Cah. O.R.S.T.O.M., Sér. Hydrobiol. 10: 201-229.

Gras, R., A. Iltis & S. Lévêque-Duwat, 1967. Le plancton du Bas- Chari et de la partie est du lac Tchad. Cah. O.R.-S.T.O.M., Sér. Hydrobiol. 1: 25-100.

Green, J., 1960. Zooplankton of the river Sokoto. The Rotifera. Proc. zool. Soc. Lond. 135: 491-523.

Green, J., 1960a. Zooplankton of the river Sokoto. The freshwater Medusa *Limnocnida*. Proc. zool. Soc. Lond. 135: 613-618.

Green, J., 1962. Zooplankton of the river Sokoto (Nigeria). The crustacea. Proc. zool. Soc. Lond. 138: 415-453.

Green, J., 1963. Zooplankton of the river Sokoto. The Rhizopoda Testacea. Proc. zool. Soc. Lond. 141: 491-514.

Green, J., 1967. The distribution and variation of *Daphnia lumholtzi* (Crustacea: Cladocera) in relation to fish predation in Lake Albert, East Africa. J. Zool. Lond. 151: 181-197.

Green, J., 1971. Latitudinal variation in associations of planktonic Rotifera. J. zool. Soc. Lond. 167: 31-39.

Green, J., 1981. Associations of Rotifera in Australian crater lakes. J. zool. Lond. 193: 469-486.

Green, J., S. A. Corbet, E. Watts & O. B. Lan, 1976. Ecological studies on Indonesian lakes. Overturn and restratification of Ranau Lamongen. J. zool. Soc. Lond. 180: 315-354.

Gurney, R., 1928. Some Copepoda from Tanganyika collected by Mr. S. R. B. Pask. Proc. zool. Soc. 22: 317-332.

Harding, J. P., 1955. The Percy Sladen Trust Expedition to Lake Titicaca in 1937, 15. Crustacea, Copepoda. Trans. linn. Soc. Lond. 1: 219-247.

Hart, R. C. & B. R. Allanson, 1976. The distribution and diel vertical migration of *Pseudodiaptomus hessei* (Mràzek) (Calanoida: Copepoda) in a subtropical lake in southern Africa. Freshwat. Biol. 6, 2: 183-198.

Hauer, J., 1937. Die Rotatorien von Sumatra, Java und Bali nach der Ergebnissen der Deutschen Limnologischen Sunda-Expedition. Arch. Hydrobiol., Suppl. 15: 296-384.

Hauer, J., 1938. Die Rotatorien von Sumatra, Java und Bali nach der Ergebnissen der Deutschen Limnologischen Sunda-Expedition. Arch. Hydrobiol., Suppl. 15: 507-602.

Hauer, J., 1941. Rotatorien aus dem Zwischengebiet Wallacea. Int. Revue ges. Hydrobiol. 41: 177-389.

Hauer, J., 1956. Rotatorien aus Venezuela und Kolombien. Ergebn. dt. Limnol. Venezuela-Exped. 1952. 1: 277-313.

Hauer, J., 1965. Rotatorien fauna des Amazonasgebietes. Int. Revue ges. Hydrobiol. 50: 341-389.

Heberer, G., 1923. Uber eine population von *Daphnia cephalata* aus Flores. Verh. dt. zool. Ges., Basel 32: 70-78.

Hebert, P. D. N., 1977. A revision of the taxonomy of the genus *Daphnia* (Crustacea: Daphniidae) in south east Australia. Aust. J. Zool. 25: 371-398.

Herbst, H. V., 1959. Brasilianische Susswasser Cyclopoiden (Crustacea, Copepoda). Gewäss. Abwäss. 24: 49-73.

Herbst, H. V., 1960. Copepoden (Crustacea, Entomostraca) aus Nicaragua und Sudperu. Gewäss. Abwäss. 27: 27-54.

Herbst, H. V., 1962. Crustacea aus dem Amazonasgebiet gesammelt von Professor Dr. H. Sioli und Dr. R. Braun, 1. Litorale und substratgebundene Gnathostoma (Copepoda). Crustaceana 3: 259-278.

Herbst, H. V., 1967. Copepoda und Cladocera (Crustacea) aus Sudamerika. Gewäss. Abwäss. 45: 96-108.

Holden, M. J. & J. Green, 1960. The hydrology and plankton of the river Sokoto. J. anim. Ecol. 29: 65-84.

Hossain, M. A., 1982. A taxonomical study of freshwater zooplankton (Rotifera, Cladocera and Copepoda) of Bangladesh. M. Sc. Thesis, Univ. Waterloo.

Idris, B. A. G. & C. H. Fernando, 1981. Cladocera of Malaysia and Singapore, with new records, redescription and remarks on some species. Hydrobiologia 77: 233-256.

Juday, C., 1915. Limnological studies on some lakes in Central America. Trans. Wis. Acad. Sci. Arts Lett. 18: 214-250.

Kanduru, A. I., 1981. The Cladocera (Crustacea: Branchiopoda) of the Indian sub-continent and their latitudinal distribution. M. Sc. Thesis, Univ. Waterloo.

Karunakaran, L. & A. Johnson, 1978. A contribution to the rotifer fauna of Singapore and Malaysia. Malayan Nat. J. 32: 173-208.

Kiefer, F., 1926. Uber einige susswasser Cyclopoiden aus Peru. Arch. Hydrobiol. 16: 494-507.

Kiefer, F., 1928b. Crustacea, 1. Copepoda aquae-dulcis. In Th. Monod (ed.), Contribution a la faune du Cameroun. Faunes Colonies françaises 1: 535-570.

Kiefer, F., 1932. Neue Diaptomiden und Cyclopiden aus Franzosisch-West-Afrika. Voyage de Ch. Alluaud et P. A. Chappuis en Afrique occidentale française. Bull. Soc. Stiinte Cluj 6: 523-528.

Kiefer, F., 1933. Freilebende Binnengewassercopepoden Diaptomiden und Cyclopiden. Voyage de Ch. Alluaud et P. A. Chappuis en Afrique occidentale française. Arch. Hydrobiol. 26: 121-142.

Kiefer, F., 1933. Die freilebende Copepoden Binnengewasser von Insulinde. Arch. Hydrobiol., Suppl. 12: 519-629.

Kiefer, F., 1933. Subwassercopepoden aus Brasilien. Zool. Anz. 105: 38-43.

Kiefer, F., 1933. Zoologische Ergebnisse einer Reise nach Bonaire Curacao und Aruba im Jahre 1930, 7 - Sub und Brackwasser - Copepoden von Bonaire Curacao und Aruba II. Cyclopoida. Zool. Jb. 64, 317.

Kiefer, F., 1934. Neue Ruderfusskrebse von der Insel Haiti. Zool. Anz. 108: 227-230.

Kiefer, F., 1935. Uber drei Cyclopiden (Crustacea, Copepoda) aus Brasilien. Zool. Anz. 112: 23-31.

Kiefer, F., 1936. Freilebende Suss und Salzwassercopepoden von der Insel Haiti. Arch. Hydrobiol. 30: 263-268, 271-301, 310-311.

Kiefer, F., 1936. Brasilianische Ruderfusskrebse (Crustacea, Copepoda) gesammelt von Herrn Dr. Otto Schubart, V. Mitteilung. Zool. Anz. 116: 31-35.

Kiefer, F., 1938. Ruderfusskrebse (Crust. Cop.) aus Mexico. Zool. Anz. 123: 277-280.

Kiefer, F., 1942. Freilebende Copepoda (Crust.). Beitr. Fauna Perus 2: 213-215.

Kiefer, F., 1952. Copepoda Calanoida und Cyclopoida. Exploration Parc National Albert. Mission H. Damas 1935-1936 21: 136 pp.

Kiefer, F., 1954. Drei neue Ruderfusskrebse (Crust. Cop.) aus Venezuela. Zool. Anz. 153: 170-174.

Kiefer, F., 1956. Freilebende Ruderfusskrebse (Crustacea, Copepoda) 1. Calanoida und Cyclopoida. Ergebn. dt. Limnol. Venezuela Exped. 1: 233-268.

Kiefer, F., 1956. Ruderfusskrebse aus dem Tanganyikasee (Crustacea Copepoda). Ergebnisse der Mission J. J. Symoens. Revue Bot. afr. 54: 241-266.

Kiefer, F., 1957. Freilebende Ruderfusskrebse (Crustacea, Copepoda) des Titicacasees. Veröff. zool. Staatssamml. Munchen 4: 125-150.

Kiefer, F., 1981. Beitrag zur kenntnis von Morphologie, Taxonomie und Geographische Verbreitung von *Mesocyclops leuckarti* auctorum. Arch. Hydrobiol., Suppl. 62: 148-190.

Kiefer, F., 1982. Vergleichende Untersuchungen über Morphologie, Taxonomie und Geographischer Verbreitung der Arten der Gattung *Tropodiaptomus* Kiefer (Copepoda: Calanoida) aus asiatischen Binnengewasser. Hydrobiologia 93: 223-253.

Kiss, R., 1960. Entomostracés de la plaine de la Ruzizi (Congo et Ruanda-Urundi). Mem. 8 Acad. Sci. Outre-Mer Bruxelles, Cl. Sci. nat. med., N.S. 11, 5: 1-48.

Kleerekoper, H., 1944. Introducao ao estudo da limnologia. Servico de Informacao Agricola, Ministerio de Agriculture, 4: 329 pp.

Klie, W., 1932. Die Ostracoden der Deutschen limnologischen Sunda-Expedition. Arch. Hydrobiol., Suppl. 11: 447-502.

Korovchinsky, N. M., 1981. Taxonomic and faunistic revision of Australian *Diaphanosoma* (Cladocera: Sididae). Aust. J. mar. Freshwat. Res. 32: 813-831.

Koste, W., 1972. Rotatorien aus Gewässern Amazoniens. Amazoniana 3: 258-505.

Koste, W., 1978. Rotatoria. Die Rädertiere Mitteleuropas. Begrundet von Max Voigt, 1 & 2. Borntraeger, Berlin, 63 pp, 234 pp.

Koste, W., 1981. Zur Morphologie systematik und Okologie von neuen monogononten Radertieren (Rotatoria) aus dem uberschwemmungsgebiet des Magela Creek in der Alligator River region Australiens N.T. Osnabruck. Naturw. Mitt. 8: 97-126.

Koste, W. & R. J. Shiel, 1980. New Rotifera from Australia. Trans. r. Soc. S. Aust. 104: 133-149.

Koste, W. & R. J. Shiel, 1982. Morphology, systematics and ecology of new monogonont Rotifera from the Alligator River region, N.T. 2. Trans. r. Soc. S. Aust. 106.

Lai, H. C. & C. H. Fernando, 1978a. The freshwater Calanoida (Crustacea: Copepoda) of Singapore and Peninsular Malay-

sia. Hydrobiologia 61: 113-127.
Lai, H. C. & C. H. Fernando, 1978b. Redescription of *Neodiaptomus botulifer* Kiefer and one of its related species (Calanoida: Copepoda). Hydrobiologia 59: 229-235.
Lai, H. C. & C. H. Fernando, 1979. The Malaysian *Tropodiaptomus* (Copepoda: Calanoida) and its distribution. Hydrobiologia 65: 75-81.
Lai, H. C. & C. H. Fernando, 1980. Zoogeographical distribution of Southeast Asian Freshwater Calanoida. Hydrobiologia 74: 53-66.
Lai, H. C. & C. H. Fernando, 1981. The freshwater Calanoida (Crustacea: Copepoda) of Thailand. Hydrobiologia 76: 161-178.
Lai, H. C., A. C. Mamaril & C. H. Fernando, 1979. The freshwater Calanoida (Copepoda) of the Philippines. Crustaceana 37: 225-240.
Lewis, W. M., 1974. Primary production in the plankton community of a tropical lake. Ecol. Monogr. 44: 377-409.
Lewis, W. M., 1979. Zooplankton Community Analysis. Springer-Verlag, N.Y.: 163 pp.
Lindberg, K., 1950a. Cyclopides (Crustacés, Copépodes) du Maroc. Bull. Soc. Sci. nat. Maroc 30: 23-32.
Lindberg, K., 1950b. Notes sur quelques Cyclopides (Crustacés, Copépodes) de l'Angola. Publ. Cult. comp. Diamentes Angola: 49-56.
Lindberg, K., 1950c. Cyclopides (Crustacés, Copépodes) de la Nigeria (Afrique occidentale). Bull. Soc. zool. Fr. 75: 145-148.
Lindberg, K., 1951a. Exploration du Parc National de l'Upemba. Cyclopides (Crustacés, Copépodes). Explor. Parc nat. Upemba Miss. de Witte 2: 1-13.
Lindberg, K., 1951b. Cyclopides (Crustacés, Copépodes). Rev. Sci. Explor. Hydrobiol. Lac Tanganyika 1946-47, 3, 2: 45-91.
Lindberg, K., 1954. Cyclopides (Crustacés, Copépodes) d'iles du Pacifique sud (Melanesie et Micronesie) et de Borneo. K. fysiogr. Sällsk. Lund Förh. 24: 161-174.
Lindberg, K., 1955. Cyclopides (Crustacés, Copépodes) des Mexique. Ark. Zool. 7: 459-489.
Löffler, H., 1963. Zur Ostrakoden und Copepoden Fauna Ekuadors. Arch. Hydrobiol. 59: 196-234.
Löffler, H., 1973. Tropical high mountain lakes of New Guinea and their zoogeographical relationship compared with other tropical high mountain lakes. Alpine & Arctic Res. 5: A193-A198.
Lowndes, A. G., 1930. Freshwater Copepoda from Abyssinia collected by M. J. Omer Cooper. Proc. zool. Soc. Lond.: 161-179.
Lowndes, A. G., 1934. Report of an expedition to Brazil and Paraguay in 1926-7 supported by the trustees of the Percy Sladen Memorial Fund and the Executive Committee of Carnegie Trust of Scotland. J. linn. Soc. Zool. 39: 83-131.
Lowndes, A. G., 1936. Scientific results of the Cambridge Expedition to the East African lakes 1930. No. 16. The smaller Crustacea. J. linn. Soc. 40 (269): 1-31.
Magis, N., 1967. Le zooplancton des lacs artificiels du Haut Katanga méridional, étude faunistique et écologique. F.U.L.-R.E.A.C., Univ. Liège: 203 pp.
Mandel, B. K., 1980. Limnological studies of a freshwater fish pond at Burdwan, West Benga, India. Jap. J. Limnol. 41: 10-18.
Mamaril, A. C. & C. H. Fernando, 1978. Freshwater zooplankton of the Philippines (Rotifera, Cladocera and Copepoda). Bull. Nat. appl. Sci. Univ. Philipp., Manila 30: 109-221.
Marques, E., 1958. Copepodes marinhos de Angola. 2a campanha 1952-1953. Mems. Jta Invest. Ultramar, Lisboa 21, 4: 199-222.
Marques, E., 1961. Copepodes dulciaquicolas de Mocambique. Mems. Jta Invest. Ultramar, Lisboa 23: 167-172.
Marques, E., 1963. Copepodes dulciaquicolas de Mocambique (lago Chicunga). Mens. Jta Invest. Ultramar 43, 2: 13-27.
Marques, E., 1966. Copépodes des eaux de Boma et de l'embouchure du fleuve Congo. Revue Zool. Bot. afr. 73, 1-2: 1-16.
Marsh, C. D., 1913. Report on freshwater Copepoda from Panama with descriptions of new species. Smithson. misc. Collns 61: 1-30.
Marsh, C. D., 1919. Report on a collection of Copepoda from Panama made in Honduras by F. J. Dyer. Proc. U.S. natn. Mus. 55: 545-548.
Marsh, C. D., 1931. On a collection of Copepoda made in El Salvador by Samuel F. H. Hildebrand and Fred J. Foster of the U.S. Bureau of Fisheries. J. Wash. Acad. Sci. 21: 207-209.
Matsumura-Tundisi, T. & J. G. Tundisi, 1976. Plankton studies in a lacustrine environment. Oecologia, Berl. 25: 265-270.
Matsumura-Tundisi, T. & O. Rocha (in press). Occurrence of copepod species (Calanoida, Cyclopoida and Harpacticoida) from Broa reservoir (São Carlos, Sao Paulo, Brazil). Revta bras. Biol.
McKenzie, K. G., 1966. Freshwater Ostracoda from north-western Australia. Aust. J. mar. Freshwat. Res. 17: 259-279.
McKenzie, K. G., 1971. Ostracoda from Lake Peunde, near Mt Wilhelm, New Guinea. Zool. Anz. 186: 391-403.
McKenzie, K. G., 1980. A new subfamily from Gorong, Seram, Moluccas, with the description of *Renaudcypris*, new genus (Crustacea: Ostracoda). Bull. Mus. natn. Hist. nat., Paris 2: 507-515.
Mendis, A. S. & C. H. Fernando, 1962. A guide to the freshwater fauna of Sri Lanka. Bull. Fish. Res. Stat. Ceylon 12: 1-160.
Michael, R. G., 1980. A historical resume of Indian limnology. Hydrobiologia 72: 15-20.
Michael, R. G. & B. K. Sharma, in press. Indian Cladocera (Crustacea: Branchiopoda: Cladocera). Fauna India Ser.
Moghrabi, A. I. el, 1972. The zooplankton of the Blue Nile. Ph.D. Thesis, Univ. Khartoum.
Moitra, S. K. & S. K. Mukherjee, 1972. Studies on the freshwater plankton of a fish pond at Kalyani, West Bengal. Vest. csl. Spol. zool. 36: 24-28.
Morton, D. W., 1977. A revision of the Australian Cyclopoida of the families Cyclopinae and Eucyclopinae (Crustacea: Copepoda). M.Sc. Thesis, Monash Univ.
Nasar, S. A. K., 1977. Investigations on the seasonal periodicity of zooplankton in a freshwater pond in Bhagalpur, India. Arch. hydrochim. hydrobiol. 6: 577-584.
Nasar, S. A. K., 1978. Diurnal movements of zooplankton in a freshwater pond in Bhagalpur, India. Bangladesh J. Zool. 6: 9-96.
Onabamiro, S. D., 1952a. Four new species of *Cyclops* sensu lato (Crustacea, Copepoda) from Nigeria. Proc. zool. Soc. London, 122: 153-166.
Onabamiro, S. D., 1952b. On the diurnal migration and seasonal fluctuation in the numbers of *Thermocyclops nigerianus*

Kiefer in a nigerian pond. Ann. trop. Med. Parasit. 46: 38-47.

Onabamiro, S. D., 1957. Some new species of *Cyclops* sensu lato (Crustacea, Copepoda) from Nigeria. J. linn. Soc. Zool. 43: 123-133.

Pearse, A. S., 1911. Report on the Crustacea collected by the University of Michigan-Walker Expedition in the State of Vera Cruz, Mexico. 13 Rep., Mich. Acad. Sci.: 108-113.

Pearse, A. S., 1915. An account of the Crustacea by the Walker Expedition to Santa Marta, Columbia. Proc. U. S. natn. Mus. 49: 531-556.

Pearse, A. S., 1921. Crustacea from Lake Valencia, Venezuela. Proc. U.S. natn. Mus. 59: 459-462.

Pearse, A. S., 1938. Copepoda from Yucatan caves. Carnegie Inst. Wash. Publ. 491: 153-154.

Pejler, B., 1977. On the global distribution of the family Brachionidae (Rotatoria). Arch. Hydrobiol., Suppl. 53: 255-306.

Poppe, S. A., 1981. Ein neuer *Diaptomus* aus Brasilien. Zool. Anz. 14: 248-250.

Proszynska, M., 1967. Bibliography of Cladocera and Copepoda of African inland waters. Ghana J. Sci. 7, 1 & 2: 37-49.

Pourriot, R., 1968. Rotifères du lac Tchad. Bull. I.F.A.N. 30, 2: 471-296.

Pourriot, R., 1980. Rotifères. In Flore et faune aquatiques de l'Afrique Sahelo-Soudanienne. O.R.S.T.O.M. Paris: 219-244.

Rajapaksa, R. & C. H. Fernando, 1981. The Cladocera of Sri Lanka (Ceylon), with remarks on some species. Hydrobiologia 94: 49-69.

Rajapaksa, R. & C. H. Fernando, in press. Freshwater zooplankton. In C. H. Fernando (ed.), Ecology and Biogeography of Sri Lanka. Monographiae Biologicae.

Richard, J., 1895. Sur quelques entomostracés d'eau douce d'Haiti. Mem. Soc. zool. Fr. 8: 189-199.

Richard, J., 1897. Entomostracés de l'Amérique du sud receuillis par Mme. U. Deiters, H. von Ihering, G. W. Müller et C. O. Poppe. Mem. Soc. zool. Fr. 10: 263-301.

Rioja, F., 1940. Observaciones acerca del plancton del Lago Patzcuaro. Ann. Inst. biol. Mex. 11: 421-422.

Rioja, F., 1940. Notas acerca de los Crustaceos del Lago Patzcuaro. Ann. Inst. biol. Mex. 11: 473-474.

Robertson, B. A., 1980. Composicas abundancia e distribuicao de Cladocera (Crustacea) na regiao de aqua livre da Represa de Curua-Una, Para. M.Sc. Thesis, Inst. nac. Pesq. Manaus, 105 pp.

Robinson, A. H. & P. K. Robinson, 1971. Seasonal distribution of zooplankton in the northern basin of Lake Tchad. J. Zool. Lond. 163: 25-61.

Russell, C. R., 1957. Some rotifers from the South Pacific Islands and Northern Australia. Trans. r. Soc. N. Z. 84: 897-902.

Russell, C. R., 1961. The Rotatoria of Queensland, Australia. Trans. r. Soc. N.Z. 1: 235-239.

Ruttner, F., 1943. Beobachtungen über die tägliche Vertikalwanderung des Planktons in tropischen Seen. Arch. Hydrobiol. 40: 474-492.

Ruttner, F., 1952. Planktonstudien der Deutschen limnologischen Sunda-Expedition. Arch. Hydrobiol., Suppl. 21: 1-274.

Rzoska, J., 1968. Observations on zooplankton distribution in a tropical river dam-basin (Gebel Aulia, White Nile, Sudan). J. anim. Ecol. 37: 185-198.

Rzoska, J., 1974. The upper Nile swamps, a tropical wetland study. Freshwat. Biol. 4: 1-30.

Rzoska, J., A. J. Brooke & C. A. Prowse, 1955. Seasonal plankton development in the White and Blue Nile near Khartoum. Verh. int. Ver. Limnol. 12: 337-334.

Sars, G. O., 1901. Contribution to the knowledge of the freshwater entomostraca of south America, 2. Copepoda-Ostracoda. Arch. Math. Naturv. 24: 1-46.

Sars, G. O., 1909. Zoological results of the Third Tanganyika expedition, conducted by Dr. W. A. Cunnington, F. Z. S., 1904-1905. Report on the Copepoda. Proc. zool. Soc. Lond.: 31-77.

Sendacz, S. & E. Kubo, 1982. Copepoda (Calanoida e Cyclopoida) de reservatorios de Estado de São Paulo. Bolm. Inst. Pesca, São Paulo, 9: 51-89.

Seymour-Sewell, R. B., 1935. Studies on the bionomics of freshwaters of India. II. On the fauna of the tanks in the Indian Museum Compound and the seasonal changes observed. Int. Revue ges. Hydrobiol. 31: 203-238.

Seymour-Sewell, R. B., 1957. A review of the sub-genus *Thermocyclops* Kiefer of the genus Mesocyclops Sars, with a description of a new form of *Mesocyclops (Thermocyclops) schmeili* Poppe and Mrazek. Rec. indian Mus. 55: 69-119.

Sharma, B. K. & R. G. Michael, 1980. Synopsis of taxonomic studies on Indian Rotatoria. Hydrobiologia 73: 229-236.

Shiel, R. J., 1981. Plankton on the Murray-Darling River system, with particular reference to the zooplankton. Ph.D. Thesis, Univ. Adelaide.

Shiel, R. J., 1983. A new species of Brachionus (Rotifera: Brachionidae) and new records from Australia. Proc. r. Soc. Vict.

Shiel, R. J. & W. Koste, 1979. Rotifera recorded from Australia. Trans. r. Soc. S. Aust. 103: 57-68.

Shiel, R. J. & W. Koste, 1983. Rotifer communities in billabongs of northern and south eastern Australia. Hydrobiologia.

Shirota, A., 1966. The plankton of South Vietnam. Overseas tech. Cooperation Ag., Japan, 489 pp.

Smirnov, N. N., 1971. Fauna U.S.S.R. N.S. 101 Crustacea 1, 2. Chydoridae fauna of the world. Akad. Nauk, Leningrad, 664 pp. (in Russian).

Smirnov, N. N., 1976. Fauna U.S.S.R. N. S. 112 Crustacea 1, 3. The Macrothricidae and Moinidae of the World. Akad. Nauk., Leningrad, 236 pp. (in Russian).

Smirnov, N. N. & B. V. Timms, 1983. A revision of the Australian Cladocera (Crustacea). Rec. aust. Mus.

Smith, K. E. & C. H. Fernando, 1981. The freshwater calanoid and cyclopoid copepod Crustacea of Cuba. Can. J. Zool. 56: 2015-2023.

Stingelin, T., 1900. Beitrag zur Kentnis der Susswasserfauna von Celebes. Revue suisse Zool. 8: 193-207.

Stingelin, T., 1915. Cladocera von Neu Caledonien. Nova Caledonia Zool. 2: 197-208.

Straškrabá, M. *et al.*, 1969. Primeira contribucion al conocimiento limnologico de las lagunas y embalses de Cuba. Acad. Cienc. Cuba, Ser. Biol. 4: 1-44.

Straškrabá, M. *et al.*, 1969. Lista de los Crustaceos dulciacuicolas de Cuba y sus relaciones zoogeograficas. Acad. Cienc.

Cuba, Ser. Biol. 8: 1-37.

Swar, D. B. & C. H. Fernando, 1979a. Cladocera from Pokhara Valley, Nepal, with notes on distribution. Hydrobiologia 66: 113-128.

Swar, D. B. & C. H. Fernando, 1979b. Seasonality and fecundity of *Daphnia lumholtzi* in Lake Phewa, Nepal. Hydrobiologia 64: 261-268.

Swar, D. B. & C. H. Fernando, 1980. Some studies on the ecology of limnetic crustacean zooplankton in lakes Begnas and Rupa, Pokhara Valley, Nepal. Hydrobiologia 70: 235-245.

Tafall, B. F. O., 1941. Materiales para el estudio del microplancton del Lago de Patzcuaro (Mexico). An. Esc. nac. Cienc. biol. 1940, 2: 343.

Tafall, B. F. O., 1943. Observaciones sobre la fauna aquatica de la cuevas de la region de Valles, San Luis Potosi. Rev. Soc. mex. Hist. nat. 4: 49-54, 57.

Tafall, B. F. O., 1944. Biodinamica del Lago de Patzcuaro. Rev. Soc. mex. Hist. nat. 5: 207.

Tait, R. D., 1981. Natural seasonal changes in the zooplankton in billabong of the Magela Creek, Northern Territory, Australia prior to development in the area. In M. Sudzuki (ed.), Some Approaches to Saprobiological Problems. Sanseido, Tokyo: 29-35.

Tait, R. D., 1982. The ecology of plankton of selected Magela catchment billabongs, Northern Territory. M.Sc. Thesis, Macquarie University.

Tait, R. D., R. J. Shiel & W. Koste, 1983. Structure and dynamics of zooplankton communities, Alligator Rivers region, N.T., Australia. Hydrobiologia.

Thiebaud, A., 1914. Copépodes de Colombie et des Cordillères de Mendoza. Mem. Soc. neuchat. Sci. nat. 5: 160-175.

Ueno, M., 1939. Zooplankton of Lago de Patzcuaro, Mexico. Ann. Zool. Jap. 18: 108.

Van de Velde, I., 1978. Cladocera and Copepoda from the valley of the river Senegal. Dodonaea 46: 192-201.

Victor, R. & C. H. Fernando, 1981. An illustrated key to the freshwater Ostracod genera of the Oriental region. Univ. Waterloo, Biol. Ser. 23: 1-92.

Von Daday, E., 1898. Microskopische Süsswasserthiere aus Ceylon. Termesz. Fuz. (Budapest) 21: 1-123.

Von Daday, 1901. Mikroskopische Süsswasserthiere aus Deutsch Neu Guinea. Termesz. Fuz. (Budapest) 26: 1-56.

Welch, P. S., 1952. Limnology. McGraw Hill, N. Y., 538 pp.

Worthington, E. B., 1931. Vertical movements of freshwater macroplankton. Int. Revue ges. Hydrobiol. 25: 394-436.

Wright, S., 1927. A revision of the South American species of *Diaptomus*. Trans. am. microsc. Soc. 66: 73-121.

Wright, S., 1935. Three new species of *Diaptomus* from northeast of Brazil. Ann. Acad. Bras. Sci. 7: 221-233.

Wright, S., 1936. Preliminary report of six new species of *Diaptomus* from Brazil. Ann. Acad. Bras. Sci. 8: 79-85.

Wright, S., 1937. A review of some species of *Diaptomus* from São Paulo. Ann. Acad. Bras. Sci. 9: 65-82.

Wright, S., 1938. A review of the *Diaptomus bergi* group with description of two new species. Trans. am. microsc. Soc. 57: 297-315.

Wright, S., 1938. Distribuicao geografica des especies de *Diaptomus* na America do Sul. Livro Jubilar Prof. Travassos 3: 561-566.

Zago, M. S. A., 1976. The planktonic Cladocera (Crustacea) and aspects of the eutrophication of the Americana reservoir, Brazil. Bolm. Zool. 1: 105-145.

Zago, M. S. A., 1976. A preliminary investigation on the cyclomorphosis of *Daphnia gessneri* Herbst 1967, in a Brazilian reservoir. Bolm. Zool. 1: 147-160.

Zaret, T. M., 1980. Predation and freshwater communities. Yale University Press, 187 pp.

Zooplankton associations in the swamps of southern Sudan

J. Green
Zoology Department, Westfield College, Hampstead, London NW3 7ST, U.K.

Kewyords: tropical zooplankton, swamps, Sudan

Abstract

Samples from the main Nile channel yield a poor zooplankton, deficient in species and heavily laden with organic debris. Samples from side arms and floodplain lakes yield a richer zooplankton, often containing 15–20 species of rotifers and ten or more species of Crustacea. When the floodplain lakes are heavily vegetated the number of species increases. For the Crustacea the index of diversity (α) normally varies between 0.6 and 2.4, but in a side arm of the Bahr el Zeraf it reaches 6.1 and in Lake Ambadi on the Bahr el Ghazal it reaches 8.4. The diversity of the rotifers is somewhat higher, normally ranging between 2 and 4.5, but reaching 23 in Lake Ambadi. The zooplankton of Lake Ambadi is the most distinctive and contains several species not found at other stations. The reasons for the peculiar zooplankton of Lake Ambadi are discussed.

Introduction

The aim of this paper is to give some account of the associations of zooplankters found in the swamps of southern Sudan, often simply called the Sudd. This is a vast area, with about 10 000 km^2 of permanent swamp, and a further area of about 90 000 km^2 which flood in the wet season (Equatorial Nile Project, 1954). A detailed account of the vegetation in this area has been given by Migahid (1947, 1948), and a general ecological account has been given by Rzoska (1974) summarising the literature up to that date.

Interest in the area has recently been revived by the construction of the Jonglei canal which is now in progress, and when this is completed it will certainly reduce the area flooded during the wet season, and perhaps reduce the area of permanent swamp.

The material for the present study was collected on two journeys. The first was made by boat in December and January 1976–77 from Kosti down to Juba and included a diversion along the Bahr el Ghazal to Lake Ambadi. The second journey in April 1978 included travel by boat from Malakal to Lake No and back, and by road from Malakal to Juba.

Methods

Seventeen stations were sampled on the first journey, and a further seven stations on the second journey. Only seven of the sites proved to have sufficiently large populations of zooplankton to merit detailed examination. These are shown diagrammatically in Fig. 1.

Samples were collected with nylon plankton nets having a mesh of 55 μm. This mesh is small enough to retain most of the rotifers and all the microcrustacea. Some samples were also taken with a hand net with a mesh size of 120 μm. Most of the samples taken with this net were from among vegetation, particularly the roots of *Eichhornia*. Some samples

Hydrobiologia 113, 93–98 (1984).

were also taken with a nylon net having a mesh of 250 μm. This retains only the larger members of the microcrustacea and a very few large rotifers. It is however a more efficient net for catching these forms because it allows a faster throughflow than the finer net. The samples were preserved in approximately 5% formalin and were examined in detail in London.

Sampling stations

The locations of the seven selected sampling stations are shown in Fig. 1. Some of their characteristics are given below.

1. Lake No. This lies at the junction of the Bahr el Ghazal and the Bahr el Jebel. The conductivity of the water is about 200 μmho cm^{-1} 20 °C (Rzoska, 1974). We found the water to be well aerated, with oxygen levels between 72 and 90% of air saturation at 26 °C. The water is fringed with *Cyperus papyrus*

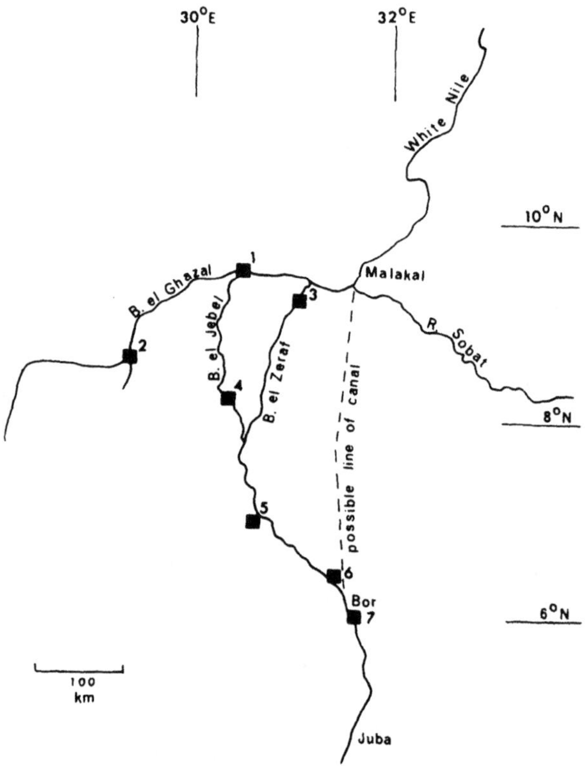

Fig. 1. Sketchmap to show locations of sampling stations. 1 – Lake No; 2 – Lake Ambadi; 3 – side arm of Bahr el Zeraf; 4 – Bahr el Jebel at R.P. 50; 5 – Lake Shambe; 6 – Lake Bur Akok; 7 – Lake Malwal.

and a band of *Eichhornia crassipes*. There are also extensive beds of *Najas pectinata*, on which *Hippopotamus* sometimes grazes. In 1976 a group of Arab fishermen were operating from a camp on the northern shore, and were catching large numbers of *Heterotis nilotica*.

2. Lake Ambadi. This lies some 190 km along the Bahr el Ghazal from Lake No. It lies in a flat featureless landscape, and the dominant grass immediately around its shores is *Vossia cuspidata*. The shallow water contains abundant *Utricularia* and a dwarfed form of *Eichhornia crassipes*. The conductivity of the water in December 1976 was 45 μmho cm^{-1} 20 °C. We found a steep stratification of oxygen in quite shallow water. Where the depth was 1.5 m the surface waters were 58% air saturated, but at a depth of 1 m this had fallen to 14% and at 1.5 m to only 4%. In more open water we found 54% at 2 m, with a water temperature of 25 °C.

3. Side arm of the Bahr el Zeraf. In this part of the Zeraf there is a narrow fringe of *Vossia* backed by *Echinochloa*. The side arm was located about 15 km south of the northern end of the Zeraf. It was shallow with a dense vegetation including *Utricularia* and *Vallisneria*. The conductivity of the water in the Zeraf in this area is about 350 μmho cm^{-1} 20 °C (Bishai, 1962).

4. Side arm of the Bahr el Jebel at R.P. 50. The entrance to this side arm is narrow, running between papyrus plants and a luxuriant growth of *Eichhornia*. Once this narrow passage has been negotiated there is an extensive lagoon with *Hippopotamus* and crocodiles. Parts of the shore are lined with *Vossia* and parts with *Typha*. The water was quite well oxygenated, with about 58% air saturation at a depth of 1 m and at a temperature of 26.5 °C. The conductivity in this region is about 250 μmho cm^{-1} 20 °C (Bishai 1962).

5. Lake Shambe. This lake has several km^2 of open water, and opens to the main channel through a wide mouth. The conductivity of the water is about 250 μmho cm^{-1} 20 °C (Bishai, 1962).

6. Lake Bur Akok. This is a complex and extensive side arm, about 20 km north of Bor. The northern end of the lake is wider than in the south and connects via a curved channel to the main river. The water was well oxygenated, with about 75% air saturation at all depths down to 2 m. In the main body of the lake the maximum depth we found was 2.7 m, but in the northern curved channel we found

a depth of 4 m. The vegetation around the lake is complex, with a mixture of *Cyperus papyrus, Vossia, Typha* and *Phragmites*. In the open water there are patches of *Potamogeton schweinfurthi*, and *Eichhornia* is present in various parts of the lake. A herd of *Hippopotamus* was present, and a varied fish fauna was found in nets set overnight.

7. Lake Malwal. This lake lies to the south of Bor. It is about 1 km long and 0.5 km wide. We could not get a boat on to the lake, and so have only samples from the shore. There does not appear to be any clearly open connection with the main river, but a swampy area, rich in *Utricularia*, lies between the river and the lake.

The zooplankton associations

1. Lake No. The crustacean zooplankton is dominated by *Thermodiaptomus galebi* (Table 1), and there is a rich association of planktonic rotifers (Table 2). The most important rotifers are *Keratella tropica, Brachionus calyciflorus* and *B. caudatus*.

Table 1. Percentage composition of the crustacean zooplankton in Lake No, April 1978. Sample taken with 250 μm net.

Thermodiaptomus galebi	62 %
Thermocyclops neglectus	11
Eucyclops sp.	4
Diaphanosoma excisum	10
Ceriodaphnia dubia	1
Daphnia barbata	1
Moina micrura	8
Bosmina longirostris	2
Ostracod indet.	1

Table 2. Percentage composition of the planktonic rotifers in Lake No, December 1976 and April 1978.

	Dec. 1976	Apr. 1978
Anuraeopsis fissa	1 %	-
Brachionus calyciflorus	13	35 %
Brachionus caudatus	7	41
Brachionus falcatus	7	3
Brachionus patulus	2	4
Filinia longiseta	4	-
Filinia opoliensis	1	-
Hexarthra mira	2	-
Horaella brehmi	1	-
Keratella tropica	60	6
Keratella lenzi	1	-
Lecane bulla	-	10
Lecane luna	-	1
Rotaria neptunia	1	-

2. Lake Ambadi. The zooplankton in this lake has a high periphytic component. The number of species is higher than at any of the other stations. The Crustacea are shown in Table 3, and the rotifers in Table 4. The latter table is remarkable for the number of species represented by only a single specimen.

3, 4. The Crustacea in the side arms of the Bahr el Zeraf and the Bahr el Jebel are compared in Table 5. It is obvious that the association at station 3 is essentially a periphytic association, whilst at station 4 it is a true planktonic association.

5. Lake Shambe. The crustacean zooplankton in this lake is very simple; only six species: three cladocerans, dominated by *Moina micrura*, and three copepods, dominated by *Thermocyclops neglectus*. In contrast, the rotifers are diverse (Table 7) with 17 species dominated by *Brachionus caudatus* (47%) and *Keratella tropica* (25%).

6, 7. Lakes Bur Akok and Malwal. These two lakes are compared in Tables 6 and 7. Both have simple associations of Crustacea; that from Malwal being the result of having access only to the shore of the lake. The rotifer associations are more complex

Table 3. Percentage composition of Crustacea in plankton samples from Lake Ambadi.

Cladocera	
Alona monacantha	2
Biapertura verrucosa	7
Bosminopsis deitersi	1
Camptocercus uncinatus	2
Chydorus eurynotus	20
Chydorus pubescens	2
Euryalona orientalis	1
Graptoleberis testudinaria	1
Grimaldina brazzai	5
Guernella raphaelis	1
Indialona globulosa	2
Kurzia longirostris	5
Macrothrix spinosa	25
Macrothrix triserialis	2
Moina micrura	3
Pleuroxus sp.	1
Pseudosida szalayi	2
Copepoda	
Mesocyclops sp.	2
Microcyclops davidi	2
Microcyclops varicans	4
Thermocyclops neglectus	10
Decapoda	
Cridina nilotica	1

Table 4. Composition of the planktonic and periphytic rotifer association in Lake Ambadi. Note that only 57 individuals were found.

Brachionus calyciflorus	1
Brachionus falcatus	2
Dipleuchlanis macrodactyla	1
Euchlanis triquetra	1
Filinia opoliensis	1
Horaella brehmi	1
Keratella tropica	8
Lecane aculeata	1
Lecane bulla	17
Lecane curvicornis	1
Lecane doryssa	1
Lecane hamata	1
Lecane leontina	3
Lecane papuana	1
Lecane rhytida	1
Lecane stichaea	1
Lecane ungulata	2
Lepadella cristata	1
Macrochaetus collinsi	3
Manfredium eudactylotum	1
Notommata triangulata	1
Ploesoma truncatum	1
Pseudoploesoma greeni	2
Rotaria neptunia	2
Testudinella incisa	1
Trichocerca insulana	1

Table 5. Percentage composition of the planktonic Crustacea in side arms of the Bahr el Zeraf (Stn. 3) and Bahr el Jebel (Stn. 4).

	Zeraf	Jebel
Cyclestheria hislopi	4	-
Ceriodaphnia cornuta	-	3
Ceriodaphnia dubia	-	15
Daphnia lumholtzi	-	1
Dunhevedia serrata	1	-
Euryalona orientalis	12	-
Indialona globulosa	8	-
Macrothrix triserialis	19	-
Moina micrura	-	3
Oxyurella tenuicaudis	2	-
Pleuroxus laevis	2	-
Pseudosida szalayi	5	-
Pseudochydorus globosus	1	-
Simocephalus latirostris	23	-
Thermodiaptomus galebi	-	26
Cyclopoids	13	52
Caridina nilotica	7	-
Acocypris hirsuta	1	-
Strandesia sp.	2	-

Table 6. Percentage composition of planktonic Crustacea in Lakes Bur Akok and Malwal.

	Bur Akok	Malwal
Ceriodaphnia cornuta	40	-
Daphnia barbata	2	-
Diaphanosoma excisum	14	37
Moina micrura	16	60
Tropodiaptomus orientalis	16	-
Cyclopoida	12	3

Table 7. Percentage composition of planktonic rotstations 4, 5, 6 & 7.

	(4) Jebel	(5) Shambe	(6) Bur Akok	(7) Malwal
Anuraeopsis fissa	-	-	2	11
Anuraeopsis navicula	-	-	-	1
Asplanchna brightwelli	-	1	-	-
Asplanchna priodonta	-	1	-	-
Brachionus budapestinensis	-	1	-	-
Brachionus calyciflorus	4	1	7	18
Brachionus caudatus	-	47	2	16
Brachionus falcatus	24	3	7	15
Brachionus patulus	1	3	-	1
Brachionus quadridentatus	2	-	-	1
Colurella sp.	1	-	-	-
Euchlanis triquetra	1	-	-	-
Filinia longiseta	11	2	1	-
Filinia opoliensis	-	-	6	1
Filinia pejleri	-	3	1	-
Hexarthra mira	3	5	1	6
Horaella brehmi	6	-	-	-
Keratella cochlearis	-	-	8	-
Keratella lenzi	-	1	-	-
Keratella tropica	46	25	58	7
Lecane bulla	-	3	1	4
Lecane curvicornis	-	2	-	-
Lecane papuana	-	-	-	8
Lecane ploenensis	-	-	-	1
Polyarthra vulgaris/ dolichoptera	-	-	1	10
Rotaria neptunia	1	2	-	-
Trichocerca similis	-	-	3	-

and quite similar. There are 13 species in the samples from Lake Bur Akok, and 14 in Lake Malwal; of these 9 are common to both lakes.

Comparisons of the associations

Tables 8 and 9 give some parameters of the associations found at the seven stations. As is normal there are more species of rotifers at each station

Table 8. Some parameters of the rotifer associations in the Sudd.

Station	Number of species	Dominance ratio	Index of diversity (α)
1. L. No (Dec. 1976)	12	60	3.8
(Apr. 1978)	7	41	2.0
2. L. Ambadi	26	34	23.0
3. Zeraf	-	-	-*
4. Jebel (R.P. 50)	10	46	3.1
5. Shambe	15	47	4.9
6. Bur Akok	13	58	4.0
7. Malwal	14	18	4.5

* Samples lost.

Table 9. Some parameters of the crustacean associations in the Sudd.

Station	Number of species	Dominance ratio	Index of diversity (α)
1. L. No (Dec. 1976)			
(Apr. 1978)	9	62	2.4
2. L. Ambadi	20	25	8.6
3. Zeraf	14	23	6.1
4. Jebel (R.P. 50)	6	52	1.8
5. Shambe	6	47	1.4
6. Bur Akok	7	40	2.0
7. Malwal	3	60	0.6

than there are species of Crustacea. The diversity (α) of both groups is highest in Lake Ambadi. There is a strong correlation between the number of species and diversity ($r = 0.94$ for rotifers), but a weaker negative correlation between species numbers and dominance ratios ($r = 0.29$), so that the associations with fewer species tend to have higher dominance ratios, but it is not a strong relationship.

The peculiar nature of the associations in Lake Ambadi is emphasised when Sorensen indices of similarity are calculated. This index is calculated from the following formula (Sorensen, 1948):

$$S = \frac{2c}{a+b} \times 100$$

where c is the number of species common to both associations, a is the number of species in one association and b is the number of species in the other association. Tables 10 and 11 show that Lake Ambadi has the most divergent association of rotifers and the second most divergent association of crustaceans. The total number of specimens obtained in the samples from Lake Ambadi was very small. A ten litre sample yielded only 7 rotifers and 23 crustaceans, making a total of 30 individuals belonging to 10 taxa. The species numbers and diversities for this lake were obtained from tows which sampled a much larger volume.

Table 10. Sørensen indices of similarity for rotifer associations in the Sudd.

	L. No	Ambadi	Jebel	Shambe	Bur Akok	Malwal	Mean
L. No	-	38	64	69	67	64	60
L. Ambadi	35	-	32	29	26	30	30
Jebel	64	35	-	54	42	48	49
Shambe	69	32	54	-	50	48	51
Bur Akok	67	28	42	50	-	67	51
Malwal	64	32	48	48	67	-	52

Table 11. Sørensen indices of similarity for crustacean associations in the Sudd.

	L. No	Ambadi	Zeraf	Jebel	Shambe	Bur Akok	Malwal	Mean
L. No	-	21	8	63	67	63	50	45
L. Ambadi	21	-	33	22	23	22	17	23
Zeraf	8	33	-	17	17	17	11	17
Jebel	63	22	17	-	92	57	40	49
Shambe	67	23	17	92	-	77	67	57
Bur Akok	63	22	17	57	77	-	60	49
Malwal	50	17	11	40	67	60	-	41

The peculiar nature of Lake Ambadi has been noted by previous authors. Grönblad *et al* (1958) give data showing that the pH is lower than in areas downstream, and the CO_2 content is much higher (about 30 mg l^{-1}). They describe a rich flora of desmids, with 205 species and forms, including 21 new species. The conductivity of water in Lake Ambadi is much lower than in other parts of the Sudd. Rzoska (1974) quotes a value of 40 μmho cm^{-1} 20 °C measured in 1954, while our measurement made in 1976 gave a value of 48 μmho cm^{-1} 20 °C. Talling (1957) also notes some features of the Bahr el Ghazal, particularly its high silica content and high transparency.

A noticeable feature as one travels along the Bahr el Ghazal towards Lake Ambadi is the reduction in size and abundance of *Cyperus papyrus*. There is a similar reduction in the size of *Eichhornia*, and when one arrives at Lake Ambadi all the plants of *Eichhornia* are dwarfed, having leaves only 10 or 12 cm long, whereas in the main channel of the White Nile the leaves are often seven or eight times this length. The simplest explanation for this diminution in the growth of both papyrus and *Eichhornia* is that there is some nutrient deficiency in Lake Ambadi and the Bahr el Ghazal. Some support for this view comes from the data of Talling (1957) who found that the phosphate content of the Bahr el Ghazal was only one ninth of that of the White Nile as it leaves Lake Albert.

A nutrient deficiency combined with low pH will affect the whole food chain, and could, at least in part, explain the low density of zooplankton in Lake Ambadi, but it does not explain the high diversity. Although oligotrophic water bodies of low pH often have diverse associations of rotifers and crustaceans, there is as yet no convincing explanation as to why this should be so.

Acknowledgement

The samples on which this study is based were collected whilst travelling with the Hydrobiology Research Unit of the University of Khartoum. My best thanks are due to Asim Moghraby for organising these journeys and making this study possible.

References

Bishai, H. M., 1962. The water characteristics of the Nile in the Sudan with a note on the effect of Eichhornia crassipes on the hydrobiology of the Nile. Hydrobiologia 19: 357–382.

Equatorial Nile Project, 1954. Its effects in the Anglo-Egyptian Sudan. Rep. Jonglei Invest. Team. Vol. 1–4.

Grönblad, R., G. A. Prowse & A. M. Scott, 1958. Sudanese Desmids. Acta bot. fenn. 58: 1–82.

Migahid, A. M., 1947. An ecological study of the Sudd swamps of the Upper Nile. Proc. Egypt. Acad. Sci. 3: 57–86.

Migahid, A. M., 1948. Report on a botanical excursion to the Sudd region. Cairo University Press: 1–159.

Rzoska, J., 1974. The Upper Nile swamps, a tropical wetland study. Freshwat. Biol. 4: 1–30.

Talling, J. F., 1957. The longitudinal succession of water characteristics in the White Nile. Hydrobiologia 11: 73–89.

Ecological studies of Cladocera in the ricefields of Tanjung Karang, Malaysia, subjected to pesticide treatment

R. P. Lim[1], M. F. Abdullah[1] & C. H. Fernando[2]
[1] *Zoology Department, University of Malaya, Kuala Lumpur, Malaysia*
[2] *Department of Biology, University of Waterloo, Waterloo, Ontario, Canada N2L 3G1*

Keywords: tropical zooplankton, ricefield Cladocera, pesticide effects

Abstract

Temporal changes in the population densities of four dominant Cladocera (*Moina micrura, Diaphanosoma excisum, Alona guttata* and *Macrothrix spinosa*) were studied in untreated and pesticide treated rice plots over a growing season. *M. micrura* was the first dominant species to occur in all the plots which were initially devoid of vegetation except for the newly transplanted rice seedlings. With the growth of the rice seedlings and the appearance of aquatic macrophytes, open water ricefields were transformed into vegetated littoral conditions precipitating the disappearance of *M. micrura* and the appearance of the littoral *A. guttata* and *M. spinosa*, and the eurytopic *D. excisum*. The cladoceran community was affected by spraying of the pesticide FMC35001, an analogue of Furadan®. The response of the four dominant species to the pesticide treatment is discussed.

Introduction

Ricefields form an important component in the landscape and economic life of the rural people along the west coast of Peninsular Malaysia. As an ecosystem it consists of two main biotopes – the rice plants and the aquatic environment. The latter biotope is temporary, littoral and heavily manipulated. The flora and fauna are often affected not only by natural changes that occur over the growing season (i.e. growth of rice plants and aquatic macrophytes) but also by pesticide applications (Soong, 1954; Yunus & Lim, 1971; Tan *et al.*, 1973; Moulton, 1973a, b; Lee & Furtado, 1977; Lim, 1980). Although the zooplankton form an important component of the aquatic fauna of ricefields they are little studied in tropical areas. Notable studies are those of Fernando (1977), Fernando *et al.* (1979), Heckman (1979), and Pont (1977).

This study aims at looking at the seasonal succession and responses of the dominant Cladocera in relation to the application of the organophosphorus pesticide FMC 35001, an analogue of Furadan®.

Materials and methods

This study was conducted in the rice plots of the Malaysian Agricultural Development Institute (MARDI) Rice Research Station in Tanjung Karang. Each plot is 0.044 ha in area and was cultivated following the usual procedures used by the farmers in the surrounding area. The pesticide was applied on three occasions following the recommendations of the manufacturer i.e. 25% EC formulated at 0.2% ai. in 786 l water: ha sprayed aerially 15, 40 and 65 days after transplanting. Samples of zooplankton were collected at weekly intervals over the growing season. A 1 l plastic beaker was used to collect the samples. Two replicate samples of 10 l each were taken randomly from each plot and concentrated in a plankton net of mesh size 10 (120 μm) aperture. The concentrated sample was preserved

Hydrobiologia 113, 99–103 (1984).

in 400 g l^{-1} sucrose-formalin solution (Haney & Hall, 1973) for analysis in the laboratory.

Physical and chemical variables were concurrently measured. Water temperature and dissolved oxygen content were measured at 15 cm below the surface of the water with a Yellow Spring Oxygen meter (Model 54), pH, conductivity and alkalinity were measured at the site.

Log (X + 1) transformations were applied to normalise the population data before statistical analysis was carried out. The Student's t-test was used to determine differences in population densities of cladocerans between the control and the pesticide treated plots.

Results

Table 1 summarises some of the physical and chemical conditions of the ricefield throughout the growing season. In general, temperature varies over a wide range with a mean of 30.1 °C. This is due to the shallow depth of the ricefields which when exposed to strong sunlight heats up readily, especially in the early stages of the growing season when the water is devoid of vegetational cover. Dissolved oxygen also varies widely with a mean of 8.6 mg l^{-1}. Low dissolved oxygen content occurs in the latter half of the growing season and is attributed to the cover provided by the growth of *Lemna* and *Salvinia*. This forms a barrier between the atmosphere and the water, limiting gaseous diffusion. Together with the rice plants they also limit primary production through increasing light attenuation.

pH is circumneutral while conductivity is high indicating the fertility of the water. The very high value of 777.0 μ S cm^{-1} is due to the application of fertilizers. Alkalinity also varies over a wide range. In general the aquatic habitat undergoes great changes over a growing season.

Table 1. Maximum, minimum and mean values of the physical and chemical parameters in the ricefields of Tanjung Karang, Malaysia.

Parameter	Max.	Min.	Mean
Temperature (° C)	33.4	26.9	30.1
Dissolved oxygen (mg/l)	14.1	3.2	8.6
pH	7.67	6.12	6.89
Conductivity (μ S/cm)	777.0	303.0	540.0
Alkalinity (mg/l) ($CaCO_3$)	34.3	7.5	20.9

Table 2 shows a list of cladoceran species collected in the ricefields of Tanjong Karang. Of these four species were dominant namely *Moina micrura*, *Alona* cf. *guttata*, *Macrothrix spinosa* and *Diaphanosoma excisum*. This is typical of most littoral cladoceran communities (e.g. Pennak, 1957; Lim, 1976; Smyly, 1952).

Table 2. Species list of Cladocera collected from the ricefields of Tanjung Karang, Malaysia.

*Diaphanosoma excisum**
Diaphanosoma sarsi
Pseudosida bidentata
Scaphaloberis kingi
*Moina micrura**
*Macrothrix spinosa**
Acroperus harpae
*Alona guttata**
Alonella excisa
Chydorus barriosi
Chydorus cf. *sphaericus*

*Dominant species.

The population densities of the four dominant species also fluctuated greatly over the growing season (Fig. 1). In both the untreated and pesticide treated plots population densities reached a maximum of around 7000 individuals per 10 l. In terms of seasonal trends, the population density in the control plot showed three peaks, while that of the

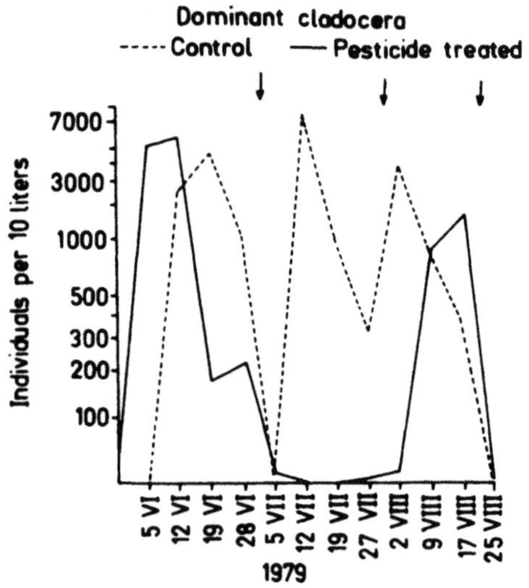

Fig. 1. Seasonal changes in population density of the dominant Cladocera in the control and pesticide treated plots. Arrows indicate pesticide application.

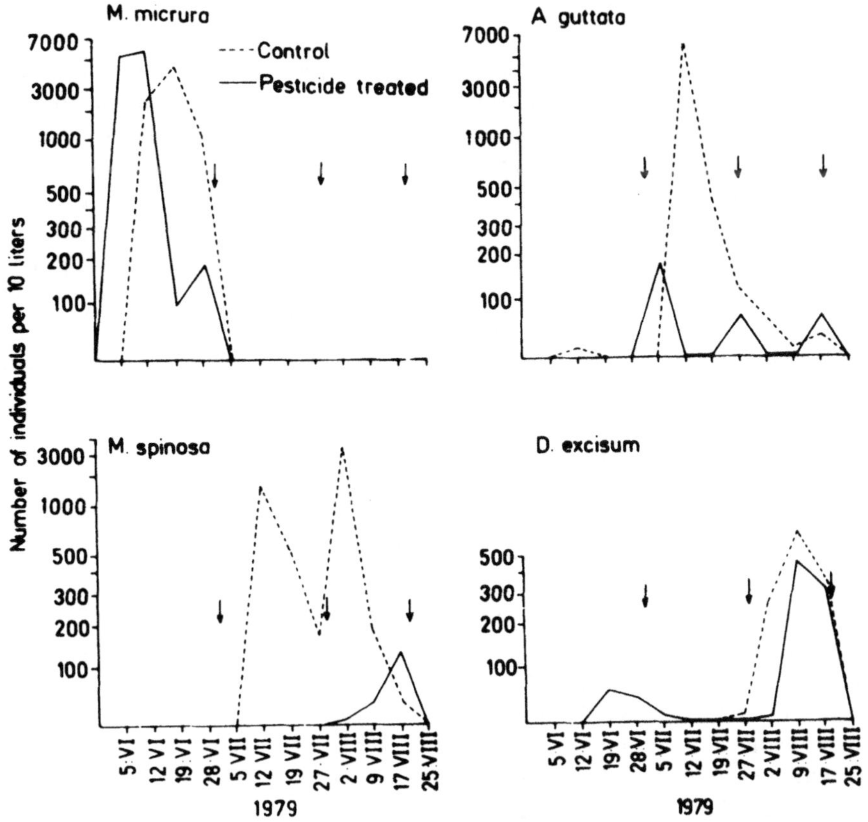

Fig. 2. Seasonal changes in population densities of *Moina micrura, Alona guttata, Macrothrix spinosa* and *Diaphanosoma excisum* in the control and pesticide treated plots. Arrows indicate pesticide application.

pesticide treated plot had two peaks with low densities in the middle in the growing season (Fig. 1). The trough in population density may be due to the effects of pesticide application.

Figure 2 shows the changes in population densities of each dominant species. *Moina micrura* was the first cladoceran species to appear after the flooding and transplanting the ricefields. This is due to open water conditions which prevailed for the first four weeks. *Moina micrura* is a limnetic species and hence its dominance during the vegetation free period. In both the control and the pesticide treated plots their populations responded similarly. Effectively they were subjected to similar conditions as pesticide spraying was carried out only after the 4th week. Thus little difference is seen in their seasonality. The sharp decrease in population density after the 5th week can be attributed to the growth of aquatic macrophytes especially the rice plants, *Salvinia, Lemna* and *Ceratophyllum*. The transition from open water to vegetated littoral conditions resulted in the appearance of littoral Cladocera dominated by *Alona* cf. *guttata* and *Macrothrix spinosa*, and the eurytopic *Diaphanosoma excisum* (Fig. 2). However, in the control plot *A. guttata* was dominant in the middle of the growing season, while *M. spinosa* dominated most of the latter half of the growing season. *D. excisum* was most dominant in the last four weeks of the growing season. In contrast, the population of *A.* cf. *guttata*, and *M. spinosa* were substantially lower in the pesticide treated plot when compared to those of the control plot ($P < 0.01$). The population density of *D. excisum* was not significantly different from that of the control ($P > 0.05$).

Discussion

The responses of the cladoceran populations in relation to pesticide application do not indicate direct kills through acute toxicity immediately after spraying. Seasonal abundances of populations of each species could be controlled by one or several factors: (i) sublethal toxicity, (ii) habitat location and food habits of the cladocerans, (iii) predator-prey relationships and inter-specific competition, and (iv) growth and development of aquatic macrophytes.

The pesticide under investigation was sprayed aerially and hence only a fraction enters the water. It is likely that the amounts entering the water are not concentrated enough to cause direct kills but can result in chronic toxicity affecting the physiological functioning of the cladocerans, especially their reproductive capacity. This depressed population densities of the cladocerans in the middle of the growing season. Sublethal effects are less towards the end of the growing season because of the development of aquatic macrophytes which act as an effective screen in preventing or greatly reducing the entry of the pesticide into the water.

Cladocerans are also variable in their habitat locations and their food preferences. Pesticides have a tendency to be adsorbed to organic particles in water and in the sediment, making the benthos a high risk habitat to be located in. *A.* cf. *guttata* and *M. spinosa* are both true littoral Cladocera spending a fair portion of their time in or near the sediment, grazing on detrital matter. This could have resulted in the two species being exposed to greater concentrations of the pesticide and hence their depressed population densities when compared to the population densities of the control plot. In contrast the eurytopic *D. excisum* was not significantly affected by the second application of the pesticide. Smith & Ison (1967) also concluded that benthic organisms are generally exposed to higher concentrations of pesticide residues and for longer periods than those inhabiting limnetic regions.

Besides sublethal effects, the more complex role of predator-prey relationships and inter-specific competition can be altered by pesticide application (Hurlbert *et al.*, 1972; Hurlbert, 1975). Aquatic organisms respond differentially to the same dosage of pesticide and also to different pesticides (Lee, 1979). Thus a predator or grazer may not be able to develop due to a lack of prey while a prey may be able to increase its density due to a decrease in a particular predator, e.g. migration of potential hemipteran predators (Lim, 1980). Such complex relationships make the overall effect of a pesticide on a species difficult to predict.

In light of the above discussion it is speculated that differential effects on the various species of Cladocera are due to their habitat and food preferences and also the role of macrophyte development in preventing the entry of pesticides into the water. Seasonal succession was not disrupted by the application of pesticides indicating further that natural environmental conditions still play an important role, over a growing season. The dosage of FMC 3001 applied does affect the cladoceran community in ricefields.

Acknowledgements

The authors would like to thank Mr. G. Liew Kim-Seng, Encik Ramli and Encik Junid for field assistance. This study was carried out with the financial assistance of grants Vote F 107/80 and UNESCO/MAB 30111C awarded to R. P. Lim.

References

Fernando, C. H., 1977. Investigations on the aquatic fauna of tropical ricefields with special reference to South East Asia. Geo-Eco-Trop 1: 169-188.

Fernando, C. H., J. I. Furtado & R. P. Lim, 1979. The aquatic fauna of the world's ricefields: A general introduction and bibliography of references to the aquatic fauna of ricefields with an introduction to the ecology of the ricefield fauna. Wallaceana, Suppl., (Kuala Lumpur) 2: 105 pp.

Haney, J. F. & D. J. Hall, 1973. Sugar coated *Daphnia*. Limnol. Oceanogr. 18: 331-333.

Heckman, C. W., 1979. Ricefield ecology in northeastern Thailand. Dr. W. Junk, The Hague, 228 pp.

Hurlbert, S. H., 1975. Secondary effects of pesticides on aquatic ecosystems. Residue Rev. 57: 81-148.

Hurlbert, S. H., M. S. Mulla & H. R. Wilson, 1972. Effects of an organophosphorus insecticide on the phytoplankton, zooplankton and insect populations of freshwater ponds. Ecol. Monogr. 42: 269-299.

Lee, K. W., 1979. The effects of Thiodan, 2,4-D and Sevin 85 on the plankton community and fish in a ricefield system. M.Sc. Thesis, University of Malaya, Kuala Lumpur, 272 pp.

Lim, R. P., 1976. Community description, Population Dynamics and Production of Cladocera with special reference to the littoral region of Pinehurst Lake, Ontario. Ph.D. Thesis, University of Waterloo, Can., 350 pp.

Lim, R. P., 1980. Population changes of some aquatic invertebrates in ricefields. In J. I. Furtado (ed.) Tropical Ecology & Development. Proc. 5 int. Soc. Trop. Ecol.: 971-980.

Moulton, T. P., 1973a. The effects of various insecticides (especially Thiodan and BHC) on fish in the paddy-fields of West Malaysia. Malaysian Agric. J. 49: 224-253.

Moulton, T. P., 1973b. More rice and less fish - some problems of the 'Green Revolution'. Aust. Nat. Hist.: 322-327.

Pennak, R. W., 1966. Structure of zooplankton populations in littoral macrophyte zone of some Colorado Lakes. Trans. am. micros. Soc. 85: 329-349.

Pont, D., 1977. Structure et évolution saisonnière des populations de copepodes, cladocères et ostracodes des rizières de Camergue. Ann. Limnol. 13: 15-28.

Smith, G. E. & B. G. Isom, 1967. Investigation of effects of large-scale applications of 2,4-D on aquatic fauna and water quality. Pest. Monitor J. 1: 16.

Smyly, W. J. P., 1952. The Entomostraca of a moorland pond. J. anim. Ecol. 21: 1-11.

Soong, M. K., 1954. Fish culture in paddy-fields in the Federation of Malaya. F.A.O. Int. Rice Comm. 4 Session, Tokyo, Japan, 5 pp. (mimeo).

Tan, C. E., B. J. Chong, H. K. Sier & T. P. Moulton, 1973. A report on paddy and paddfield fish production in Krian, Perak. Bull. 128 Minist. Agric. & Fish. Malaysia, 58 pp.

Yunus, A. & G. S. Lim, 1971. A problem in the use of insecticides in paddy-fields in West Malaysia - a case study. Malaysian Agric. J. 48: 167-178.

The freshwater zooplankton of Central America and the Caribbean

Carmen Collado, C. H. Fernando & Dawn Sephton
Department of Biology, University of Waterloo, Waterloo, Ontario, Canada, N2L 3G1

Keywords: tropical zooplankton, Central America, Caribbean

Abstract

So far mainly sporadic studies have been made on the freshwater zooplankton of this region. We studied material from Costa Rica, Cuba, Bahamas, El Salvador, Haiti and Trinidad and listed unpublished species data from Jamaica. In all 183 species of Rotifera; 104 of Cladocera; 64 Calanoida and Cyclopoida and a few Ostracoda are known from the region which includes Central America, the Caribbean Islands from the Bahamas to Trinidad and the islands off South America and Central America. Records from individual countries are generally low except for Cuban Cladocera and Copepoda. The total number of Copepoda and Cladocera recorded for the whole regions appears to be reasonably comprehensive. *Daphnia* is rare or absent from the equatorial regions and it is likely that the low species diversity may be due partly to the lack of a range of habitat types.

Introduction

The freshwater zooplankton of Central America and the Caribbean has been investigated sporadically since the first decade of this century. The area consists of a relatively narrow bridge of land joining Mexico to Colombia, a large number of islands stretching from Trinidad to Puerto Rico (Caribbean Islands), the Bahamas in the North and Trinidad, the islands Curacao, Aruba and Bonaire near the South American sub-continent and the large islands of Cuba, Hispaniola (Haiti, Dominican Republic) and Puerto Rico. Running across the central region of the Central American portion is a mountain massif. Most of the land is at a relatively low level. There are few but a varied range of lakes and more recently reservoirs have been constructed. The range of freshwater habitats does not appear to be very varied though and the riverine habitat is by far the most common.

Our study has included the identification of zooplankton from samples collected by us (C.H.F. and C.C.) or sent to us by various scientists. We have also compiled the literature on the Rotifera, Cladocera and Copepoda (Calanoida, Cyclopoida). It is quite clear from these two sources of information that very little is known about the species composition in the region as a whole. The best known country is Cuba and the least known is Belize. (Other Cyclopoida from Caves have been recorded by C. Plesa, 1981.)

This study has shown very clearly the wide gaps that exist in our knowledge of the zooplankton species and their distribution. There is a great need for intensive and extensive studies on systematics, distribution and ecology.

Previous literature

The previous literature for this area plus Mexico have been compiled by Pourriot (1982), Rotifera; Frey (1982), Cladocera and Fernando & Smith (1982), Copepoda. Some of the literature consists

Hydrobiologia 113, 105–119 (1984).

of reports based on relatively few and sporadic collections from miscellaneous habitats. To this category belong the following papers: Biochina (1976), Brehm (1932) Dodds (1926) Herbst (1960), Kiefer (1933, 1934), Marsh (1913, 1920, 1931), Orghidan & Negrea (1970, 1973), Richard (1895), Ridder (1966), Straškrabá (1969), Van de Velde *et al* (1978), Wilson (1942) and Wilson (1941). The relatively few lakes in the region have received attention from a number of workers: Brehm (1939), Brinson & Nordlie (1974), Deevey (1957), Deevey *et al* (1980), Edmondson (1933), Goulden (1966), Juday (1915) Packham & Dineen (1931) and Zaret (1975). These studies give us an idea of the composition of the limnetic zooplankton. There are a few detailed studies based on a substantial number of zooplankton samples collected over a wide geographic area, albeit in one country sometimes. These are the studies of Candelas (1956), Harring (1915), Kiefer (1969), Lindberg (1954b), Pourriot (1975) Ridder (1977) and Smith & Fernando (1978). Frey (1980) has recently described a non-swimming chydorid cladoceran from the rainforests in the region. Harring (1915), Kiefer (1936), Lindberg (1954a, 1954b), Smith & Fernando (1978) and Straškrabá (1936) have given data on the zoogeography of the species found in the region.

Materials and methods

Zooplankton samples were collected by two of the authors (C. C. and C. H. F.) in Costa Rica, Cuba and Haiti. Samples were received from the Bahamas, El Salvador and Trinidad. A list of species identified in zooplankton samples collected in Jamaica was supplied by Mr. A. C. Mamaril. Cladocera from Cuban samples were listed by Dr. V. Kořínek. All these samples were collected with a 25 mesh (64 μm) net and preserved in 5–10% formalin. The areas where our samples were collected in Costa Rica, Cuba and Haiti are shown in Fig. 1. All the localities in the region where previous samples have been collected are shown in Fig. 2.

Results and discussion

Lists of Rotifera, Cladocera and Copepoda (Cyclopoida and Calanoida) recorded by us and previous workers are given in Tables 1–3. Since Ostracoda are often found in tropical zooplankton we have listed the forms identified in the region (Table 4). Species numbers identified by us and other workers are given in Tables 5 and 6. It is clear that the number of species recorded from each country or group of islands is relatively low when compared with records from tropical and subtropical regions which have been intensely studied, e.g. Malaysia (Fernando & Ponyi 1981, Fernando & Zankai 1981, Idris & Fernando 1981). The only country where intensive and extensive collecting has been done is Cuba. Here, except for the Rotifera which have not been studied in detail, the numbers of Cladocera (54) and cyclopoid and calanoid Copepoda (26) are roughly comparable to Malaysian records. However, the Cuban fauna is not tropical but subtropical in composition, with four *Daphnia* spp. and temperate region cyclopoids (Smith & Fernando, 1980). The sampling in Costa Rica was intense though restricted to a relatively limited area. Over 200 samples were examined. The relatively low number of species of Rotifera, Cladocera and Copepoda recorded is surprising. There is a distinct possibility that the zooplankton of Central America and the Carribean is not as diverse as tropical areas such as Malaysia, which has a much greater diversity of freshwater habitats. The low number of calanoids recorded (Table 3) is noteworthy. Only one species, *Prionodiaptomus colombiensis* is South American. The others are either endemic e.g. *Mastigodiaptomus amatitlanensis, Diaptomus gatunensis* or found in Mexico too, e.g. *Mastigodiaptomus albuquerquemis, Leptodiaptomus siciliodes*. Lai & Fernando (1981) suggested that the low number of Calanoida is indicative of oligotrophic conditions in South East Asia. Whether this is so in Central America and the Caribbean or due to historical reasons is not clear. The data on Calanoida species and their distribution is too meagre for any definite conclusions.

The total number of cladoceran species in the region (107), is a respectable total considering the total latitude and the total land mass. In comparison Europe has 150 species, Australia 130 species and the subcontinent of India 130 species. However the records in any single country or group of islands are low in species. This is probably due largely to low intensity of collecting. The virtual absence of *Daphnia* in the more tropical regions is

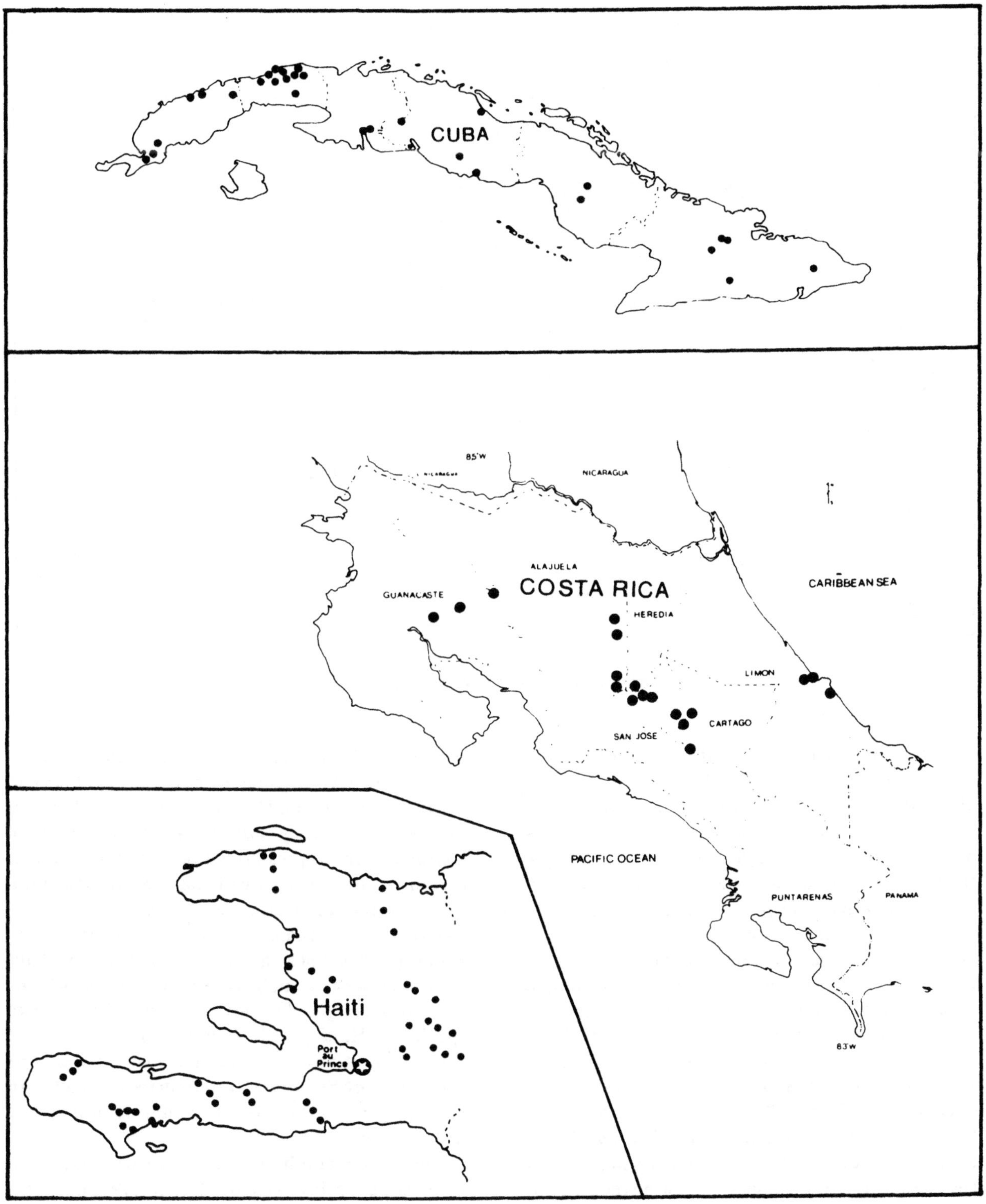

Fig. 1. Sampling localities (●) of material studied by us.

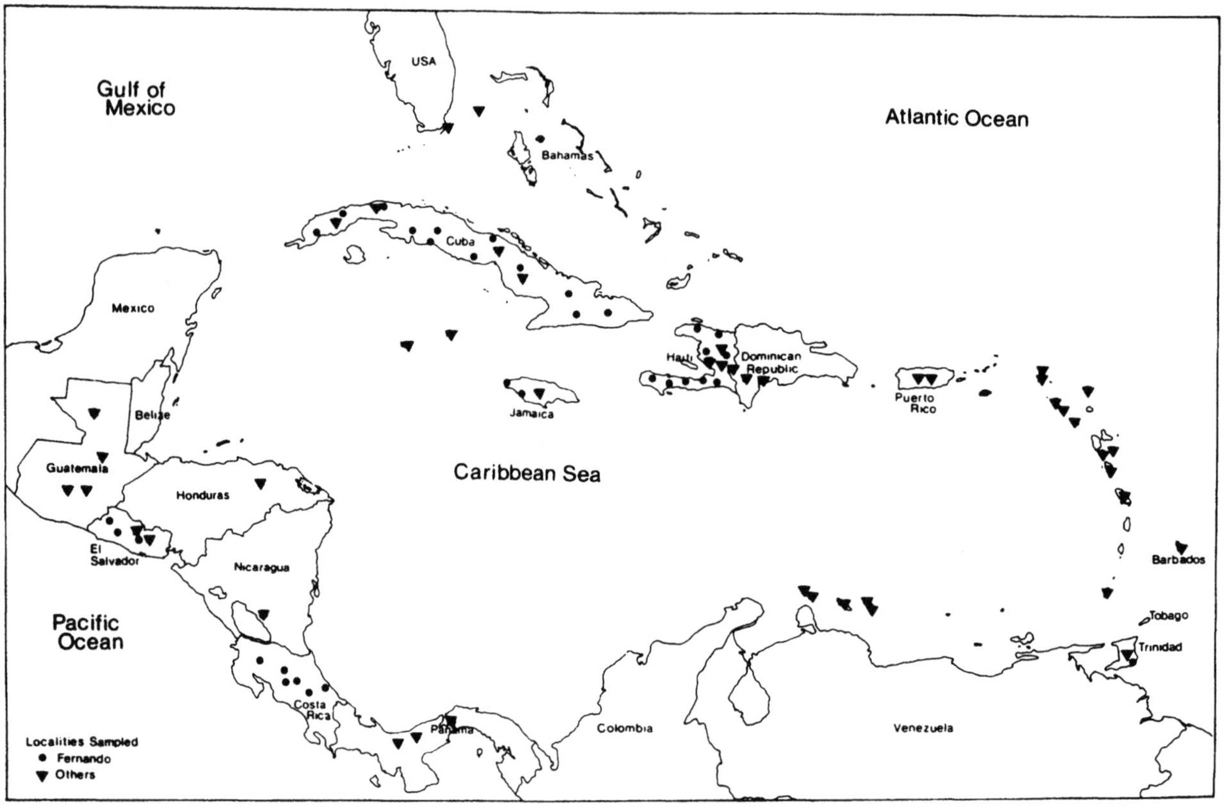

Fig. 2. All sampling localities for freshwater zooplankton in the region.

not surprising. In tropical rainforest regions *Daphnia* is absent to all intents and purposes (Dumont, 1979; Idris & Fernando, 1981). Also missing are the carnivorous Cladocera *Leptodora* and *Polyphemus*. Fernando (1980a, b) has analysed the composition of tropical freshwater zooplankton and pointed out the rarity of *Daphnia* spp, the low diversity of limnetic crustaceans and the absence of carnivorous Cladocera. These features hold true for the tropical low lands of the area studied in the present investigation. Some probable reasons for the low diversity were also explored (Fernando, 1980b).

In the Rotifera the genus *Brachionus* is well represented (13 species) as in other tropical regions (Pejler, 1977). Especially marked is the diversity of the genus *Lecane* (53 species). The low number of species recorded, except in Panama where 101 of the 185 species for the region have been recorded, is probably due to the use of large mesh plankton nets and the difficulties of identification.

These are a few especially interesting features in the zooplankton composition. This includes the occurrence of *Cypria pelagica* as a dominant species in lakes of the Peten District, Guatemala throughout the year (Deevey *et al*, 1980). The occurrence of Ostracoda in tropical freshwater zooplankton is certainly more widespread than the records indicate. Another interesting feature concerns the Hispaniolan lakes studied by Edmondson (1933) and Kiefer (1938). These are relict salt lakes and have halophile species such as *Apocyclops dengizicus* and *Halicyclops*. The low number of Cladocera species in Haiti (Table 4) is probably a result of saline conditions in these and costal lakes. The occurrence of *Pseudodiaptomus* in two Costa Rican habitats, one of them far from marine influence is noteworthy.

The zoogeography of freshwater crustaceans of Cuba has been discussed by Harring (1915), Straškrabá (1969) and Smith & Fernando (1978). The three components of the fauna – North American, South American and endemic – can be easily distin-

Table 1. List of Rotifera recorded in Central America by Fernando, and co-workers (F) and by others (O).

	Species	Aruba	Bonaire	Curacao	Bahamas	Belize	Caribbean Islands	Costa Rica	Cuba	Dominican Republic	El Salvador	Guatemala	Haiti	Honduras	Jamaica	Nicaragua	Panama	Puerto Rico	Trinidad
1.	*Anuraeopsis fissa*	—	—	—	—	—	—	—	—	—	—	—	—	—	—	—	O	—	—
2.	*A. coelata*	—	—	—	F	—	—	F	—	—	—	—	—	—	—	—	—	—	—
3.	*Ascomorpha saltans*	—	—	—	—	—	O	—	—	—	—	—	—	—	—	—	—	—	—
4.	*Asplanchna brightwelli*	O	—	—	—	—	O	F	F	—	F	O	F,O	—	—	—	O	—	—
5.	*A. priodonta*	—	—	—	—	—	—	—	—	—	—	—	—	—	—	O	—	—	—
6.	*A. sieboldi*	—	—	—	—	—	—	F	—	—	—	—	—	—	—	—	—	—	—
7.	*Beauchampiella eudactylota*	—	—	—	—	—	—	—	—	—	—	—	—	—	—	—	O	—	—
8.	*Brachionus angularis*	O	O	O	—	—	O	F	F	—	F	—	—	—	—	—	—	—	—
9.	*B. budapestinensis*	—	—	—	—	—	—	—	F	—	F	—	—	—	—	—	—	—	—
10.	*B. calyciflorus*	—	O	—	—	—	O	F	F	—	F	—	F	—	—	O	—	—	—
11.	*B. caudatus*	—	—	—	—	—	O	F	—	—	—	—	—	—	—	O	—	—	—
12.	*B. dolabratus*	—	—	—	—	—	—	—	—	—	—	—	—	—	—	—	O	—	—
13.	*B. havanensis*	—	O	—	—	—	O	F	F	—	F	O	—	—	—	O	—	—	—
14.	*B. mirabilus*	—	—	—	—	—	O	—	—	—	—	—	—	—	—	—	—	—	—
15.	*B. patulus*	—	—	O	—	—	O	—	—	—	F	O	F	—	F	—	—	—	—
16.	*B. plicatilis*	O	O	O	—	—	—	F	—	O	—	—	O	—	—	O	—	—	—
17.	*B. quadridentatus*	O	—	—	—	—	—	F	F	—	—	—	—	—	—	—	—	—	—
18.	*B. rubens*	O	O	—	—	—	—	F	—	—	—	—	—	—	—	—	—	—	F
19.	*B. urceolaris*	—	—	—	—	—	—	F	—	—	—	—	—	—	F	—	—	—	—
20.	*B. urceus*	—	—	—	—	—	—	—	—	—	—	—	O	—	—	—	O	—	—
21.	*Cephalodella auriculata*	—	—	—	—	—	—	—	—	—	—	—	—	—	—	—	O	—	—
22.	*C. forficata*	—	—	—	—	—	—	—	—	—	—	—	—	—	—	—	O	—	—
23.	*C. gibba*	—	—	—	—	—	O	—	—	—	—	—	—	—	—	—	O	—	—
24.	*C. gracilis*	—	—	—	—	—	—	—	—	—	—	—	—	—	—	—	O	—	—
25.	*C. tenuiseta*	—	—	—	—	—	O	—	—	—	—	—	—	—	—	—	—	—	—
26.	*Chromogaster ovalis*	—	—	—	—	—	O	—	—	—	—	—	—	—	—	—	—	—	—
27.	*Collotheca coroneta*	—	—	—	—	—	—	—	—	—	—	—	O	—	—	—	—	—	—
28.	*C. polyphema*	—	—	—	—	—	—	—	—	—	—	—	—	—	—	—	O	—	—
29.	*Colurella adriatica*	—	—	O	—	—	O	—	—	—	—	—	—	—	—	—	—	—	—
30.	*C. althausae*	—	—	O	—	—	—	—	—	—	—	—	—	—	—	—	—	—	—
31.	*C. anodonta*	—	—	O	—	—	O	—	—	—	—	—	—	—	—	—	—	—	—
32.	*C. colurus*	O	O	O	—	—	O	—	—	—	—	—	—	—	—	—	—	—	—
33.	*C. obtusa*	—	—	O	—	—	O	—	—	—	—	—	—	—	—	—	—	—	—
34.	*C. unicinata*	—	—	—	—	—	O	—	—	—	—	—	—	—	—	—	—	—	—
35.	*Conochilus dossuarius*	—	—	—	—	—	—	F	—	—	—	—	—	—	—	—	—	—	F
36.	*C. hippocrepis*	—	—	—	—	—	—	—	—	—	—	—	—	—	—	—	O	—	—
37.	*C. lyspoenepis*	—	—	—	—	—	—	F	—	—	—	—	—	—	—	—	—	—	—
38.	*C. natans*	—	—	—	—	—	—	F	—	—	—	—	—	—	—	—	—	—	—
39.	*C. unicornis*	—	—	—	—	—	—	F	—	—	—	—	—	—	—	—	—	—	—
40.	*Dicranophorus epicharis*	—	—	—	—	—	O	—	—	—	—	—	—	—	—	—	—	—	—
41.	*D. forcipatus*	—	—	—	—	—	—	—	—	—	—	—	—	—	—	—	O	—	—
42.	*D. robustus*	—	—	—	—	—	—	F	F	—	—	—	—	—	—	—	—	—	—
43.	*D. tegillus*	—	—	—	—	—	—	F	—	—	—	—	—	—	—	—	—	—	—
44.	*Dipleuchlanis propatula*	—	—	—	—	—	O	—	F	—	—	—	O	—	—	—	O	—	F
45.	*Dissotrocha macrostyla*	—	—	—	—	—	—	—	—	—	—	—	—	—	—	—	O	—	—
46.	*Encentrum felis*	—	—	—	—	—	—	—	—	—	—	—	—	—	—	—	O	—	—
47.	*Epiphanes clavulata*	—	—	—	—	—	O	—	—	—	—	—	—	—	—	—	O	—	—
48.	*Euchlanis dilatata*	O	O	O	—	—	O	F	—	—	—	—	O	—	—	—	O	—	F
49.	*E. lyra*	O	—	O	—	—	—	—	—	—	—	—	—	—	—	—	—	—	—
50.	*E. oropha*	—	—	—	—	—	—	—	—	—	—	—	O	—	—	—	O	—	—

Table 1. Continued.

	Species	Aruba	Bonaire	Curacao	Bahamas	Belize	Caribbean Islands	Costa Rica	Cuba	Dominican Republic	El Salvador	Guatemala	Haiti	Honduras	Jamaica	Nicaragua	Panama	Puerto Rico	Trinidad
51.	*E. parva*	—	O	—	—	—	—	—	—	—	—	—	—	—	—	—	—	—	—
52.	*E. perpusilla*	O	—	—	—	—	—	—	—	—	—	—	—	—	—	—	—	—	—
53.	*E. plicata*	—	—	—	—	—	—	—	—	—	—	—	—	—	—	—	O	—	—
54.	*E. triquetra*	—	—	—	—	—	—	—	—	—	—	—	—	—	—	—	O	—	—
55.	*Filinia longiseta*	—	—	—	—	—	O	F	F	—	F	O	—	—	—	—	—	—	—
56.	*F. opoliensis*	—	—	—	—	—	—	F	—	—	—	—	—	—	—	—	—	—	—
57.	*F. pejleri*	—	—	—	—	—	—	F	—	—	F	—	—	—	—	—	—	—	F
58.	*F. terminalis*	—	—	—	—	—	—	F	—	—	—	—	—	—	—	—	—	—	—
59.	*Floscularia ringens*	—	—	—	—	—	—	—	—	—	—	—	—	—	—	—	O	—	—
60.	*Hexarthra fennica*	O	—	O	—	—	O	—	—	—	O	—	—	—	—	O	—	—	—
61.	*H. intermedia*	O	O	—	—	—	O	F	F	—	—	—	—	—	—	—	—	—	—
62.	*H. mira*	—	—	—	—	—	—	—	F	—	—	—	—	—	—	—	—	—	—
63.	*Keratella americana*	—	—	—	—	—	—	F	F	—	—	—	—	—	—	O	—	—	
64.	*K. cochlearis*	—	—	—	—	—	—	—	F	—	F	—	—	—	—	—	—	—	—
65.	*K. lenzi*	—	—	—	—	—	O	—	—	—	—	—	—	—	—	—	—	—	—
66.	*K. procurva*	—	—	—	—	—	O	—	—	—	—	—	—	—	—	—	—	—	—
67.	*K. quadrata*	—	—	—	—	—	—	—	—	—	—	—	—	—	—	—	O	—	—
68.	*K. stipitata*	—	—	—	—	—	—	—	—	—	—	O	—	—	—	—	O	—	—
69.	*K. taurocephala*	—	—	—	—	—	—	—	—	—	F	—	—	—	—	—	—	—	—
70.	*Lacinularia flosculosa*	—	—	—	—	—	—	F	—	—	—	—	—	—	—	—	—	—	—
71.	*Lecane aculeata*	—	O	O	—	—	O	—	—	—	—	—	—	—	—	—	—	—	—
72.	*L. amorpha*	—	—	—	—	—	—	—	—	—	—	—	—	—	—	—	O	—	—
73.	*L. arcula*	—	—	—	—	—	—	—	—	—	—	—	—	—	—	—	O	—	—
74.	*L. arcuata*	—	O	O	—	—	O	—	—	—	—	—	—	—	—	—	—	—	—
75.	*L. bifurca*	—	—	—	—	—	—	—	—	—	—	—	O	—	—	—	O	—	—
76.	*L. bondi*	—	—	—	—	—	—	—	—	—	—	—	O	—	—	—	—	—	—
77.	*L. bulla*	—	O	—	—	—	O	F	F	—	—	—	O	—	F	O	O	—	F
78.	*L. closterocerca*	—	—	—	—	—	—	—	—	—	—	—	—	—	—	—	O	—	—
79.	*L. compta*	—	—	—	—	—	—	—	—	—	—	—	—	—	—	—	O	—	—
80.	*L. cornuta*	O	O	O	—	—	O	—	—	—	—	—	—	—	—	—	O	—	—
81.	*L. crenata*	—	—	O	—	—	—	—	—	—	—	—	O	—	—	—	—	—	—
82.	*L. crepida*	—	—	—	—	—	—	—	—	—	—	—	—	—	—	—	O	—	—
83.	*L. curvicornis*	—	—	—	—	—	O	—	—	—	—	—	O	F	—	—	O	—	—
84.	*L. decipiens*	—	—	—	—	—	—	—	—	—	—	—	—	—	—	—	O	—	—
85.	*L. doryssa*	—	—	—	—	—	—	—	—	—	—	—	—	—	—	—	O	—	—
86.	*L. elegans*	—	—	—	—	—	—	—	—	—	—	—	—	—	—	—	Ó	—	—
87.	*L. ercodes*	—	—	—	—	—	—	—	—	—	—	—	—	—	—	—	O	—	—
88.	*L. flexilis*	—	—	—	—	—	O	—	—	—	—	—	O	—	—	—	O	—	—
89.	*L. furcata*	—	—	—	—	—	O	—	—	—	—	—	—	—	—	—	O	—	—
90.	*L. grandis*	O	O	—	—	—	O	—	—	—	—	—	—	—	—	—	—	—	—
91.	*L. hamata*	—	—	—	—	—	—	—	—	—	—	—	—	—	—	—	—	—	—
92.	*L. harringi*	O	O	O	—	—	O	—	—	—	—	—	—	—	—	—	—	—	—
93.	*L. hornemanni*	—	—	—	—	—	O	—	—	—	—	—	O	—	—	—	O	—	—
94.	*L. hummelincki*	—	—	O	—	—	—	—	—	—	—	—	—	—	—	—	—	—	—
95.	*L. inermis*	O	O	O	—	—	O	—	—	—	—	—	—	—	—	—	O	—	—
96.	*L. leontina*	—	—	—	—	—	O	—	F	—	—	—	F,O	—	F	—	O	—	—
97.	*L. ludwigii*	—	—	—	—	—	—	—	—	—	—	—	—	—	—	—	O	—	—
98.	*L. luna*	O	O	—	—	—	O	F	F	—	—	—	O	—	—	—	O	—	—
99.	*L. lunaris*	O	O	O	—	—	O	—	—	—	—	—	O	—	—	—	O	—	—
100.	*L. marshi*	—	—	—	—	—	—	—	—	—	—	—	—	—	—	—	O	—	—

Table 1. Continued.

	Species	Aruba	Bonaire	Curacao	Bahamas	Belize	Caribbean Islands	Costa Rica	Cuba	Dominican Republic	El Salvador	Guatemala	Haiti	Honduras	Jamaica	Nicaragua	Panama	Puerto Rico	Trinidad
101.	*L. monostyla*	—	—	—	—	—	—	—	—	—	—	—	—	—	—	—	O	—	—
102.	*L. nana*	—	—	—	—	—	O	—	—	—	—	—	O	—	—	—	O	—	—
103.	*L. obtusa*	—	—	—	—	—	O	—	—	—	—	—	—	—	—	—	O	—	—
104.	*L. ohionensis*	—	—	—	—	—	—	—	—	—	—	—	O	—	—	—	—	—	—
105.	*L. papuana*	—	—	—	—	—	O	—	—	—	F	—	—	—	F	—	O	—	—
106.	*L. plesia*	—	—	—	—	—	O	—	—	—	—	—	—	—	—	—	—	—	—
107.	*L. pleonensis*	—	—	—	—	—	—	—	F	—	—	—	O	—	—	—	O	—	—
108.	*L. punctata*	—	O	—	—	—	—	—	—	—	—	—	—	—	—	—	—	—	—
109.	*L. pusilla*	—	—	—	—	—	—	—	—	—	—	—	—	—	—	—	O	—	—
110.	*L. pyriformis*	—	O	—	—	—	O	—	—	—	—	—	—	—	—	—	O	—	—
111.	*L. quadridenta*	—	—	—	—	—	O	—	—	—	—	—	—	—	F	—	O	—	—
112.	*L. rotundata*	—	—	—	—	—	—	—	—	—	—	O	—	—	—	—	—	—	
113.	*L. rugosa*	—	—	—	—	—	—	—	—	—	—	—	—	—	—	—	O	—	—
114.	*L. scutata*	O	O	O	—	—	O	—	—	—	—	—	O	—	—	—	—	—	—
115.	*L. sibina*	—	—	—	—	—	—	—	—	—	—	—	O	—	—	—	O	—	—
116.	*L. spinfera*	—	—	—	—	—	—	—	—	—	—	—	O	—	—	—	—	—	—
117.	*L. strenroosi*	—	—	—	—	—	—	F	—	—	—	—	—	—	—	—	—	—	—
118.	*L. styrax*	O	—	—	—	—	—	—	—	—	—	—	—	—	—	—	—	—	—
119.	*L. tenuiseta*	—	—	—	—	—	O	—	—	—	—	—	—	—	—	—	O	—	—
120.	*L. thalera*	O	O	O	—	—	O	—	—	—	—	—	—	—	—	—	—	—	—
121.	*L. ungulata*	O	O	—	—	—	O	—	—	—	—	—	—	—	—	—	—	—	—
122.	*L. unquitata*	—	—	—	—	—	—	—	—	—	—	—	—	—	—	—	—	—	F
123.	*L. virga*	—	—	—	—	—	—	—	—	—	—	—	—	—	—	—	O	—	—
124.	*Lepadella acuminata*	—	—	O	—	—	—	—	—	—	—	—	—	—	—	—	—	—	—
125.	*L. cryptopus*	—	—	—	—	—	—	—	—	—	—	—	—	—	—	—	O	—	—
126.	*L. imbricata*	—	—	O	—	—	—	—	—	—	—	—	—	—	—	—	O	—	—
127.	*L. myersi*	—	—	—	—	—	—	—	—	—	—	—	O	—	—	—	—	—	—
128.	*L. ovalis*	—	—	—	—	—	O	—	—	—	—	—	—	—	—	—	O	—	—
129.	*L. patella*	O	O	O	—	—	O	—	—	—	—	—	—	—	—	—	O	—	—
130.	*L. persimilis*	—	O	O	—	—	—	—	—	—	—	—	—	—	—	—	—	—	—
131.	*L. triptera*	—	—	—	—	—	—	—	—	—	—	—	—	—	—	—	O	—	—
132.	*Macrochaetus collinsi*	—	—	—	—	—	—	—	—	—	—	—	O	—	—	—	O	—	—
133.	*Limnias ceratophylli*	—	—	—	—	—	—	—	—	—	—	—	—	—	—	—	O	—	—
134.	*Monommata orbis*	—	—	—	—	—	—	—	—	—	—	—	—	—	—	—	O	—	—
135.	*Mytilina acanthophora*	—	—	—	—	—	O	—	—	—	—	—	—	—	—	—	—	—	—
136.	*M. mucronata*	—	—	—	—	—	O	—	—	—	—	—	—	—	—	—	—	—	—
137.	*M. trigona*	—	—	—	—	—	—	—	—	—	—	—	—	—	—	—	O	—	—
138.	*M. ventralis*	—	—	—	—	—	—	—	—	—	—	—	—	—	—	—	O	—	—
139.	*Notholca longispina*	—	—	—	—	—	—	—	—	—	—	—	—	—	—	—	O	—	—
140.	*Notommata aurita*	—	—	—	—	—	—	—	—	—	—	—	—	—	—	—	O	—	—
141.	*N. cerberus*	—	—	—	—	—	—	—	—	—	—	—	—	—	—	—	O	—	—
142.	*N. copeus*	—	—	—	—	—	—	—	—	—	—	—	—	—	—	—	O	—	—
143.	*N. pseudoceberus*	—	—	—	—	—	—	—	—	—	—	—	—	—	—	—	O	—	—
144.	*N. torulosa*	—	—	—	—	—	—	—	—	—	—	—	—	—	—	—	O	—	—
145.	*Philodina macrostyla*	—	—	—	—	—	—	—	—	—	—	—	—	—	—	—	O	—	—
146.	*Platyias leloupi*	—	—	—	—	—	O	—	—	—	—	—	—	—	—	—	—	—	—
147.	*P. patulus*	—	—	—	—	—	O	—	F	—	—	—	O	—	—	—	O	—	—
148.	*P. quadricornis*	—	—	—	—	—	—	F	—	—	—	—	F,O	—	F	—	O	—	F
149.	*Pleurotrocha robusta*	—	—	—	—	—	—	—	—	—	—	—	O	—	—	—	—	—	—
150.	*P. sordida*	—	—	—	—	—	—	—	—	—	—	—	—	—	—	—	O	—	—

Table 1. Continued.

	Species	Aruba	Bonaire	Curacao	Bahamas	Belize	Caribbean Islands	Costa Rica	Cuba	Dominican Republic	El Salvador	Guatemala	Haiti	Honduras	Jamaica	Nicaragua	Panama	Puerto Rico	Trinidad
151.	*Philodina sp.*	—	—	—	F	—	—	—	—	—	—	—	—	—	—	—	—	—	—
152.	*Polyarthra remata*	—	—	—	—	—	O	—	—	—	—	—	—	—	—	O	—	—	—
153.	*P. vulgaris*	—	—	—	—	—	O	F	F	—	F	—	—	—	—	—	O	—	—
154.	*P. trigla*	—	—	—	—	—	—	—	—	—	—	—	—	—	—	—	O	—	—
155.	*Pompholyx complanata*	—	—	—	—	—	—	F	F	—	F	—	F	—	—	—	—	—	—
156.	*Proales sordida*	—	—	—	—	—	—	—	—	—	—	—	—	—	—	—	O	—	—
157.	*Ptygura longipes*	—	—	—	—	—	—	—	—	—	—	—	—	—	—	—	O	—	—
158.	*P. pectinifer*	—	—	—	—	—	—	—	—	—	—	—	—	—	—	—	O	—	—
159.	*Resticula melandocus*	—	—	—	—	—	—	F	—	—	—	—	—	—	—	—	—	—	—
160.	*Rotaria macrura*	—	—	—	—	—	—	—	—	—	—	—	—	—	—	—	O	—	—
161.	*R. neptunia*	—	—	—	—	—	—	—	—	—	—	—	—	—	—	—	O	—	—
162.	*Scaridium longicaudum*	—	—	—	—	—	—	—	—	—	—	—	—	—	—	—	O	—	—
163.	*Sinantherina spinosa*	—	—	—	—	—	—	—	—	—	—	—	—	—	—	—	O	—	—
164.	*Squatinella mutica*	—	—	—	—	—	—	—	—	—	—	—	—	—	—	—	O	—	—
165.	*S. tridentata caribbea*	—	—	—	—	—	—	—	F	—	—	—	—	—	—	—	—	—	—
166.	*Testudinella incisa*	—	—	—	—	—	—	—	—	—	—	—	—	—	—	—	O	—	—
167.	*T. patina*	O	O	O	F	—	O	—	F	—	—	—	O	—	—	—	—	—	—
168.	*T. trilobata*	—	—	—	—	—	—	—	—	—	—	—	—	—	—	—	O	—	—
169.	*Trichocerca bricristata*	—	—	—	—	—	—	—	—	—	—	—	—	—	—	—	O	—	—
170.	*T. brachyura*	—	—	O	—	—	—	—	—	—	—	—	—	—	—	—	O	—	—
171.	*T. dixon-nuttalii*	—	—	—	—	—	—	F	—	—	—	—	—	—	—	—	O	—	—
172.	*T. gracilis*	—	—	—	—	—	—	F	—	—	—	—	—	—	—	—	—	—	—
173.	*T. nitida*	—	—	—	—	—	—	—	—	—	—	—	—	—	—	—	O	—	—
174.	*T. pusilla*	—	—	—	—	—	O	—	—	—	—	—	—	—	—	—	—	—	—
175.	*T. rattus*	—	—	—	—	—	O	—	—	—	—	—	O	—	—	—	O	—	—
176.	*T. similis*	—	—	O	—	—	O	F	—	—	—	—	—	—	—	—	—	—	—
177.	*T. stylata*	—	—	—	—	—	—	—	—	—	—	—	—	—	—	—	O	—	—
178.	*T. tenuior*	—	—	—	—	—	—	—	—	—	—	—	—	—	—	—	O	—	—
179.	*T. tigris*	—	—	—	—	—	O	—	—	—	—	—	—	—	—	—	O	—	—
180.	*T. voluta*	—	—	—	—	—	—	—	—	—	—	—	—	—	—	—	O	—	—
181.	*T. weberi*	—	—	—	—	—	O	—	—	—	—	—	—	—	—	—	O	—	—
182.	*Tripleuchlanis plicata*	O	O	O	—	—	O	—	F	—	—	—	O	—	—	—	O	—	—
183.	*Trichotria tetractis*	—	—	—	—	—	—	—	—	—	—	—	—	—	—	—	O	—	—

Table 2. List of Cladocera recorded in Central America by Fernando, and co-workers (F) and by others (O).

	Species	Aruba	Bonaire	Curacao	Bahamas	Belize	Caribbean Islands	Costa Rica	Cuba	Dominican Republic	El Salvador	Guatemala	Haiti	Honduras	Jamaica	Nicaragua	Panama	Puerto Rico	Trinidad
1.	*Alona affinis*	—	—	—	—	—	—	—	—	—	—	O	—	—	—	—	O	—	—
2.	*A. cambouei*	—	—	—	—	—	—	—	—	—	—	O	—	—	—	—	—	—	—
3.	*A. costata*	—	—	—	—	—	—	—	—	—	—	O	—	—	—	—	—	—	F
4.	*A. davidi*	—	—	—	F	—	—	F	O,F	—	—	—	O,F	—	F	—	—	—	O
5.	*A. dentifera*	—	—	—	—	—	—	—	—	—	—	O	—	—	—	—	O	—	—

Table 2. Continued.

	Species	Aruba	Bonaire	Curacao	Bahamas	Belize	Caribbean Islands	Costa Rica	Cuba	Dominican Republic	El Salvador	Guatemala	Haiti	Honduras	Jamaica	Nicaragua	Panama	Puerto Rico	Trinidad
6.	*A. guttata*	—	—	—	—	—	—	F	—	—	—	—	—	—	—	—	—	—	—
7.	*A. intermedia*	—	—	—	—	—	—	—	—	—	—	—	—	—	—	—	O	—	—
8.	*A. karua*	—	—	—	—	—	—	—	—	—	—	O	—	—	—	—	—	—	—
9.	*A. monocantha*	—	—	—	—	—	—	—	O	—	—	O	—	—	—	—	—	—	—
10.	*A. pulchella*	—	—	—	—	—	—	—	O	—	—	O	—	—	—	—	—	—	—
11.	*A. rectangula*	—	—	—	—	—	—	—	—	—	—	—	—	—	—	—	O	O	—
12.	*A. rustica*	—	—	—	—	—	—	—	O	—	—	—	—	—	—	—	—	—	—
13.	*A. verrucosa*	—	—	—	—	—	—	—	O	—	—	O	—	—	—	—	O	O	—
14.	*Alonella dadayi*	—	—	—	—	—	—	—	—	—	—	—	—	—	—	—	O	—	—
15.	*A. excisa*	—	—	—	—	—	—	—	O	—	—	O	—	—	F	—	—	—	—
16.	*A. hamulata*	—	—	—	—	—	—	—	—	—	—	O	—	—	—	—	—	—	—
17.	*A. lineolata*	—	—	—	—	—	—	—	—	—	F	—	—	—	—	—	—	—	—
18.	*A. pulchella*	—	—	—	—	—	—	—	—	—	—	O	—	—	—	—	—	—	—
19.	*Bosmina hagmanni*	—	—	—	—	—	—	F	—	—	—	O	—	—	—	—	—	—	—
20.	*B. longirostris*	—	—	—	—	—	—	F	O,F	—	O,F	O	—	—	—	—	O	—	—
21.	*B. longispina*	—	—	—	—	—	—	—	—	—	—	—	—	—	—	—	O	—	—
22.	*B. obtusirostris*	—	—	—	—	—	—	—	—	—	—	O	—	—	—	—	—	—	—
23.	*B. tubicen*	—	—	—	—	—	—	F	O	—	—	O	—	—	—	—	O	—	—
24.	*Bosminopsis deitersi*	—	—	—	—	—	—	F	—	—	—	O	—	—	—	—	O	—	—
25.	*Bryospilus repens*	—	—	—	—	—	—	—	—	—	—	—	—	—	—	—	—	O	—
26.	*Camptocercus australis*	—	—	—	—	—	—	—	O	—	—	—	—	—	—	—	—	—	—
27.	*C. rectirostris*	—	—	—	—	—	—	—	O	—	—	O	—	—	—	—	—	—	—
28.	*C. similis*	—	—	—	—	—	—	—	—	—	—	O	—	—	—	—	—	—	—
29.	*C. uncinatus*	—	—	—	—	—	—	F	—	—	—	O	—	—	F	—	—	—	—
30.	*Ceriodaphnia affinis*	—	—	—	—	—	—	—	—	—	F	O	—	—	—	—	—	—	—
31.	*C. cornuta*	—	—	—	F	—	—	F	O,F	—	F	O	F	—	—	—	O	O	F
32.	*C. dubia*	—	—	—	—	—	—	—	O,F	—	—	O	F	—	—	—	—	—	—
33.	*C. lacustris*	—	—	—	—	—	—	—	—	—	—	O	—	—	—	—	—	—	—
34.	*C. pulchella*	—	—	—	—	—	—	—	—	—	O	O	—	—	—	—	—	—	—
35.	*C. quadrangula*	—	—	—	—	—	—	—	—	—	—	O	—	—	—	—	—	—	—
36.	**Chydorus barroisi*	—	—	—	—	—	—	—	O	—	—	O	—	—	F	—	O	—	—
37.	*C. dentifera*	—	—	—	—	—	—	—	—	—	—	—	F	—	—	—	—	—	—
38.	*C. eurynotus*	—	—	—	F	—	—	F	O,F	—	F	O	—	—	—	—	O	—	—
39.	*C. hermanni*	—	—	—	—	—	—	F	—	—	—	—	—	—	—	—	—	—	—
40.	*C. hybridus*	—	—	—	—	—	—	—	O	—	—	O	—	—	—	—	O	—	—
41.	*C. nitidulus*	—	—	—	—	—	—	—	—	—	—	O	—	—	—	—	—	—	—
42.	*C. parvus*	—	—	—	—	—	—	F	—	—	—	—	—	—	—	—	—	—	—
43.	*C. poppei*	—	—	—	—	—	—	—	—	—	—	O	—	—	—	—	—	—	—
44.	*C. pubescens*	—	—	—	—	—	—	—	O	—	F	O	F	—	—	—	—	—	—
45.	*C. reticulatus*	—	—	—	—	—	—	—	—	—	—	—	F	—	—	—	—	—	—
46.	*C. sphaericus*	—	—	—	—	—	—	—	—	—	—	O	—	—	—	—	O	—	—
47.	*Dadaya macrops*	—	—	—	—	—	—	—	—	—	—	—	F	—	—	—	O	—	—
48.	*Daphnia ambigua*	—	—	—	—	—	—	—	O	—	—	—	—	—	—	—	—	—	—
49.	*D. hyalina*	—	—	—	—	—	—	—	—	—	O	O	—	—	—	—	—	—	—
50.	*D. laevis*	—	—	—	—	—	—	—	O,F	—	—	O	—	—	—	—	—	—	—
51.	*D. longispina*	—	—	—	—	—	—	—	—	—	—	O	—	—	—	—	—	—	—
52.	*D. parvula*	—	—	—	—	—	—	—	O	—	—	—	—	—	—	—	—	—	—
53.	*D. pulex*	—	—	—	—	—	—	—	—	—	—	O	—	—	—	—	—	—	—
54.	*D. pulicaria*	—	—	—	—	—	—	—	O	—	—	—	—	—	—	—	—	—	—
55.	*Diaphanosoma brachyurum*	—	—	—	—	—	—	—	O	—	F	O	F	—	—	—	O	—	F
56.	*D. brevireme*	—	—	—	—	—	—	—	O,F	—	—	—	—	—	—	—	—	—	—
57.	*D. fluviatile*	—	—	—	—	—	—	—	—	—	—	O	O	—	—	—	—	—	—

* Some of these species have now been moved into the genus *Ephemeroporus* Frey.

Table 2. Continued.

	Species	Aruba	Bonaire	Curacao	Bahamas	Belize	Caribbean Islands	Costa Rica	Cuba	Dominican Republic	El Salvador	Guatemala	Haiti	Honduras	Jamaica	Nicaragua	Panama	Puerto Rico	Trinidad
58.	*D. spinulosum*	—	—	—	—	—	—	F	O	—	F	O	—	—	—	—	—	—	F
59.	*Disparalona dadayi*	—	—	—	—	—	—	—	O	—	—	O	—	—	—	—	O	—	—
60.	*Dunhevedia crassa*	—	—	—	—	—	—	—	O	—	—	—	F	—	—	—	O	—	—
61.	*D. odontoplax*	—	—	—	—	—	—	—	O	—	—	—	—	—	—	—	O	—	—
62.	*D. serrata*	—	—	—	—	—	—	—	O	—	—	—	F	—	F	—	—	—	—
63.	*Echinisca elegans*	—	—	—	—	—	—	—	O	—	—	—	—	—	—	—	—	—	—
64.	*E. rosea*	—	—	—	—	—	—	—	O,F	—	—	O	F	—	—	—	O	—	—
65.	*E. triserialis*	—	—	—	—	—	—	—	O	—	—	O	—	—	—	—	—	—	—
66.	*Euryalona occidentalis*	—	—	—	—	—	—	—	O,F	—	—	—	—	—	—	—	O	—	—
67.	*E. orientalis*	—	—	—	—	—	—	—	O	—	—	O	—	—	—	—	—	—	—
68.	*Graptoleberis testudinaria*	—	—	—	—	—	—	—	—	—	—	O	—	—	—	—	—	—	—
69.	*Grimaldina brazzai*	—	—	—	—	—	—	F	O	—	—	—	—	—	—	—	O	—	—
70.	*Indialona globulosa*	—	—	—	—	—	—	—	O	—	F	O	—	—	F	—	—	—	—
71.	*Ilyocryptus spinifer*	—	—	—	—	—	—	F	O,F	—	—	O	—	—	F	—	O	O	F
72.	*Kurzia latissima*	—	—	—	—	—	—	—	O	—	—	O	F	—	—	—	O	O	—
73.	*Latonopsis australis*	—	—	—	—	—	—	—	—	—	—	O	—	—	F	—	—	—	—
74.	*L. fasciculata*	—	—	—	—	—	—	—	O	—	—	—	—	—	—	—	—	—	—
75.	*L. occidentalis*	—	—	—	—	—	—	—	O,F	—	—	—	—	—	—	—	—	—	—
76.	*L. serricauda*	—	—	—	—	—	—	—	—	—	—	—	—	—	—	—	—	O	—
77.	*Leydigia acanthocercoides*	—	—	—	—	—	—	—	O	—	F	O	—	—	—	—	O	—	—
78.	*L. leydigi*	—	—	—	—	—	—	—	O	—	—	O	—	—	—	—	—	—	—
79.	*L. megalops*	—	—	—	—	—	—	—	—	—	—	O	—	—	—	—	—	O	—
80.	*Leydigia parva*	—	—	—	—	—	—	—	—	—	—	O	—	—	—	—	—	—	—
81.	*Leydigiopsis ornata*	—	—	—	—	—	—	—	O	—	—	O	—	—	—	—	—	—	—
82.	*Macrothrix goeldi*	—	—	—	—	—	—	—	O,F	—	—	—	—	—	—	—	—	—	—
83.	*M. laticornis*	—	—	—	—	—	—	—	—	—	—	O	—	—	—	—	—	—	—
84.	*M. spinosa*	—	—	—	—	—	—	F	—	—	F	—	—	—	—	—	—	—	F
85.	*M. squamosa*	—	—	—	—	—	—	—	—	—	—	—	—	—	—	—	O	—	—
86.	*M. triserialis*	—	—	—	F	—	—	F	—	—	—	—	—	—	—	—	—	—	—
87.	*Moina affinis*	—	—	—	—	—	—	—	O	—	—	—	—	—	—	—	—	—	—
88.	*M. macrocopa*	—	—	—	—	—	—	—	—	—	—	—	O	—	—	—	—	—	—
89.	*M. micrura*	—	—	—	—	—	—	F	O,F	—	F	O	F	—	—	—	O	O	F
90.	*M. minuta*	—	—	—	—	—	—	—	—	—	—	—	—	—	—	—	O	—	—
91.	*M. reticulata*	—	—	—	—	—	—	F	O	—	—	—	—	—	—	—	—	—	—
92.	*M. wierzejskii*	—	—	—	—	—	—	—	—	—	—	—	O	—	—	—	—	—	—
93.	*Moinodaphnia macleayi*	—	—	—	—	—	—	F	O	—	F	O	F	—	—	—	O	—	F
94.	*Oxyurella ciliata*	—	—	—	—	—	—	—	—	—	—	—	O	—	—	—	—	—	—
95.	*O. longicaudis*	—	—	—	—	—	—	—	—	—	—	O	—	—	—	—	—	—	—
96.	*O. tenuicaudis*	—	—	—	—	—	—	—	O	—	—	O	—	—	—	—	O	—	—
97.	*Pseudochydorus globosus*	—	—	—	—	—	—	—	O	—	—	O	—	—	—	—	—	—	—
98.	*Pseudosida bidentata*	—	—	—	—	—	—	—	O	—	—	O	—	—	—	—	O	—	—
99.	*P. ramosa*	—	—	—	—	—	—	—	O	—	—	O	—	—	—	—	—	—	—
100.	*P. tridentata*	—	—	—	—	—	—	—	—	—	—	O	—	—	—	—	—	—	—
101.	*Scapholeberis mucronata*	—	—	—	—	—	—	—	—	—	—	—	—	—	—	—	—	O	—
102.	*Simocephalus acutirostratus*	—	—	—	—	—	—	F	O	—	—	—	—	—	—	—	—	—	—
103.	*S. elizabethae*	—	—	—	—	—	—	—	O	—	—	—	—	—	—	—	—	—	—
104.	*S. exspinosus*	—	—	—	—	—	—	—	—	—	—	—	—	—	—	—	O	—	—
105.	*S. latirostris*	—	—	—	—	—	—	F	—	—	—	—	—	—	—	—	—	—	—
106.	*S. serrulatus*	—	—	—	—	—	—	F	O,F	—	—	O	—	—	—	—	O	—	—
107.	*S. vetulus*	—	—	—	—	—	—	—	O,F	—	—	—	—	—	O	—	—	—	—

* *E. triserialis* = *M. triserialis; C. dentifera* = *A. dentifera; Alonella pulchella* = *Alona pulchella.*

Table 3. List of Copepoda recorded in Central America by Fernando, and co-workers (F) and by others (O).

	Species	Aruba	Bonaire	Curacao	Bahamas	Belize	Caribbean Islands	Costa Rica	Cuba	Dominican Republic	El Salvador	Guatemala	Haiti	Honduras	Jamaica	Nicaragua	Panama	Puerto Rico	Trinidad
1.	*Acanthocyclops vernalis*	—	—	—	—	—	—	—	F	—	—	—	—	—	—	—	—	—	—
2.	*Arctodiaptomus dorsalis*	—	—	—	—	—	—	F	F	—	—	—	—	—	—	—	—	—	—
3.	*A. asymmetricus*	—	—	—	—	—	—	—	O,F	—	—	—	—	—	—	—	—	—	—
4.	*Cyclops vicinus*	—	—	—	—	—	—	—	F	—	—	—	—	—	—	—	—	—	—
5.	*Diatomus dampfi*	—	—	—	—	—	—	—	—	—	—	O	—	—	—	—	O	—	—
6.	*D. gatunensis*	—	—	—	—	—	—	—	—	—	—	—	—	—	—	—	O	—	—
7.	*D. leoninicollis*	—	—	—	—	—	—	—	—	—	—	—	—	—	—	—	O	—	—
8.	*D. marshi*	—	—	—	—	—	—	—	—	—	O,F	—	—	O	—	—	O	—	—
9.	*D. silvaticus*	—	—	—	—	—	—	—	—	—	—	—	—	—	—	—	—	—	O
10.	*Ectocyclops phaleratus*	—	—	—	—	—	—	—	F	—	—	O	F	—	—	—	O	—	—
11.	*Eucyclops ariguanabensis*	—	—	—	—	—	—	—	O	—	—	—	—	—	—	—	—	—	—
12.	*E. bondi*	—	—	—	—	—	—	—	F	O	—	—	O	—	—	—	—	—	F
13.	*E. ensifer*	—	—	—	—	—	—	—	—	—	—	—	—	—	—	O	—	—	—
14.	*E. extensus*	—	—	—	—	—	—	—	—	—	—	—	O	—	—	—	—	—	—
15.	*E. prasinus*	—	—	—	—	—	—	—	—	—	—	O	—	O	—	—	O	—	—
16.	*E. serrulatus*	—	—	—	—	—	—	F	F	—	—	O	F	O	—	—	O	—	—
17.	*E. speratus*	—	—	—	—	—	—	—	F	—	—	—	—	—	—	—	—	—	—
18.	*Halicyclops aequoreus*	—	—	—	—	—	—	—	—	—	—	—	—	—	—	—	O	—	—
19.	*H. exiguus*	—	—	—	—	—	—	—	—	—	—	—	O	—	—	—	—	—	—
20.	*H. similis*	—	—	—	—	—	—	—	—	—	—	—	O	—	—	—	—	—	—
21.	*Leptodiatomus siciloides*	—	—	—	—	—	—	—	—	—	O	—	—	—	—	—	—	—	—
22.	*Macrocyclops albidus*	—	—	—	F	—	—	—	F	O	O	O	F	O	—	O	O	—	—
23.	*M. ater*	—	—	—	—	—	—	—	F	—	—	—	—	—	—	—	O	—	—
24.	*Mastigodiaptomus amatitlanensis*	—	—	—	—	—	—	—	—	—	—	O	—	—	—	—	—	—	—
25.	*Mastigodiaptomus albuquerquensis*	—	—	—	—	—	—	—	F	—	—	O	F	—	—	—	—	—	—
26.	*M. purpureus*	—	—	—	—	—	—	—	O,F	—	—	—	F	—	—	—	—	—	—
27.	*Mesocyclops brasilianus*	—	—	—	—	—	—	F	—	—	—	—	—	—	—	O	—	—	—
28.	*M. dybowski*	—	—	—	—	—	—	—	—	—	—	—	—	—	—	—	—	—	F
29.	*M. edax*	—	—	—	—	—	—	—	F	—	—	—	—	—	—	—	—	—	—
30.	*M. ellipticus*	—	—	—	—	—	—	—	F	—	—	—	—	—	—	—	—	—	—
31.	**M. leukarti*	—	—	—	—	—	—	F	F	—	O,F	—	F	O	—	—	O	—	F
32.	*M. longisetus*	—	—	—	—	—	—	—	F	—	F	—	O	—	—	—	—	—	—
33.	*M. meridianus*	—	O	—	—	—	O	—	—	—	—	—	—	—	—	—	—	—	—
34.	*M. nicaraguensis*	—	—	—	—	—	—	—	—	—	—	—	—	—	—	O	—	—	—
35.	*Microcyclops anceps*	—	—	—	—	—	—	—	—	—	—	—	—	—	—	—	O	—	—
36.	*M. bicolor*	—	—	—	—	—	—	—	—	—	—	O	—	—	—	—	—	—	—
37.	*M. ceibaensis*	—	—	—	—	—	—	—	F	—	—	—	—	O	—	—	—	—	—
38.	*M. dengizicus*	—	—	—	—	—	—	—	—	—	—	—	O	—	—	—	—	—	—
39.	*M. diversus*	—	—	O	—	—	—	—	—	—	—	—	O	—	—	—	—	—	—
40.	*M. dubitabilis*	—	—	O	—	—	—	—	—	O	—	—	O	—	—	—	—	—	—
41.	*M. mendocinus*	—	—	—	—	—	—	—	F	—	—	—	O	—	—	O	—	—	—
42.	*M. panamensis*	—	—	O	—	—	—	—	F	O	—	—	O	O	—	—	O	—	—
43.	*M. varicans*	—	—	—	—	—	—	F	F	—	F	—	O,F	O	—	—	O	—	F
44.	*Metacyclops hartmanni*	O	—	—	—	—	—	—	—	—	—	—	—	—	—	O	—	—	—
45.	*M. viduus*	O	—	—	—	—	—	—	—	—	—	—	—	—	—	—	—	—	—
46.	*Oithona nana*	—	—	—	—	—	—	—	—	—	—	—	—	—	—	—	O	—	—

* Not found outside Europe. Records refer mainly to *T. thermocyclopoides*.

Table 3. Continued.

	Species	Aruba	Bonaire	Curacao	Bahamas	Belize	Caribbean Islands	Costa Rica	Cuba	Dominican Republic	El Salvador	Guatemala	Haiti	Honduras	Jamaica	Nicaragua	Panama	Puerto Rico	Trinidad
47.	*O. plumifera*	—	—	—	—	—	—	—	—	—	—	—	—	—	—	—	O	—	—
48.	*O. setiger*	—	—	O	—	—	—	—	—	—	—	—	—	—	—	—	—	—	—
49.	*O. similis*	—	—	—	—	—	—	—	—	—	—	—	—	—	—	—	O	—	—
50.	*O. spinirostris*	—	—	—	—	—	—	—	—	—	—	—	—	—	—	—	O	—	—
51.	*Paracyclops fimbriatus*	—	—	—	—	—	—	—	F	—	—	O	—	—	—	O	—	—	—
52.	*Pronodiaptomus colombiensis*	—	—	—	—	—	—	—	—	—	F	O	—	—	—	—	—	—	—
53.	*Pseudodiaptomus cristobelensis*	—	—	—	—	—	—	—	—	—	—	—	—	—	—	—	O	—	—
54.	*P. culebrensis*	—	—	—	—	—	—	F	—	—	—	—	—	—	—	—	—	—	—
55.	**Thermocyclops crassus*	—	—	—	—	—	—	F	F	—	F	—	F	—	—	—	—	—	F
56.	*T. decipiens*	—	O	—	—	—	O	F	—	—	—	—	—	—	—	—	—	—	—
57.	*T. inversus*	—	—	—	—	—	—	F	F	—	F	—	—	—	—	—	—	—	—
58.	*T. orthonoides*	—	—	—	—	—	—	—	—	—	—	—	O	—	—	—	—	—	—
59.	*T. tenuis*	—	—	—	—	—	—	F	F	—	O	—	—	—	—	—	O	—	F
60.	*Tropocyclops prasinus*	—	—	—	—	—	—	—	F	—	—	—	F	—	—	—	O	—	—

* Not found outside Europe. Records refer mainly to *T. thermocyclopoides*.

Table 4. List of Ostracoda recorded in Central America by Fernando, and co-workers (F) and by others (O).

	Species	Aruba	Bonaire	Curacao	Bahamas	Belize	Caribbean Islands	Costa Rica	Cuba	Dominican Republic	El Salvador	Guatemala	Haiti	Honduras	Jamaica	Nicaragua	Panama	Puerto Rico	Trinidad
1.	*Chrissia humilis indica*	—	—	—	—	—	—	—	—	—	F	—	—	—	—	—	—	—	—
2.	*Cypretta sp.*	—	—	—	—	—	—	—	—	—	—	—	F	—	—	—	—	—	—
3.	*C. margalefi*	—	—	—	—	—	—	—	O	—	—	—	—	—	—	—	—	—	—
4.	*Cypria sp.*	—	—	—	—	—	—	—	—	—	—	—	F	—	—	—	—	—	—
5.	*Cypria pelagica*	—	—	—	—	—	—	—	—	—	—	O	—	—	—	—	—	—	—
6.	*Cypria petenensis*	—	—	—	—	—	—	—	—	—	F	—	—	—	—	—	—	—	—
7.	*Cypridopsis*	—	—	—	—	—	—	—	—	—	—	—	F	—	—	—	—	—	—
8.	*Cytheridella ilosvayi*	—	—	—	—	—	—	—	O	—	—	—	—	—	—	—	—	—	—
9.	*Dolerocypris maya*	—	—	—	—	—	—	—	—	—	—	O	—	—	—	—	—	—	—
10.	*Entocythere hamata*	—	—	—	—	—	—	—	O	—	—	—	—	—	—	—	—	—	—
11.	*Entocythere heterodonta cubensis*	—	—	—	—	—	—	—	O	—	—	—	—	—	—	—	—	—	—
12.	*Entocythere ruibali*	—	—	—	—	—	—	—	O	—	—	—	F	—	—	—	—	—	—
13.	*Heterocypris sp.*	—	—	—	—	—	—	—	—	—	F	—	F	—	—	—	—	—	—
14.	*Ilyocypris sp.*	—	—	—	—	—	—	—	—	—	—	O	—	—	—	—	—	—	—
15.	*Limnocythere opesta*	—	—	—	—	—	—	—	—	—	—	—	F	—	—	—	—	—	—
16.	*Physocypria sp.*	—	—	—	—	—	—	—	—	—	—	—	F	—	—	—	—	—	—
17.	*Potamocypris sp.*	—	—	—	—	—	—	—	—	—	—	—	F	—	—	—	—	—	—
18.	*Strandesia sp.*	—	—	—	—	—	—	—	—	—	—	—	F	—	—	—	—	—	—
19.	*Stenocypris malcolmsoni*	—	—	—	—	—	—	—	O	—	—	—	—	—	—	—	—	—	—

Table 5. Total number of species recorded in Central America by all workers.

Species	Aruba	Bonaire	Curacao	Bahamas	Belize	Caribbean Islands	Costa Rica	Cuba	Dominican Republic	El Salvador	Guatemala	Haiti	Honduras	Jamaica	Nicaragua	Panama	Puerto Rico	Trinidad
Rotifera	24	28	29	3	-	63	40	22	1	13	5	35	-	7	11	101	-	8
Cladocera	-	-	-	4	-	-	22	54	-	15	63	18	-	9	-	35	10	9
Copepoda	2	2	-	1	-	2	9	56	5	10	10	23	8	-	8	22	-	7
Ostracoda	-	-	-	-	-	-	-	-	6	3	3	9	-	-	-	-	-	-
Bryozoa (*Plumatella*)	-	-	-	-	-	-	-	X	-	-	-	-	-	-	-	-	-	X

Table 6. Total number of species recorded in Central America by Fernando and co-workers.

Species	Aruba	Bonaire	Curacao	Bahamas	Belize	Caribbean Islands	Costa Rica	Cuba	Dominican Republic	El Salvador	Guatemala	Haiti	Honduras	Jamaica	Nicaragua	Panama	Puerto Rico	Trinidad
Rotifera	-	-	-	3	-	-	40	22	-	13	-	6	-	7	-	-	-	8
Cladocera	-	-	-	4	-	-	22	15	-	13	-	14	-	8	-	-	-	8
Copepoda	-	-	-	1	-	-	9	26	-	7	-	10	-	-	-	-	-	-
Ostracoda	-	-	-	-	-	-	-	-	-	3	-	9	-	-	-	-	-	-
Bryozoa (*Plumatella*)	-	-	-	-	-	-	-	X	-	-	-	-	-	-	-	-	-	X

guished. As expected from the longer historical connection, the North American element of the fauna is far more prominent than the South American element except in Panama where tropicopolitan forms derived from S. America dominate (Harring, 1915). More work needs to be done before the zoogeography of entire Carribean region can be discussed.

The apparent low diversity of species in the zooplankton contrasts strongly with the recent findings of Brandorff *et al* (1982), that in a tropical Amazonian river a high species diversity exists. Certainly the ecological conditions present now and the history of the freshwater habitats in Central America and the Caribbean must account for the differences. The equatorial effect on some elements of the fauna, e.g. the absence of carnivorous Cladocera and the low diversity of *Daphnia* spp. are common to both these situations. The importance of organized, long-term studies must be stressed. In their absence the composition, distribution and ecology of a very important component of a tropical freshwater ecosystem will remain relatively unknown. Answers to important basic aspects of global zooplankton ecology may well lie in questions posed as a result of a better knowledge of tropical materials.

Acknowledgements

We wish to thank the following for sending us zooplankton samples: Dr. W. K. Besch, Karlsrue, Western Germany (El Salvador); Mr. Tyrone Gopee, Entomologist, Caroni, Trinidad (Trinidad) and Dr. G. Hanek, Bahamas, (Bahamas, Haiti). Mr. Augustus Mamaril, Biology Department, Mississippi State University, Mississippi sent us data on zooplankton species in Jamaica and Dr. M. Dickman, Brock University, St. Catherines, Canada generously donated a collection of zooplankton made from different habitats in Costa Rica. Mr. R. Chengalath, National Museum of Natural Sciences, Ottawa, Canada helped us with the identification of Rotifera. Dr. V. Kořínek Charles University Prague loaned us material from Cuba and supplied a list of Cladocera recorded from Cuba.

References

Biochina, G. I., 1976. Caracterización de la fauna de la vegetación de los embalses de Cuba. Acad. Cienc. Cuba, Depto. Ecología Forestal. Ser. Forestal 33: 1-6.

Brandorff, G.-O., W. Koste & N. N. Smirnov, 1982. The composition and structure of rotiferan and crustacean communities of the lower Rio Nhumunda, Amazonas, Brazil. Stud. Neotrop. Fauna & Envir. 17: 67-121.

Brehm, V., 1932. Notizen zur Suswasserfauna Guatemalas und Mexico. Zool. Anz. 99: 63-66.

Brehm, V., 1939. La fauna microscópica del lago Petén, Guatemala. An. Esc. nac. Cienc. biol., Mex. 1: 173-202.

Brinson, M. M. & F. G. Nordlie, 1974. Lake Izabal Guatemala. Verh. int. Ver. Limnol. 19: 1408-1479.

Candelas, G. A., 1956. Studies on the freshwater plankton of Puerto Rico. Ph.D. Thesis, Univ. Minnesota.

Deevey, E. S., 1957. Limnological studies in middle America. Trans. Conn. Acad. Arts Sci. 39: 213-328.

Deevey, E. S., G. B. Deevey & M. Brenner, 1980. Structure of zooplankton communities in the Peten District, Guatemala. In: W. C. Kerfoot (ed.), The Evolution and Ecology of Zooplankton Communities. University Press Hanover, New England: 669-678.

De Ridder, M., 1966. Rotifers from Nicaragua. Hydrobiologia 27: 238-247.

De Ridder, M., 1977. Rotatoria of the Caribbean region. Studies on the fauna of Curaçao and other Caribbean Islands. 52: 137 pp.

Dodds, G. S., 1926. Entomostraca from the Panama Canal Zone with descriptions of one new species. Univ. Michigan Occ. Pap. 174: 2-27.

Dumont, H. J., 1979. Limnologie van Sahara en Sahel. Rijksuniv. Gent, 557 pp.

Edmondson, W. T., 1933. Investigations of some Hispanialan lakes. I: The Rotifera. Arch. Hydrobiol. 26: 465-471.

Fernando, C. H., 1980a. The freshwater zooplankton of Sri Lanka with a discussion of tropical freshwater zooplankton composition. Int. Revue ges. Hydrobiol. 65: 85-125.

Fernando, C. H., 1980b. The species and size composition of tropical freshwater zooplankton with special reference to the Oriental region (South East Asia). Int. Revue ges. Hydrobiol. 65: 411-426.

Fernando, C. H. & J. E. Ponyi, 1981. The free living freshwater cyclopoid Copepoda (Crustacea) of Malaysia and Singapore. Hydrobiologia 78: 113-123.

Fernando, C. H. & K. W. Smith, 1982. Copepoda. In S. H. Hurlbert (ed.), Aquatic Biota of Mexico, Central America and the Caribbean. San Diego State University Press: 192-195.

Fernando, C. H. & N. P. Zankai, 1981. The Rotifera of Malaysia and Singapore with remarks on some species. Hydrobiologia 78: 205-219.

Frey, D. G., 1980. The non-swimming chydorid Cladocera of wet forests, with descriptions of a new genus and two new species. Int. Revue ges. Hydrobiol. 65: 613-641.

Frey, D. G., 1982. Cladocera. In S. H. Hurlbert (ed.), Aquatic Biota of Mexico, Central America and the Caribbean. San Diego State University Press: 177-186.

Goulden, C. E., 1966. The history of Laguna de Petenxil. Mem. Conn. Acad. Arts Sci. 17: 1-126.

Harring, H. K., 1915. Report on Rotatoria from Panama with descriptions of new species. Proc. U. S. nat. Mus. 47: 525-564.

Idris, B. A. G. & C. H. Fernando, 1981. The Cladocera of Malaysia and Singapore, with new records, redescriptions and remarks on some species. Hydrobiologia 77: 233-256.

Juday, C., 1915. Limnological studies on some lakes in Central America. Trans. Wis. Acad. Sci. Arts Lett. 18: 214-250.

Kiefer, F., 1933. Suss und Brackwasser Copepoden von Bonaire, Curacao und Aruba 2. Cyclopoida. Zool. Jb. Syst. Okol. Geogr. Tiere 64: 405-415.

Kiefer, F., 1934. Neue Ruderfusskrebse von der Insel Harte. Zool. Anz. 108: 227-233.

Kiefer, F., 1936. Freilebende, Suss- und Salzwassercopepoden von der Insel Haiti; mit eine Revision der Gattung Halicyclops Norman. Arch. Hydrobiol. 30: 263-317.

Lai, H. C. & C. H. Fernando, 1980. Zoogeographical distribution of Southeast Asian Calanoida. Hydrobiologia 74: 53-66.

Lindberg, K., 1954a. Cyclopoides (Crustacés, Copépodes) de l'Amérique du Sud. Ark. Zool. 7: 193-222.

Lindberg, K., 1954b. Cyclopoides (Crustacés, Copépodes) du Mexique. Ark. Zool. 7: 459-489.

Marsh, C. D., 1913. Report on freshwater Copepoda from Panama. Smithson. misc. Collns 61: 12-19.

Marsh, C. D., 1920. Report on a collection of Copepoda made in Honduras by F. J. Dyer. Proc. U.S. Nat. Mus. 55: 545-548.

Marsh, C. D., 1931. On a collection of Copepoda made in El Salvador by Samuel F. Hilderbrand and Fred. J. Foster of the U.S. Bureau of Fisheries. J. Wash. Acad. Sci. 21: 207-209.

Orghidan, T. & St. Negrea, 1970. Quelques cladocères des eaux Souterraines et epigées de Cuba. Trav. Inst. Spéol. Emile Racavitza 9: 113-120.

Orghidan, T. & St. Negrea, 1973. Cladocères des eaux souterraines et epigées de Cuba. In Resultes de expéditions biospe-'ologiques Cubano-Roumaines à Cuba. Bucharest: 105-115.

Pejler, B., 1977. On the global distribution of the family Brachionidae (Rotatoria). Arch. Hydrobiol., Suppl. 53: 255-306.

Peckham, R. S. & C. F. Dineen, 1931. Summer plankton of lake Amatitlan, Guatemala. Am Midl. Nat. 50: 377-381.

Plesa, C., 1981. Cyclopides (Crustacea, Copepoda) de Cuba. In Resultes des expéditions biospéologiques Cubano-Roumaines à Cuba, Bucarest: 17-34.

Pourriot, R., 1975. Rotifères des Antilles. Cah. O.R.S.T.O.M. Sér. Hydrobiol. 9: 81-90.

Pourriot, R., 1982. Rotifera. In S. H. Hurlbert (ed.), Aquatic Biota of Mexico, Central America and the West Indies. Edit. San Diego State University Press: 140-143.

Richard, J., 1895. Sur quelques Entomostracés d'eau douce d'Haiti. Mem. Soc. zool. Fr. 8: 189-199.

Smith, K. E. & C. H. Fernando, 1978. The freshwater calanoid and cyclopoid Crustacea of Cuba. Can. J. Zool. 56: 2015-2023.

Smith, K. E. & C. H. Fernando, 1980. Guía para los copépodos (Calanoida y Cyclopoida) de las aguas dulces de Cuba. Acad. Cienc. Cuba, Habana: 1-28.

Straškrabá, M., 1969. Lista de los crustaceos dulceacuícolas de Cuba y sus relaciones zoogeográficas. Acad. Cienc. Cuba. Inst. Biol. Ser. biol. 8: 1-37.

Van de Velde, I., H. J. Dumont & P. Grootaert, 1978. Report on a collection of Cladocera from Mexico and Guatamala. Arch. Hydrobiol. 83: 391-404.

Wilson, C. B., 1942. The copepods of the plankton gathered during the last Cruise of the Carnegie. Carnegie Inst. Washington Publ. 536: 1-237.

Wilson, M. S., 1941. New species and distribution records of diaptomid copepods from the Marsh collection in the United States Natural Museum. J. Wash. Acad. Sci. 31: 509-515.

Zaret, T. M., 1975. Strategies for existance of zooplankton prey in homogeneous environments. Verh. int. Ver. Limnol. 19: 1484-1489.

A study of the zooplankton community of Billings Reservoir – São Paulo

S. Sendacz
Instituto de Pesca, Divisão de Pesca Interior, São Paulo, Brasil

Keywords: tropical zooplankton, seasonal fluctuation, reservoir, eutrophic

Abstract

The composition of the zooplankton of the Billings Reservoir and its variation in an eutrophic environment, subject to frequent blooms of algae (chiefly Cyanophyceae) was studied during one year (from October, 1977 to September, 1978) in two stations in the littoral and in the limnetic zone.

The zooplankton community in the limnetic zone was dominated by cyclopoid copepods (*Thermocyclops crassus* and *Metacyclops mendocinus*) and by rotifers (*Brachionus, Polyarthra* and others) which represented, respectively 38.5 and 35.5% of the total zooplankton. At the littoral zone, cyclopoids were the most abundant (42.3%).

The cladocerans were the least significant group at both stations, and calanoid copepods were found only at the littoral zone.

A higher production of small filtrators, such as rotifers, cyclopoid nauplii and *Bosmina* sp was observed.

Introduction

The purpose of this article is to give information on the composition of the zooplankton of Billings Reservoir, which is located near the city of São Paulo (23° 42′ S and 46° 24′ W).

The reservoir, whose area is 127 million m^2, was built in 1927; it has an irregular outline, and its maximum depth is 19 m. It was first used for electric power generation, and lately has also been used for water supply and as sewage clarifiers, as sewage and industrial effluents from the city of São Paulo are received by this water body through Pinheiros River.

The eutrophication of the Billings Reservoir, through the years, has been studied by Branco (1959, 1966), Kawai & Branco (1969) and Kawai (1977), always with an emphasis on its sanitary conditions.

The zooplankton analysis was undertaken at two branches of Billings Reservoir (Riacho Grande and Rio Pequeno). Riacho Grande is used for the water supply of small cities near São Paulo (São Bernardo, Santo André, São Caetano e Diadema), and in spite of its location relatively far from the catchment area, is not free from the influence of the sources of pollution.

A description of physical and chemical variables is given by Sendacz (1978) and Xavier (1981). Mean air temperatures are high from December to March (21 and 22 °C), and the lowest values occur in June and July (14 and 15 °C).

The mean annual pluviometric values for the region are 1 300 and 1 500 mm; the rainy season takes place from October to March, and the dry season from April to September.

The progressive eutrophication of Billings Reservoir justifies the continuous studies of its physical, chemical and biological characteristics.

Some aspects of the zooplankton community, such as its composition and seasonal fluctuation during the course of one year are analyzed.

Hydrobiologia 113, 121–127 (1984).

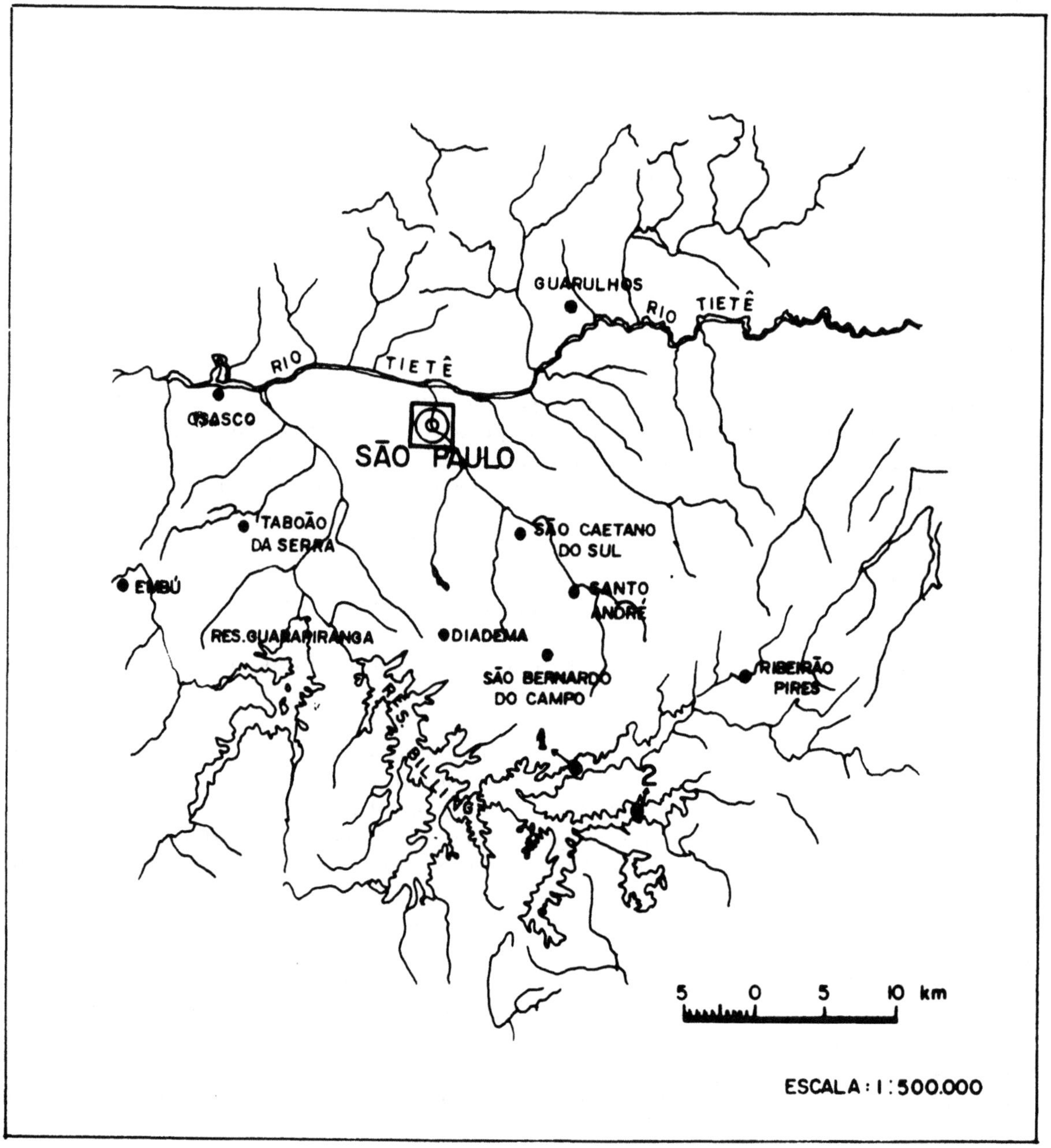

Fig. 1. Billings Reservoir – location of the sampling stations.

Material and methods

Sampling was carried out fortnightly, from September, 1977 to August, 1978, at two stations, one at the limnetic zone (Riacho Grande), and the other at the littoral zone (Rio Pequeno) (Fig. 1).

Station 1 is located approximately 100 m from the edge of Riacho Grande, with a mean depth of 6 m. During most of the studied period, a water bloom of blue-green algae, mainly *Anabaena spiroides* was observed. Station 2 is located at the border of Rio Pequeno, showing a slight slope; there is no aquatic vegetation, and the substrate consists of stones covered by clay material.

The zooplankton population was sampled with a plankton net with 25 cm diameter and 68 μm mesh size, through vertical hauls at station 1 and horizontal hauls at station 2. The samples were preserved by the addition of formalin 4%.

Samples were submitted to total counting or subsample methods, depending on the volume of plankton. For subsampling, a 1 ml Stempel pipette was used. All organisms were counted under stereoscopic microscope.

Results

Figure 2 illustrates the seasonal variation in the total zooplankton population and its major components at station 1. It is apparent that these variations were largely due to those of the Copepoda Cyclopoida and Rotifera, which represented, respectively, 38.5 and 35.5% of the total zooplankton population; the Cladocera population was less significant, representing 23%, and no calanoid copepods occurred.

The peaks of abundance, in October and November of 1977, and August and September of 1978, were due to the large number of cyclopoid nauplii.

A decrease in the population density was noted from January to July, and during the cold and dry period, a less pronounced peak was observed in June. This one was due to the Rotifera and Cladocera population.

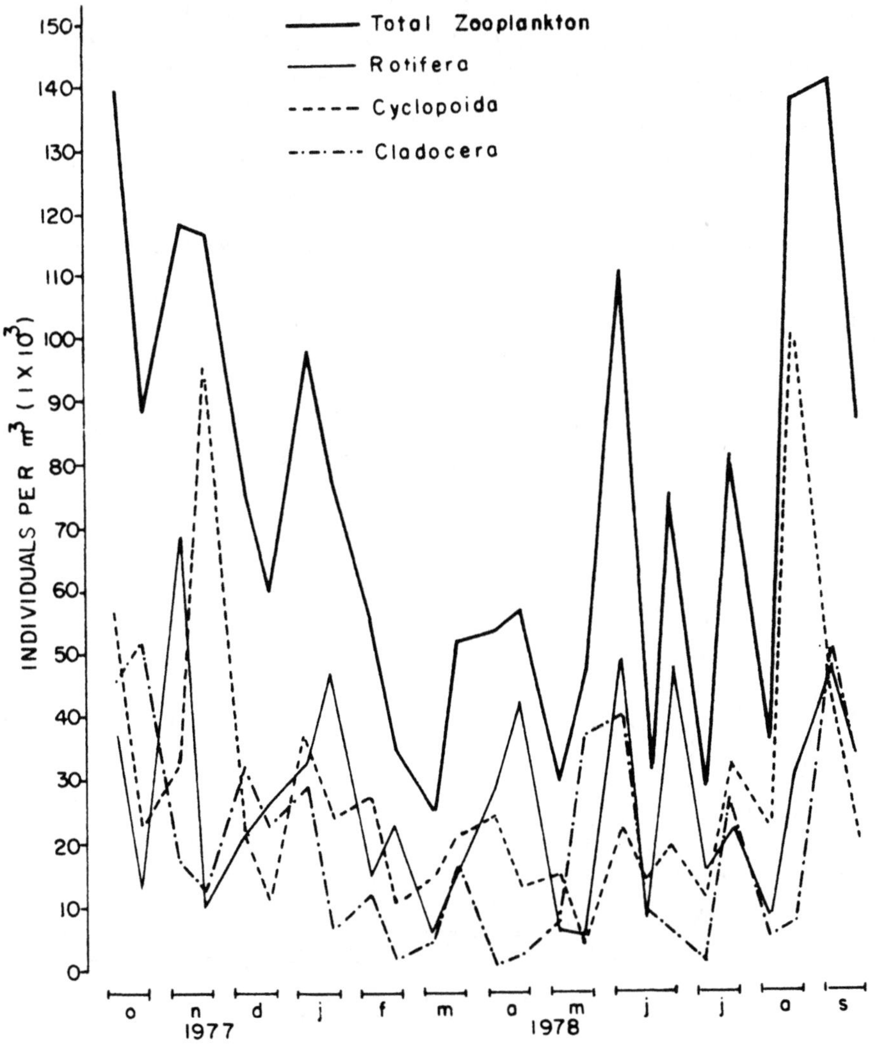

Fig. 2. Seasonal variation in numerical density at station 1 (limnetic zone) – Billings Reservoir.

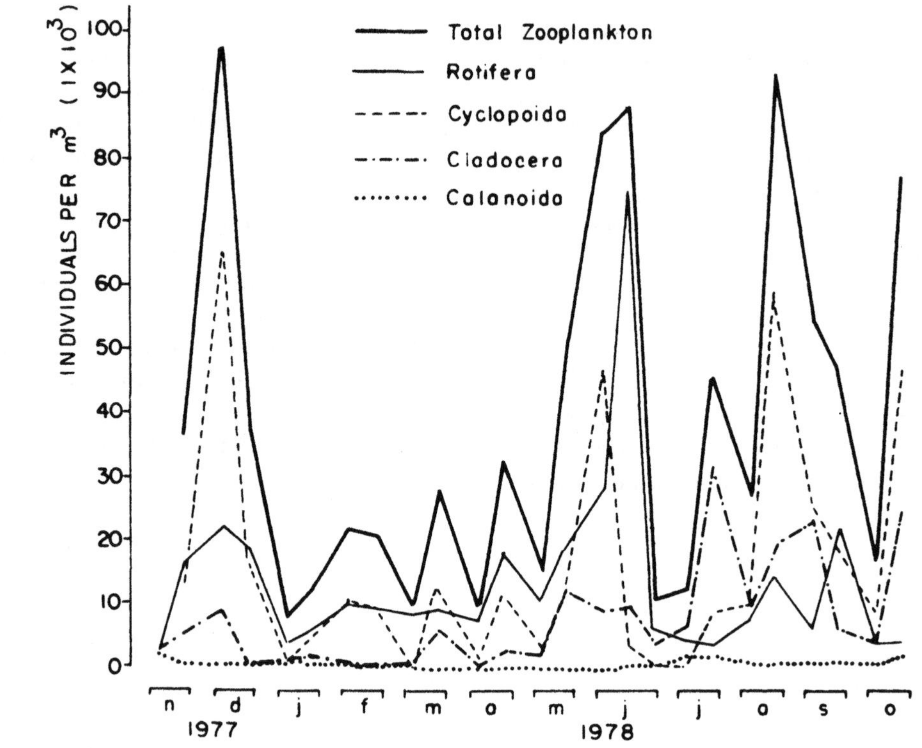

Fig. 3. Seasonal variation in numerical density at station 2 (littoral zone) - Billings Reservoir.

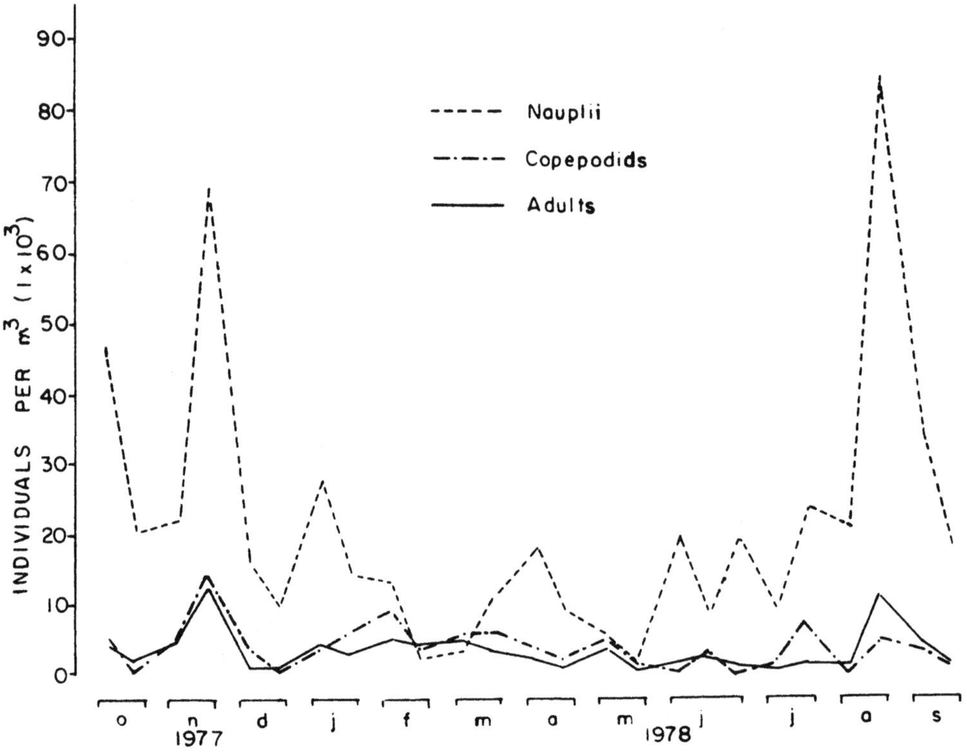

Fig. 4. Seasonal variation in numerical density of the developmental stages of Copepoda Cyclopoida at station 1 (limnetic zone) Billings Reservoir.

The seasonal variation of the zooplankton population at station 2, located at the littoral zone, in a different branch of Billings Reservoir, is shown in Fig. 3.

Here the total zooplankton per m^3 was usually lower than in station 1, and generally the numbers of cyclopoid copepods were higher than those of rotifers (42.3% and 35.7% of total zooplankton, respectively). Cladocerans represented 20.5%, and a calanoid copepod *(Diaptomus corderoi)* was registered, although hardly significant in the samples for most of the year.

The total zooplankton numbers showed a gradual rise from January to May, and there was a steep increase in May and June, and then in August. The peak of abundance in June was largely due to the great number of Rotifera, and in August to nauplii of Cyclopoida. An analysis of the cyclopoid population shows that it was numerically dominated by *Thermocyclops crassus* and *Metacyclops mendocinus; Tropocyclops prasinus* f *meridionalis* was found in relatively small numbers, and *Eucyclops serrulatus* was scarcely represented.

The dominant species of cyclopoid copepod changed according to the dry and rainy seasons. *Thermocyclops crassus* was dominant during the rainy season (October to March), and *Metacyclops mendocinus* during the dry season (April to September).

Table 1 shows the relative abundance of the Copepoda Cyclopoida at both stations.

At both stations, there is no succession of the different stages of copepod development (nauplii, copepodids, adults) as shown in Fig. 4 and 5. The peaks of abundance were always due to the nauplii.

About 10 genera of Rotifera were present, in larger number and frequency, the most abundant

Table 1. Relative abundance (%) of Copepoda Cyclopoida at both stations.

	Thermocyclops crassus	*Metacyclops mendocinus*	*Tropocyclops prasinus*
Station 1			
Rainy season	50.34	25.50	24.16
Dry season	32.43	52.44	15.13
Station 2			
Rainy season	68.05	26.22	5.73
Dry season	40.18	55.55	4.27

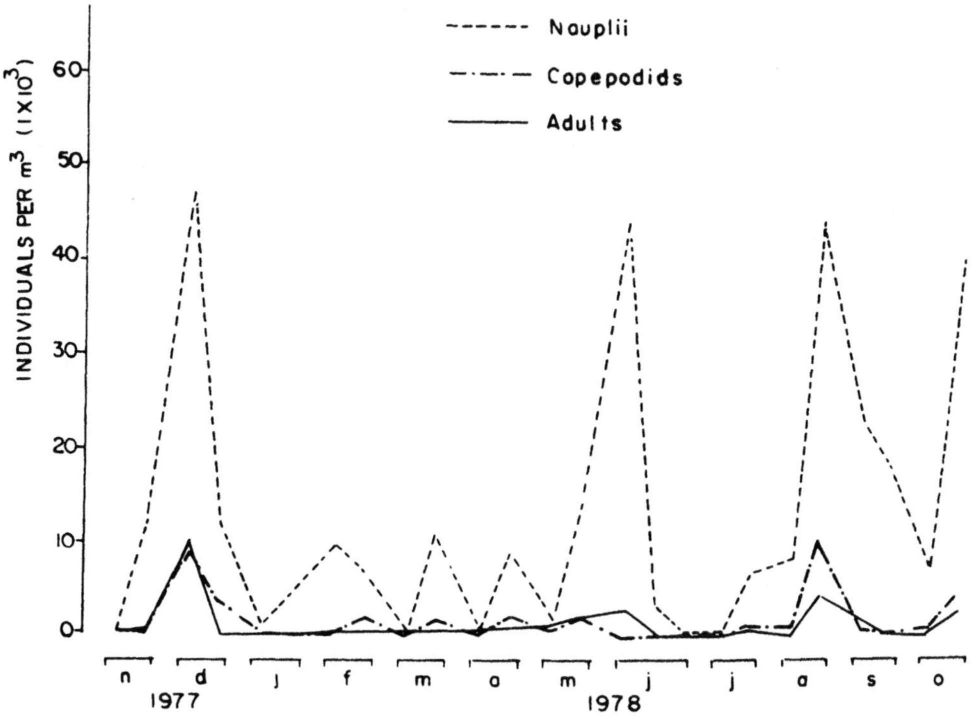

Fig. 5. Seasonal variation in numerical density of the developmental stages of Copepoda Cyclopoida at station 2 (littoral zone) - Billings Reservoir.

being *Brachionus, Polyarthra, Keratella, Thricocerca, Hexarthra* and *Asplanchna.* At station 1, *Brachionus, Polyarthra, Keratella* and *Thricocerca* were abundant, and at station 2 *Polyarthra* and *Keratella* occurred in great numbers, *Brachionus* and *Thricocerca* showing lower densities.

The Cladocera were represented by *Bosmina* sp, *Diaphanosoma* sp, *Daphnia gessneri* and *Ceriodaphnia cornuta.*

Bosmina sp was the dominant species, although *Diaphanosoma* sp reached great numbers. *Daphnia gessneri* and *Ceriodaphnia cornuta* were present in relatively small numbers, and their occurrence was irregular throughout the period of study.

Discussion

It is known that the eutrophication affects the specific composition of the zooplankton through physical and chemical alterations of the environment which also change the phytoplankton composition, promoting changes in the quality and quantity of available food for the zooplankton population. In this way, there are different communities associated with different trophic conditions.

Billings Reservoir, an environment in an advanced stage of eutrophication (Branco, 1966; Branco & Kawai, 1969), showed high contents of chlorophyl, nutrients, as well as low transparency values (Sendacz, 1978; Xavier, 1981). Intense photosynthesis at the surface can produce oxygen tensions of up to 124.26% oxigen saturation.

The obtained data show that in the zooplankton composition of Billings Reservoir, cyclopoid copepods were hardly more significant than rotifers, cladocerans were less significant, and calanoid copepods were scarcely represented only at the littoral zone.

Several authors, mainly at temperate regions, have associated the degree of eutrophication with the composition of the zooplankton community (Bradshaw, 1964; Patalas, 1972; Hillbricht-Ilkowska & Weglenska, 1970); they observed a dominance of cyclopoid copepods and cladocerans in eutrophic environments.

In tropical and subtropical regions, Zago (1972) and Carvalho (1975) observed, in a reservoir undergoing a progressive eutrophication, first a dominance of cladocerans then of cyclopoid copepods, being calanoid scarce at the beginning of the study, and totally absent later. On the other hand, in an oligotrophic reservoir, Matsumura-Tundisi & Tundisi (1976) have found a dominance of rotifers followed by cladocerans and copepods.

In recent studies of the zooplankton communities in 17 reservoirs in São Paulo State, southern Brazil, Sendacz *et al.* (1982) observed that rotifers were dominant in most of the reservoirs, followed by copepods in eutrophic lakes, and by cladocerans in oligotrophic ones.

The dominant group of the zooplankton of Billings Reservoir, the cyclopoid copepods, showed the association of two species, *Thermocyclops crassus* and *Metacyclops mendocinus,* although the frequent association in tropical and subtropical region is *Thermocyclops* and *Mesocyclops,* with higher densities of the first (Burgis, 1969; Roa, 1972; Carvalho, 1975; Matsumura-Tundisi & Tundisi, 1976; Brandorff, 1977). It seems probable that *Metacyclops* replaces *Mesocyclops* in eutrophic lakes, which was confirmed by Sendacz *et al.* (1982).

The species of calanoid copepod, *Diaptomus corderoi,* scarcely represented at the littoral zone is probably also adapted to more eutrophic environments.

The seasonal variation in the zooplankton population was largely due to the nauplii of cyclopoid copepods; there is no succession of the different stages of the development as already pointed out for tropical regions by Burgis (1969) and Carvalho (1975).

Roa (1972) and Matsumura-Tundisi & Tundisi (1976) observed that peaks of abundance in the zooplankton population were registered at the beginning of the warm and rainy period; in this study, it was not observed an increase in number of the population associated to high temperatures and precipitation.

The occurrence of lower densities of cladocerans compared to other zooplanktonic groups, is due to the lack of suitable food, since large particles predominate in the phytoplankton of Billings Reservoir. In spite of this, *Bosmina* sp was present in large numbers and frequency; although observation about crustaceans as indicators of trophic conditions is rare, *Bosmina* is associated with eutrophic conditions. The size of the particles ingested by *Bosmina* is smaller than those ingested by other cladocerans (Burns, 1968 *apud* Green, 1976), allowing this organism to feed on bacteria and detritus.

Diaphanosoma sp was also abundant and frequent, the opposite occurring with *Daphnia gessneri*. Probably the first is well adapted to eutrophic environments; Zago (1976) observed at Americana Reservoir, a replacement of *Daphnia gessneri* by *Diaphanosoma* sp as the dominant species of Cladocera with the eutrophication of the reservoir. On the other hand, Brandorff (1977) verified for the oligotrophic Lake Castanho, that *Daphnia gessneri* was the dominant cladoceran, *Diaphanosoma sarsi* being found in reduced numbers.

The Rotifera showed higher diversity compared with the other groups, reaching high densities throughout the studied period. The organisms of this group are considered opportunists (Allan, 1976), reaching high or low densities according to the tolerance to environmental conditions. This fact is demonstrated by the seasonal fluctuation of this group, which followed an irregular pattern.

The rotifers of Billings Reservoir at the limnetic zone, did not show a marked dominance over cyclopoid copepods, which dominated even at the littoral zone.

References

Allan, J. D., 1976. Life history patterns in zooplankton. Am. Nat. 110: 165-180.

Bradshaw, A. S., 1964. The crustacean zooplankton picture. Lake Erie 1939-49-59; Kayuga 1910-51-61. Verh. int. Ver. Limnol. 15: 700-708.

Branco, S. M., 1959. Estudo dos fatores biológicos responsáveis pelas más qualidades que caracterizam as águas de abastecimento do ABC. S. Paulo, Div. Tratamento.

Branco, S. M., 1966. Estudo das condições sanitárias da Represa Billings. Arq. Fac. Hig., Univ. S. Paulo 20: 57-86.

Brandorff, G. O., 1977. Untersuchungen zur populationsdynamik des Crustaceenplanktons in tropischen Lago Castanho (Amazonas, Brasilien). PhD Thesis, Univ. Kiel, 109 pp.

Burgis, M. J., 1969. A preliminary study of ecology of zooplankton in Lake George, Uganda. Verh. int. Ver. Limnol. 17: 297-302.

Carvalho, M. A. J., 1975. A Represa de Americana: aspectos físicoquímicos e a variação das populações Copepoda Cyclopoida de vida livre. PhD Thesis, Inst. Bioc., Univ. S. Paulo, 157 pp.

Green, J. D., 1976. Plankton of Lake Ototoa, a sand-dune lake in Northern New Zealand. N.Z. J. mar. Freshwat. Res. 10: 43-59.

Hillbricht-Ilkowska, A. & T. Weglenska, 1970. Some relations between production and zooplankton structure of two lakes of a varying trophy. Polskie Arch. Hydrobiol. 17: 233-240.

Kawai, H., 1977. Modelo matemático para avaliação do desenvolvimento da eutrofização na Represa Billings. In Congresso Brasileiro de Engenharia Sanitária, 9. Belo Horizonte: 1-20.

Kawai, H. & S. M. Branco, 1969. Estudo sobre as relações entre despejos domésticos e industriais da região da Grande São Paulo e a eutrofização do Reservatório Billings. Revta. DAE, S. Paulo 29: 57-68.

Matsumura-Tundisi, T. & J. G. Tundisi, 1976. Plankton studies in a lacustrine environment, 1. Preliminary data on zooplankton ecology of Broa Reservoir. Oecologia 25: 265-270.

Patalas, K., 1972. Crustacean plankton and the eutrophication of the St. Lawrence Great Lakes. J. Fish. Res. Bd Can. 29: 1451-1462.

Roa, E. Z., 1972. Zooplankton de la laguna de Campoma, Edo. Sucre, Venezuela. Cuad. Oceanogr. Univ. Oriente 3: 49-53.

Sendacz, S., 1978. Alguns aspectos do zooplancton da Represa Billings. MSc Thesis, Inst. Bioc., Univ. S. Paulo: 84 pp.

Sendacz, S., E. Kubo & M. A. Cestarolli, in press. Composição do zooplancton de 17 reservatórios do sudeste do Estado de São Paulo.

Xavier, M. B., 1981. Represa Billings - São Paulo I. Fitoplancton e características físicas da água. B. Inst. Pesca. 8: 23-46.

Zago, M. S. A., 1976. The planktonic Cladocera (Crustacea) and aspects of the eutrophication of Americana Reservoir, Brazil. Bolm. Zool. Univ. S. Paulo 1: 105-145.

Species composition and seasonal abundance of zooplankton in two Ethiopian Rift Valley lakes - Lakes Abiata and Langano

Kassahun Wodajo & Amha Belay
Department of Biology, Science Faculty, Addis Abeba University, P.O. Box 1176, Addis Abeba, Ethiopia

Keywords: tropical zooplankton, species composition, seasonal abundance, Ethiopian Rift Valley lakes

Abstract

The composition and annual cycle of the zooplankton of two Ethiopian Rift Valley soda lakes is described. Lake Langano has a conductivity of 1 400 to 1 800 μS cm^{-1} and a permanent mineral turbidity. Lake Abiata is more concentrated (conductivity 19 000 to 23 000 μS cm^{-1}) and more alkaline but less turbid; it is characterised by dense phytoplankton blooms, mainly cyanophytes.

The zooplankton assemblage is typically tropical, with relatively few species of Cladocera and Copepoda. There was a marked difference in zooplankton between the two lakes, Lake Abiata showing much higher concentrations and greater wet season/dry season differences. The species composition was also different. Lake Abiata lacked Cladocera, and calanoid copepods occurred only during the wet season with lower conductivities. These two phenomena were attributed to high sodium bicarbonate concentration and to dense cyanophyte blooms. Eleven species of rotifers occurred in Lake Abiata, including six *Brachionus* spp. but *B. rubens* was the only rotifer found in Lake Langano. The seasonal variation of the zooplankton is discussed in relation to seasonal fluctuations in conductivity and Chl a concentration.

Introduction

The knowledge of zooplankton cycles in tropical African lakes is meagre. The only detailed data comes from Lake George, Uganda (Burgis, 1974) and Lake Chad (Carmouze *et al*, 1983). Very little work has been done on the zooplankton of the Horn of Africa (see Dumont, 1980).

In recent years, several useful investigations have been carried out on Lakes Abiata and Langano (Baxter *et al*, 1965; Wood *et al*, 1978; Wodajo, 1982), while some older studies on the zooplankton of Lake Abiata (Lowndes, 1930) and Lake Langano (Cannicci & Almagia, 1947) have only limited significance. The aim of the present study is to supply information of a preliminary nature on the seasonal variability of the zooplankton of these two lakes. First, the species composition of the limnetic zooplankters was determined and, second, we assessed the seasonal changes in composition and abundance of major zooplankton groups (Cladocera, Cyclopoida, Calanoida and Rotifera) in relation to changes in conductivity and pH, and in relation to major algal groups and algal biomass.

Study area

Lake Abiata and Lake Langano are situated in the northern, Ethiopian section of the Great African Rift Valley (Fig. 1). Some of the physical characteristics of the lakes are listed below (Table 1).

The origin and the geology of these closed graben lakes are documented by Mohr (1962), DiPaola (1972) and Gasse & Street (1978). Grove & Gondie (1971) believe that about 9 200 ± 190 years B.P.

Hydrobiologia 113, 129–136 (1984).

Table 1. Some physical characteristics of Lakes Abiata and Langano.

Lake	Altitude[a] (m)	Surface (km²)	Volume[c] (km³)	Max. depth[a] (m)	$\bar{Z}$ (m)	Max. length[a] (km)	Max. width[a] (km)	Perimeter[a] (km)	Catchment[a] (km²)
Abiata	1573	207[a] 205[b]	1.7	14.2	7.6[a] 8.9[b]	20.9	12.4	61.9	848
Langano	1585	230[a] 228[b]	5.0	46.2	16.9[a] 21.9[b]	25.4	16.0	77.5	832

[a] Cannicci & Almagia (1947).
[b] Wood, Prosser & Baxter (1978).
[c] Loffredo & Maldura (1941).

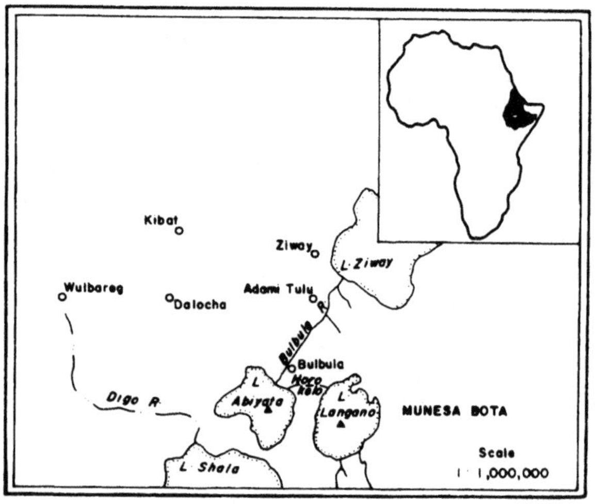

Fig. 1. Map showing the location and drainage of Lake Abiata and Lake Langano. Sampling sites (▲) are shown.

Lakes Abiata, Langano and Shala were united with Lake Ziwai (Fig. 1), but were isolated later by faulting and other crustal movements, and by desiccation (Mohr, 1962; Grove & Gondie, 1971; Di-Paola, 1972).

Lake Abiata is typically greenish, while Lake Langano maintains a stable colloidal suspension of silt and is light reddish brown in colour. Light attenuation due to non-algal causes especially from silt is considerable in the two lakes, about 78% in Lake Abiata and between 94 and 98% in Lake Langano (Wood *et al,* 1978).

Their waters show a marked predominance in bicarbonate and carbonate among the anions. The range of alkalinity is between 10 and 15 meq l^{-1} for Lake Langano and between 180 and 300 meq l^{-1} for Lake Abiata (Talling & Talling, 1965; Wodajo, 1982). Sulphates are present in low concentrations, 20–50 mg l^{-1}, in Lake Langano and at higher concentrations, 175–260 mg l^{-1}, in Lake Abiata (Loffredo & Maldura, 1941; Talling & Talling, 1965; Wodajo, 1982). Chlorides are present in a larger concentration (3 000–4 000 mg l^{-1}) in Lake Abiata compared to Lake Langano (100–200 mg l^{-1}), and show a parallel increase with conductivity and total ions (Wodajo, 1982). Lakes Abiata and Langano contain 5 000–7 000 mg l^{-1} Na and 400–500 mg l^{-1} Na, respectively. A high proportion of sodium to potassium (20 to 1) exists in both lakes (Wodajo, 1982). The divalent cations (Ca^{2+} and Mg^{2+}), are often found in low concentrations, <3 and 10 mg l^{-1} Ca in Lake Abiata and Lake Langano, respectively, and less than 8 mg l^{-1} is found in both lakes (Loffredo & Maldura, 1941; Talling & Talling, 1965; Wodajo, 1982).

Climate and vegetation

The two lake basins lie within the sub-humid region of Ethiopia (Gamatchu, 1977). Most rain falls between late June and mid-September, with a monthly total of more than 160 mm. Normal weather for this period involves heavy rain for an hour or so, often followed by bright sunshine. The least diurnal variation in air temperature occurs at this time (± 6 °C). The driest period of the year is November to February, when a monthly total of less than 5–25 mm rainfall is recorded. The lowest daily minimum temperature (14 °C) occurs at this time. A shorter period of fairly heavy rainfall is observed between March and May (Wodajo, 1982) when the highest daily maximum temperatures (27–29 °C) are reached.

The natural vegetation around the lakes is composed of *Acacia* spp. and scrub grassland. Around Lake Langano *Scirpus inclinatus* and *Juncellus laevigatus* are present, but in Lake Abiata macrophytes are entirely absent.

Hydrology

The water balance of Lake Abiata is controlled by river discharge, rainfall and evaporation. Water levels in Lake Abiata were very low during most of the period of this study, and the water receded more than 400 m from its average shore line. The two inflows, the Horakalo and Bulbula rivers (Fig. 1), had been dry for more than eight months (October, 1980 to June, 1981), but high rainfall in their headwaters between June and September, 1981, resulted in a 6 cm rise (October, 1981) in the water level of the lake. The water level of Lake Abiata has been know to fluctuate with alternating wet and dry periods (Baxter *et al*, 1965; Grove & Gondie, 1971; DiPaola, 1972; Wood *et al*, 1978; Wodajo, 1982).

Lake Langano lies in a topographically closed basin but hydrologically, it is characterized by having an outflow. A portion of its inflow is derived from hot spring waters. The annual differences in the water level of this lake are governed by rainfall and evaporation (Wodajo, 1982). The level rises between July and September after the onset of rainfall, and falls progressively during the dry months (November and February).

Materials and methods

Vertical zooplankton samples were collected monthly with No. 10 mesh (157 μm) net between November, 1980 and October, 1981. Five samples (ca. 250 ml) were obtained from a single station on each lake (Fig. 1) using vertical hauls after Edmondson & Winberg (1971). All samples were preserved in 4–5% Formalin. For subsampling the sample was poured into a 500 ml beaker and stirred; subsamples were taken with a bulb pipette and the animals were counted in a small petri-dish (ca. 10 ml) with equal transects on the bottom, after staining with 1–3 drops of Lugol's solution to facilitate observation. The subsamples were examined and counted under a stereoscopic microscope (magnification 20–80×). Adult Cladocera, Cyclopoida, Calanoida, Rotifera, and immature Copepoda (nauplii and copepodids) were counted as separate groups. The final estimate of the population (individuals m^{-3} of lake water) was computed using the formula of Edmondson & Winberg (1971),

$$V = \pi r^2 d$$

where, V is the volume of water filtered, r is the radius of the net, and d is the length of the course of the net through the water.

Besides the above groups of zooplankton other invertebrates collected were identified to various levels using appropriate keys in Edmondson (1959).

Electrical conductivity and pH were measured in the field following the methods of Talling & Talling (1965). The concentrations of the major anions and cations incorporated in this paper were determined by standard methods (for detailed information, see Wodajo, 1982). Chlorophyll a concentration was estimated spectrophotometrically after concentration of the phytoplankton on glass fibre filters (Whatman GF/C), extraction of the pigment in hot 90% methanol, and clearing for 2 minutes by centrifugation (4 000 r.p.m.), and using the formula in Talling & Driver (1963). Degradation products were not distinguished.

Results

Table 2 shows that Cladocera occurred only in Lake Langano. These were *Daphnia barbata, Daphnia longispina* and *Ceriodaphnia cornuta. Lovenula africana* was the only calanoid present in either lake, in Lake Langano throughout the year but in Lake Abiata only between November, 1981, and May 1982 (Fig. 2). Harpacticoids were present in Abiata in small numbers but not in Langano. The Abiata species was probably *Nitocra lacustris.*

There were only two or three species of cyclopoid species in both lakes. *Mesocyclops aequatorialis, Eucyclops* cf. *macrurus*, and *Eucyclops* sp. were found in Lake Abiata; *Mesocyclops* sp. and *Eucyclops* cf. *serrulatus* were found in Lake Langano.

Eleven rotifer species were identified from lake Abiata, mostly *Brachionus* spp. *B. caudatus, B. caliciflorus* and *B. angularis* could be found at any time of the year. *B. rubens* was the only rotifer encountered in limnetic samples from Lake Langano, throughout the whole study period.

In Lake Abiata high ostracod densities were encountered between September and November. Ostracods also occurred during April in both lakes.

Table 2. List of zooplankton species from Lake Abiata and Lake Langano. The zooplankton species not described in earlier works (Lowndes, 1930; Cannicci & Almagia, 1947) are indicated (*).

Lake	Calanoida	Cyclopoida	Cladocera	Rotifera	Others
Abiata	*Lovenula africana* (Daday)*	*Mesocyclops aequatorialis* Kiefer. *Eucyclops* spec. (cf. *macrurus*)* *Eucyclops* spec.		*Brachionus caudatus* (Hauer)* *B. calyciflorus* (Pallas) *B. nilsoni* (Ahlstrom)* *B. budapestinensis* (Daday)* *B. plicatilis* (O. F. Muller)* *B. angularis* (Grosse)* *Filinia terminalis* (Plate)* *Horaella brehmi* (Donner)* *Lecana bulla* (Gosse) *Lecana luna* *(O.F. Muller)* *Trichocerca similis* Wierzejski)*	Ostracoda* Chironomidae Larvae Corixidae *Micronecta* sp.* Ectroprocta *Plumatella repens* L.* Hemiptera *Anisops** Protozoa *Arcella** Harpacticoida *Nitocra lacustris** (Schmankewitsch)*
Langano	*Lovenula africana* (Daday)	*Mesocyclops* sp. *Eucyclops* cf *serrulatus* (Fischer)	*Daphnia barbata* Weltner) *Ceriodaphnia cornuta* (Sars)* *Daphnia longispina* (O. F. Muller)*	*Brachionus rubens* (Ehrenberg)*	Ostracoda* Sididae* Unidentified

Large numbers of corixids and *Anisops* sp. (Notonectidae) were seen in inshore waters of Lake Abiata between March and April. Larval Chironomidae were found in bottom sediments of Lake Abiata in November, 1980, and October, 1981, but did not appear to enter the plankton. Pseudoplanktonic Protozoa (Arcella) were found in April, 1981 in limnetic samples from Lake Abiata.

Figures 2 and 3 compare the annual cycle of different zooplankters and contrast these with phytoplankton fluctuations (Chl a, mg m^{-3}), conductivity, and pH. Zooplankton density was much greater in Lake Abiata and increased to a maximum between May and June when there was also a rapid increase in phytoplankton biomass, as indicated by the Chl a figures (Fig. 2). In Lake Langano a maximum was observed between February and March (Fig. 3).

The zooplankton displayed marked seasonal changes in composition in both lakes:

Calanoida. The population density of *L. africana* peaked at about 1 000–3 000 m^{-3} between November and February in Lake Abiata, subsequently numbers declined progressively and none were observed after May, 1981 (Fig. 2). In Lake Langano (Fig. 3), *L. africana* was found in variable numbers throughout the study period, its density peaking between February and March with low values between December and January, and April and May. The declines occurred during blooms of blue-green algae (Fig. 2), while during the maximum large chlorophycean populations were present (Wodajo, 1982).

Cyclopoida. Cyclopoid copepods were numerous between March and June and between August and September in Lake Abiata (Fig. 2) and high densities were recorded between March and May. However, their annual variation was rather irregular. During periods of high cyclopoid density, large populations of rotifers and a high algal biomass was also observed.

In Lake Langano, cyclopoid density was low (average of about 800 m^{-3}) and a minor peak was observed between February and April (Fig. 3). The

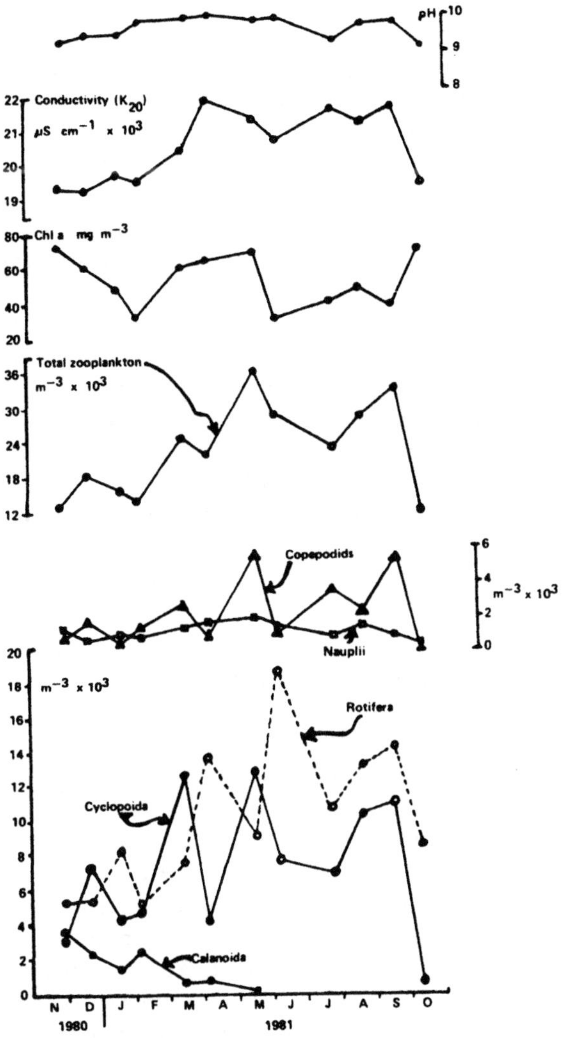

Fig. 2. Seasonal distribution of the main groups of zooplankters and algal biomass in Lake Abiata. Changes in conductivity and pH are also shown.

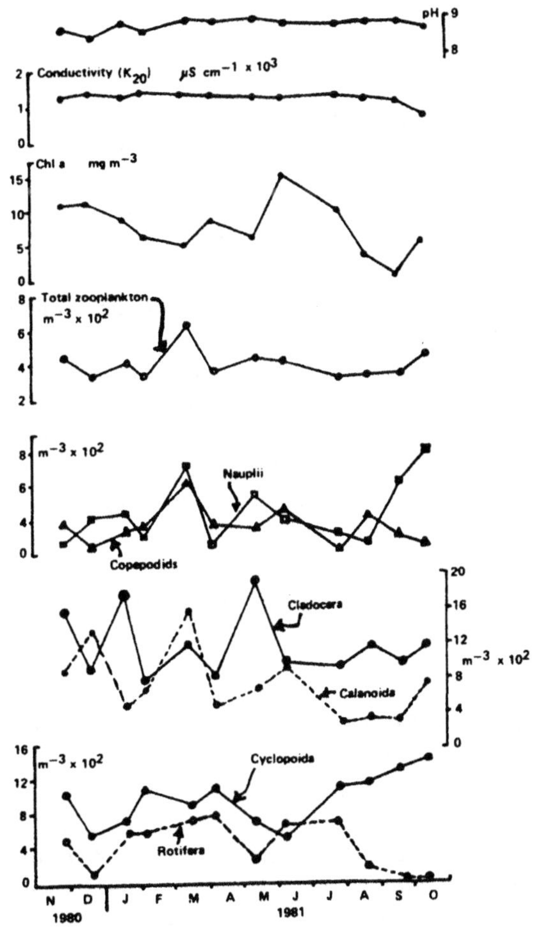

Fig. 3. Seasonal distribution of the main groups of zooplankters and algal biomass in Lake Langano. Changes in conductivity and pH are also shown.

cyclopoid population increased in number from 1 200 to 1 600 m^{-3} between July and October, while the other component of the zooplankton, the rotifers, declined in numbers.

In both lakes the decrease in nauplii appears to have been accompanied by the appearance of copepodids (Figs. 2 & 3).

Rotifera. The relative importance of rotifers varied greatly between the two lakes. Their contribution to the total zooplankton was high (ca. 44%) in Lake Abiata, whereas in Lake Langano they constituted only about 9% of the total zooplankton.

They were the most seasonal of all groups. In Lake Abiata there were blooms between March and May and August and September (Fig. 2) and in Lake Langano the highest densities were recorded between January and April, and between June and July (Fig. 3).

Cladocera. These were present only in Lake Langano and were found in large numbers (> 800 m^{-3}) throughout the study (Fig. 3). Two pulses were obvious, one in January and one in May (> 1 800 m^{-3}). Low population densities (< 1 200 m^{-3}) corresponded to relatively high cyanophycean concentrations (> 10 mg Chl a m^{-3}) and to large cyclopoid populations (> 1 200 m^{-3}).

Discussion

Species numbers and composition

The two lakes are similar in mode of origin (Mohr, 1962; Grove & Gondie, 1971) and in some physical characteristics (Loffredo & Maldura, 1941; Cannicci & Almagia, 1947). However, they show marked differences in zooplankton density and species composition (Table 2), fewer species being recorded in Lake Langano. Nevertheless, their zooplankton as a whole has a typical tropical, limnetic, species composition (cf. Fernando, 1980a, b).

Cladoceran and calanoid species form a major component of the limnetic zooplankton in Lake Langano. Two Cladoceran species, *Daphnia longispina* and *Ceriodaphnia cornuta*, previously not recorded (Cannicci & Almagia, 1947), were found during the present study. These could have been introduced with fish (*Clarias* spec. and *Barbus* spec.) from Lake Ziwai. They are eurytopic species occurring in a wide range of freshwater habitats in Africa (e.g. Burgis, 1973; Rzoska, 1976) and in South East Asia (Fernando, 1980a, b). Nevertheless, Cladocera are conspicuously absent from the limnetic zooplankton of Lake Abiata. This, and the scarcity of Calanoida need explanation. A number of factors combined may account for the situation. High populations of blue-green algae dominate in Lake Abiata (Baxter *et al*, 1965; Wodajo, 1982), which could make it uninhabitable by Cladocera. Burgis (1973) suggested that the lack of small algal particles restrict Cladocera in blue-green algal-rich Lake George, Uganda, and this would apply to most Cladocera (Fernando, 1980b). Generally, Cladocera have problems with manipulation and consumption of blue-green algal bundles (Dumont, 1977; Edmondson, 1974). These are not eaten by *Daphnia barbata* in Lake George (Moriarty *et al*, 1973). Arnold (1971) has shown that blue-green algal cultures fed to Cladocera have an adverse effect on their reproductive rate and longevity.

Lake Abiata has high concentrations of sodium bicarbonate and chloride and seasonal and long-term changes occur because of variation in evaporation and inflows (Wood *et al*, 1978; Wodajo, 1982); it is also occasionally deoxygenated (Wodajo, 1982). These may also limit cladoceran survival and growth. Herbst (1954) showed that Cladocera are sensitive to low oxygen concentration, and Heisey & Porter (1977) found that oxygen deficiency occurring in highly productive lakes can change the quality of the deeper water refuge for Daphnia species. However, Talling & Talling (1965) report an absence of most Entomostraca from African Rift Lakes with alkalinities above 80 meq l^{-1}. Thus the high alkalinity (> 180 meq l^{-1}) of Lake Abiata may be the principal factor affecting cladocera.

The decline of the calanoid *Lovenula africana* after May, 1982 in Lake Abiata may be related to a rapid increase in salinity (conductivities about 22 000 $\mu S\ cm^{-1}$) (Fig. 2). Such high conductivity values appear to exceed the range of salinity in which calanoids can exist (LaBarbara & Kilham, 1974). Other factors, such as elevated pH values (pH 9.8–10.0) and a large cyanophyte population (>60 mg Chl a m^{-3}) (Fig. 2) could add to this adverse effect. O'Brien & deNoyelles (1972) report that high pH values can cause zooplankton mortality. Blue-green algal inhibition (Edmondson, 1965; Fernando, 1980a, b) can also be considerable. In Lake Langano, the conductivity changes were within the limit of tolerance of these calanoids (cf. LaBarbara & Kilham, 1974).

The species composition of cyclopoid Copepoda in the lakes is in line with Fernando's (1980a, b) findings for other tropical lakes. They are represented by few species of two genera, *Eucyclops* and *Mesocyclops* (Table 2). Their systematics has not been authenticated (Dumont, pers. commun.), in particular *E.* cf. *serrulatus* from Lake Langano may not be this species. Dumont (1980) suggests that *E. serrulatus* Fischer (= agilis Koch) occurs neither in Africa nor in America because of considerable endemicity in the genus *Eucyclops*.

The harpacticoid copepod, *Nitocra lacustris*, common in Lake Abiata, appears to be characteristic of the more alkaline, saline Lake Abiata, as Lowndes (1930) recorded it from other soda lakes in Ethiopia.

The rotifer fauna fits the assemblage noted on a world-wide scale by Green (1972) and Pejler (1977). In Lake Langano, one species, *B. rubens* was found. It was not reported by previous workers (Cannicci & Almagia, 1947). The paucity of rotifers in this lake may be due to the low productivity associated with the permanent turbidity. The richness of the rotifers in Lake Abiata is probably related to an abundant food supply.

Seasonal variation

Seasonal variation in zooplankton density and composition (Figs. 2 and 3) was related to some extent to alternating wet and dry season conditions and to changes in the density and composition of the phytoplankton (Wodajo, 1982).

In Lake Langano there was a negative relationship between Cladocera and Cyanophyte densities, but a positive relationship between the latter and rotifer densities. Cyclopoids showed a weak and probably not significant increase with phytoplankton biomass.

The alternation between Cladocera and rotifers may have been due to their preference for different types of phytoplankton. The cladoceran maximum coincided with blooms of chlorophytes whereas the rotifer maximum coincided with a bloom of cyanophytes.

In Lake Abiata the algae (cyanophytes) bloomed (Fig. 3) during the short rainy period and there was a concomitant increase in densities of rotifers and cyclopoids. The latter increased more slowly, probably because of their lower reproductive rates (Edmondson, 1965; Dumont *et al*, 1975). The densities of all three groups of organisms decreased after the big rains (October–February), with the dilution of the lakes and an increase in lake volumes.

Acknowledgements

We are indebted to Prof. C. H. Fernando, Dr. H. J. Dumont and Prof. J. Green for identification and supply of relevant literature. The project has been financially supported by the Swedish Agency for Research Cooperation with Developing Countries (SAREC) through the School of Graduate Studies, Addis Abeba University. Prof. A. D. Harrison read the manuscript and made critical comments.

References

Arnold, D. E., 1971. Ingestion, assimilation, survival and reproduction by *Daphnia pulex* fed to seven species of blue-green algae. Limnol. Oceanogr. 16: 906–920.

Baxter, R. M., M. V. Prosser, J. F. Talling & R. B. Wood, 1965. Stratification in tropical African lakes at moderate altitudes (1 500 to 2 000 m). Limnol. Oceanogr. 10: 514–520.

Burgis, M. J., 1973. Observation on Cladocera of Lake George, Uganda. J. zool. Soc. Lond. 173: 339–349.

Burgis, M. J., 1974. Revised estimates for the biomass and production of zooplankton in Lake George, Uganda. Freshwat. Biol. 4: 535–541.

Cannicci, G. & F. Almagia, 1947. Notizie sulla 'Facies' Planctonica di Alcuni Laghi Della Fossa Galla. Boll. Pesca Piscicult. Idrobiol. 23: 54–74.

Carmouze, J. P., J. R. Durand & C. Lévêque, 1983. Lake Chad. Monographiae Biologicae, 53. Junk, The Hague, 575 pp.

DiPaola, G. H., 1972. The Ethiopian Rift Valley (between 7°00′ and 8°40′ Lat. North). Bull. Volcanol. 36: 517–560.

Dumont, H. J., I. Van de Velde & S. Dumont, 1975. The dry weight estimate of biomass of Cladocera, Copepoda and Rotifera from the plankton, periphyton and benthos of continental waters. Oecologia 19: 75–97.

Dumont, H. J., 1977. Biotic factors in the population dynamics of rotifers. Arch. Hydrobiol., Beih. 8: 98–122.

Dumont, H. J., 1980. Zooplankton and the science of biogeography: the example of Africa. In W. C. Kerfoot (ed.) Evolution and Ecology of Zooplankton Communities. The University Press of New England, Hanover (N.H.); Lond.: 685–696.

Edmondson, W. T. (ed.), 1959. H. B. Ward and G. C. Whipple's Fresh-water Biology, 2nd edn. John Wiley & Sons, N.Y. Lond., 1248 pp.

Edmondson, W. T., 1965. Reproductive rate of planktonic rotifers as related to food and temperature in nature. Ecol. Monogr. 35: 61–111.

Edmondson, W. T. & G. C. Winberg (eds.), 1971. A Manual on Methods for the Assessment of Secondary productivity in Freshwaters. Blackwell Scientific Publications, Oxford.

Edmondson, W. T., 1974. Secondary production. Mitt. int. Ver. Limnol. 20: 229–272.

Fernando, C. H., 1980a. The species and size composition of tropical freshwater zooplankton with special reference to the oriental region (South East Asia). Int. Revue ges. Hydrobiol. 65: 411–426.

Fernando, C. H., 1980b. The freshwater zooplankton of Sri Lanka, with a discussion on tropical freshwater zooplankton composition. Int. Revue ges. Hydrobiol. 65: 85–125.

Gamatchu, D., 1977. Aspects of climate and water budget in Ethiopia. Addis Abeba University Press, Addis Abeba.

Gasse, F. & F. A. Street, 1978. Late quaternary lake-level fluctuations and environments of the Northern Rift Valley and Afar Depression (Ethiopia and Djibouti). Palaeogeography, Palaeoclimatology, Palaeoecology 24: 279–325.

Green, J., 1972. Latitudinal variation in associations of planktonic Rotifera. J. Zool., Lond. 167: 31–34.

Grove, A. T. & A. S. Gondie, 1971. Late quaternary lake-levels in the Rift Valley of Southern Ethiopia and elsewhere in tropical Africa. Nature 234: 403–405.

Herbst, H. R., 1954. The tolerance of oxygen deficiency in the water by certain Cladocera. Mem. Ist. ital. Idrobiol. 8: 97–107.

Heisey, D. & K. V. Porter, 1977. The effect of ambient oxygen concentration on filtering and respiration rates of *Daphnia geleata mendotae* and *Daphnia magna*. Limnol. Oceanogr. 22: 839–845.

LaBarbara, M. C. & P. Kilham, 1974. The Chemical ecology of copepod distribution in the lakes of east and central Africa. Limnol. Oceanogr. 10: 459–465.

Loffredo, S. & C. M. Maldura, 1941. Risultati generale delle ricerche di chimica limnologica sulle acqua dei laghi dell' Africa orientale italiana esplorati della Missione ittiologica. In G. Brunelli (ed.), Esplorazione dei Laghi della Fossa Galla, i. Collezione scientifica e documentaria dell' Africa Italiana III. Airoldi, Verbania: 181-200.

Lowndes, A. G., 1930. Freshwater Copepoda from Abyssinia collected by J. Omer Cooper. Proc. zool. Soc. 11: 161-179.

Moriarty, D. J. W., J. P. E. Darlington, I. G. Dunn, M. Moriarty & M. P. Telvin, 1973. Feeding and grazing in Lake George, Uganda. Proc. r. Soc. Lond. B 184: 299-319.

Mohr, P., 1962. The Ethiopian Rift System. Bull. Geophys. Obs. 3: 33-62.

O'Brien, W. J. & F. deNoyelles, Jr., 1972. Relationship between nutrient concentration, phytoplankton density, and zooplankton density in nutrient enriched experimental ponds. Hydrobiologia 44: 105-125.

Pejler, B., 1977. On the global distribution of the family Brachionidae (Rotatoria). Arch. Hydrobiol. Suppl. 53: 255-306.

Rzoska, J. (ed.), 1976. The Nile, Biology of an Ancient River. Monogaphiae Biologociae, 29. Junk, The Hague: 344 pp.

Talling, J. F. & D. Driver, 1963. Some problems in the estimation of chlorophyll a in phytoplankton. Proc. Conf. primary productivity measurement, mar. & freshwat., U.S. Atomic Energy Commission. TID 7633: 142-147.

Talling, J. F. & I. B. Talling, 1965. The chemical composition of African Lake Waters. Int. Revue ges. Hydrobiol. 50: 421-463.

Wodajo, K., 1982. Comparative Limnology of Lake Abiata and Lake Langano in relation to primary and secondary production. M.Sci. Thesis, Addis Abeba Univ.

Wood, R. B., M. V. Prosser & R. M. Baxter, 1978. Optical characteristics of the Rift Valley Lakes, Ethiopia. Sinet: ethiop. J. Sci: 1: 73-85.

Zooplankton composition of ten reservoirs in southern Brazil

Marlene Sofia Arcifa
Departamento de Biologia F.F.C.L., Universidade de São Paulo, Ribeirão Preto, Brazil

Keywords: tropical zooplankton, reservoirs, Brazil

Abstract

The zooplankton of ten reservoirs of São Paulo State was analyzed as part of a larger project, 'Typology of Reservoirs of São Paulo State'.

Twenty-four genera of Rotifera, six species of Copepoda and at least nine species of Cladocera were found in samples collected on four occasions in 1979. In general, Rotifera dominated in most reservoirs, although fluctuations occurred during the year.

The reservoirs were arranged in four groups, according to zooplankton density, whose range was 10 to 500 i l^{-1}.

The average composition of Crustacea, in number of species at any one time is comparable to those of other water bodies, being a little higher than that of Colorado lakes.

The number of species of limnetic Cladocera in Brazil is between those of Holarctic Region and Tropical Asia.

Ceriodaphnia cornuta and *Bosminopsis deitersi*, and a few species of *Daphnia* are typical of Brazilian zooplankton. *Thermocyclops crassus* is common in the southern reservoirs but *T. minutus* seems to be more widely distributed in Brazil. Calanoida occurred in relatively few reservoirs in São Paulo and usually one species at one time. *Brachionus* and *Keratella* were more abundant closer to the Equator then to the Tropics, where other genera seem to be more abundant.

The range in size of the planktonic Crustacea is relatively small when compared to temperate lakes, being similar to that of other tropical lakes.

Introduction

As part of a larger project 'Typology of Reservoirs of São Paulo State', which comprised the study of 52 reservoirs (Tundisi, 1981), this paper deals with the zooplankton of ten of these reservoirs.

A summary of the characteristics of the reservoirs can be found in Arcifa *et al* (1981), and detailed papers will be published in a near future.

The purpose of this paper is to show the following:
a) The year-round composition of the zooplankton in our reservoirs and a comparison with other water bodies.
b) The momentary composition of the zooplankton in number of species and a comparison with other temperate and tropical lakes.
c) The range in size of the planktonic Crustacea in Brazilian water bodies in comparison with temperate and tropical environments.

Materials and methods

The ten reservoirs studied were: Americana (Am), Atibainha (At), Cabuçu (Cb), Cachoeira (Ca), Ja-

Hydrobiologia 113, 137–145 (1984).

guari (Jg), Paraibuna [3 Stations: Félix Guisard (FG), Paraibuna (Pa), Paraitinga (Pt)], Pedro Beicht (PB), Ribeirão do Campo (RC), Santa Branca (SB), and Taiaçupeba (Ta). Distinction was not made between reservoirs and Paraibuna Stations.

Four samplings were taken in the reservoirs in 1979: March, May–June, August, and December, totaling 47 samples.

Samples of 200 l were taken with a pump and the water was filtered through a 68 μm net. Also vertical hauls were made with a 70 μm net. Samples were fixed with 4% formalin in the field.

As most Rotifera could not be recognized and numbers were underestimated in the pump samples, net samples were utilized for the countings. Crustacea were not as difficult.

For Crustacea, 1 ml aliquots were taken with a piston pipette, and countings were made in a transparent plastic box with a square ruling at the bottom, until 300 individuals were totalled; also a Petri dish with 8.7 cm in diameter divided in 4 quadrants was employed for samples for which 1 ml aliquots were not adequate. For Rotifera, a Sedgwick-Rafter chamber was used and up to 5 aliquots were counted. Whole samples were examined for rare species.

Results and discussion

The following genera and species were found in the 47 samples:

COPEPODA
- Calanoida
  - *Odontodiaptomus paulistanus* (Wright, 1936)
  - *Diaptomus corderoi* Wright, 1936
- Cyclopoida
  - *Thermocyclops crassus* Fischer, 1853
  - *Mesocyclops longisetus* (Thiebaud, 1914)
  - *Tropocyclops prasinus* (Fischer, 1960)
  - *Metacyclops mendocinus* Wierzejski, 1893

CLADOCERA
- *Daphnia gessneri* Herbst, 1967
- *Daphnia ambigua* Scourfield, 1947
- *Ceriodaphnia cornuta* Sars, 1886
- *C. reticulata* (Jurine, 1820)
- *Bosmina longirostris* (O.F. Müller, 1785) and probably other species
- *Bosminopsis deitersi* Richard, 1895
- *Diaphanosoma neotropicum* Brehm & Thomsen, 1936
- *Diaphanosoma* sp
- *Moina* sp

ROTIFERA
- *Brachionus* Pallas
- *Keratella* Bory de St. Vincent
- *Macrochaetus* Perty
- *Ptygura* Ehrenberg
- *'Ptygura'*
- *Sinantherina* Bory de St. Vincent
- *Floscularia* Cuvier
- *Polyarthra* Ehrenberg
- *Synchaeta* Ehrenberg
- *Conochilus* Hlava
- *Conochiloides* Hlava
- *Collotheca* Harring
- *Stephanocerus* Ehrenberg
- *Hexarthra* Schmarda
- *Trichocerca* Lamarck
- *Gastropus* Imhof
- *Ascomorpha* Perty
- *Testudinella* Bory de St. Vincent
- *Filinia Bory* de St. Vincent
- *Pompholyx* Gosse
- *Lecane* Nitzsch
- *Monostyla* Ehrenberg
- *Asplanchna* Gosse
- Fam. Philodinidae

The percentage composition of the zooplankton groups considering each sampling period (Table 1) showed the dominance of Rotifera in most samplings in the reservoirs. On a few occasions Copepoda and Cladocera were predominant due to a 'population explosion' of one or two species as, for instance, *Bosmina* spp in Cabuçu and *Bosminopsis deitersi* in Santa Branca, in August.

Rotifera also predominated in most of the 17 other reservoirs of São Paulo studied at the same time as ours (Sendacz *et al*, in press).

Noteworthy in Table 1 are the high percentages of nauplius and copepodid stages when compared to the percentages of adults.

The most frequent organisms were *Bosmina* spp, *Bosminopsis deitersi, Diaphanosoma* spp (mainly D. sp.; *D. neotropicum* is rare), *Ceriodaphnia cornuta, Thermocyclops crassus, Tropocyclops prasi-*

Table 1. Percentage composition of zooplankton groups in the four sampling periods.

	1	2	3	4	1	2	3	4
	Americana				Atibainha			
Copepoda N + C	14	9	75	34	14	28	46	15
AD.	<1	<1	2	2	<1	1	7	16
Cladocera	10	42	2	17	6	2	2	27
Rotifera	75	48	20	47	79	68	45	42
	Cabuçu				Cachoeira			
Copepoda N + C	<1	0	<1	<1	25	32	10	54
AD.	<1	<1	0	0	2	4	1	11
Cladocera	5	2	94	23	7	15	35	14
Rotifera	94	97	5	76	65	48	54	21
	F. Guisard				Jaguari			
Copepoda N + C	59	12	32	67	46	33	22	13
AD.	<1	6	1	2	3	9	<1	1
Cladocera	8	32	26	13	4	12	9	8
Rotifera	32	49	41	18	47	45	68	78
	Paraibuna				Paraitinga			
Copepoda N + C	<1	13	14	41	25	15	25	4
AD.	<1	2	2	4	2	6	9	3
Cladocera	<1	9	71	32	11	32	52	1
Rotifera	99	76	13	23	62	47	14	92
	Pedro Beicht				Ribeirão do Campo			
Copepoda N + C	5	5	47	9	8	37	71	
AD.	<1	4	5	4	1	22	7	
Cladocera	<1	3	6	21	1	20	9	
Rotifera	94	88	42	66	90	21	13	
	Sta Branca				Taiaçupeba			
Copepoda N + C	<1	17	2	4	1	6	<1	6
AD.	<1	2	0	3	<1	1	<1	1
Cladocera	<1	29	77	25	50	27	<1	15
Rotifera	99	52	21	68	48	66	99	78

N = nauplius, C = copepodid, Ad. = adult.

Table 2. Frequency of occurrence of Crustacea and Rotifera in 47 samples taken in the four sampling periods.

	1	2	3	4	Total	%
Collotheca	11	11	12	11	45	96
Bosmina spp	9	12	11	11	43	91
Diaphanosoma spp	10	11	6	10	37	79
Conochilus	8	11	7	10	35	76
Keratella	12	8	7	11	35	74
Bosminopsis deitersi	8	11	8	8	35	74
Ptygura	11	9	6	8	34	72
Polyarthra	6	10	10	6	32	68
Ceriodaphnia cornuta	10	9	5	8	32	68
Thermocyclops crassus	10	10	4	8	32	68
Gastropus	10	6	8	7	31	66
Daphnia gessneri	7	10	5	8	30	64
Hexartha	8	7	2	9	26	55
Brachionus	6	7	6	5	24	51
Tropocyclops prasinus	6	6	4	5	21	44
Trichocerca	6	8	2	3	19	40
Ceriodaphnia reticulata	6	8	2	2	18	38
Odontodiaptomus paulistanus	2	5	5	3	15	32
Lecane	5	1	6	3	15	32
Mesocyclops longisetus	5	4	3	2	14	29
Filinia	3	5	2	2	12	25
Asplanchna	3	4	3	1	11	23
Metacyclops mendocinus	4	2	1	2	9	19
Macrochaetus	4	3	2	0	9	19
Moina sp	4	3	1	1	9	19
Diaptomus corderoi	2	1	2	2	7	15
Testudinella	0	2	2	2	6	15
Others					1-5	2-10

nus, among Crustacea, and *Collotheca*, *Conochilus*, *Keratella*, and *Ptygura* among Rotifera (Table 2).

Fluctuations in population density took place from one sampling period to another and for this reason it is difficult to discuss dominant species in general. A table was constructed with four samplings (Table 3), indicating dominant organisms in each zooplankton group. *Bosmina* spp and *Bosminopsis deitersi* predominated in most reservoirs followed by *Ceriodaphnia cornuta* and *Diaphanosoma* sp. *Daphnia gessneri* appeared in high numbers only in Americana, although it is relatively frequent (Table 2). Among Copepoda, *Thermocyclops crassus* was the most abundant followed by *Tropocyclops prasinus*. In two reservoirs, Ribeirão do Campo and Paraibuna (St. Paraibuna), *Odontodiaptomus paulistanus* and *Diaptomus corderoi*, respectively, were dominant over Cyclopoida. *Ptygura*, *Conochilus*, and *Collotheca* were the most abundant Rotifera.

In other reservoirs of São Paulo (Sendacz *et al*, *op. cit.)*, *C. cornuta*, *Diaphanosoma* sp, *Bosmina* sp, *T. crassus*, *Tropocyclops prasinus*, *Collotheca*, *Keratella*, *Polyarthra* and *Trichocerca* were among the most frequent and sometimes the most abundant organisms.

As regards the genera *Thermocyclops* and *Mesocyclops*, pointed out as a typical association in tropical waters (Hutchinson, 1967), they occurred together in most of our reservoirs. Generally *M. longisetus* was not abundant.

According to the average densities of zooplankton (Table 4), it is possible to arrange the reservoirs in four groups:

Table 3. Dominant organisms (in %, above 10%) in each zooplankton group considering the means of the four samplings (excluding nauplius and copepodid stages).

	Am	At	Cb	Ca	FG	Ja	Pa	Pt	PB	RC	SB	Ta
Bosmina spp	14	32	89	16	11	11			36	16		48
Bosminopsis deitersi		11		33	19		46	34	63	66	87	20
Daphnia gessneri	46											
Ceriodaphnia cornuta	32	17		12	37	43	15	27				
C. reticulata						22		14				
Diaphanosoma sp		30		24	24	19	26	14				25
Thermocyclops crassus	46	57	42	70	71	63	17	20			59	62
Metacyclops mendocinus	54			19								
Tropocyclops prasinus		36	57	11				99		20		20
Mesocyclops longisetus					25			15				
Odontodiaptomus paulistanus										80	40	
Diaptomus corderoi						31	76	63				14
Conochilus	77	12	18	16	25	27		24			14	54
Keratella		20		10								
Ptygura		38	37	57		37	37	29	29		27	
Collotheca		15	14		13	20	34	15	69	84	20	
'Ptygura'			14									
Hexarthra					33							
Polyarthra												18

Table 4. Average densities of zooplankton in 10 reservoirs in 1979 (nauplius and copepodid stages included).

Reservoirs	$i\,l^{-1}$
Taiaçupeba (Ta)	581
Americana (Am)	504
Cabuçu (Cb)	172
Pedro Beicht (PB)	157
Atibainha (At)	59
Cachoeira (Ca)	57
Sta. Branca (SB)	47
Félix Guisard (FG)	41
Rib. do Campo (RC)	31
Jaguarí (Ja)	24
Paraitinga (Pt)	12
Paraibuna (Pa)	10

I. Americana, Taiaçupeba: ca. 500 $i\,l^{-1}$
II. Cabuçu, Pedro Beicht: 100–200 $i\,l^{-1}$
III. Atibainha, Cachoeira, Santa Branca, Paraibuna (St. F. Guisard), Ribeirão do Campo, and Jaguari: 20–60 il^{-1}
IV. Paraibuna (St. Paraitinga and Paraibuna): ca. 10 $i\,l^{-1}$

Americana is the most eutrophic of the reservoirs and its richness in zooplankton coincided with high values of conductivity, chlorophyll a, and primary productivity (Arcifa *et al*, 1981). Taiaçupeba was still filling up in 1979 showing large quantities of plant fibres in suspension. Although the densities of zooplankton were similar in both reservoirs, the nature of the zooplankton communities separated them as the dominance of Rotifera as regards Crustacea (mainly Copepoda) was evident in the latter one (Table 1).

Besides zooplankton densities, Cabuçu and Pedro Beicht have in common the fact of being the oldest of the reservoirs (built in 1907 and 1913, respectively). P. Beicht, despite having one of the lowest values of conductivity, showed relatively high contents of chlorophyll a, and primary productivity rates (Arcifa *et al, op. cit.*), suggesting a kind of 'steady tate' if it is possible to speak about stability in a reservoir.

The scarcity of zooplankton in the two stations of Paraibuna (Pa and Pt) coincided with poverty in chlorophyll a, setting them apart of the other station of the same reservoir (FG), whose features justified its position in group III.

The wide distribution of the organisms in general and taxonomic problems make difficult the grouping of the reservoirs according to zooplankton composition. Still unsolved taxonomic problems made useless the application of any similarity index; on the other hand it would be more interesting to deal with the data of the 52 reservoirs of the 'Typology Project', which will be done in the future.

In any case some observations can be made which will probably apply to the other reservoirs in São Paulo. Calanoida, due to their restricted distribution characterize the 5 reservoirs were they occurred – Jaguari, Paraibuna (3 stations), Ribeirão do Campo, Santa Branca and Taiaçupeba. Another feature of Calanoida distribution was that the two species never occurred together, except in one reservoir. The same fact was observed by Sendacz *et al (op cit.)* in other reservoirs of São Paulo, where the association of two species of Calanoida occurred in three of the seventeen reservoirs.

Americana, which can be isolated by several features, is also unique among the reervoirs by the fact that only in it *Daphnia gessneri* occurred in relatively large numbers.

The composition of the limnetic zooplankton at any one time in the ten reservoirs differs a little from those of other water bodies (Table 5). Our averages are higher than those of the Colorado lakes, known by their simplicity of momentary planktonic composition (Pennak, 1957). The averages for Nhamundá lakes are higher, but according to the authors this could be explained by the many littoral species included (Brandorff *et al,* 1982).

The average for Rotifera in our reservoirs would most probably be higher if species were taken into consideration. Dra. Nora Ponyi identified the Rotifera in one sample, from Americana Reservoir, and found 2 species of *Brachionus (angularis* and *calyciflorus dorcas)*, 2 species of *Keratella (cochlearis* and *tropica)*, 2 *Polyarthra (vulgaris* and *major)* and 1 species of *Conochilus (unicornis)*.

Patalas (1971) found lakes with simple to complex zooplankton composition, but for purpose of comparison I calculated the means, which showed relatively larger number for Copepoda mainly due to several species of Calanoida. This is a clear difference with our reservoirs, where the number of species of Calanoida was low.

Lake Lanao (Lewis, 1979) has a comparative simpler zooplankton community considering a year-round composition which according to the author is not much different from its momentary composition.

It would not be wrong to say that the momentary composition of zooplankton of São Paulo reservoirs and of other tropical lakes was similar to those of some temperate lakes.

Based on a survey in the available Brazilian literature on water bodies located from the North to the Tropic of Capricorn, comprising lakes and reservoirs with variable characteristics, it was possible to construct a table with the supposedly planktonic Crustacea (Table 7). Based on that table, on data of this work and in Fernando (1980a), a table was made to compare the number of species of Cladocera present in the Holarctic Region, in Tropical Asia, and Brazil (Table 6).

The number of limnetic Cladocera in the Holarctic Region is higher than in Brazil, as well as in Tropical Asia, due largely to the high numbers of *Daphnia* species. The higher numbers of species of Bosminidae here are due chiefly to the addition of a

Table 5. Composition of limnetic zooplankton in the 10 reservoirs, considering the 47 samples taken in 1979, and comparison of their means with data of other authors.

Taxonomic category	Number of species per sample																Total number of species and genera of Rotifera
	0	1	2	3	4	5	6	7	8	9	10	11	12	13	Mean	Median	
Copepoda	3	11	17	13	2	1									2.0	2	6
Cladocera		3	4	7	7	7	16	3							4.5	6	9
Rotifera[a]	1	1		4		2	3	12	8	5	5	4	1	1	7.5[a]	7[a]	24[a]

	Brandorff *et al* (1982)	Pennak (1957)	World (in Pennak)	Patalas[b] (1971)
Copepoda	3.2	1.3	2.7	4.5
Cladocera	5.8	1.6	2.8	3.8
Rotifera	21.1	4.8	5.5	–

[a] Genera only (including *Lecane*)
[b] Calculations based on Patalas' data.

Table 6. Comparison of the composition of limnetic Cladocera in the Holarctic Region, Tropical Asia, and Brazil.

Families *Genera*	Holarctic[a]	Trop. Asia[a]	Brazil[b]
	Number of species		
	30	12	19
Holopedidae	1	0	1
Polyphemidae	1	0	0
Leptodoridae	1	0	0
Daphnidae			
Daphnia	18	1	3
Ceriodaphnia	4	2	3
Sididae			
Diaphanosoma	2	5	4
Bosminidae	3	3	5
Moinidae	0	1	3

[a] Fernando (1980a).
[b] See Table 7.

new species – *Bosminopsis negrensis* Brandorff (1976a). Apparently there is another endemic species of *Bosminopsis* in Amazonia reported by Brandorff *et al* (1982).

In Tropical Asia there is one species of planktonic Moinidae – *Moina micrura* – whose importance is smaller here, where *M. minuta* and *M. reticulata* assume a greater significance in terms of occurrence. *M. micrura*, although not usually planktonic in temperate lakes was found to be euplanktonic in two lakes in Europe (Goulden, 1968).

To conclude, the Cladocera of the reservoirs in São Paulo and other regions of Brazil are tropical: few species of *Daphnia, Ceriodaphnia cornuta* widely distributed and *Diaphanosoma* relatively frequent *(brachyurum* and *sarsi* are the most frequent species, the latter apparently restricted to Amazonian waters). *Bosminopsis deitersi*, which Fernando (1980a) points out as occurring sometimes in tropical limnetic zooplankton is very common here, sometimes being the dominant species. Species of *Bosmina* are also frequent and sometimes abundant in some of the southern reservoirs.

Concerning Copepoda, *Thermocyclops crassus* is one of the most common Cyclopoida in São Paulo reservoirs, as well as in South East Asia and African lakes (frequently reported as *T. hyalinus* in the latter ones), but it seems that *T. minutus* is more widely distributed in Brazil.

Calanoida were not as frequent in São Paulo reservoirs as in temperate lakes. Some families with large species were not represented in Brazil, considering the available literature. The diversity of Calanoida seems to be larger in Amazonia than in southern reservoirs (see Brandorff, 1976b for S. American Diaptomidae).

The genus *Brachionus* is considered to assume a greater importance towards the Equator (Green, 1972b; Pejler, 1977). Fernando (1980a) also indicates the dominance of *Brachionus* and *Keratella tropica* in Sri Lanka. In Brazil, in three of the five lakes studied by Green (1972c), *Brachionus* was dominant. Brandorff *et al* (1982) found *Brachionus zahniseri* (two varieties) and *Keratella americana* by far the predominant Rotifera. In our reservoirs, although *Keratella* was frequent, in general it was not abundant. Perhaps this is distinctive for the tropics, i.e. *Brachionus* or Brachionidae could be dominant near the Equator but not closer to the limit of the tropics, where other genera and families may be dominant.

The size composition of the planktonic Crustacea in Brazilian water bodies has a smaller range than in temperate ones due to the absence of large Cladocera and Copepoda. The lower end of the spectrum (0.35–0.40 mm) is represented by small species such as *Bosmina* spp, *Bosminopsis deitersi*, and *Ceriodaphnia cornuta* and the upper end by *Daphnia gessneri* (mean – 1.3 mm; maximum size – 1.5 mm). Concerning Copepoda the size range can be enlarged in some cases with the occurrence of *Argyrodiaptomus furcatus*, whose female can attain 1.65 mm, on average (Sendacz & Kubo, in press).

The size range is similar to that found in other tropical water bodies in Africa, South East Asia, Central America (Burgis *et al*, 1973; Fernando, 1980b; Zaret, 1980). Some exceptions can take place as the occurrence of large Cladocera and Copepoda in Vietnam (Shirota, in Fernando, 1980b), considered by the latter author as subtropical, although located in the tropics.

Temporary and weedy ponds can shelter larger species of Cladocera and Copepoda than the limnetic zone of tropical lakes and reservoirs. Thus, large Calanoida can be found in temporary ponds of Sri Lanka (Fernando, 1980a). Species of *Simocephalus* are not uncommon in São Paulo; large and coloured forms (approximately 2 mm) can be found among macrophytes in ponds.

Table 7. Occurrence of Cladocera and Copepoda in the plankton of lakes and reservoirs in Brazil.

	Res. Broa [1, 2]	Res. Billings [3]	Res. Americana [4, 5]	Res. Jacaré-Pepira [6]	Lagoa Belgo-Mineira [7]	Lago D. Helvécio [7, 8]	Lagoa Bonita [7]	Lagoa Carioca [7]	Lago Castanho [9, 10]	Lago Joanicu [11]	Res. Curúa Una [12]	Lago Redondo [10]	Lago Jacaretinga [10, 13]	Lago Cristalino [10]	Lago Tarumã Mirim [10]	5 Lagos Mato Grosso [14]	Lakes - Rio Nhamundá [15]	17 Res. S. Paulo [16]
Cladocera																		
Daphnia gessneri		+	+	+		+			+	+	+		+					+
D. dubia					+													
+ *Ceriodaphnia cornuta*	+	+	+	+	+	+	+	+	+	+	+	+	+	+	+	+	+	+
C. reticulata			+	+					+		+							
C. pulchella										+								
Bosmina coregoni	+				+	+	+	+										
B. longirostris			+															
* *B. chilensis*				+					+			+	+	+	+			
* *B. hagmanni*											+					+		
* *Neobosmina hagmanni*										+								
Bosminopsis deitersi	+		+	+		+	+		+	+	+	+	+	+	+	+	+	+
B. negrensis														+	+			
Moina reticulata					+				+		+	+	+	+	+		+	
Moina minuta	+					+			+		+						+	+
Moina micrura				+														
Diaphanosoma brachyurum	+			+	+	+	+									+		
D. sarsi									+		+		+		+	+	+	
D. neotropicum			+															
D. fluviatile													+					
Holopedium amazonicum									+					+	+	+	+	
Copepoda					Only Cladocera			Only Cladocera			Only Cladocera					Only Cladocera		
'Diaptomus' negrensis																	+	
Diaptomus corderoi		+				+												+
'D. linus'																	+	
'Diaptomus ohlei'									+									
Notodiaptomus amazonicus									+			+	+					
N. coniferoides									+	+			+					
N. conifer																		+
N. iheringi																		+
Argyrodiaptomus azevedoi									+								+	
A. furcatus	+			+		+												+
Dactylodiaptomus pearsei									+								+	
Odontodiaptomus paulistanus																		+
Rhacodiaptomus retroflexus																	+	
R. calamensis																	+	
Aspinus acicularis														+			+	
Trichodiaptomus coronatus																	+	
Oithona amazonica									+			+	+	+	+		+	
Thermocyclops crassus		+	+															+
T. minutus	+					+			+	+			+		+		+	+
Mesocyclops longisetus			+	+		+			+								+	+
M. leuckarti									+									
M. meridianus																	+	
Tropocyclops prasinus			+	+		+												+
Microcyclops anceps																		+
Metacyclops mendocinus		+	+															+

References: 1. Matsumura-Tundisi & Tundisi (1976); 2. Heredia-Seixas (1981); 3. Sendacz (1978); 4. Zago (1976); 5. Carvalho (1975); 6. Claro (1981); 7. Santos (1980); 8. Okano (1980); 9. Brandorff (1977); 10. Hardy (1978); 11. Herbst (1967); 12. Junk *et al.* (1981); 13. Brandorff & Andrade (1978); 14. Green (1972a); 15. Brandorff *et al* (1982); 16. Sendacz *et al* (in press).

* Probably same species.

\+ *Ceriodaphnia cornuta* includes forms *rigaudi* and *cornuta*.

The question, 'why is the size range of tropical zooplankton smaller in comparison to the temperate one?' still remains unanswered. If the influence of predation by planktivorous, mainly fish, intensively studied by several authors and compiled by Zaret (1980), can not explain certain cases, what can we say about factors such as temperature? If temperature, at first sight, seems to be the preponderant factor, how could the occurrence of larger forms of Copepoda and Cladocera in ponds be explained, or the fact that the largest species of other invertebrates as many insect orders and arachnids, for example, are inhabitants of the tropics?

The idea that food is not suitable in the tropics, chiefly for Cladocera, is a general one. But only in eutrophic water bodies do blue-green algae reach densities high enough to inhibit filter feeders. The problem with blue-greens is not restricted to the tropics; it also extends to temperate eutrophic water bodies.

Fernando's question (1980b: 422), if low oxygen concentration is responsible for the low diversity of Cladocera, implies that in tropical water bodies, oxygen is low, what is not always true. Oxygen can be low in the water column in some water bodies under certain circumstances or in some cases in the hypolimnion, which can be avoided by zooplankters.

Acknowledgements

My thanks to Dr. M. A. J. Carvalho for Copepoda and Rotifera identifications, and countings of the latter, and M. C. Marino and R. Spada for counting the Crustacea, S. Sendacz for helping with Brazilian literature, Dr. C. G. Froehlich for revision of the English, and FAPESP for financial support.

References

Arcifa, M. S., M. A. J. Carvalho, S. M. F. Gianesella-Galvão, G. Y. Shimizu, C. G. Froehlich & R. M. C. Castro, 1981. Limnology of ten reservoirs in Southern Brazil. Verh. int. Ver. Limnol. 21: 1048–1053.

Brandorff, G. O., 1976a. A new species of *Bosminopsis* (Crustacea, Cladocera) from the Rio Negro. Acta Amazônica 6: 109–114.

Brandorff, G. O., 1976b. The geographic distribution of the Diaptomidae in South America (Crustacea, Copepoda). Revta. bras. Biol. 36: 613–627.

Brandorff, G. O., 1977. Untersuchungen zur Populationsdynamick des Crustaceenplanktons im tropischen Lago Castanho (Amazonas, Brasilien). PhD Thesis, Univ. Kiel, 109 pp.

Brandorff, G. O. & E. R. Andrade, 1978. The relationship between the water level of the Amazon River and the fate of the zooplankton population in Lago Jacaretinga, a vázea lake in the Central Amazon. Stud. neotrop. Fauna Env. 13: 63–70.

Brandorff, G. O., W. Koste & N. N. Smirnov, 1982. The composition and structure of Rotiferan and Crustacean communities of the lower Rio Nhamundá, Amazonas, Brazil. Stud. neotrop. Fauna Env. 17: 69–121.

Burgis, M., J. P. E. C. Darlington, I G. Dunn, G. C. Ganf, J. J. Gwohaba & L. M. McGowen, 1973. The biomass and distribution of organisms in Lake George, Uganda. Proc. r. Soc. Lond. 184: 271–298.

Carvalho, M. A. J., 1975. A Represa de Americana: aspectos físico-químicos e a variação das populações de Copepoda Cyclopoida de vida livre. Tese Dout., Univ. S. Paulo, 80 pp.

Claro, S. M., 1981. Aspectos limnológicos da Represa do Jacaré-Pepira (Brotas-S.P.) com ènfase na comunidade zooplanctônica. Diss. Mestr., Univ. Fed. S. Carlos, 206 pp.

Fernando, C. H., 1980a. The freshwater zooplankton of Sri Lanka, with a discussion of tropical freshwater zooplankton composition. Int. Revue ges. Hydrobiol. 65: 85–125.

Fernando, C. H., 1980b. The species and size composition of tropical freshwater zooplankton with special reference to the Oriental Region. Int. Revue ges. Hydrobiol. 65: 411–426.

Goulden, C. E., 1968. The systematics and evolution of the Moinidae. Trans. am. phil. Soc. Philadelphia 58: 1–101.

Green, J., 1972a. Freshwater ecology in the Mato Grosso, Central Brazil, 2. Associations of Cladocera in meander lakes of the Rio Suiá Missú. J. nat. Hist. 6: 215–227.

Green, J., 1972b. Latitudinal variation in associations of planktonic rotifera. J. Zool., Lond. 167: 31–39.

Green, J., 1972c. Freshwater ecology in the Mato Grosso, Central Brazil, 3. Associations of Rotifera in meander lakes of the Rio Suiá Missú. J. nat. Hist. 6: 229–241.

Hardy, E. R., 1978. Composição do zooplâncton em cinco lagos da Amazônia Central. Diss. Mestr., Univ. Fed. S. Carlos, 149 pp.

Herbst, H. V., 1967. Copepoda und Cladocera (Crustacea) aus Südamerika. Gewäss. Abwäss. 44, 45: 96–108.

Heredia-Seixas, M., 1981. Aspectos ecológicos das populações de Cladocera (Crustacea) na Represa do Lobo ('Broa'), São Carlos, SP. Diss. Mestr., Univ. Fed. S. Carlos, 116 pp.

Hutchinson, G. E., 1967. A treatise on limnology. 2: Introduction to lake biology and the limnoplankton. John Wiley & Sons, N.Y., 1115 pp.

Junk, W. J., B. A. Robertson, A. J. Darwich & I. Veira, 1981. Investigações limnológicas e ictiológicas em Curuá-Una, a primeira represa hidrelética na Amazônia Central. Acta Amazonica 11: 689–716.

Lewis, W. M., 1979. Zooplankton community analysis: studies on a tropical system. Springer-Verlag, N.Y., 163 pp.

Matsumura-Tundisi, T. & J. G. Tundisi, 1976. Plankton studies in a lacustrine environment, 1. Preliminary data on zooplankton ecology of Broa Reservoir. Oecologia 25: 265-270.

Okano, W. Y., 1980. Padrão de migração vertical e flutuação sazonal das principais espécies de Copepoda (Crustacea) do Lago Dom Helvécio. Parque florestal do Rio Doce-MG. Diss. Mestr., Univ. Fed. S. Carlos, 168 pp.

Patalas, K., 1971. Crustacean plankton communities in forty-five lakes in the Experimental Lakes Area, northwestern Ontario. J. Fish. Res. Bd Can. 28: 231-244.

Pejler, B., 1977. On the global distribution of the family Brachionidae (Rotatoria). Arch. Hydrobiol., Suppl. 53: 255-306.

Pennak, R. W., 1957. Species composition of limnetic zooplankton communities. Limnol. Oceanogr. 2: 222-232.

dos Santos, L. C., 1980. Estudo das populações de Cladocera em cinco lagos naturais (Parque Florestal do Rio Doce-MG) que se encontram em diferentes estágios de evolução. Diss. Mestr., Univ. Fed. S. Carlos, 260 pp.

Sendacz, S., 1978. Alguns aspectos do zooplâncton da Represa Billings. Diss. Mestr., Univ. S. Paulo, 84 pp.

Sendacz, S. & E. Kubo in press. Copepoda (Calanoida e Cyclopoida) de reservatórios do Estado de São Paulo. Bol. Inst. Pesca.

Sendacz, S., E. Kubo & M. A. Cestarolli, in press. Composição do zooplâncton de 17 reservatórios do sudeste do Estado de São Paulo. Monogr. Ecol.

Tundisi, J. G., 1981. Typology of reservoirs in Southern Brazil. Verh. int. Ver. Limnol. 21: 1031-1039.

Zago, M. S. A., 1976. The planktonic Cladocera (Crustacea) and aspects of the eutrophication of Americana Reservoir, Brazil. Bolm. Zool., Univ. S. Paulo 1: 105-145.

Zaret, T. M., 1980. Predation and freshwater Communities. Yale University Press, New Haven & London, 187 pp.

The distribution of brackish water Calanoida (Copepoda) along the coasts of Brazil

M. S. de Almeida Prado Por & F. A. Lansac Tôha
Instituto Oceanográfico, Universidade de São Paulo, CEP-05008 São Paulo, Brazil

Keywords: tropical zooplankton, estuarine Calanoida, marine zoogeography, Brazilian coast

Abstract

The distribution of 18 species of Calanoida Copepoda found in the estuaries of the Brazilian coast is analyzed. A slight tendency of a latitudinal decrease in diversity, from North to South is found. Salinity ranges from the different species are discussed and in some cases interspecific vicariance patterns are found. Two basic types of estuaries of the Brazilian coast are characterized on the basis of environmental parameters and population dynamics.

Introduction

The distribution and the communities of brackish water Calanoida along the coasts of Brazil have been analyzed on the basis of data from the literature and some personal, unpublished data.

From North to South, Calanoida of brackish waters have been reported from the following localities (see also Fig. 1):

1. Estuary of River Tocantins (Pará)
   Lat. 1°0′S (Dahl, 1894; Cipolli & Carvalho, 1973; pers. data)
2. Rio Capim and Rio Guamá (Pará)
   Lat. 1°5′S (Cipolli & Carvalho, 1973)
3. Rio Paraiba do Norte (Paraiba)
   Lat. 7°0′S (Nordi, 1982; Pekala, 1982)
4. Rio Botafogo, Itamaracá (Pernambuco)
   Lat. 7°49′S (Nascimento, 1980)
5. Rio Pomonga (Sergipe)
   Lat. 10°40′S (Alcantara *et al*, 1979)
6. Baia de Vitória (Espirito Santo)
   Lat. 20°30′S (pers. data)
7. Rio Una do Prelado (São Paulo)
   Lat. 24°40′S (pers. data)
8. Estuary and lagoon of Cananéia (São Paulo)
   Lat. 25°0′S (J. Tundisi & T. M. Tundisi, 1968; T. M. Tundisi, 1977; Almeida Prado & Fonseca, unpubl.)

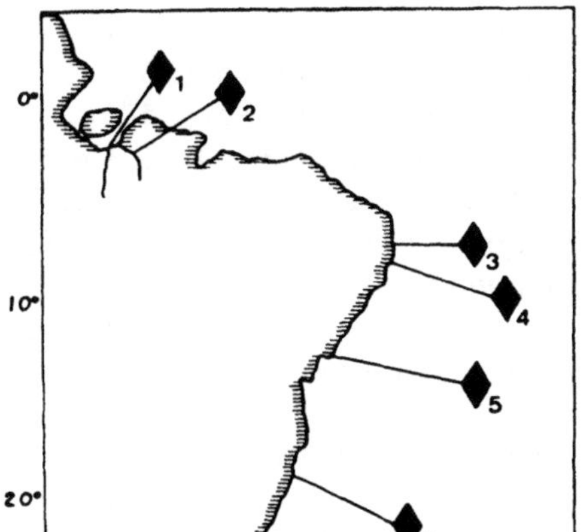

Fig. 1. Communities of brackish water Calanoida along the coast of Brazil. Numbers refer to the list of localities mentioned in the text.

Several other localities and salinity data for species of *Pseudodiaptomus* are given by Wright

(1936). Information on the riverine Diaptomidae is summarized by Cipolla & Carvalho (1973). General data on distribution have been taken from Björnberg (1963).

Results

Eighteen species of calanoids are up till now on record from the Brazilian coastal brackish waters:

	Localities (see Fig. 1)
Eucalanus monachus (Giesbrecht)	1
Eucalanus pileatus Giesbrecht	3, 5
Paracalanus aculeatus Giesbrecht	1, 5
Paracalanus crassirostris Dahl	1, 3, 4, 5, 6, 7, 8
Paracalanus quasimodo Bowman	3, 5
Pracalanus campaneri Björnberg	5
Clausocalanus furcatus (Brady)	5, 7
Centropages velificatus (Dana)	1, 3, 4, 5, 7, 8
Temora stylifera (Dana)	1, 3, 5, 7, 8
Pseudodiaptomus acutus (Dahl)	1, 3, 4, 5, 7, 8
Pseudodiaptomus richardii (Dahl)	1, 2, 3, 4, 5, 6, 7, 8
Pseudodiaptomus marshi Wright	3
Pseudodiaptomus gracilis (Dahl)	1, 2
Labidocera fluviatilis Dahl	1, 3, 4, 5, 8
Calanopia americana Dahl	1, 3, 4, 5
Acartia lilljeborgi Giesbrecht	1, 3, 4, 5, 6, 7, 8
Acartia giesbrechti Dahl	1
Acartia tonsa Dana	8

The central complex of species which appears in all the estuaries along the East coast of Brazil as far South as 25°S consists of *Paracalanus crassirostris, Pseudodiaptomus acutus, P. richardi* and *Acartia lilljeborgi*. Beyond the tidal influence in the rivers, the euryhaline calanoids are replaced by the complex of species belonging to genus *Notodiaptomus* Kiefer (Cipolla & Carvalho, 1973). Towards the open sea the estuarine species complex is replaced by the dominant neritic species *Centropages velificatus* and *Temora stylifera*; these two neritic species, together with a few other frequently mix with the typical estuarine fauna, as discussed below.

There is some decrease in the species diversity from the equatorial North to the temperate South: *Eucalanus monachus, Acartia giesbrechti* and *Pseudodiaptomus gracilis* have till now been reported only from the Amazonian system. *Pseudodiaptomus marshi* also seems to be restricted to North Brazilian waters. *Calanopia americana* and *Paracalanus quasimodo* are found in estuaries of the Northeast but in temperate waters they appear only in open, neritic waters. *Acartia tonsa*, a species considered by Björnberg (1981) to be typical for the temperate-subantarctic brackish waters of the southern coasts of South America, appeared recently in the well investigated plankton of Cananéia, after the closure of the Valo Grande canal and the general increase of salinities there. The synonymity of this species with the geographically distant *A. giesbrechti* has still to be settled (Björnberg, 1981).

In Table 1 an attempt is made to present the salinity preferences of some of the most important species of the Brazilian estuarine calanoids. The data should be considered only approximative, because of the large temperature differences between the widely spaced localities; the lack of experimental work on salinity:temperature spectra; and the well-known liability of estuarine field conditions.

There are few data on the oligohaline, low salinity zone. Only *Pseudodiaptomus gracilis* is reported to appear jointly with the more euryhaline freshwater diaptomids. From mesohaline values to seawater concentration there are several successional series such as that of *Acartia lilljeborgi*→*Labidocera fluviatilis*→*Centropages velificatus*; the intrageneris succession of *Pseudodiaptomus gracilis*→ *P. marshi*→*P. richardi*→*P. acutus*. Likewise *Paracalanus crassirostris*, dominant in the estuaries, is replaced toward the open sea by other species of this genus.

Two basic types of estuaries can be recognized along the Brazilian coasts in accordance to their physiography and faunal dynamics (Fig. 2). *Type A* estuaries are more complex systems where a lagoon or a marine coastal waterway is intercalated between the river estuary proper and the open sea. This type is best represented by the Cananéia system but another example is that of the Santa Cruz channel of Itamaracá (Pernambuco) with its inflowing rivers (among them Rio Botafogo). In this estuary

Table 1. Preferences of the most important estuarine Calanoida of Brazil.

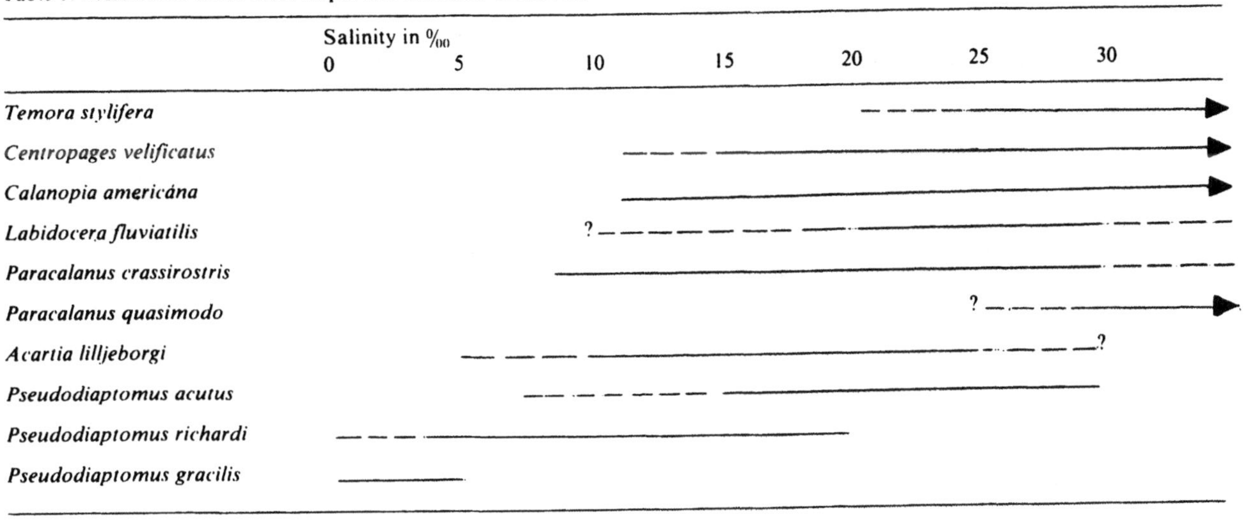

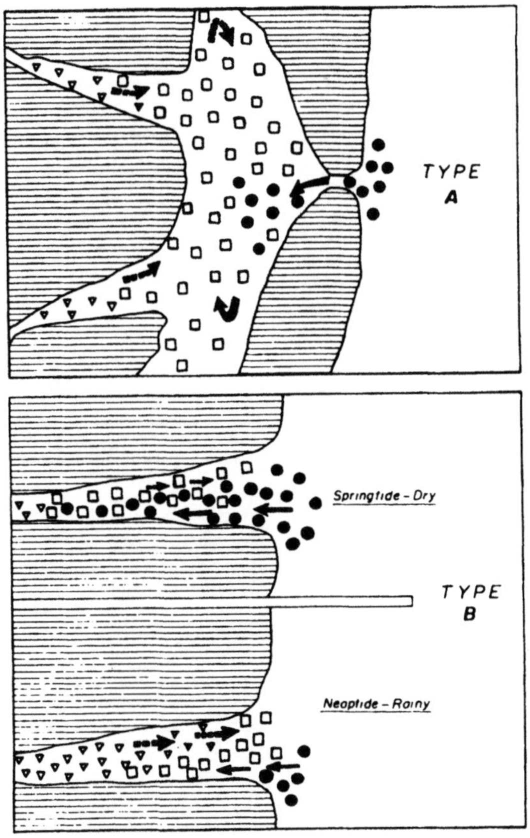

Fig. 2. Two types of estuaries of the Brazilian coast. Symbols as follows: □ brackish biota, ● marine biota, ▿ freshwater biota, ➔ marine inflow, - - ➔ freshwater inflow.

type a more stable estuarine community can be found over the year, relatively little influenced by the marine coastal plankton and by the upland freshwater plankton.

Type B estuaries are generally situated on late post-pliocenic lowlands and they empty directly into the sea. In such estuaries, like that of Rio Una do Prelado and of Rio Pomonga, at spring-high tides many marine coastal species penetrate far up the estuary. Sometimes marine species may be found many kilometers upstream. Conversely, at neap tides, especially after heavy rains, freshwater plankton invades the whole estuary. The typical estuarine calanoids, wedged in between the two extreme communities are dislocated accordingly over wide distances within the estuary. Marine species that are reported to advance over wide distances into the estuaries are *Clausocalanus furcatus, Paracalanus aculeatus* and *P. quasimodo.*

References

Alcântara, A. V., C. E. Falavigna de Rocha & M. A. Santos, 1979. Caracterização hidrológica e biológica do Rio Sergipe (mimeo). Aracaju, Sergipe, 123 pp.

Björnberg, T. K. S., 1963. On the marine free-living Copepoda off Brazil. Bolm. Inst. Oceanogr. São Paulo 13: 3–142.

Björnberg, T. K. S., 1981. Copepoda. In D. Boltovskoy (ed.), Atlas del zooplancton del Atlantico Sudoccidental y métodos de trabajo con el zooplancton marino I.N.I.D.E.P., Mar del Plata, 936 pp.

Cipollli, N. M. & M. A. Juliano de Carvalho, 1973. Levantamento de Calanoida e Cyclopoida (Copepoda, Crustacea) das éguas da região do Guamá, Capim e Tocantins, com nota sobre a fauna acompanhante. Pap. Avulsos Zool. S. Paulo 27: 95–110.

Dahl, F., 1894. Copepoden des unteren Amazona. Ber. nat. Ges. (Zool.) 8: 1–14.

Do Nascimento, D. A., 1980. Composição do Zooplâncton no estuario do Rio Botafogo, Itamaracá, PE. M. Thesis, (mimeo). Univ. Curitiba, 180 pp.

Nordi, N., 1982. Ecologia do Zooplancton no estuario do Rio Paraiba do Norte, Paraiba, Brasil, Master's Thesis, Univ. São Carlos S.P., 130 pp.

Pekala, G. A., 1982. Distribução espacial e fluctuação anual dos Copépodes (Crustacea) no estuario do Rio Paraiba do Norte (Paraiba Brasil). Master Thesis Univ. São Carlos S.P., 111 pp. (mimeo).

Tundisi, J. & T. M. Tundisi, 1968. Plankton studies in Mangrove environments. V. The salinity tolerances of some planktonic crustaceans. Bolm. Inst. Oceanogr. São Paulo 17: 57–65.

Tundisi, T. M., 1972. Aspectos ecológicos do zooplancton da região lagunar de Cananéia com especial referência aos Copepoda (Crustacea). Ph.D. Thesis (mimeo). Univ. São Paulo, 191 pp.

Wright, S., 1936. A Revision of the South American Species of *Pseudodiaptomus*. Ann. Acad. Bras. Sci. Rio de Janeiro 8: 1–24.

An outline of the distribution patterns of the freshwater Copepoda of Israel and surroundings

F. D. Por
Department of Zoology, The Hebrew University of Jerusalem, 91904 Jerusalem, Israel

Keywords: tropical zooplankton, freshwater Copepoda, freshwater zoogeography, Levantine province, Israel, Sinai Peninsula

Abstract

The distribution of the freshwater Copepoda in the Levantine province is analyzed. The characteristics of the five subprovinces of the freshwater fauna in the area are briefly presented and the representative species of Copepoda, mainly of the Harpacticoida, are mentioned. The presence of a palearctic enclave in the mountains of South Sinai and the presence of Ethiopian species in the Jordan–Dead Sea valley, is emphasized.

Introduction

The freshwater Copepoda of Israel have been investigated by many renowned copepodologists such as Brady, Richard, Brehm, Gurney, Harding, Chappuis and Kiefer. However, only recently with the work carried out under the auspices of the 'Fauna Palaestina Committee' of the Israel Academy of Sciences and by the Israel Inland Water Ecological Service, has more complete distributional data become available. Some 50 species from the area are known today; however, a complete listing is still premature, mainly because of the much needed revisions in the cyclopoid genera *Eucyclops, Paracyclops, Mesocyclops* and *Microcyclops*. Information existing on the species of Harpacticoida, as well as on the poorly represented Diaptomidae, is more reliable. While a comprehensive listing of the species is still impossible, a fairly clear zoogeographical and ecological picture already emerges. This will be presented below.

General zoogeographic framework

As explained in an earlier paper (Por, 1975), Israel is part of the Levantine province – a transitional area between the Palearctic and Paleoeremic regions, situated in a narrow belt of the Eastern Mediterranean shoreland. The presence of a longitudinal geomorphological feature, the Jordan–Dead Sea rift valley and the accompanying mountain ridges on both sides of this valley, imparts an additional element of heterogeneity to this province.

More specifically, there is a northern area that still has a predominantly palearctic fauna (Lebanon and the Nothern Galilee) – delimited to the south by the 'Nehring line', corresponding more or less to the annual isohyet of 800 mm. Most of the territory of Israel represents an intermediate area where there is a mixture of palearctics and paleoeremians, an area that underwent many subrecent climatic fluctuations. This area extends south to the isohyet of 200 mm, the so-called 'Bodenheimer line'. South of this line, in the Negev and Sinai deserts, there is a solid predominance of desertic, paleoeremic animals, with some small enclaves of palearctic species in the high mountains.

To complicate this picture, in the rift valley, a number of outrunners of the East African wet-tropical Ethiopian fauna are encountered.

In the above-mentioned paper (Por, 1975) a subdivision of the Levantine continental water fauna into five subprovinces has been proposed. The distribution of the freshwater Copepoda generally corresponds with the limits set in the 1975 paper. Nothing is known about the fauna of the Transjordanian mountains.

Characteristics of the five freshwater subprovinces (Fig. 1)

Several ecological factors intervene in defining this subprovinciality. The presence of cold-stenothermic and richly oxygenated freshwater defines *subprovince I*. The transitional *subprovince II*, between the Nehring and Bodenheimer lines, is very much dominated by human influence such as deforestation and spring management. The presence of artificially deepened and protected springs, since antiquity, the existence of classical waterworks connecting many springs into continuous water-supply systems for the cities, are extremely important. The substrate and profundity of the water pools, some excavated and maintained by the villagers and more recently the expansion of the residual water reservoirs for agricultural use, are important for the lenitic fauna.

In *Subprovince III*, the soûthern desertic one, diversity of biota is extremely low and chiefly dependent on extremely short life cycles and on a considerable euryhalinity. Deep and protected or spring-fed pools are extremely important environments. Such environments, together with a few spring-fed streamlets, are important environments in the mountains of the Negev and especially of Southern Sinai. Unfortunately, these areas are insufficiently known. *Subprovince IV*, of the Jordan–Dead Sea system, that also contains Lake Kinneret, is basically characterized by oligohaline warm-stenothermic permanent waters. Finally, *subprovince V*, the area of the coastal streams, has suffered enormously from stream pollution and from an increase in salinity, due to overexploitation of the freshwater aquifers of the coastal plain; to the south, this subprovince is limited by the increasing seasonality of the streams.

The distribution of the freshwater Copepoda

The northern subprovince (No. I) is chiefly characterized by the presence of stenothermic harpacticoids. The genus *Moraria* as reported by Chappuis (1955) has its southern limit in Lebanon. *Elaphoidella putealis* has been reported from Lebanon (Chappuis, op. cit.), *E. gracilis* (Sars) and *E. pani* (Por, 1983) live in the headwaters of the Jordan and some streams of the Golan Heights. The Anatolian species *Attheyella (Mrazekiella) osmana* (Kiefer) also reaches its southern limits in the mountains of Gallilee (Por, op. cit.). Of the Diaptomidae, *Hemidiaptomus guerneyi canaanita* Dimentman and Por inhabits the rainpools in close association with *Arctodiaptomus similis similis* (Kiefer). In a few pools of the Golan Heights another subspecies is found, namely *A. similis irregularis* (Kiefer) which is reported also from Syria and Transjordania (Dimentman & Por, in press). The cyclopoid fauna of this subregion is negatively characterized by the total absence of southern species.

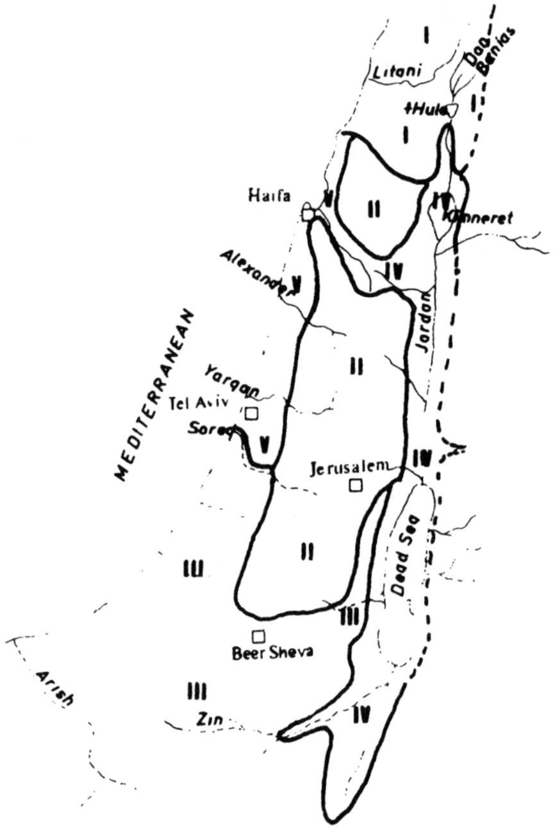

Fig. 1. The subprovinces of freshwater fauna of Israel and adjacent areas (see text) (modified after Por, 1975).

The transitional subprovince (No. II) is, as mentioned above, the area where the palearctic species reach the southernmost limit of their continuous distribution. The predominant cyclopoid of this subprovince is *Eucyclops serrulatus* (Fischer), followed by *Paracyclops fimbriatus* auct. and *P. affinis* auct. *Cyclops strenuus* is found in mountain temporary pools. Species of *Acanthocyclops* as well as *Cyclops vicinus* are found in pools and reservoirs of standing water. A complex of three harpacticoids characterizes the springs and brooks of the Cisjordanian mountains: *Canthocamptus microstaphylinus* Wolf, *Atheyella (A.) crassa* (Sars) and *Bryocamptus minutus* (Claus). A newly described species, *Attheyella (Mrazekiella) napthalica* Por (Por, 1983) has been reported till now only from a spring near the town of Tiberias. *Arctodiaptomus similis similis* is widespread in the subprovince but *Hemidiaptomus guerneyi canaanita* appears only in a restricted area of the Judean mountains, where deep and rocky rainpools last for more than three months (Dimentman & Por, in press).

The mountains of southern Sinai, some 400 km south of the Bodenheimer line, are a small enclave of a residual palearctic fauna as documented at present by the presence there of *Atheyella (A.) crassa* and of *A. (M.) naphtalica.*

The desertic subprovince (No. III), south of the 200 mm isohyet and also in the Judean Desert, has an extremely impoverished copepod fauna. In the short-lived temporary pools or residual waters of the flash floods only species of the genera *Microcyclops* and *Cryptocyclops* are found. An interesting exception is the presence of the diaptomid *Lovenula (Neolovenula) alluaudi* (Guerne & Richard), which appears in spring-fed pools or other semi-permanent or perennial waterbodies. These have chlorinity values of up to 788 mg l^{-1} (Dimentman & Por, in press). It is interesting that *Lovenula* does not coexist with any of the two northern diaptomid species. A harpacticoid, *Nitocra incerta* (Richard) found in the Nahal Zin, desertic river system, indicates temporary or historical connections between this system and the Dead Sea valley.

The Jordan-Dead Sea valley subprovince (No. IV). This is a subprovince which has a very diversified copepod fauna: in the brackish, mainly oligohaline waters, palearctic, paleotropic and marine-relict species intermingle. Unfortunately, man-made modifications have been very considerable in this subprovince: Lake Hula has been drained thirty years ago; Lake Kinneret had its hydrological regime severely changed; River Jordan had its water debit reduced by a whole order of magnitude and its water quality deteriorated; and springs of the Dead Sea area were captured.

Several African cyclopoids appear in the subprovince: *Afrocyclops gibsoni* Kiefer in the Jordan (Ortal & Por, 1978), *Tropocyclops confinis* Kiefer in the spring of En Gedi (unpubl.) and *Mesocyclops ogunnus* (Van der Velde, 1984), the dominant copepod of Lake Kinneret. The palearctic lacustrine *Eucyclops macrurus* (G. O. Sars) appears exclusively in Lake Kinneret (Por, 1968). A species of *Halicyclops* has recently been found in the Dead Sea oasis of En Feshka (unpubl.). The harpacticoids are a typically oligohaline dominant complex: *Nitocra incerta* (Richard), *Cletocamptus deitersi* (Richard) and *Onychocamptus mohammed* (Blanchard & Richard). *N. incerta* inhabits all the Jordan system up to the Dan and Banias springs. Other species of genus *Nitocra* are: *N. hibernica* Brady, a northern species, *N. balnearia* Por, an endemic of the highly saline Hamei Zohar hot springs and *N. lacustris*, a cosmopolitic species (Por, 1964). *Cletocamptus deitersi* is interesting, owing to its circumtropical distribution. Lake Kinneret is further characterized by two endemic marine species: *Pseudobradya barroisi* Por and *Nannopus palustris tiberiadis* Por; *Schizopera taricheana* Por, in a spring on the shore of this lake, is the only representative of the genus in the subprovince (Por, 1968).

Little is known of the fauna of the old Lake Hula. It seems that is was the only locality where *Attheyella (M.) trispinosa* (Brady) lived in the area (Por, 1967).

The subprovince of the coastal streams (No. V) extends to River Soreq in the south, the last permanent running water of the southern coastal plain. The basic harpacticoid fauna of this subprovince is that of *Nitocra incerta, Onychocamptus mohammed* similar to the Jordan valley and *Canthocamptus microstaphylinus* of the Cisjordanian mountains. *C. microstaphylinus*, a species with resistance cysts, is probably the most eurytopic freshwater harpacticoid of our province. Long-lasting rainpools form along the coast, where drainage is made difficult by a line of subrecent dunes. Where water persists for more than two months, *Hemidiaptomus guerneyi canaanita* appears together with the ubiquitous *Arctodiaptomus* (Dimentman & Por, in press). This subprovince, probably the least defined one, cannot be characterized on the basis of cyclopoids.

Subterranean fauna. Little is known on this subject from Lebanon (Lindberg, 1949; Chappuis, 1955). Typically subterranean harpacticoids were reported only recently from a cave near Jerusalem (Por, 1981): *Bryocyclops absalomi* Por, the only representative of this genus north of the Sahara, and *Parastenocaris nomiae* Por, the only *Parastenocaris* known till now from the Levantine province.

References

Chappuis, P. A., 1955. Crustacés d'eau douce cavernicoles. Mission Henri Coiffait au Liban. Biospeologica 75. Arch. zool. exp. Grale, 91: 533-636.

Dimentman, Ch. & F. D. Por, in press. Distribution and habitat characteristics of the Diaptomidae (Copepoda, Calanoida) in the inland waters of Israel and Northern Sinai and some comments on their taxonomy and biology. Hydrobiologia.

Kiefer, F., 1974. Zur Kenntnis von Morphologie und Systematik einiger Arten der Gattung *Arctodiaptomus* Kiefer (Crustacea, Copepoda). Zool. Scr. 3: 11-22.

Lindberg, K., 1949. Cyclopoïdes (Crust. Cop.) du Liban et de Syrie. Kungl. Fysiogr. Salls. Lund 19(10): 1-11.

Ortal, R. & F. D. Por, 1978. Effects of hydrological changes on aquatic communities in the Lower Jordan River. Verh. int. Ver. Limnol. 20: 1543-1551.

Por, F. D., 1964. The genus *Nitocra* Boeck (Copepoda, Harpacticoida) in the Jordan rift valley. Isr. J. Zool. 13: 78-88.

Por, F. D., 1967. Copepods from Lake Huleh; Lake Tiberias investigations 4. Sea Fish. Res. St. Haifa Bull. 48: 1-2.

Por, F. D., 1968. The benthic Copepoda of Lake Tiberias and of some inflowing springs. Isr. J. Zool. 17: 51-79.

Por, F. D., 1975. An outline of the zoogeography of the Levant. Zool. Scr. 4: 5-20.

Por, F. D., 1981. A new species of *Bryocyclops* (Copepoda: Cyclopoida) and of *Parastenocaris* (Copepoda: Harpacticoida) from a cave in Israel and some comments on the origin of the Cavernicolous Copepods. Isr. J. Zool. 30: 35-46.

Por, F. D., 1983. The freshwater Canthocamptidae (Copepoda: Harpacticoida) of Israel and Sinai. Isr. J. Zool. 32 (2-3): 113-134.

Velde, Van der, I., 1984. Revision of the African species of the genus *Mesocyclops* Sars, 1914 (Copepoda: Cyclopidae). Hydrobiologia 109: 3-66.

Some comments on tropical littoral Cladocera, with a description of *Alona incredibilis* sp. nov.

N. N. Smirnov
Institute of Evolutionary Morphology and Ecology of Animals of the U.S.S.R. Academy of Sciences, Leninsky prospect 33, Moscow, U.S.S.R. 117071

Keywords: tropical zooplankton, littoral Cladocera, *Alona incredibilis* sp. nov.

Abstract

Tropical littoral Cladocera with special reference to *Alona incredibilis* sp. nov. and *Alonella brasiliensis* Bergamin, 1935 from the Amazon basin, are commented upon.

Introduction

Tropical waters are populated by numerous littoral Cladocerans which are far from being adequately known to science. In studying them the following problems may be encountered: description of communities (biocenology); revealing of special adaptations of known and newly found species; interpretation of endemic species and genera; possibility of true identity of species in far-away lands and, if found, explanations thereof; studies in formation of Cladocera communities in new man-made lakes, in composition of faunas of particular countries.

A number of species new for science was described from tropical waters (e.g. Dumont, 1981; Brandorff, Koste & Smirnov, 1982; Idris & Fernando, 1981; Korovchinsky, 1981, 1982). Experience in sub-tropical countries also revealed many species and genera new for science, as well as distinctive features of taxocenoses (Khalaf & Smirnov, 1976; Smirnov, 1971, 1976, 1977; Smirnov & Timms, 1983). Specialized species are found in particular substrata, as it was, for example, in moss-like underwater (Fryer & Paggi, 1972) or above-water substrata (Frey, 1980).

Within our programme tropical Cladocera are studied in connection with revisions of particular Cladocera groups, studies in fauna of particular regions, in adaptations (morphological, physiological, behavioural, etc.), in historic biocenology (paleolimnology).

Description

A new species is described and another rare species is commented on, both from South America. More new species from the Rio Nhamunda have been also described (Brandorff *et al*, 1982). The slides are deposited in the Zoological Museum of the Moscow University.

Alona incredibilis sp. nov. (Figs. 1–6)

Material. 4 ♀ from the Rio Negro right of Ilha Tamaquaré, 1959.9.24, pH 4.6, on one slide 3181.

Holotype. MGU ♀ 3181-2 (i.e. the female designated 2 on the mentioned slide).

Paratypes. MGU three females NN 1, 3, 4 on the slide 3181.

Female. The general outline of the body is widely oval. The posterior corners of the shell are rounded and lack denticles. The setae of the ventral margin of the valve are short. There are minor setae along the inner side of the posterior margin of the valve. The surface of the shell has no noticeable sculpturing. The anterior (rostral) side of the head shield is blunt. There are three interconnected major head

Hydrobiologia 113, 155–158 (1984).

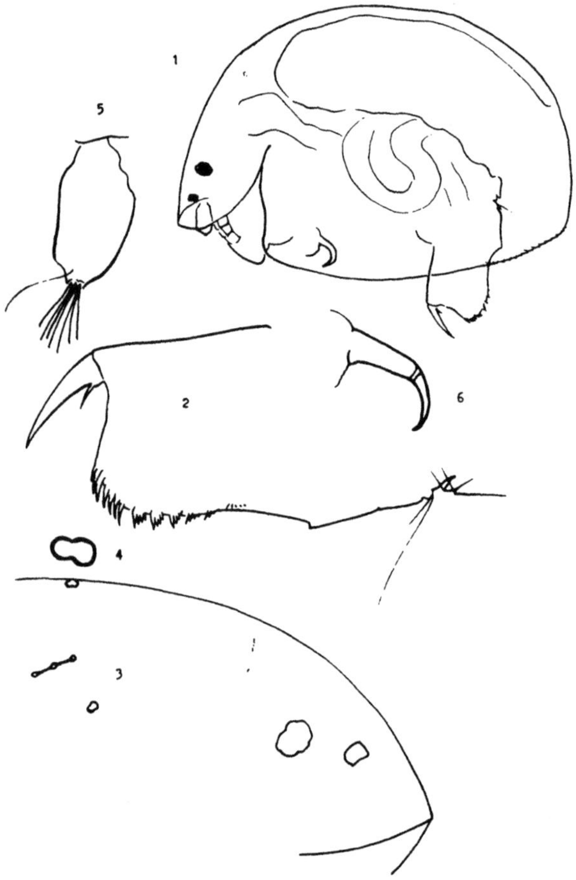

Figs. 1–6. Alona incredibilis sp. nov. The Rio Negro right of Ilha Tamaquaré 1959.9.24, pH 4.6.
(1) left-side view, length 0.44 mm, slide 3181-2 (holotype); (2) post-abdomen of the same individual; (3) location of the head pores, slide 3181-3; (4) enlarged lateral head pore, slide 3181-3; (5) antennule, slide 3181-1; (6) hook-like seta of thoracic limb I.

pores. The lateral head pores are large, distinctly outlined, in the form of a wide 8, situated at about one interpore distance (between the anterior and posterior major head pores) from the anterior major head pore.

The post-abdomen is short and wide. Its distal-dorsal corner is rounded. The anal margin is only slightly depressed, so the preanal corner is not protruding. There are three different groups of anal teeth along the postanal margin: distal simple, densely placed, about 7 in number; they are followed by three pectinate teeth; proximally few groups of small teeth are situated, the distalmost tooth being the largest in each group.

The claw is stout, shorter than the width of the post-abdomen. Its basal spine is equal in length to the width of the claw in lateral view. The antenule is short, thick, and abruptly narrowing at its apex into a smaller diameter, from where seasory papillae arise. The antenna is rather small. The labral lamina is rounded.

There is a big hook-like seta on thoracic limb I. The intestine is convoluted. The ocellus is smaller than the eye, midway between the eye and the tip of the rostrum.

Length. 0.44 mm (holotype 3181-2).

Male. Unknown.

Locality. The Rio Negro right of Ilha Tamaquaré.

Etymology. The specific name is given by reason of this species possessing several peculiar and partly unique features, and being an addition to one of the biggest genera of Chydoridae.

Differential diagnosis. Alone incredibilis sp. nov. possesses a combination of distinctive features which seem to ensure a successful crawling: a short post-abdomen having three different kinds of the anal teeth, a stout claw, a hook-like seta on thoracic limb I, a blunt rostrum, the thick antennule abruptly narrowing on its distal end. In addition it has lateral head pores as large openings.

The lateral head pores as wide openings occur also in forms having two major head pores. Such similar structure in *Biapertura verrucosa* (Sars, 1901) (*Alona verrucosa* Sars, 1901) is shown in Fig. 7.

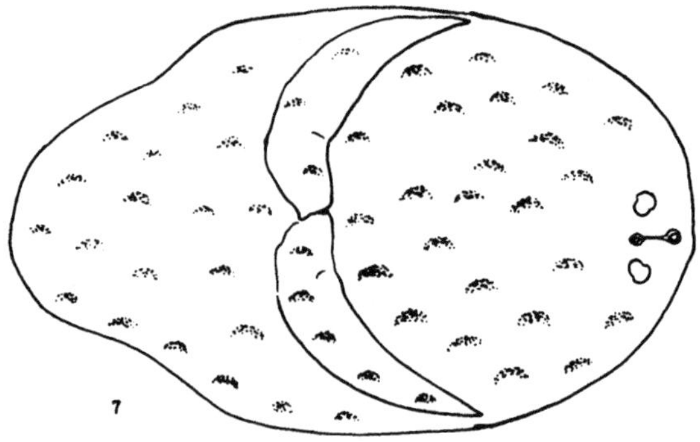

Fig. 7. Head shield of *Biapertura verrucosa* (Sars 1901). The lower Rio Nhamundá. Slide 1041.

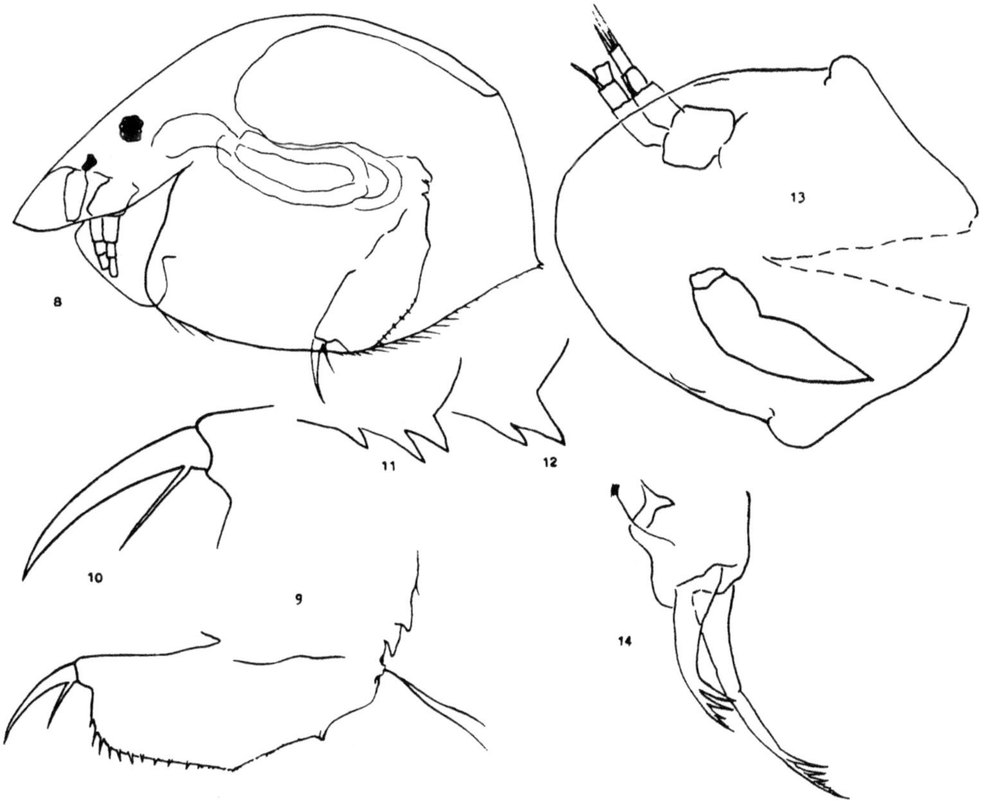

Figs. 8–14. Alonella brasiliensis Bergamin, 1935. The lower Rio Nhamunda (Amazon basin).
(8) left-side view, length 0.39 mm, slide 419; (9) post-abdomen of a female of 0.43 mm length, slide 572; (10) claw, slide 419; (11) ventro-posterior corner of the left valve, slide 572; (12) the same of the right valve, slide 572; (13) head shield, antenna, mandibula, slide 674; (14) external branch of the endite of thoracic limb I, slide 674.

Alonella brasiliensis Bergamin, 1935 (Figs. 8–14)
Bergamin, 1935: 284–285, Fig. 1 (photo), 2

Material. Several females from the lower Rio Nhamundá (Amazon basin).

Slides. MGU ♀ 419, 572, 674.

Female. The dorsal and ventral sides of the body are convex. The posterior side of the valve is almost straight. The ventro-posterior corner of the valve continues into two or three small acute denticles. The ventral margin of the valve has setules, in its posterior part sparsely distributed.

There is no noticeable sculpturing on the shell. The anterior (rostral) side of the head shield, as well as its posterior side, is widely rounded. No head pores are found.

The post-abdomen is slightly narrowing distally. There are anal teeth along both the post-anal margin (where teeth are larger) and anal margin (where teeth are small and grouped). The claw has a long basal spine which is half as long as the claw. The antennule does not reach the tip of the rostrum. The antenna has setae 0-0-3/0-1-3. There is a spine on the first segment of the endopodite reaching the end of the second segment.

There are two large setae on the external branch of the endite of thoracic limb I. These setae are unilaterally setulated; the setules are stout in the proximal part and diminish distally. The presence of only two setae in this group seemed doubtful in cases of some *Alona* and *Pleuroxus* species (as figured in Smirnov 1971) until confirmed for *Bryophilus* (Frey 1980) and *Oxyurella* (checked by me for the European *Oxyurella tenuicaudis*).

The intestine is convoluted. The ocellus is smaller than the eye, closer to the eye than to the tip of the rostrum.

Length. 0.39–0.43 mm. According to Bergamin (1935) 0.35–0.45 mm.

Male. Unknown.

Our samples correspond to Bergamin's publication, fortunately supplied with a photograph. Externally this form is similar to *Pleuroxus*; however, it differs by its very blunt rostrum. *Alonella brasiliensis* is not included in the list of species in Brandorff, Koste, and Smirnov (1982). Only few females were available and thoracic limb IV was not studied. No head pores are found. Thus the generic attribution remains problematic.

Acknowledgements

The material is kindly sent by professor H. Sioli, Dr G.-O. Brandorff, Dr M. Franken. Handling of the individual cladocerans was greatly facilitated by R. W. Pennak's micro-loop kindly supplied by himself.

References

Brandorff, G.-O., W. Koste & N. N. Smirnov, 1982. The composition and structure of Rotiferan and Crustacean communities of the lower Rio Nhamundá, Amazonas, Brazil. Stud. neotrop. Fauna & Envir. 17: 69–121 (Cladocera: 75, 76, 91–103).

Dumont, H. J., 1981. Cladocera and free-living Copepoda from the Fouta Djalon and adjacent mountain areas in West Africa. Hydrobiologia 85: 97–116.

Frey, D. G., 1980. The non-swimming Chydorid Cladocera of wet forests, with description of a new genus and two new species. Int. Revue ges. Hydrobiol. 65, 5: 613–641.

Fryer, G. & J. C. Paggi, 1972. A new cladoceran genus of the family Macrothricidae from Argentina. Crustaceana 23, 3: 255–262.

Idris, B. A. G. & C. H. Fernando, 1981. Two new species of cladoceran crustaceans of the genera Macrothrix and Alona Baird from Malaysia. Hydrobiologia 76: 81–85.

Khalaf, A. N. & N. N. Smirnov, 1976. On littoral Cladocera of Iraq. Hydrobiologia 51, 1: 91–94.

Korovchinsky, N. M., 1981. Taxonomic revision of Australian Diaphanosoma (Cladocera: Sididae). Aust. J. mar. Freshwat. Res. 32: 813–831.

Korovchinsky, N. M., 1982. Materials to systematics of Diaphanosoma genus (Cladocera: Sididae). Zoologischesky zhurnal 51, 5: 690–698 (in Russian).

Smirnov, N. N., 1971. Chydoridae of the World's fauna. Fauna SSSR. N 101. Rakoobraznye I, 2. Leningrad. Nauka Publishers, 531 pp. (in Russian).

Smirnov, N. N., 1976. Macrothricidae and Moinidae of the World's fauna. Fauna SSSR. N 112. Rakoobraznye I, 3. Leningrad. Nauka Publishers, 237 pp. (in Russian).

Smirnov, N. N., 1977. Quattuor species novae Cladocerorum Australiae. Crustaceana 32, 3: 119–128.

Smirnov, N. N. & B. V. Timms, 1983. Revision of Australian Cladocera. Memoirs of the Australian Museum, Sydney, Suppl. 1. 132 pp.

Attempt at ecological prognosis of the plankton in the man-made lake 'Parana Medio (Chapeton Transect)', Argentina

N. N. Smirnov
Institute of Evolutionary Morphology and Ecology of Animals of the U.S.S.R. Academy of Sciences, Leninsky prospect 33, Moscow, USSR 117071

Keywords: tropical zooplankton, ecological prognosis, the Parana

Abstract

Characteristics of the projected man-made lake on the Middle Parana are described, as well as environmental conditions in the river and their modifications in the future lake. The sources of the future plankton are considered. The formation of the future pelagic ecosystem, mainly of the zooplankton, during the initial and the subsequent stage is discussed.

The projected man-made lake along 300 km of the Middle Parana (from the future dam crossing the Chapeton Island above the city of Santa Fe and the city of Parana upstream to about the cities of Reconquista and of Goya) in the Republic of Argentina will radically affect a large and unique biospheric complex.

Description of the site, the future lake, and some environmental factors

This man-made lake will be a very large and long ditch, having a maximum width of more than 40 km (Fig. 1) and steep left (natural) and right (artificial) shores. Thus the expected ecosystem will be mainly pelagic (planktonic and nectonic).

While prognosis for nectonic fish fauna in such a man-made lake is a subject of special consideration, attention is focused here on the plankton and conditions of its existence at present and after damming. At present the Middle Parana ecosystem is a potamium of major rank existing in the mainstream of the Parana, its tributary channels, and numerous floodplain lakes, ponds, swamps harbouring various flourishing vegetation and many species of plants and animals (including those endemic to South America). Waters of other types, e.g. lakes (not floodplain lakes) are not abundant in northern Argentina.

Thus, the damming of the Parana will drastically change one of the major biospheric complexes of South America, producing a new large ecosystem requiring scientific and technical efforts for its formation and management to make the situation as least destructive and as much useful as possible. Some statistics of the future man-made lake are listed in Table 1.

The man-made lake 'Parana Medio (Chapeton)' will possess extremely long dams, the lateral dam preventing the formation of a lake of enormous area. The slope of the inner surface of the lateral dam is about 1:3.5. Along 150 km of the left shore from Chapeton Island to the City of La Paz there are 100 km of steep high bank which will be transformed after damming. Over 10 years the inshore shallow zone will be formed 60 m wide, from the edge of water to the step-like border of deeper waters at a depth of 3 m.

The distribution of depths (at water discharge of 1600 $m^3 s^{-1}$ average for many years) will be as indicated in Table 2. The depth of the Parana is about 10 m, often reaching 30 m. The man-made lake is projected to have a mean depth of 10–15 m

Hydrobiologia 113, 159–163 (1984).

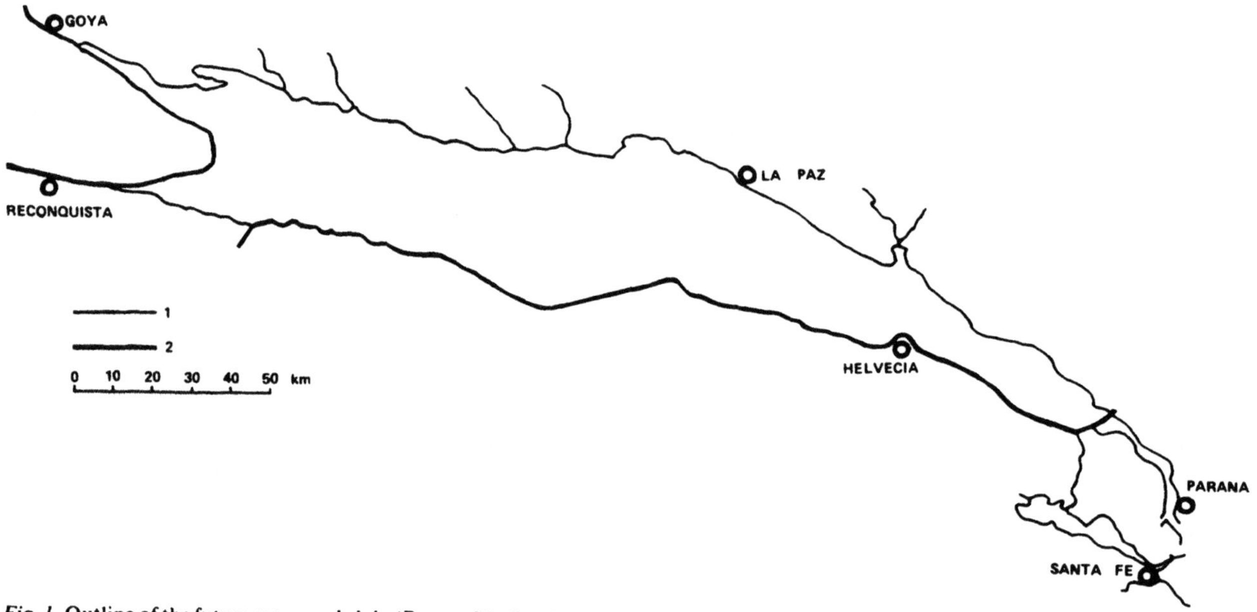

Fig. 1. Outline of the future man-made lake 'Parana Medio (Chapeton Transect)'. 1 natural shoreline, 2 dams, O cities mentioned in the text.

Table 1. Some morphometric characteristics of the man-made lake Parana Medio (Chapeton).

Areas	km^2	%
Man-made lake (at normal water level of 29 m and average many-year water discharge of 16 000 $m^3 s^{-1}$)	8320	100
Overflooded floodplain of the right bank	4500	54
Non-overflooded floodplain of the right bank outside of lateral dam in the lower region (from the transect of the dam to the City of Helvecia) (ca. 5% of the total area of the floodplain of the right shore)	240	-
Overflooded floodplain of the left bank	1760	21
Newly forming large bays of the left bank	300	3.6
Riverbed and large channels	1760	21.4
Area of the bed of the future man-made lake covered with vegetation, not less than	6600	-
including covered with *Eichhornia*, of order ca.	1000	-
Lengths	**km**	**%**
Total length of the shoreline	1130	100
Lateral and frontal dams	235	21
Generalized shoreline of large bays of the left bank	430	38
Other shoreline, including the North transect	465	41

and a maximum depth of 60 m. Water exchange will be 9 times per year.

The steepness of the future shores and the very limited area of bays are indications that the rooted vegetation will not occupy a large area. It will be confined to the zone of shallow water, for the most part not wider than 60 m, and to bays of the left bank (with a total area of about 300 km, which is less than 1% of the total area of the future lake).

About 235 km of the inshore zone (the length of the dams) will be covered with encrusting and attached filamentous algae and with other periphyton. The side of the lateral dam facing the lake will be concrete down to a depth of 7 m, with various species of algae growing on the surface.

Another limiting factor for the rooted vegetation is wave beating to reach at the mentioned open areas and known wind velocities and directions 6.56 t m^{-2} (height of wave 3.53 m). Thus a wave of 3.3 m

Table 2. Distribution of areas in the Parana Medio (Chapeton) man-made lake corresponding to different depths.

Water level, m above sea level	km^2	%
29	8320	100
27	8030	96.5
24	7740	93
29–27 (water level commonly discharged during low-water period)	290	3.5
27–24 (zone overgrowing with vegetation)	290	3.5

in height acts on 1 m^2 of inshore vegetation with a mass of about 6 t. It seems that there are no data characterizing the destruction of vegetation by waves of certain force. The analogous maximum height of waves is known for the Kuybyshev man-made lake on the Volga (U.S.S.R.) where, together with water level fluctuations, this had delimited the rooted vegetation to separate patches of relatively small area.

In the man-made lake 'Parana Medio (Chapeton Transect)' water level fluctuations will be within the limits of 27–29 m above sea level. The water discharge from this man-made lake will occur mainly by letting the deepest layers through the generators. Piling of water by winds is expected up to 90 cm.

In the future man-made lake the ecological conditions will not be diverse. The physical variables will be modified. Thus the average velocity of current in the mid-stream of the Parana is 0.8–1.2 m s^{-1}, in the future lake a system of currents will be formed having velocities of 0.1–0.3 m s^{-1}. At present the water transparency in the Parana never surpasses 67 cm, being much less during most of the year (J. Paggi & S. Paggi, 1974). Along the man-made lake the sedimentation of the particles will progress and the transparency will surpass 1 m.

At present, the water of the Middle Parana is characterized by a rather high mineralization (TDS 70.7 mg l^{-1}) containing a large quantity of Na^{1+} (7.9 mg l^{-1}), and Cl^{1-} (9.1 mg l^{-1}), about 1 mg NO_3^{1-} l^{-1} and 0.4 mg $PO_4^{2-}l^{-1}$ (Bonetto *et al*, 1981). The ground waters of the right shore will be pressed away after damming and the surface drainage will be barred by the lateral dam, so water in the man-made lake will be of the bicarbonate-calcium type (Depetris, 1976).

The dissolution of the eroding material of the high left bank will provide a new source of dissolved minerals; in 10 years 56 × 10^6 m^3 of the left bank material will be washed out (according to calculations by N. N. Kondratyev of Moscow). There are only two large tributary rivers on the left bank of the Middle Parana.

The decomposition of the overflooded vegetation and soils will provide a large source of dissolved matter, including various organic compounds. The latter processes will tend to shift pH towards 6–7, from the present-day pH 7.5, in the Parana. In the floodplain waterbodies of the Parana the acid pH of water is often recorded.

The lateral dam prevents pollution along the right bank from coming into the man-made lake. There are about 50 small factories using water along the borders of the future lake. The fertilizers are used on very limited area of the adjacent fields. The total number of cattle in the zone of the Middle Parana surpasses 1 million and is several times more than the human population.

If the figures of daily production of pollutants by cattle are taken as 247 t of N, 137 t of P, 220 t of K by 1 million individuals they illustrate that this is a noticeable source of biogenous elements.

After damming the whole floodplain ecosystem will be eliminated.

Methods

Ecosystemic variables to be considered for an ecological prognosis are extremely complex and numerous. Nevertheless they may be summarized as follows. For determination of the changes in an ecosystem it is necessary to consider all biological components, areas of biocenoses, structural and functional variables of biocenoses, environmental factors and their modifications after damming. Special features of the discussed natural object are to be taken into consideration. In making an ecological prognosis for a particular project it is preferable to use quantitative expressions that are capable of being transformed into technical solutions and norms.

The considerations and conclusions presented here follow from the data of the Project 'Parana Medio (Chapeton Transect)' and from the data by Argentinian scientists.

It should also be noted that because of the great complexity of the biological systems and considering the theory of ecological prognostication no bio-ecological prognosis can be guaranteed; in other words the prognostic conclusions in bio-ecological prognosis may be formulated only to some degree of accuracy.

Plankton

Most of the data on zooplankton of lotic and lenitic situations of the Middle Parana were obtained by Dr S. De Paggi and Dr J. Paggi (J. Paggi

& S. Paggi, 1974; S. Paggi, 1980a, b). In the mainstream of the Middle Parana 97 species are found. In 1974 these authors list for lotic water of the Middle Parana no less than 25 species of Cladocera (*Diaphanosoma* 2, Daphniidae 5, *Moina* 2, Bosminidae 4, Macrothricidae 2, Chydoridae 9), 13 species of Copepoda, and numerous Rotifera.

In 1980 J. Paggi lists for lenitic waters 17 species of Cladocera *(Diaphanosoma,* 2, *Moina* 2, *Bosmina* 2, Chydoridae 7), 16 species of Copepoda, and numerous Rotatoria. Many of these species with limited distribution, e.g. of Cladocera: *Diaphanosoma fluviatile, Daphnia ambigua, Moina minuta* Hansen, 1899, *Bosmina hagmanni, Bosmina huaronensis, Chydorus pubescens* Sars, 1901.

Rotatoria dominate in the river, comprising up to 70 per cent of the total number of animals, which is commonly 1–23 thousand individuals per 1 m^3.

. The river phytoplankton consists of diatoms, Euchlorophyceae, Conjugata, Cyanophyceae, Euglenophyta. In the floodplain waterbodies phytoplankton is very diverse and represented by species of diatoms, Euchlorophyceae, Conjugata, Cyanophyceae, Dinophyceae, Cryptophyceae, Xanthophyceae (De Emiliani 1979, 1980).

Stands of floating and rooted vegetation are populated by diverse fauna of Cladocera (of 11 genera), Copepoda, Ostracoda, Isopoda, Amphipoda, Decapoda, Rotifera (of 18 genera), and of many other groups (Dioni, 1967; Neiff & Poi De Neiff, 1978).

The above listed cenoses will provide sources of plankton formation in the man-made lake.

In the man-made lake, as commonly occurs, the percentage of Copepoda and Cladocera will increase, temperatures above the level of about 20 °C at the water surface also contributing to the reproduction of the latter. At the available level of nutrients, namely about 1 mg NO_3^{1-} l^{-1}, and their income (including those from the eroding shore, from animal husbandry) green algae will flourish. pH shifted below 7 (acid) will be favourable for the development of desmids.

The abundant development of algae will also benefit by the increased water transparency, slow water movements. Green algae are a favourite food for planktonic Cladocera (Sididae, Daphniidae, Moinidae, Bosminidae). Thus, mass development of these Cladocera is to be expected, as well as of some Chydoridae species depending on abundant dying off material.

The formation of the pelagic ecosystem will undergo two main stages. During the initial period of about three years the catastrophic transformation of the overflooded cenoses and intensive decomposition of the overflooded vegetation and soil, entailing an oxygen shortage will occur. There will be no equilibrium between explosive-like development of particular kinds of life using abundant food from decomposing plants, decomposing solid organic soil matter, eroding shore.

Kills of species requiring high oxygen concentration are probably due to oxygen depletion or to liberated toxic organic compounds. Most of the species of the lotic zooplankton will perish. Zooplankton of this period will be composed of species of the former lenitic situations capable of withstanding extreme conditions, and multiplying in abundance one after another.

During the subsequent period an equilibrium of limnological processes will be formed. Most of the initial vegetative material will decompose and oxygen supply will improve. The algal abundances will be more moderate and more evenly distributed in time. In the zooplankton the species may develop differently from the species of the initial period, and permanent for consequent years. The sources of its formation will be: the survived species of the former floodplain and the species brought by the river and its tributaries.

The zooplankton will mainly determine the existence of fishes and of the other preying animals of this level in the Parana Medio man-made lake.

Extensive growths of periphyton on the great surface of uniquely long dams will harbour species of demersal plankton and crawling species of Chydoridae and Macrothricidae. The latter two will thus develop in abundance at about 500 000 m^2 of concrete slopes overgrown by attached algae.

The abundant development of floating filamentous green algae is not excluded, as vast free resources of nutrients will be available and vascular plants will be scarce. In this case areas of filamentous algae will be populated by Chydoridae and by species of many other invertebrate groups crawling among them. Extensive growth of filamentous algae (of Conjugata) was observed over drowned forest in warm-water man-made Lake Brokopondo in the northern South America (Leentvaar, 1966).

The zooplankton of the vegetation will be confined to small areas of macrophytes.

The formation of the zooplankton and other communities in the future Parana Medio (Chapeton) Dam deserves thorough recording and scientific study as a new case under these limnological, climatic and biogeographic conditions. A more complete listing of the existing planktonic and demersal species of Cladocera, Copepoda, Rotifera, etc. is desirable.

Concentrations of marine Cladocera opposite the estuary of the Rio De La Plata are indicated for the present situation (Ramirez & De Vreese, 1974). The situation in the sea in years with very low discharge of river water should be studied as a model of the situation after damming.

Taking the above into consideration, measures may be elaborated for regulating the newly-to-be-formed, predominantly pelagic ecosystem towards a situation favouring propagation of more valuable fishes and better quality of water. The elimination of an existing ecosystem along 300 km of the river by this man-made lake is to be compensated for by the creation of a large natural reserve, before damming, extending along the Parana below the Chapeton dam, of comparable length – obviously not less than 200 km and floodplain-wide.

Acknowledgements

The author was privileged to have worked as an expert for the 'Parana Medio (Chapeton Transect)' project, to have had the opportunity to visit a number of technical and scientific institutions in Argentina, to have enjoyed gracious hospitality and broad possibilities for scientific investigation and discussions, and to have had access to libraries and other facilities. The author is grateful to Gerente del Projecto ing. M. A. Beretta, Head of the Department of Ecology of this project; T. Gutierrez, Director of Instituto de Limnologia in Santo Tome; señora Dr C. Pignalberi de Hassan, Director of Centro de Ecologia Aplicada del Litoral; Dr A. A. Bonetto, Director of Museo Provincial de Ciencias Naturales 'Florentino Ameghino' in Santa Fe; C. A. Virasoro, Mrs Dr S. De Paggi and Mr Dr J. Paggi, head of the USSR experts; A. K. Fink who efficiently and pleasantly fulfilled technical and coordinative activity, and to many individuals of the Argentinian institutions and group of USSR experts.

My two communications could be presented at the International Workshop on Tropical Zooplankton due to the generosity of Universidade Federal de São Carlos which paid for my subsistence in Brazil and of the USSR Academy of Sciences which paid for my travel by air between Moscow and São Paulo.

References

Bonetto, A. A. & H. G. Lancelle, 1981. Calidad de las aquas del Rio Parana Medio, principales caracteristicas fisicas y quimicas. Cent. Ecol. Aplicada del Litoral, 35 pp.

De Emiliani, G. M. O., 1979. Campana 'Keratella I' a lo largo del Rio Parana Medio, 3. Fitoplancton de ambientes leniticos. Rev. Asoc. Cienc. nat. Litoral N 10: 73–84.

De Emiliani, G. M. O., 1980. Fitoplacton de una laguna del valle alluvial del Parana Medio ('Los Matadores', Santa Fe, Argentina), 1. Estructura y distribucion en relacion a factores ambientales. Ecologia N 4: 127–140.

De Paggi, S. J., 1980a. Campana limnologica 'Keratella I' en el Rio Parana Medio: zooplankton de ambientes loticos. Ecologia 4: 69–75.

De Paggi, S. J., 1980b. Zooplankton del Rio Parana Medio: estudios en un perfil transversal, 8. Reunion Argentina de ecologia, Santa Fe. Mayo 65: 11–17.

Depetris, P. J., 1976. Hydrochemistry of the Parana River. Limnol. Oceanogr. 21, 5: 736–739.

Dioni, W., 1967. Investigacion preliminar de la estructura basica de las asociaciones de la micro y meso fauna de las raices de las plantas flotantes. Acta zool. 23: 111–137.

Leentvaar, P., 1966. The Brokopondo Lake in Surinam. Verh. int. Ver. Limnol. 16, 2: 680–684.

Mordukhai-Boltovskoy Philar, D. (ed.), 1979. The River Volga and its life. Dr W. Junk, The Hague, Boston, Lond., 473 pp.

Neiff, J. J. & A. Poi de Neiff, 1978. Estudios succesionales en los camalotales chaqunos y su fauna asociada, 1. Etapa seral Pistia stratiotes – Eichhornia crassipes. Physis 37, 95: 29–39.

Paggi, J. C., 1980. Campana limnologica 'Keratella I' en el Rio Parana Medio (Argentia): zooplankton de ambientes leniticos. Ecologia N 4: 77–88.

Paggi, J. C. & S. J. De Paggi, 1974. Primeros estudios sobre el zooplankton del Parana Medio. Physis 33, N 86: 91–114.

Ramirez, F. C. & P. De Vreese, 1974. Taxonomia y distribucion de los Cladoceros (Crustacea, Phyllopoda) de un sector de la Plataforma Bonaerense y adyacentias. Physis, A, 33, N 87: 511–526.

Table 1. (cont.)

Species	Habitat							
	River	Large reservoirs				Tank	Pond	Field
		High	Deep	Mod.	Shallow			
Moina macrocopa				+				+
Moina micrura	+	+		+	+	+		+
Sididae								
Diaphanosoma sp.				+	+			
Diaphanosoma excisum	+	+	+	+	+			+
* *Diaphanosoma* cf. *modigliani*				+	+	+	+	
Diaphanosoma sarsi	+	+		+				+
* *Diaphanosoma* cf. *volzi*				+				
Latonopsis australis				+			+	
* *Pseudosida* cf. *bidentata*								+
III. **Copepoda**								
CYCLOPOIDA								
Cyclopidae								
Acanthocyclops vernalis				+				
Cyclops nanus				+				+
Ectocyclops sp.	+							
* *Macrocyclops* cf. *distinctus*	+							
* *Mesocyclops* cf- *thermocyclopoides*		+		+	+	+		
Microcyclops bicolor			+	+				
Microcyclops varicans		+		+				
* *Paracyclops* cf. *fimbriatus*	+			+				
Thermocyclops cf. *crassus*	+	+	+	+	+	+	+	+
Tropocyclops prasinus	+			+				
Ergasilidae								
Ergasilus sp.				+	+			
CALANOIDA								
Diaptomidae								
Allodiaptomus raoi var. *membranigera*				+				
Diaptomus chaffonjoni	+	+						
Diaptomus javanicus		+						
Neodiaptomus botulifer			+	+	+			
Neodiaptomus malaysianus	+							
Neodiaptomus memphistopheles	+							
Neodiaptomus schmackeri			+	+	+	+	+	+
Phyllodiaptomus annae	+							
Temoridae								
Eurytemora sp.						+		
HARPACTICOIDA	+			+		+	+	
IV. **Ostracoda**								
Cypridae								
gen. sp. div.					+			+
V. **Brachiura**								
Argulidae								
Argulus sp.				+				
Total	35	18	25	65	35	30	26	68

High = High altitude reservoir (> 759 m a.s.l.)
Deep = Deep reservoirs
Mod. = Moderately deep reservoirs
Shallow = Shallow reservoirs
Tank = Irrigation Tanks
Field = Trichogaster fish fields
* = Not previously reported from Thailand

in the early zoeae of *P. pandaliformis* and *P. northropi* and evaluates rates of oxygen uptake by the first zoea of *P. pandaliformis* when subjected to a wide salinity range. These results, combined with the metabolic data of Moreira *et al.* (1982) for larval *P. northropi*, are an attempt to assess to what extent larval physiological responses to salinity variation might reflect the general distribution patterns of the adult shrimps.

Material and methods

Ovigerous *Palaemon northropi* were collected by hand from intertidal rock pools either at the southern extremity of the Guaecá Beach (ca. 23°49′18″S; 45°27′18″W), about 10 km south of São Sebastião, or on the isthmus connecting Cabelo Gordo de Fora Beach with Baleeiro Promontory nearby. Ovigerous *P. pandaliformis* were taken with a sieve from the marginal vegetation at the lower reaches of a small river discharging onto the southern end of the Enseada Beach (ca. 23°43′34″S; 45°25′14″W), some 12 km south of Caraguatatuba; both towns are located on the coast of the State of São Paulo, Brazil.

Water temperatures taken during collections varied from 20 to 29 °C; salinities, measured by refractometer, were from 0 to 28‰ in the stream depending on tidal influence and 16 to 35‰ in the rock pools. Shrimps were transported to the laboratory in São Paulo where each species was maintained in an aerated aquarium in water from the collection site. Pieces of raw fish, mussel and live *Tubifex* worms were given as food.

Selected females of both species were maintained individually until eclosion of the larvae which were then divided into groups of 10 in culture dishes, each containing 100 ml of experimental medium (0, 7, 14, 21, 28 or 35‰ S). Three replicates were performed for each salinity. Larvae were then kept at 20 °C in a constant temperature chamber under a 12 h light: 12 h dark photo-regime. Test salinities, prepared from São Sebastião Channel seawater and river water, were renewed daily and a record of moulting activity and mortality taken. Experiments were conducted for 7 days.

Cartesian diver microrespirometers (Holter, 1943), each having a total volume of between 8 and 13 μl, were used to measure the rates of oxygen uptake (QO_2) by individual, intermoult, stage I *Palaemon pandaliformis* zoeae, at 20 °C, in salinities of 0, 7, 14, 21, 28 and 35‰. Determinations were made each 30 min over a 90 min period, the first reading for each diver being disregarded. Such measurements represent 'routine metabolism' (Prosser, 1973) since the zoeae could move but not swim inside the divers. For each salinity, a minimum of seven repetitions was performed, each zoea being used only once. Results are expressed as microlitres of oxygen consumed per milligramme dry weight per hour.

To determine the larval dry weight (calculated as 49.8 μg), 3 samples, each of 10 larvae, were rinsed with distilled water, dried overnight at 80 °C, placed in a desiccator for 2 h and weighed on a Cahn Gram Electrobalance (0.1 μg sensitivity).

Significant differences between means for respiratory data were calculated using a t-test according to Zar (1974).

Results

Figures 1 and 2 show the percentage survival and moulting of *Palaemon pandaliformis* and *P. northropi* zoeae over a 7 day period at 20 °C. While all *P. northropi* zoeae died in 0‰ S after 24 h, *P. pandaliformis* zoeae exhibited far better tolerance of this medium, 25% surviving after 4 days. Survival for both species was also very poor in 7‰ S but increased in 14 and 21‰ S. Best survival for both species was registered in 28‰ S (about 90% after 7 days). Exposure to São Sebastião Channel seawater (35 ‰ S) increased mortality in zoeae of both species, no survival being registered for *P. northropi* zoeae, and only 45% for *P. pandaliformis* zoeae, after the experimental period.

Moulting did not occur in zoeae of either species in 0‰ S, nor in *P. northropi* zoeae in 7‰ S. Although *P. northropi* zoeae I generally commenced moulting before *P. pandaliformis* zoeae, on the fourth day the percentage of zoeae II was greater for *P. pandaliformis* than for *P. northropi* in all salinities where moulting occurred, with the greatest differences being observed in high salinities (28 and 35‰ S). On the last day of the experimental period, some *P. pandaliformis* zoeae had reached stage III in salinities of 14 and 21‰ while all *P. northropi* zoeae remained in stage II.

Lai, H. C. & C. H. Fernando, 1978. Redescription of *Neodiaptomus botulifer*. Kiefer and one of its related species. (Calanoida: Copepoda). Hydrobiologia 59: 229–235.

Lai, H. C. & C. H. Fernando, 1981. The freshwater Calanoida Crustacea of Thailand. Hydrobiologia 76: 161–178.

Mamaril, A. C. & Fernando, C. H., 1978. Freshwater zooplankton of the Philippines, Rotifera, Cladocera and Copepoda. Nat. appl. Sci. Bull. Manilla 30: 109–221.

Mizuno, T., 1968. Illustration of freshwater plankton of Japan, Hoikusha, Osaka, 351 pp.

Office of the National Economic and Social Development Board, Kingdom of Thailand, 1981. Rehabilitation and Development of Bung Boraped and Nong Han, 1. Bung Boraped.

Office of the National Economic and Social Development Board, Kingdom of Thailand, 1982. Rehabilitation and Development of Bung Boraped and Nong Han, 2. Nong Han.

Pennak, R. W., 1953. Freshwater Invertebrates of the United State. The Ronal Press Company, N. Y.: 350–421.

Pothipitak, K., 1971. Hydrobiological and Fisheries Surveys of Meklong and Kwae River. Fish. Res. Unit annu. Rep. Inld Fish. Div., Dep. Fish., Bangkok, Thailand, 63–78 (in Thai).

Royal Irrigation Department, 1970. Lam Pra Plerng Project. Nakhon Rajasima Province. Planning Div. R. I. Dep. Bangkok (in Thai).

Royal Irrigation Department, 1970. Special paper on the dedication ceremony of the Lam Takong Dam in Nakhon Rajasima Province, 16 pp (in Thai).

Royal Irrigation Department, 1974. Nam Oon Project. Sakelnakorn Province. Planning Div. R. I. Dept, Bangkok (in Thai).

Royal Irrigation Department, 1974. Special paper on the dedication ceremony of the Vajiralongkorn Dam in Kanchanaburi Province, 11 pp. (in Thai).

Sangkhakul, S., 1971. Hydrobiological and Fisheries surveys in Lum Takong reservoir. Annu. Rep., Inld Fish. Div., Dep. Fish., Bangkok, Thailand: 51–62 (in Thai).

Sangkhakul, S., V. Srisuwannataj, W. Pokaew & W. Martjootaseranee, 1972. Hydrobiological and fisheries survey in Sirindhorn reservoir. Fish. Res. Unit annu. Rep., Inld Fish. Div., Dep. Fish., Bangkok: 12–29 (in Thai).

Sangkhakul, S., 1974. Zooplankton in Bhumibol Reservoir, Tak Province. M. Sci. Thesis, Kasetsart Univ., Bangkok, Thailand, 51 pp. (in Thai).

Sangkhakul, S., 1975. Plankton in Kang Krachan Reservoir. Fish. Res. Unit annu. Rep., Inld. Fish. Div., Dep. Fish., Bangkok: 40-56 (in Thai).

Scourfield, D. J. & J. P. Harding, 1966. A Key to the British Species of Freshwater Cladocera, 3rd edn. Scient. Publ. 5. Freshwat. biol. Ass., Ambleside, Westmorland, pp. 55.

Sibke, A., 1971. Hydrobiological and fisheries surveys of Rong Hua Chang reservoir. Fish. Res. Unit annu. rep., Inld Fish. Div., Dep. Fish., Bangkok, Thailand: 79-98 (in Thai).

Sibke, S. & A. Nujsombat, 1974. Hydrobiological and fisheries surveys in Chulabhorn reservoir. Fish. Res. Unit annu. Rep. Inld Fish. Div., Dep. Fish., Bangkok, Thailand: 50-60 (in Thai).

Srisuwanataj, V., 1971. Hydrobiological and Fisheries surveys in Lum Dom Noi reservoir. Fish. Res. Unit annu. Rep., Inld Fish. Div., Dep. Fish., Bangkok, Thailand: 37–50 (in Thai).

Tressler, W. L., 1963. Ostracoda. In: W. T. Edmondson (ed.), Freshwater Biology, 2nd edn. John Wiley & Sons, N. Y.: 657–734.

Victor, R. & C. H. Fernando, 1979. The freshwater ostracods (Crustacea: Ostracoda) of India. Rec. zool. Sci. India 74: 147–242.

Waew-Ngam, B., 1973. Limnological and Fisheries survey in Sirikit reservoir. Fish. Res. Unit annu. Rep., Inld Fish. Div., Dep. Fish., Bangkok, Thailand: 24-41 (in Thai).

Wilson, M. S. & H. C. Yeatman, 1963. Free-living copepods. In: W. T. Edmondson (ed.), Freshwater Biology, 2nd edn. John Wiley & Sons, N.Y.: 735–737.

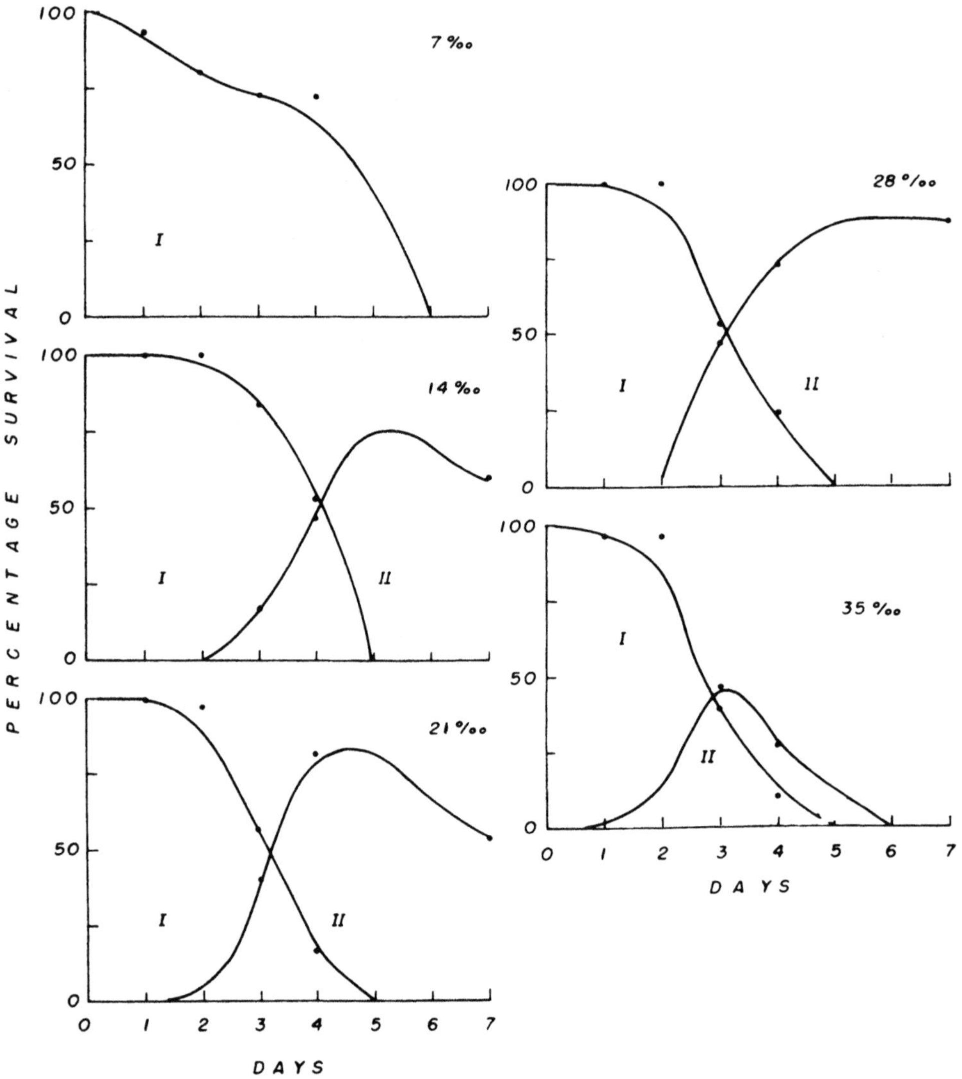

Fig. 2. Eye-fitted curves representing percentage survival and moulting in early zoeae of *Palaemon northropi* exposed to different salinities over a seven day period. I, first zoea; II, second zoea.

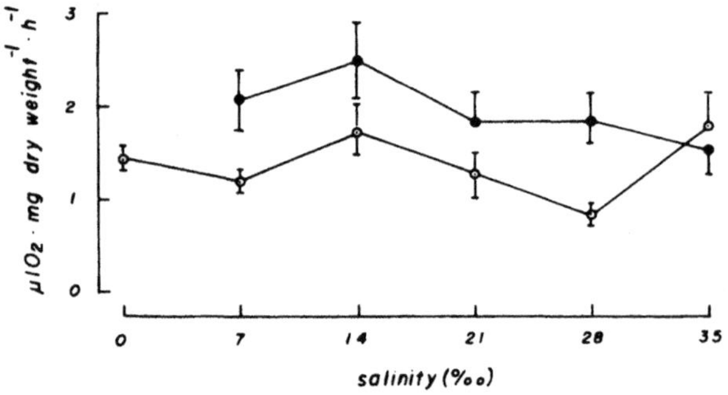

Fig. 3. Rate of oxygen consumption by stage I *Palaemon pandaliformis* (○) and *Palaemon northropi* (●) zoeae in different salinities. (Curve for *P. northropi* from Moreira *et al.*, 1982).

dependency on moderately saline media, at least in the early developmental phases. Since adult *P. pandaliformis* typically inhabit fresh to low salinity waters, the tolerance of the first zoea to freshwater should enhance survival during transport to more saline media. *Palaemon northropi* zoea I, however, appears to have little chance of survival on encountering freshwater. Such differences in larval tolerance of freshwater may be important in determining the rate and extent of adult penetration into the freshwater biotope.

Time and extent of moulting are also markedly affected by salinity exposure in the two species. The lower limit for moulting in *P. northropi* zoea I is 14‰ S while *P. pandaliformis* zoeae I are able to moult in 7‰ S. Moulting to zoea III occurs in moderate salinities in *P. pandaliformis* but was not observed in *P. northropi*. Thus, *P. pandaliformis* zoeae appear to develop more rapidly over a wider salinity range than do *P. northropi* zoeae. However, such data suggest that while survival for *P. pandaliformis* zoeae may be optimal in 28‰ S, lower salinities are necessary for initiation of the developmental processes in this species. Clearly, the salinities required for development are far more restricted than those ensuring survival.

The metabolism-salinity responses of neotropical palaemonid larvae may be classified into two groups according to Moreira *et al.* (1982). The zoeae of *Macrobrachium acanthurus* and *M. olfersii* increase metabolic rates in concentrated and dilute media while zoeal *M. carcinus* and *M. heterochirus* tend to have decreased metabolic rates in both high and low salinities. *Palaemon pandaliformis* zoeae I exhibit a U-shaped metabolism–salinity curve, thus showing similarities with the responses for zoeal *Macrobrachium acanthurus, M. olfersii* and *M. holthuisi* (Moreira *et al.*, 1980), all predominantly freshwater species in the adult form. *Palaemon northropi* zoeal responses, however, are more typical of marine-brackish water crustaceans, i.e., showing a gradual increase in oxygen consumption rate with increasing dilution of the medium, within the limits of tolerance, as occurs for example in marine populations of the shrimp *Metapenaeus monoceros* (Rao, 1958).

Although adult *Palaemon* species have often been employed as the subject of physiological studies, the effects of salinity on survival and respiratory metabolism of these common components of the littoral ecosystem have not been well examined. In *Palaemon serratus*, Scheer & Scheer (1954) and Siva & Regnault (1980) studied hormonal control and influence of the intermoult cycle on respiratory rate, respectively. Reeve (1969) presented data on oxygen consumption in relation to size, temperature and oxygen tension, and recently, Hagerman & Weber (1981) have studied the influence of water oxygen tension on metabolic rate in *P. adspersus*.

While many factors complicate any direct correlation between oxygen consumption and salinity exposure (see Lockwood, 1976 and Gilles, 1979 for reviews), the larval metabolism–salinity responses obtained in the present study do tend to reflect the general distribution patterns of the adult shrimps. Adult *P. pandaliformis* are usually encountered in freshwater and the freshly hatched larvae of this species proved to be more resistant than those of *P. northropi* to fresh and diluted seawater, exhibiting very stable metabolic rates in these media. However, while not showing the euryhalinity of *P. pandaliformis* larvae, zoeae of the marine *P. northropi* did maintain more stable metabolic rates in the higher salinities (21 to 35‰ S), suggesting an inherent adaptation to more saline media.

References

Gilles, R., 1979. Intracellular organic osmotic effectors. In R. Gilles (ed.), Mechanisms of Osmoregulation in Animals. John Wiley & Sons, Chichester: 111-154.

Hagerman, L. & R. E. Weber, 1981. Respiration rate, haemolymph oxygen tension and haemocyanin level in the shrimp Palaemon adspersus Rathke. J. exp. mar. Biol. Ecol. 54: 13-20.

Holter, H., 1943. Technique of the Cartesian diver. C. r. Lab. Carlsberg, Sér. Chim. 24: 399-478.

Holthuis, L. B., 1952. A general revision of the Palaemonidae (Crustacea, Decapoda, Natantia) of the Americas, 2. The sub-family Palaemoninae. Allan Hancock Found. Publ., occ. Pap. 12: 1-396.

Lockwood, A. P. M., 1976. Physiological adaptation to life in estuaries. In R. C. Newell (ed.), Adaptation to Environment. Butterworth & Co, Lond.: 315-392.

Moreira, G. S., J. C. McNamara & P. S. Moreira, 1982. The effect of salinity on the metabolic rates of some palaemonid shrimp larvae. Aquaculture 29: 95-100.

Moreira, G. S., J. C. McNamara, P. S. Moreira & M. Weinrich, 1980. Temperature and salinity effects on the respiratory metabolism of the first zoeal stage of Macrobrachium holthuisi Genofre & Lobão (Decapoda: Palaemonidae). J. exp. mar. Biol. Ecol. 47: 141-148.

Prosser, C. L., 1973. Oxygen: respiration and metabolism. In C. L. Prosser (ed.), Comparative Animal Physiology. W. B. Saunders Co., Philad.: 163-211.

Rao, K. P., 1958. Oxygen consumption as a function of size and salinity in Metapenaeus monoceros Fab. from marine and brackish water environments. J. exp. Biol. 35: 307-313.

Reeve, M. R., 1969. The laboratory culture of the prawn Palaemon serratus. Fish. Invest., Lond. 26: 1-38.

Scheer, B. T. & M. A. Scheer, 1954. The hormonal control of metabolisms in crustaceans. 8. Oxygen consumption in Leander serratus. Publ. Staz. zool. Napoli 25: 419-426.

Silva, S. C. & M. Regnault, 1980. Variations de la consommation d'oxygène chez la crevette Palaemon serratus Penant au cours du cycle d'intermue. Cah. Biol. mar 21: 279-286.

Zar, J. H., 1974. Biostatistical Analysis. Prentice-Hall, Englewood Cliffs, N.J., 620 pp.

Consumption of zooplanktonic organisms by *Astyanax fasciatus* Cuvier, 1819 (Osteichthyes, Characidae) in Lobo (Broa) Reservoir, São Carlos, SP, Brazil

Paulina Maria Maia Barbosa & Takako Matsumura-Tundisi
Universidade Federal de São Carlos,
Departamento de Ciências Biológicas, Lab. de Limnologia, Via Washington Luis, Km 235, Caixa Postal 676, 13.560 São Carlos, SP, Brazil

Keywords: tropical zooplankton, predation, fish, composition, preferential, food

Abstract

The consumption of zooplanktonic organisms by young and adult fish of *Astyanax fasciatus* in Lobo Reservoir was studied from October 1978 to September 1979 by analysing the abundance of zooplankton in the gut of fishes and relating it to the variation of the zooplankton community in the lake. To analyse the diet the following methods were used: numeric frequency of occurrence, and frequency rate. *Astyanax fasciatus* showed a large spectrum of feeding. The young stage feeds mainly on zooplankton and macroinvertebrates (insect larvae) but the adults also consume algae and detritus. Even in relation to the zooplankton the young stage showed a food preference different from that of the adults, selecting copepods over cladocerans, while the adults consumed more cladocerans.

Introduction

Most species of fish utilize planktonic components as food at least during the initial phases of development since the choice of items to be ingested depends on several factors related to prey (agility, visibility, availability, capacity of disguise) and predator (chase efficiency, conditioning to certain types of food), although environmental conditions such as water turbidity may interfere.

According to Nikolsky (1963), many species show organs adapted to the capture and digestion of certain items, but these adaptations do not remain constant during the fish whole life. Feeding variations within the same species can be observed among individuals at different stages of development and among sexes, in different regions or locations within the same environment and still, in distinct periods.

As stated by Lagler *et al* (1962), most fishes are highly adaptable in their feeding habits, utilizing the most readily available items, pure carnivorous or herbivorous species being few.

It has been shown by many authors that, despite the multiplicity of factors which interfere with prey selection, some fish species prey selectively on zooplanktonic organisms, interfering with the size composition of the zooplanktonic community and their spatial and temporal distribution; it also contributes to the development of different forms within the same species (Brooks & Dodson, 1965; Wells, 1970; Zaret, 1969, 1972; Hutchinson, 1971; Zaret & Kerfoot, 1975). Thus, in order to understand the zooplanktonic community dynamics of the Lobo Reservoir (Broa) which has been studied by many authors, an attempt has been made to analyze the feeding process of an abundant fish in the reservoir, *Astyanax fasciatus*, which feeds on zooplankton at least during its young phase and may influence the density fluctuation of zooplanktonic populations.

Hydrobiologia 113, 171–181 (1984).

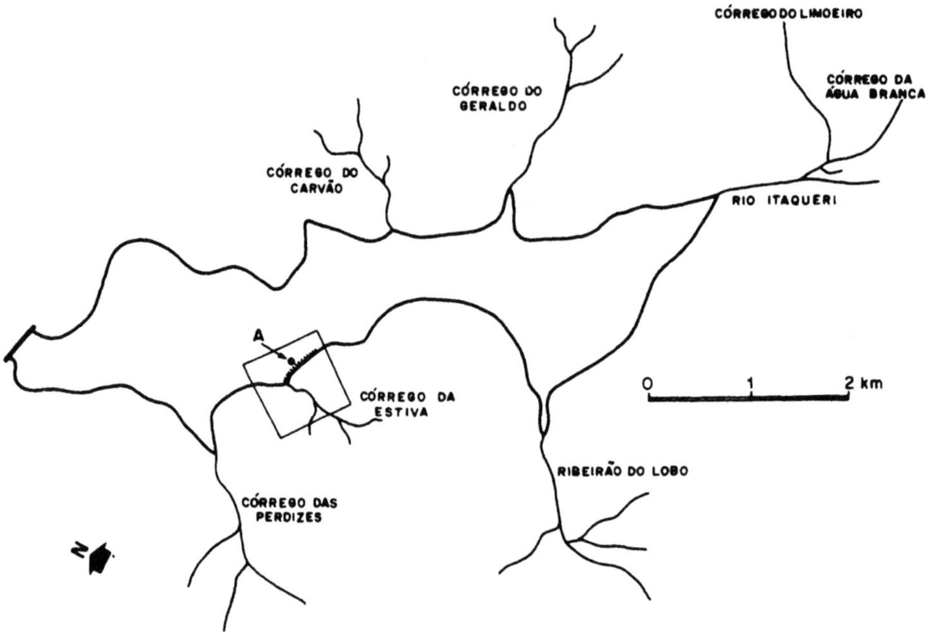

Fig. 1. Maps of the Lobo Reservoir (Broa) with the zoo-plankton collection site (A) and the haul area for fish collections.

Material and methods

Area and period of collection

Figure 1 shows a morphometric map of the Lobo Reservoir indicating the region where fish and zooplankton have been collected. This is a shallow region (1.5–2.0 m) composed of aquatic macrophytes which can protect the organisms, either zooplankton or fishes. Also in this region, the population of *A. fasciatus* can be found in large amounts during the whole year. The zooplankton collection station was set at approximately 50 meters offshore.

Fish and zooplankton collections were obtained monthly from October, 1978 to September, 1979 except in February when it was not possible to collect zooplankton.

Zooplankton collection

Samples for zooplankton quantitative and qualitative analysis were obtained with vertical hauls at the station by using a plankton net of 68 μm mesh aperture. The samples were fixed with 4% neutralized formalin.

For quantitative analysis, the samples were concentrated to a known volume and after homogenization, sub-samples were withdrawn and counted; samples with low density of organisms were totally counted. In order to calculate the volume filtered by the net we used a formula described by Edmondson & Winberg (1971).

In qualitative analysis, organisms belonging to Copepoda and Cladocera groups were identified, whenever possible, to species while organisms belonging to Rotifera were identified to genus. Although protozoans were present in environmental samples they were not considered since they were not found in the digestive tract of fishes, perhaps due to fixation problems.

Fish collection

The fishes used in this work were divided into two classes of length: young from 2.0 to 4.0 cm and adults from 8.0 to 10.0 cm (in a total of 205 specimens).

Young fishes were captured with a haul net of 0.3 cm mesh aperture at approximately 4 p.m. and adult fishes with a 0.8 cm haul net after 7 p.m. These schedules were established after a test sampling which verified the occurrence of young specimens only during day-time in contrast with adults which were found after sunset.

After collection the fishes were narcotized with a benzocaine solution for 10 minutes in order to avoid regurgitation and then fixed with 10% formalin.

To analyze the digestive tract content 10 specimens of each class were utilized although in some months a smaller number of young fish were analysed. Particularly in October and November 1978 and September 1979, no young fish was captured.

For qualitative and quantitative analysis of the ingested material we used the content present in the whole digestive tract.

The counting and identification of this material were mostly made by examining the total content recurring whenever necessary, to sub-sampling. Only entire organisms were counted; those that, although digested, could be identified were also counted.

Methods utilized for fish feeding analysis

The following methods were considered in the present study: numerical, occurrence frequency and frequency index. In the numerical method all components of each feeding item were recorded for the two classes of length analyzed during the different months, the results being expressed in percentage.

In the occurrence frequency method the number of specimens monthly examined where feeding components were found, was recorded independent of size or quantity of organisms present. To have a general and comparable idea of the feeding process between young and adult specimens, the total percentage of the different food items was calculated by considering as 100% the number of occurrence records of each item during the different months (Hynes, 1950).

The frequency index expresses a relation between the number of samples which contained a given item and the total analyzed. The results obtained allow different prey classifications: preferential when $f > 0.5$, secondary when $0.1 < f < 0.5$ and accidental when $f < 0.1$.

Results

Diet composition of A. fasciatus considering two classes of length

The digestive tract content analysis of young and adult fishes permitted the identification of the following items: zooplankton, insects, acari, algae, vegetal detritus, inorganic detritus (sand) and, rarely, fish scales.

Among zooplanktonic organisms, cladocerans and copepods were found to be the major components in the two classes of length studied. Rotifers and ostracods, due to low abundance and frequency were grouped into the item 'others'.

Figure 2 shows the occurrence frequency (%) of the diet items of young and adult specimens. In both cases, higher values were obtained for cladocerans, copepods and macroinvertebrates (larvae and adult insects as well as hydracari) show-

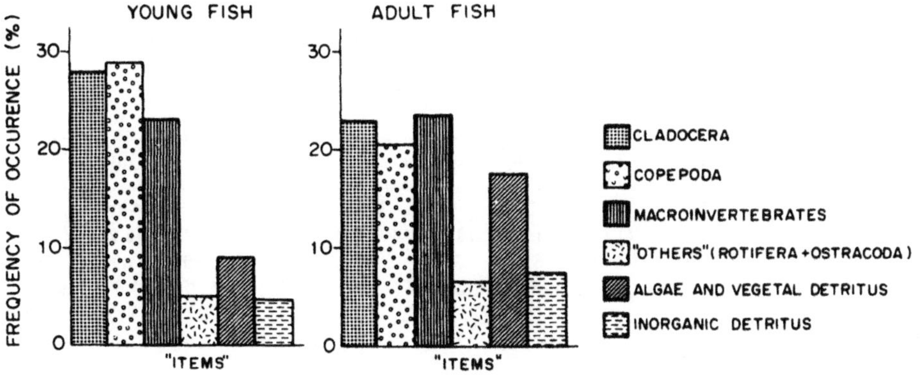

Fig. 2. Occurrence frequency (%) of the main items found in the digestive tract of young and adult specimens of *A. fasciatus* from the Lobo Reservoir during the period October, 1978–September, 1979.

Table 1. Frequency index (f) of the feeding items found in the digestive tract of young and adult specimens of *Astyanax fasciatus*.

Feeding items	Frequency index (f)	
	Young fish	Adult fish
Zooplankton	1.00	0.90
Macroinvertebrates	0.72	0.93
Algae and vegetal detritus	0.28	0.69
Inorganic detritus	0.14	0.29
Others[a]	0.18	0.27

[a] Rotifera and Ostracoda.

ing for young fishes 28.08%, 29.23% and 23.46% respectively and for adults 23.23%, 20.65% and 23.87%. It can be observed that the frequency of algae and vegetal detritus was relatively higher for adults (17.85%) than for young specimens (9.23%). When comparing the frequency of items in the two classes of length, a slightly higher occurrence of cladocerans and copepods is observed in young fishes while the occurrence of macroinvertebrates is practically the same. Inorganic detritus and 'others' had a low frequency in either class of length although a slightly higher frequency was observed in adults.

Table 1 gives the frequency index obtained for the different items in young and adult fishes. Both stages selected zooplankton and macroinvertebrates as preferential preys with indexes of 1.0 and 0.72 in young and 0.9 and 0.93 in adults respectively. Algae and vegetal detritus were prefered by adults and secondary by young specimens. Inorganic detritus and 'others' could be considered secondary for both classes with higher indexes in adults.

Analysis of absolute and relative numerical values with respect to main feeding items of the digestive tract content of young and adult fishes

Table 2 provides the absolute and relative numerical values obtained for cladocerans, copepods, macro-invertebrates and 'others' (rotifers and ostracods) in young fish digestive tracts for a year-long period. It can be observed that Cladocera showed the highest values in December, April and July making 61.66%, 92.29% and 96.28% of the total respectively while the lowest values were found in January and June corresponding to 1.69% and 0.70%. Copepoda were abundant during almost the whole year with low percentages only during April, May, and July (2.25%, 6.636% and 1.38% respectively). Their maximum percentage was observed in June (98.78%) when the lowest amount of cladocerans was found in young fishes. It can also be observed in relation to young fishes that macroinvertebrates were important only in May when their relative abundance was 44.89%, similar to cladocerans at the same period (47.88%).

Rotifera and Ostracoda had a relative abundance always inferior to 5% whenever present.

Table 3 shows the absolute and relative numerical values of the main diet items of *A. fasciatus* in a year-long period. Cladocera contributed with 72.79% of the total ingested while Copepoda with 17.75% and macroinvertebrates with 7.12%. The rel-

Table 2. Absolute (n) and relative (%) values of the major diet components of young *A. fasciatus* collected from October 1978 to September 1979 in Lobo (Broa) Reservoir.

Feeding items		1978			1979									Total annual
		Oct.	Nov.	Dec.	Jan.	Feb.	Mar.	Apr.	May	Jun.	Jul.	Aug.	Sept.	
Cladocera	(n)	–	–	357	8	28	38	3504	384	8	4809	262	–	9401
	(%)			(61.66)	(1.69)	(4.21)	(8.28)	(96.29)	(47.88)	(0.70)	(96.28)	(7.72)		(58.16)
Copepoda	(n)	–	–	204	447	632	380	82	51	1129	69	3108	–	6110
	(%)			(35.23)	(94.70)	(95.04)	(82.79)	(2.25)	(6.36)	(98.78)	(1.38)	(91.57)		(37.80)
Macroinvertebrates	(n)	–	–	18	17	5	33	49	360	5	117	23	–	632
	(%)			(3.11)	(3.60)	(0.75)	(7.19)	(1.35)	(44.89)	(0.44)	(2.34)	(0.68)		(3.91)
Others[a]	(n)	–	–	–	–	–	8	4	7	1	–	1	–	21
	(%)						(1.74)	(0.11)	(0.87)	(0.09)		(0.03)		(0.13)

[a] Rotifera and Ostracoda.

Table 3. Absolute (n) and relative (%) values of the major diet components of young *A. fasciatus* collected from October 1978 to September 1979 in Lobo (Broa) Reservoir.

Feeding items		1978			1979									Total annual
		Oct.	Nov.	Dec.	Jan.	Feb.	Mar.	Apr.	May	Jun.	Jul.	Aug.	Sept.	
Cladocera	(n)	2404	–	1818	13289	7327	3152	201	8505	125	2889	236	4663	44609
	(%)	(77.20)		(87.11)	(93.67)	(90.64)	(40.74)	(6.09)	(86.72)	(5.93)	(65.30)	(29.69)	(82.60)	(72.79)
Copepoda	(n)	337	–	102	344	537	2845	2063	311	1839	1123	529	848	10878
	(%)	(10.82)		(4.89)	(2.42)	(6.64)	(36.77)	(62.50)	(3.17)	(87.28)	(25.38)	(66.54)	(15.02)	(17.75)
Macroinver-tebrates	(n)	373	–	146	457	174	656	975	908	133	396	28	114	4360
	(%)	(11.97)		(7.0)	(3.23)	(2.15)	(8.48)	(29.53)	(9.26)	(6.31)	(8.95)	(3.53)	(2.02)	(7.12)
Others[a]	(n)	–	–	21	97	46	1084	62	83	10	16	2	20	1441
	(%)			(1.01)	(0.68)	(0.57)	(14.01)	(1.88)	(0.85)	(0.48)	(0.36)	(0.26)	(0.35)	(2.35)

[a] Rotifera and Ostracoda.

ative abundance of Rotifera and Ostracoda was small (2.35% of the total ingested) except in March (14.01%). Cladocerans were present in relatively high proportions most of the year, peaks being obtained during the beginning of the spring (September and October) and most of the summer. In March and April, a decrease in Cladocera abundance was verified and from there on regular fluctuations were observed with peaks in May, July and September (86.72%, 65.30% and 82.60% respectively) and minimum values in June and August (5.93% and 29.69% respectively) In November, only algae and vegetal detritus were found.

Copepoda showed minimum values from September to February; from March on, regular fluctuations were observed with peaks in April, June and August (62.50%, 87.28% and 66.54% respectively) and the lowest percentages were observed in May, July and September (3.17%, 25.38% and 15.02% respectively).

The relative abundance of macroinvertebrates was almost always inferior to 10% expect in April when it was 29.53%. Rotifera and Ostracoda grouped into 'others' showed only one peak in March (14.01%). During the other months their values were insignificant and their relative abundance was inferior to 2%.

Analysis of relative abundance and occurrence frequency of major Cladocera and Copepoda species in the digestive tract of young and adult A. fasciatus

Table 4 shows absolute and relative abundance data of the main zooplanktonic components present in *A. fasciatus* young and adult specimens. Among cladocerans, *Bosminopsis deitersi* was the only species abundant in both young and adults. It was present in large amounts in the digestive tract of young specimens in April and July (3.235 and 4.549 individuals respectively) and in the digestive tract of adults in June and September (2,157 and 3,411 individuals respectively). Other species such as *Bosmina* sp. *Diaphanosoma brachyurum, Moina minuta* and *Chydorus globosus* were also found in young and adult specimens but their occurrence in large quantities in certain months was observed only in the digestive tract of adults as follows: *Bosmina* sp, October (1, 122 individuals), January (1, 521 individuals), March (2,406 individuals); *Diaphanosoma brachyurum,* January (4,069 individuals); *Moina minuta,* January (1,498 individuals), February (6,442 individuals); *Chydorus globosus,* May (4,015 individuals). Besides these species, *Acroperus harpae, Ilyocryptus spinifer* and *Simocephalus serrulatus* were present, although in small quantities.

Among copepods, the Calanoid *Argyrodiaptomus furcatus* was the most abundant species in the digestive tract of young fishes in June (1,129 individuals) and in August (1,594 individuals) while in adults a great abundance occurred in March (2,841 individuals), April (2,061 individuals) and June (1,831 individuals). The Cyclopoida *(Thermocyclops minutus, Mesocyclops longisetus)* and Harpacticoida *(Elaphoidella bidens)* occurred in small numbers except in August when it was recorded the occurrence of *Thermocyclops minutes* in great abundance in the digestive tract of young specimens (1,513 individuals).

Table 4. Relative abundance (%) and absolute values (n) of the major zooplanktonic components found in the digestive tract of young and adult specimens of *A. fasciatus* from October 1978 to September 1979.

Organisms		1978			1979								
		Oct.	Nov.	Dec.	Jan.	Feb.	Mar.	Apr.	May	Jun.	Jul.	Aug.	Sept.
Young fish													
Bosmina spp	(%)			6.25	0.22	–	0.24	0.12	0.24	–	0.48	–	
	(n)			(35)	(1)		(1)	(4)	(1)		(23)		
Bosminopsis deitersi	(%)			42.14	–	–	0.71	94.62	3.41	0.44	95.07	1.63	
	(n)			(236)			(3)	3235)	(14)	(5)	(4549)	(54)	
Diaphanosoma brachyurum	(%)			8.57	0.22	0.15	0.24	–	–	–	–	1.51	
	(n)			(48)	(1)	(1)	(1)					(50)	
Acroperus harpae	(%)			6.60	0.66	0.46	2.86	0.12	15.12	–	2.88	0.21	
	(n)			(37)	(3)	(3)	(12)	(90)	(62)		(138)	(7)	
Chydorus globosus	(%)			–	0.44	2.30	0.95	2.63	50.73	0.09	0.12	2.06	
	(n)				(2)	(15)	(4)	(3)	(208)	(1)	(6)	(68)	
Ilyocryptus spinifer	(%)			–	–	0.15	2.62	0.03	16.34	–	–	0.33	
	(n)			–		(1)	(11)	(1)	(67)			(11)	
Calanoida (*Argyrodiaptomus furcatus*)	(%)			36.43	98.46	96.78	90.48	2.37	11.46	99.38	1.38	48.32	
	(n)			(204)	(447)	(631)	(380)	(81)	(47)	(1129)	(66)	(1594)	
Cyclopoida (*Thermocyclops minutus*, *Mesocyclops longisetus*)	(%)			–	–	0.15	–	0.03	0.49	–	0.06	45.86	
	(n)					(1)		(1)	(2)		(3)	(1513)	
Harpacticoida	(%)			–	–	–	–	–	0.49	–	–	0.03	
	(n)								(2)			(1)	
Rotifera	(%)			–	–	–	0.71	0.12	–	0.09	–	–	
	(n)						(3)	(4)		(1)			
Ostracoda	(%)			–	–	–	1.19	–	1.71	–	–	0.03	
	(n)						(5)		(7)			(1)	
Adult fish													
Bosmina spp	(%)	41.11	–	35.63	11.84	8.36	34.13	2.75	8.58	0.20	1.40	1.98	0.80
	(n)	(1122)		(672)	(1521)	(653)	(2406)	(65)	(482)	(4)	(53)	(14)	(35)
Bosminopsis deitersi	(%)	46.24	–	52.76	8.31	0.52	3.57	0.46	0.96	4.64	57.05	22.32	78.23
	(n)	(1262)		(995)	(1068)	(41)	(252)	(11)	(54)	(91)	(2157)	(158)	(3411)
Diaphanosoma brachyurum	(%)	0.07	–	1.48	31.66	0.56	0.58	0.08	–	0.66	0.13	–	0.14
	(n)	(2)		(28)	(4069)	(44)	(41)	(2)		(13)	(5)		(6)
Chydorus globosus	(%)	0.22	–	1.06	1.63	0.34	0.07	2.88	71.44	0.25	2.56	0.56	0.69
	(n)	(6)		(20)	(210)	(27)	(5)	(68)	(4015)	(5)	(97)	(4)	(30)
Ilyocryptus spinifer	(%)	–	–	0.42	7.25	0.23	0.31	0.34	12.22	–	–	0.14	0.04
	(n)			(8)	(932)	(18)	(22)	(8)	(687)			(1)	(2)
Simocephalus serrulatus	(%)	–	–	1.85	0.86	–	0.03	3.55	–	–	8.73	–	0.18
	(n)			(35)	(111)		(2)	(84)			(330)		(8)
Moina minuta	(%)	–	–	0.27	35.00	82.50	5.56	0.04	–	–	–	–	–
	(n)			(5)	(4498)	(6442)	(392)	(1)					
Calanoida *Argyrodiaptomus furcatus*	(%)	12.24	–	5.20	2.43	6.83	40.30	87.18	5.07	93.42	25.73	74.58	18.35
	(n)	(334)	–	(98)	(312)	(533)	(2841)	(2061)	(285)	(1831)	(973)	(528)	(800)
Cyclopoida (*Thermocyclops minutus*, *Mesocyclops longisetus*	(%)	–	–	0.21	0.17	–	0.03	–	0.14	0.31	3.97	–	1.10
	(n)			(4)	(22)		(2)		(8)	(8)	(150)		(48)
Harpacticoida	(%)	0.11	–	–	0.08	0.05	0.03	0.08	0.11	–	–	0.14	–
	(n)	(3)			(10)	(4)	(2)	(2)	(6)			(1)	
Rotifera	(%)	–	–	–	0.01	0.01	12.31	0.04	0.03	0.46	0.34	0.14	0.46
	(n)				(2)	(1)	(869)	(1)	(2)	(9)	(13)	(1)	(20)
Ostracoda	(%)	–	–	1.11	0.74	0.58	3.06	2.58	1.44	0.05	0.08	0.14	–
	(n)			(21)	(95)	(45)	(216)	(61)	(81)	(1)	(3)	(1)	

Rotifers and ostracods have been found more in adults than in young specimens although in small quantities.

Relatively, *A. furcatus* was the most abundant species in fish digestive tract but this abundance has been a little higher in young fishes (40%) than in adult fishes (30%).

Occurrence frequency and frequency index

In general, the most abundant organisms showed higher values of occurrence frequency and frequency index what leads to the supposition that the type of food consumed by fish in this way depends on availability of species in the environment.

The results presented in Table 5 show, for young

Table 5. Occurrence frequency (N) and frequency index (f) of the major zooplanktonic components found in the digestive tract of young and adult specimens of *A. fasciatus* from October 1978 to September 1979.

Organisms	1978			1979								
	Oct. N-f	Nov. N-f	Dec. N-f	Jan. N-f	Feb. N-f	Mar. N-f	Apr. N-f	May N-f	Jun. N-f	Jul. N-f	Aug. N-f	Sept. N-f
Young fish												
Bosmina spp.			3-0.3	1-0.1	-	1-0.1	2-0.2	1-0.1	-	5-0.5	-	
Bosminopses deitersi			6-0.6	-	-	2-0.2	9-0.9	5-0.5	4-0.4	10-1.0	9-0.9	
Diaphanosoma brachyurum			4-0.4	1-0.1	1-0.1	1-0.1	-	-	-	-	6-0.6	
Acroperus harpae			7-0.7	3-0.3	3-0.3	5-0.5	9-0.9	4-0.4	-	10-1.0	4-0.4	
Chydorus globosus			-	2-0.2	4-0.4	3-0.3	3-0.3	7-0.7	1-0.1	4-0.4	10-1.0	
Ilyocryptus spinifer			-	-	1-0.1	5-0.5	1-0.1	7-0.7	-	-	6.-0.6	
Calanoida (*Argyrodiaptomus furcatus*			9-0.9	6-0.6	7-0.7	9-0.9	9-0.9	6-0.6	10-1.0	7-0.7	10-1.0	
Cyclopoida *Thermocyclops minutus, Mesocyclops longisetus)*			-	-	1-0.1	-	1-0.1	1-0.1	-	2-0.2	9-0.9	
Harpacticoida			-	-	-	-	-	1-0.1	-	-	1-0.1	
Rotifera			-	-	-	3-0.3	3-0.3	-	-	-	-	
Ostracoda			-	-	-	4-0.4	-	3-0.3	-	-	1-0.1	
Adult fish												
Bosmina spp	10-1.0	-	8-0.8	10-1.0	10-1.0	10-1.0	9-0.9	9-0.9	3-0.3	6-0.6	6-0.6	6-0.6
Bosminopsis deitersi	10-1.0	-	9-0.9	6-0.6	7-0.7	7-0.7	6-0.6	10-1.0	10-1.0	8-0.8	10-1.0	9-0.9
Diaphanosoma brachyurum	2-0.2	-	2-0.2	6-0.6	4-0.4	2-0.2	2-0.2	-	3-0.3	2-0.2	-	2-0.2
Chydorus globosus	3-0.3	-	5-0.5	6-0.6	7-0.7	1-0.1	5-0.5	6-0.6	2-0.2	4-0.4	2-0.2	6-0.6
Ilyocryptus spinifer	-	-	1-0.1	9-0.9	3-0.3	6-0.6	3-0.3	5-0.5	-	-	1-0.1	1-0.1
Simocephalus serrulatus	-	-	1-0.1	2-0.2	-	1-0.1	4-0.4	-	-	1-0.1	-	1-0.1
Moina minuta	-	-	2-0.2	6-0.6	8-0.8	7-0.7	1-0.1	-	-	-	-	-
Calanoida (*Argyrodiaptomus furcatus*	8-0.8	-	4-04	8-0.8	7-0.7	9-0.9	9-0.9	7-0.7	10-1.0	10-1.0	10-1.0	9-0.9
Cyclopoida (*Mesocyclops minutus Thermocyclops longisetus)*	-	-	2-0.2	4-0.4	-	2-0.2	-	7-0.7	3-0.3	3-0.3	-	8-0.8
Harpacticoida	3-0.3	-	-	3-0.3	3-0.3	2-0.2	2-0.2	2-0.2	-	-	1-0.1	-
Rotifera	-	-	-	1-0.1	1-0.1	4-0.4	1-0.1	1-0.1	2-0.2	1-0.1	1-0.1	2-0.2
Ostracoda	-	-	3-0.3	4-0.4	2-0.2	3-0.3	5-0.5	6-0.6	1-0.1	1-0.1	1-0.1	-

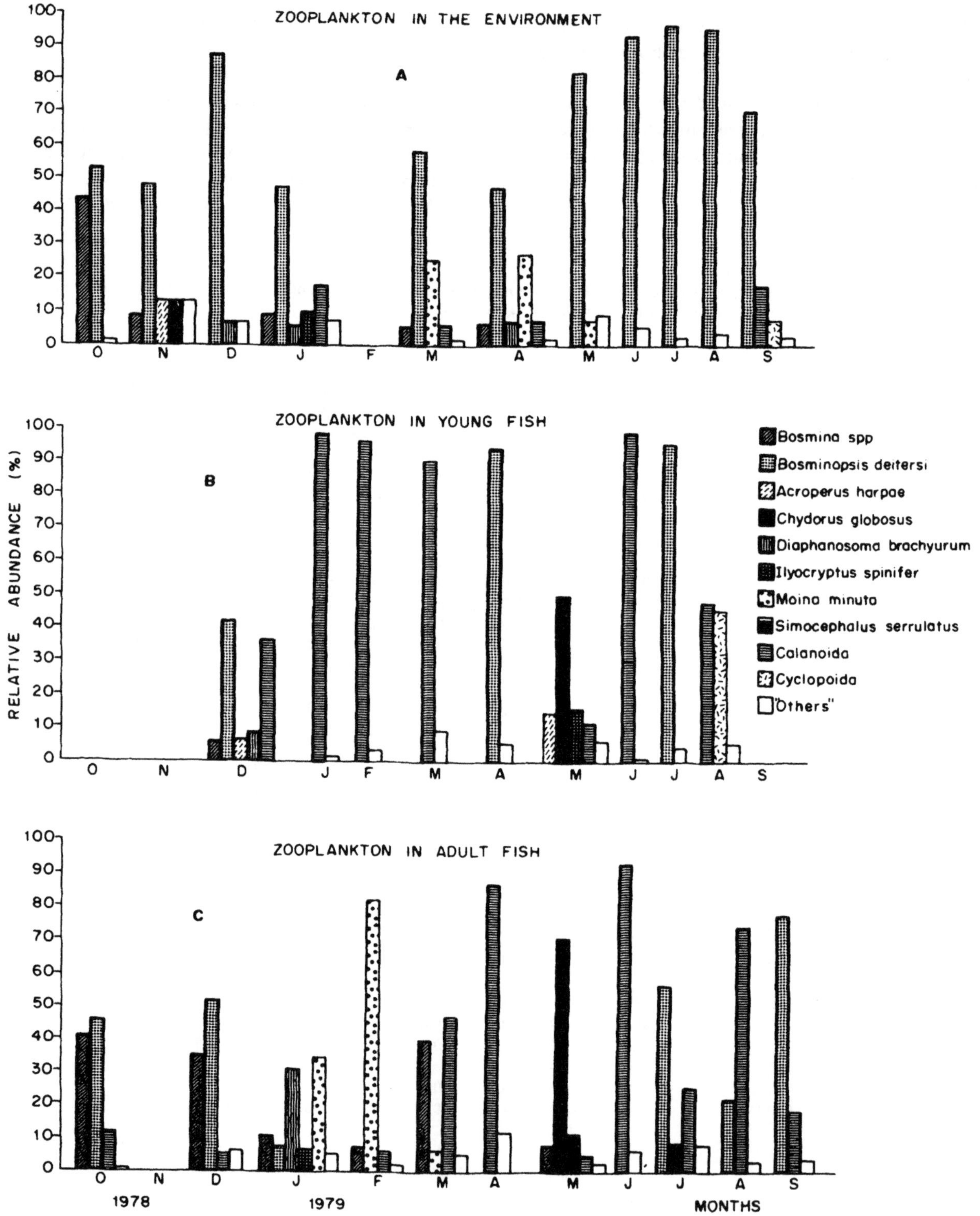

Fig. 3. Relative abundance of the major zooplanktonic organisms from the Lobo Reservoir found in the environment and in the digestive tract of young and adults specimens of *A. fasciatus*.

fishes, that only *Bosminopsis deitersi* and *Acroperus harpae*, among the cladocerans, presented indexes higher than 0.5 in most months in which they were considered preferential preys. The other species have shown indexes lower than 0.5 being considered accidental and occasionally preferential preys. Among copepods, *A. furcatus* presented high values of frequency index, being a preferential prey for young fishes during the whole year. These high values provide evidence for the importance of this species in the feeding of *A. fasciatus* in the considered environment. Cyclopoida and Harpacticoida were not found to be important in most periods of the year except in August when the Cyclopoida showed a high index (0.9). Rotifera and Ostracoda were rarely found and presented very low frequency indexes.

For adult fishes, *Bosmina* sp and *Bosminopsis deitersi* were found to be preferential preys during the whole year, presenting relatively high frequency indexes. As it was also observed for young fishes, the other cladocerans presented low indexes, being considered preferential preys only during certain periods. Among the copepods, *A. furcatus* was found to be a preferential prey, showing relatively high frequency indexes (0.8–1.0). Cyclopoida showed high indexes only in May and September (0.7 and 0.8 respectively). Harpacticoida, as rotifers and ostracods, have shown low indexes during the whole period.

Relation between zooplanktonic components found in young and adult A. fasciatus digestive tracts and at the sampling station

Figure 3 shows the monthly variation of the relative abundance of zooplanktonic components in the environment and in the digestive tract of young and adult fishes.

Only adult crustaceans have been analyzed as zooplanktonic components of the environment since they were the most abundant individuals in the digestive tract of fishes. The results show that more abundant species in the environment are not always the most consumed by fish. During the whole year, *Bosminopsis deitersi* was the most abundant species in the environment; however, it was not the most abundant species in the digestive tract of both classes (young and adults), but *A. furcatus* whose occurrence in the environment was inferior to 10%.

Adult fishes show a higher variability in zooplanktonic species consumption. The preferred species according to their dominance in different months were the following: *Bosminopsis deitersi* in October, December, July and September; *A. furcatus* in March, April, June and August; *Moina minuta* in January, February; *Bosmina* spp in October, December and May; *Chydorus globosus* in May and *Diaphanosoma brachyurum* in July.

Young fishes had *A. furcatus* as preferential prey in most months (December, January, February, May, July and August) followed by *Bosminopsis deitersi* (December, April and July), *Acroperus harpae* (May) and *Thermocyclops minutus* (August).

The importance of *A. furcatus* for the feeding of young and adult fishes was obvious and despite its low abundance in the environment it was the major feeding item found in young and adults in many months, perhaps indicating a selective feeding upon this species.

Other species were important food sources for *A. fasciatus*; however, it was not possible to establish a relation since the abundance of these species in the samples analyzed was very low. Perhaps, other relations could have been seen if other points of collection had been fixed since fishes in motion can explore large areas and consequently feed in regions different from that where they were captured.

Discussion

According to the amount of variation in the types of food ingested, fishes can be classified into: monophagic, stenophagic and euryphagic, the latter being a common character among fresh water fishes (Nikolsky, 1963). *A. fasciatus* can be considered euryphagic since it feeds on a considerable variety of items: zooplankton, macroinvertebrates, algae, vegetal and inorganic detritus. As Lowe-McConnell (1975) has observed, many species of fish show a large feeding plasticity, utilizing different preys according to the stage of development, environmental changes or availability of items.

The result of the present study show a significant difference in occurrence frequency of the main items in the two classes of length. Thus, for young and adult fishes, the following species were the most frequent items in decreasing order: Copepoda, Cladocera, Macroinvertebrates and Macroinvertebrates, Cladocera, algae and vegetal detritus respectively.

Many authors (Brooks, 1968; Zaret, 1980; Confer & Blader, 1975; Jansen, 1976; Frost, 1977) have shown that cladocerans are preferred over copepods by most planktivorous fishes. This preference has been attributed to body shape and size or to the mechanism of locomotion which, being more efficient in copepods, facilitates their escape. As stated by Zaret (1980), although Copepoda show a more elongated shape, their antennae are usually at a right angle, resulting in a larger surface which can be an important feature for fish attraction.

The abundance of copepods found in the digestive tract of young fishes indicates a feeding selectivity at least at this stage of development, preferring Copepoda over Cladocera, and vice versa for adults. It is difficult to find a convincing explanation for this difference in feeding preference, but it is probably related to the feeding mechanism.

Jansen (1976) has observed in *Alosa pseudoharengus* the occurrence of three ways of capturing food: particles in small fishes (49–114 mm); 'gulping' in medium fishes (124–152 mm) and filtration in large fishes (178–180 mm). In the first case, a higher selectivity with relation to prey size occurs, while in filtration the selection is weaker, being more important at night when preys are not visible.

Astyanax fasciatus, whose young are found in the littoral region during daytime, may present a feeding in the form of particles, selecting *A. furcatus*. This Calanoid can be an easy prey to young fishes with such feeding behavior since they are bigger than other zooplanktonic organisms and are pigmented, features Zaret (1972) and Zaret & Kerfoot (1978) consider important in contributing to prey visibility.

On the other hand, adult fishes are found near the shore of the reservoir at night, probably to feed since according to Zaret (1975), zooplanktivorous fishes usually feed at the shore of reservoirs. Maybe for these organisms (for adult fishes), the filter-feeding process is the most efficient mechanism since light conditions do not benefit visual selection. Thus, cladocerans which are more abundant than copepods in the environment (44% and 15% respectively) are consumed more. Many cladocerans such as *M. minuta, D. brachyurum, B. deitersi, Bosmina* spp, dominated the zooplankton components alternately within the 12 months of study.

Often, the comparison between prey abundance in the environment and in the digestive tract of young and adult specimens does not indicate a direct relation and, as pointed out by Werner & Hall (1974). These types of comparisons are not always satisfactory since the abundance of one prey in the environment does not always indicate that it is really found by fishes. Features such as body size, movement patterns, and environmental complexity may influence the availability and the frequency of encounter.

However, predation by fishes constitutes a very important factor in the variation of composition and zooplankton population density. In 'Broa' reservoir, although precipitation is considered to be an important variable for the seasonal fluctuation of zooplankton density (Matsumura-Tundisi & Tundisi, 1976), predation by fishes constitutes one of the key factors in controlling the zooplankton population. The amplitude of the effects of predation depends on the characteristics of the other fish populations in terms of size, spatial distribution and feeding habits.

References

Brooks, J. L., 1968. The effects of prey-size selection by lake planktivores. Syst. Zool. 17: 272–291.

Brooks, J. L. & S. I. Dodson, 1965. Predation, body size, and composition of plankton. Science 150: 28–35.

Confer, J. L. & P. I. Blades, 1975. Omnivorous zooplankton and planktivorous fish. Limnol. Oceanogr. 20: 571–579.

Edmondson, W. T. & G. G. Winberg, 1971. A manual on methods for the assessment of secondary productivity in fresh waters. Blackwell Scientific Pubications, Oxford, 358 pp.

Frost, W. E., 1977. The food of Charr, *Salvelinus willughbü* (Günther) in Windermere. J. Fish. Biol. 11: 531–547.

Hutchinson, B. P., 1971. The effect of fish predation on the zooplankton of ten Adirondack Lakes, with particular references to the alewife, *Alosa pseudoharengus*. Trans. Am. Fish. Soc. 2: 325–335.

Hynes, H. B. N., 1950. The food of fresh-water sticklebacks (*Gasterosteus aculeatus* and *Pygosteus pungitius*), with a review of methods used in studies of the food of fishes. J anim. Ecol. 19: 35–58.

Jansen, J., 1976. Feeding modes and prey size selection in the alewife (*Alosa pseudoharengus*). J. Fish. Res. Bd. Can. 33: 1972–1975.

Lagler, K. F., J. E. Bardach & R. R. Miller, 1962. Ichtyology. John Wiley & Sons, N. Y., 545 pp.

Lowe-McConnel, R. H., 1975. Fresh communities in tropical fresh-waters. Logman, city, 337 pp.

Matsumura-Tundisi, T. & J. G. Tundisi, 1976. Plankton studies in a lacustrine environment, 1. Preliminary data on zoo-plankton ecology of Broa reservoir. Oecologia 25: 265-270.

Wells, L. R., 1970. Effects of alewife predation on zoo-plankton populations in Lake Michigan. Limnol. Oceanogr. 15: 556-565.

Werner, E. E. & D. J. Hall, 1974. Optimal foraging and the size selection of prey by the Bluegill sunfish (*Lepomis macrochirus*). Ecology 55: 1024-1052.

Zaret, T. M., 1969. Predation-balanced polymorphism of *Ceriodaphnia cornuta* Sars. Limnol. Oceanogr. 14: 301-303.

Zaret, T. M., 1972. Predators, invisible prey, and the nature of polymorphism in the Cladocera (Class Crustacea). Limnol. Oceanogr. 17: 171-183.

Zaret, T. M., 1980. Predation and freshwater communities. Yale University Press, City, 187 pp.

Zaret, T. M. & W. C. Kerfoot, 1975. Fish predation on *Bosmina longirostris:* Body-size selection versus visibility selection. Ecology 56: 232-237.

Spatial structure of the zooplankton community in the Cabo Frio region (Brazil) influenced by coastal upwelling

Jean L. Valentin
Brazilian Naval Research Institute, Cabo Frio Project, 28910 Arraial do Cabo-RJ, Brazil

Keywords: tropical zooplankton, upwelling, Brazil, statistic analysis, spatial structure, copepod diversity

Abstract

Cabo Frio area (Central-southern Brazilian littoral) is characterized by a coastal upwelling due to prevailing E-NE winds, mainly during the spring-summer season.

The geographic distribution of the zooplankton was described in order to separate ecological areas in terms of specific communities. Cluster analysis by weighted pair-group method (WPGA) and principal component (PC) analysis was performed on the most important hydrological parameters and zooplankton taxa. The first PC, explaining 38% of the total zooplankton variability, represented the influence of the nearshore–offshore gradient. Zooplankton communities and ecological areas were defined as follows:

i) nearshore neritic: the richest region with more than 5 000 org. m^{-3}. In this very narrow nearshore region we observed 3 groups of taxa deliminated by upwelling effects and trophic conditions: a) *Ctenocalanus vanus, Penilia avirostris, Calanoides carinatus* and Copepod larva, all herbivorous species associated with the phytoplankton blooms in the upwelling zone; b) *Creseis acicula* and *Temora stylifera;* c) Siphonophora, Chaetognaths and a high diversity of Copepoda dominated by *Coryceus amazonicus, C. giesbrechti, Eucalanus pileatus, Clausocalanus acuicornis,* all carnivorous or omnivorous feeders representing a high degree of complexity of the ecosystem.

ii) offshore neritic: characterized by appendicularians, this region constitutes an intermediate zone between coastal and oceanic waters.

iii) oceanic: occupied by the Brazil Current where zooplankton is poor, but diversified, with typical copepods from tropical warm water: *Clausocalanus furcatus, Mecynocera clausi, Corycella gracilis, Oithona setigera, Coryceus typicus, Oncea conifera, Undinula vulgaris, Calocalanus pavo, Haloptilus longicornis, Lucicutia flavicornis.*

Introduction

Most marine zooplankton species have relatively limited and variable areas of distribution. However, two main areas dominate: the neritic and the oceanic. This scheme can occasionally be altered, notably in coastal upwelling regions implying great hydrological changes in the neritic area, which subsequently affect zooplankton distribution. Zooplankton species with narrow ecological requirements can be used as biological indicators to identify water displacements and mixing in upwelling areas.

The concept of indicator species has a statistical value in ecological work (Bougis, 1974) and underlines the necessity of using advanced statistical analysis in the evaluation of data.

The central-southern Brazilian littoral is characterized by coastal upwelling in the Cabo Frio region (23 °S latitude). This upwelling has already been investigated with regard to physical and chemical aspects (Moreira da Silva, 1968, 1973; Rodriguez,

Hydrobiologia 113, 183–199 (1984).

1973; Valentin, 1974) and zooplankton community *(Mureb et al*, 1976; Valentin & Moreira, 1978; Valentin *et al*, 1976; Valentin, 1980). Under the influence of the prevailing E-NE winds and the Coriolis deviation, the surface water flows toward the open sea and is replaced by deeper water from approximately 300 m. By penetrating the euphotic layer, this deep cold water (South Atlantic Central Water; Thomsen, 1962) influences the structure of this tropical coastal ecosystem, and gives it a subtropical appearance characterized by taxa generally met with at the higher latitudes or in deeper waters.

This report aims to define the main zooplankton associations and their contribution in drawing different ecological areas around Cabo Frio region. That was carried out by using a multivariate statistical treatment of faunistic and hydrologic data.

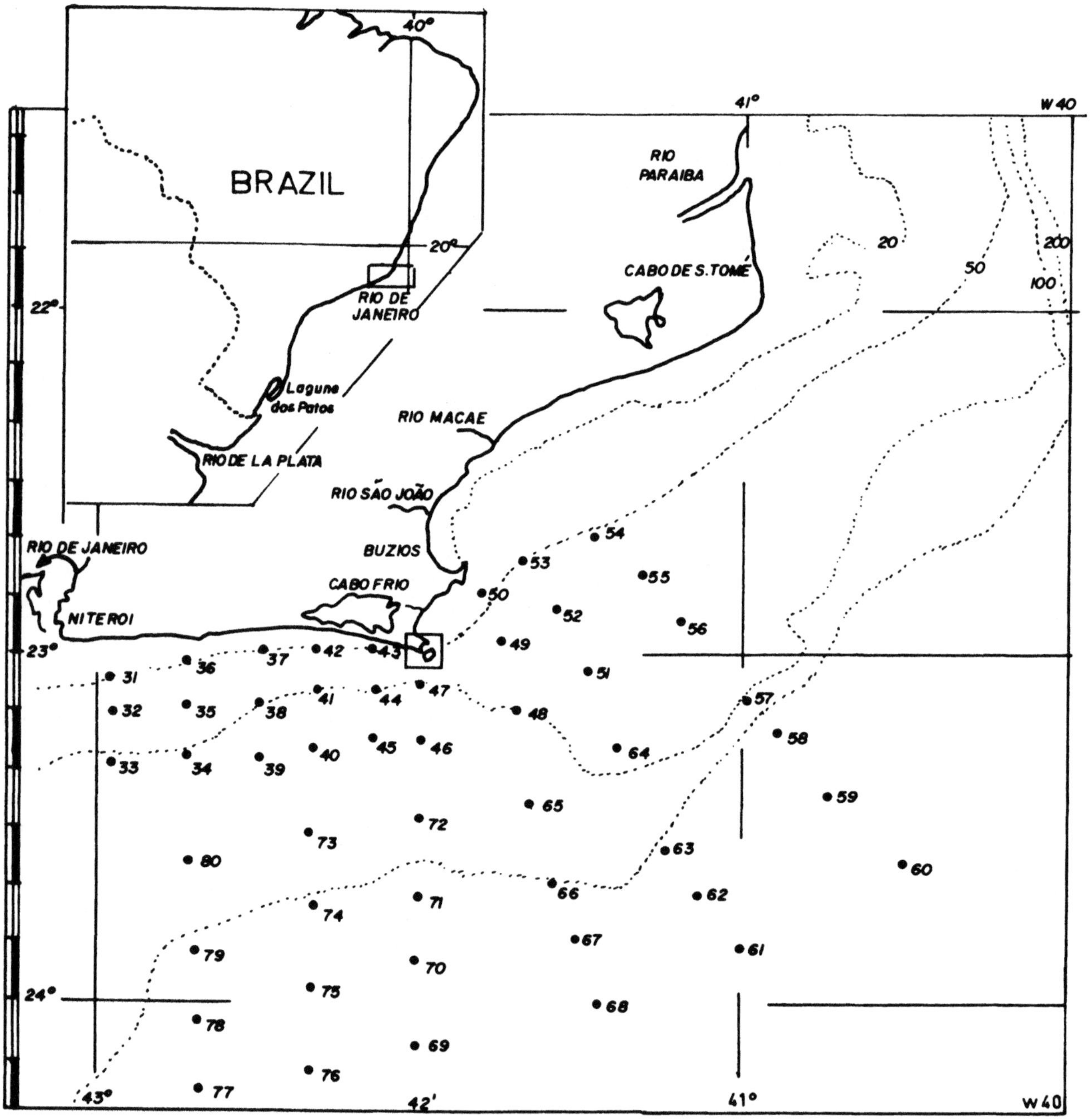

Fig. 1. Cabo Frio region with position of the collecting stations.

Methods

Zooplankton samples were collected by vertical plankton-net hauls from the bottom (upper 200 m depth) to the surface with a 250 μm mesh size, at 50 stations scattered around Cabo Frio up to 100 km offshore (Fig. 1).

Other hydrobiological variables measured included: temperature (by a reversing thermometer), salinity (using an inductance salinometer), chlorophyll *a* (by an absorbance method; UNESCO, 1966) and phytoplankton (enumerated after sedimentation, Utermohl (1958) technique). Water samples were collected at different depths (with 10 m interval) within the water column, except for primary biomass which was measured at 0, 10 and 20 m depth only.

The following statistical methods were applied to the data: (i) principal component analysis (Harmann, 1970) and (ii) cluster analysis by weighted pair-group method (using FORTRAN IV program provided by Davis (1973)).

The diversity of copepod populations was calculated using the Shannon index.

This work was conducted during the CABO FRIO IV cruise of the R/V Alte SALDANHA (Brazilian Navy) in December 1978.

Results

Hydrobiological conditions

Surface temperatures deliminated two upwelling areas close to the Cabo Frio-Rio de Janeiro shore, with temperatures below 20 °C. At 50 m depth, cold water (below 16 °C) occurred in a continuous narrow fringe in the nearshore region (Fig. 2). Offshore temperatures amounted to 24 °C in the upper oceanic water of the Brazil current.

Surface salinity ranged from 35.50‰ at the two coastal upwelling centers to 36.8‰ for the Brazil current water at around 50 km off-shore (Fig. 3).

Phytoplankton density was less than could be expected in an upwelling region: slightly exceeding 5 000 cells l^{-1} and 1.0 μg chl. *a* l^{-1} (Fig. 4). Nevertheless, the relative population characteristics, e.g. the diatom-dinoflagellate ratio, indicate a certain degree of neritic influence because tropical oceanic water of the Brazil current is inhabited by a poor phytoplankton dominated by dinoflagellate populations. On the other hand, diatoms were the most important phytoplankton group in the upwelling water around Cabo Frio (Fig. 5). There were some identical findings at some stations east of Cabo Frio showing the influence of a second eutrophic zone issue from a northern upwelling area (Valentin *et al.*, 1978). The most important phytoplankton were *Leptocylindrus danicus* (30–50% of the population) and the genus *Rhizosolenia* (25–45%) with species *R. fragilissima*, *R. delicatula*, *R. alata*, in addition to some large-size species like *Guinardia flacida*, *Hemiaulus* sp., *Cerataulina pelagica*, *Rhizosolenia imbricata*.

Zooplankton community (Fig. 6)

As anticipated, coastal stations and especially the upwelling regions were the richest with more than 1 000 ind. m^{-3}. Maxima (more than 5 000 ind. m^{-3}) occurred in the upwelling regions. Off the continental slope zooplankton density was always below 500 ind. m^{-3}. Copepods constituted the majority of the biomass (Fig. 7), with an average of 69%, 80% offshore but declining below 50% nearshore, where they were displaced by *Penilia avirostris*, *Fritillaria haplostoma* and general meroplantonic forms. *Ctenocalanus vanus*, *Temora stylifera* and *Paracalanus parvus* at the coastal stations, *Oithona setigera* and *Clausocalanus furcatus* off-shore contributed to at least 10% of the copepod populations and were the most frequent species. They were accompanied by *Calanoides carinatus*, *Coryceus amazonicus*, *C. typicus* and *C. giesbrechti* above the continental shelf and by *Corycella gracilis*, *Oncea conifera* and *Oncea venusta* at the most oceanic-influenced stations (Table 1).

Specific diversity (Shannon's index) (Fig. 8) was calculated for the copepod populations. Lowest values (below 3 bits. ind^{-1}) occurred close to Cabo Frio at stations with upwelling, except stations 41 and 42 occupied by Brazil current water. Large abundances of *Paracalanus parvus* at st. 37 and of *Ctenocalanus vanus* at st. 32 made diversity values below 2 bits. ind^{-1}. Beyond the 100 m isobathymetric line diversity ranged from 3 to 4 bits. ind^{-1}.

We observed (Table 1) that stations 64, 65 and 66, although very far from the shore (see Fig. 1), were inhabited by copepod species similar to the coastal ones. This observation and the distribution of phy-

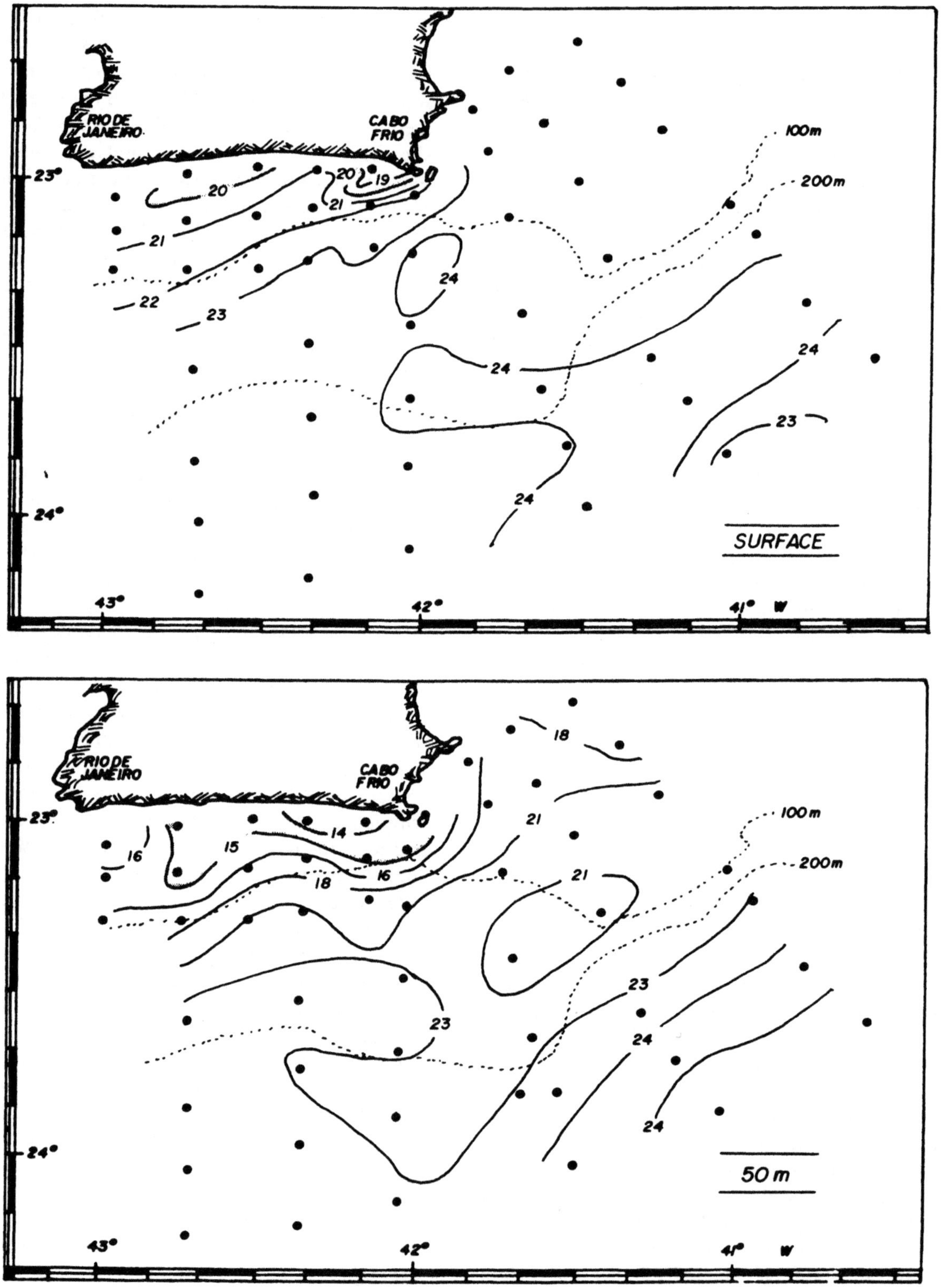

Fig. 2. Cabo Frio region. Isotherms at surface and 50 m depth. Shaded areas indicate deep cold water occurrence.

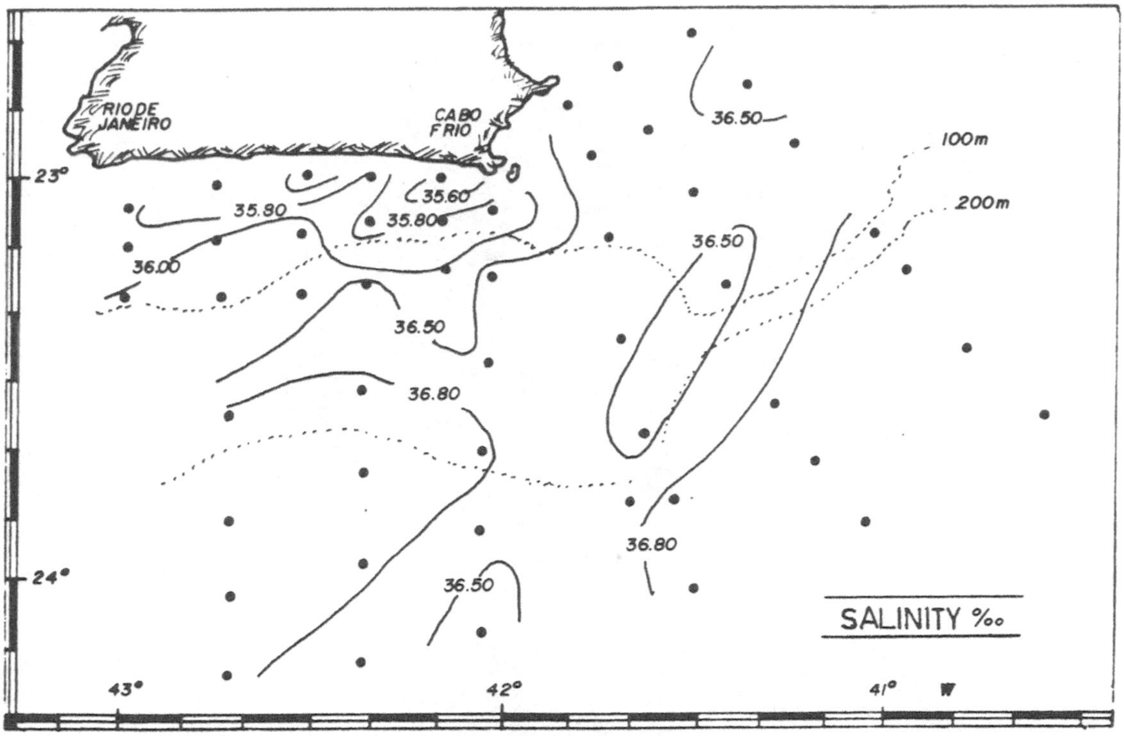

Fig. 3. Cabo Frio region. Variations of surface salinity.

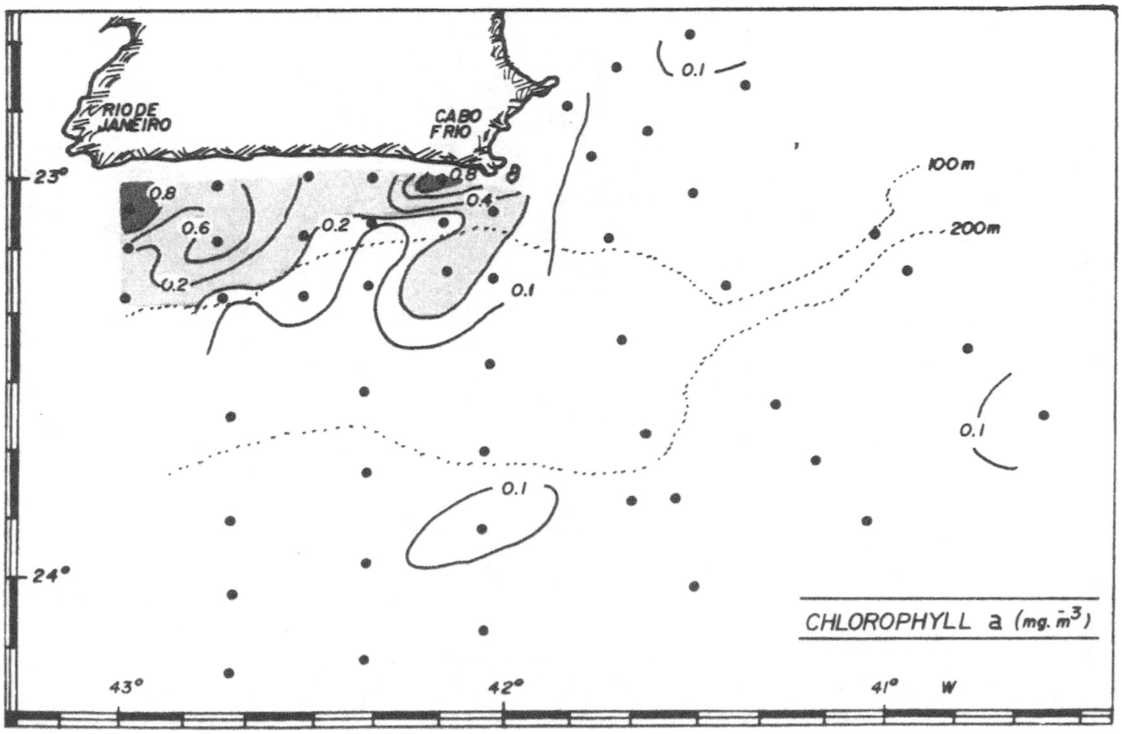

Fig. 4. Cabo Frio region. Variations of surface chlorophyll *a* ($\mu g\ l^{-1}$).

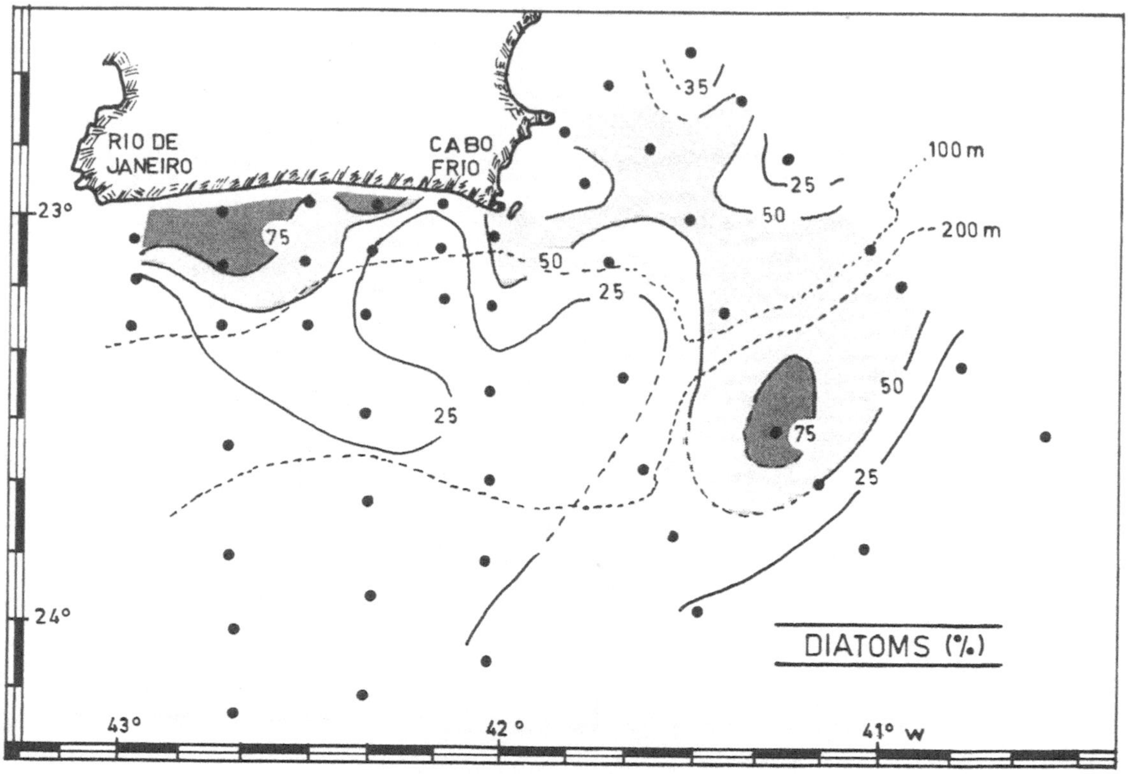

Fig. 5. Cabo Frio region. Percentage of diatoms at the surface.

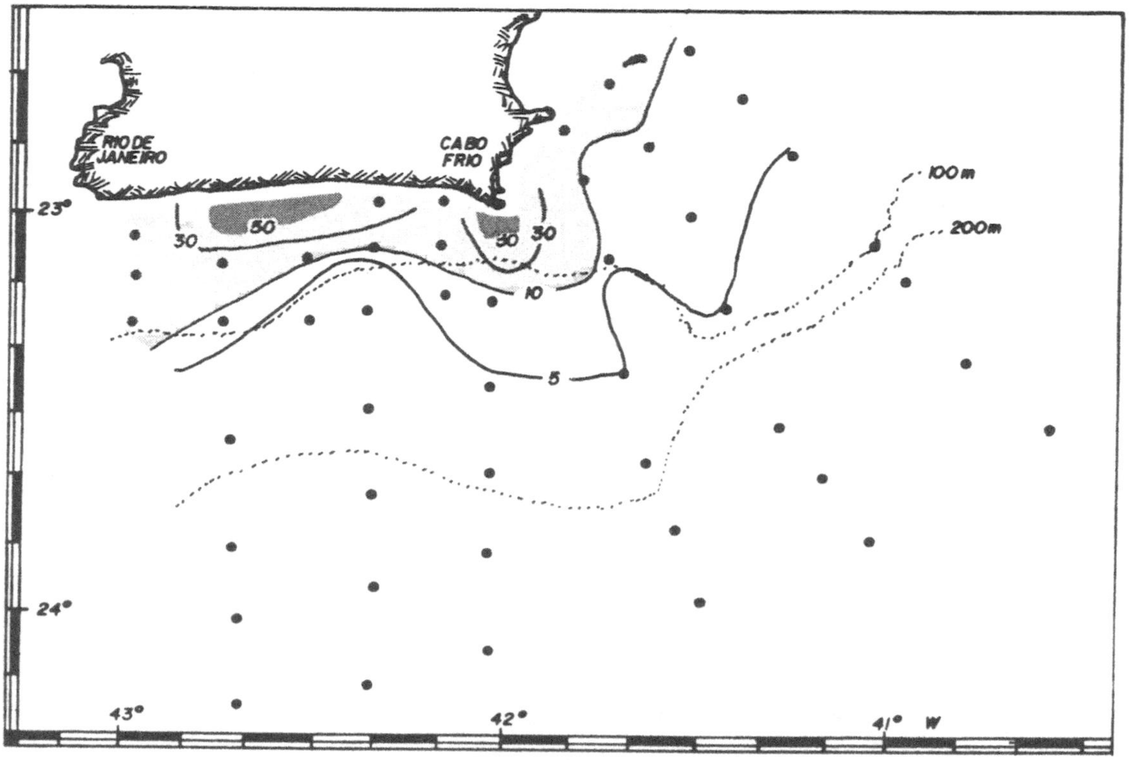

Fig. 6. Cabo Frio region. Distribution of total zooplankton (10^2 ind m^{-1}).

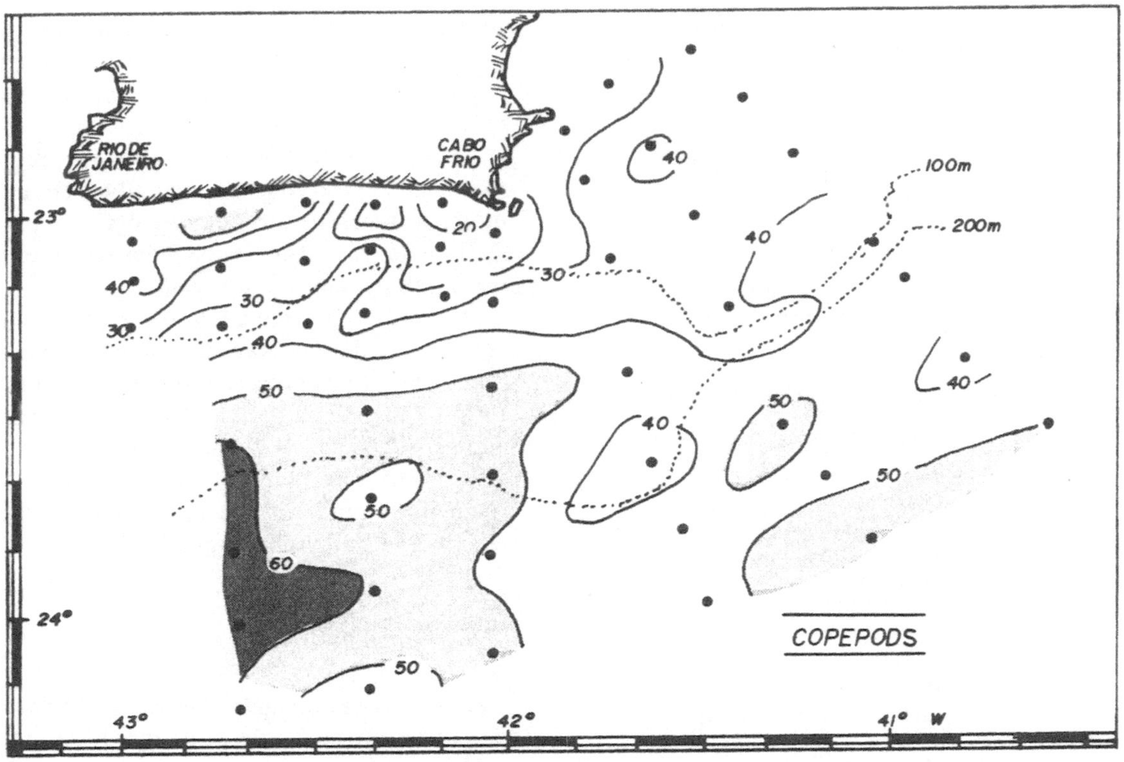

Fig. 7. Cabo Frio region. Distribution of copepods (% of total zooplankton number).

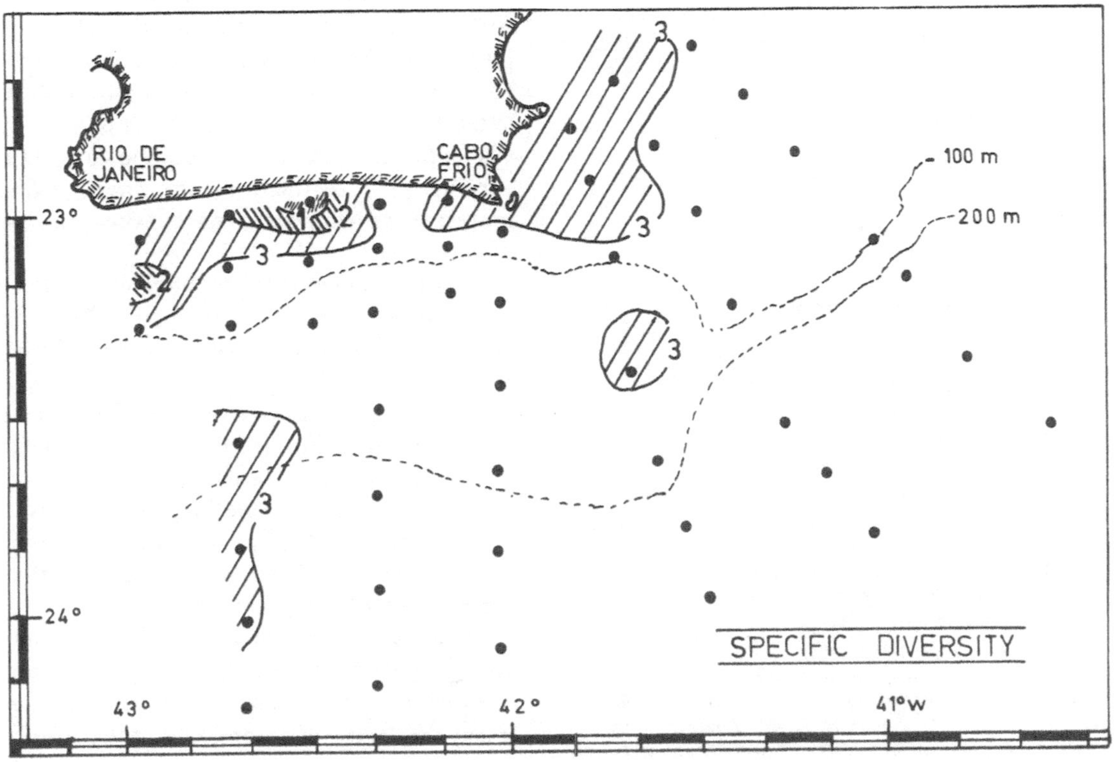

Fig. 8. Cabo Frio region. Diversity distribution of the copepod populations.

Table 1. Abundance of different species of copepods (in % of total copepods) at each station.

Station no.	Copepods (%)
31	CTV (31), CAC (26), PAP (21), TEM (14)
32	CTV (37), PAP (17), CAC (10)
33	CTV (37), TEM (19), CAC (15)
34	TEM (18), COA (17), CTV (13)
35	CTV (27), COA (16), CAC (12), TEM (11)
36	CTV (39), CAC (30), PAP (23)
37	PAP (89)
38	TEM (27), COA (16), PAP (14)
39	COG (20), TEM (18), COA (13), PAP (10)
40	TEM (22), OIS (10)
41	TEM (34), COG (13), CAC (13)
42	CTV (25), COG (18), CAC (13)
43	PAP (42), CTV (26)
44	CTV (36), PAP (16), COG (11)
45	CTV (29), CAC (18), CLA (16), COG (16)
46	CTV (32), PAP (12), CLA (11), TEM (10)
47	PAP (22), COG (21), CTV (16), CLA (11)
48	CTV (34), OIS (11), TEM (9)
49	CTV (51), ONV (14)
50	CTV (42), PAP (25)
51	TEM (27), PAP (25), CTV (9), CLA (9)
52	CLA (32), OIS (22)
53	CTV (46), PAP (14), TEM (9)
54	TEM (29), CAC (19), PAP (13)
55	PAP (30), TEM (19), CLF (9)
56	TEM (18), PAP (18), CLF (15), CTV (12)
57	OIS (32), CLF (12), ONV (10)
58	OIS (23), PAP (10)
59	OIS (23), COR (16)
60	OIS (32), COR (20)
61	OIS (40)
62	CLF (20), ONC (20), OIS (19)
63	OIS (34), ONC (10)
64	PAP (26), TEM (18), CTV (13), CLF (11)
65	PAP (49), COG (11)
66	COG (22), CTV (18)
67	OIS (19), ONC (18), COG (10)
68	OIS (21), ONC (14), CLF (14)
69	OIS (19), CLF (11), COR (9), COG (9)
70	OIS (14), PAP (13), CLF (12), ONC (10)
71	OIS (25), CLF (23)
72	OIS (22), COR (18), CLF (17)
73	CLF (24), OIS (15), COR (14)
74	OIS (29), CLF (14), COR (11)
75	CLF (26), OIS (17), COG (10), TEM (10)
76	CLF (20), OIS (19), ONC (10)
77	CLF (27), OIS (19), COG (11), ONC (10)
78	COG (30), CLF (26), OIS (11)
79	CLF (33), OIS (20), COR (16)
80	CLF (44), OIS (15), COR (12)

Abbreviations: CTV = *Ctenocalanus vanus*, CAC = *Calanoides carinatus*, PAP = *Paracalanus parvus*, TEM = *Temora stylifera*, COA = *Coryceus amazonicus*, COG = *Coryceus giesbrechti*, OIS = *Oithona setigera*, CLA = *Clausocalanus acuicornis*, CLF = *Clausocalanus furcatus*, ONV = *Oncaea venusta*, ONC = *Oncaea conifera*, COR = *Corycella gracilis*.

toplankton (Fig. 4) support the hypothesis that the surface masses originated from the northern upwelling center and thus produced eutrophic conditions as far as 80 km away.

Except for copepods, the cladoceran *Penilia avirostris* most often showed the highest zooplankton biomass, reaching 50% of the total population at stations 36, 42 and 44. It disappeared completely in the off-slope samples, however.

Other taxa that had a relatively high share in the coastal populations were Ostracoda (e.g. *Conchoecia* – 20%), Pteropoda (*Creseis acicula* – more than 10%), appendicularians (*Oikopleura longicauda* and *Fritillaria* sp. – 40% close to Cabo Frio), meroplanktonic forms (15%) and fish eggs. Over the 100 m isobathymetric line, and mainly off the slope, we observed only chaetognaths besides copepods (Fig. 9).

Statistical analysis

Inter-correlation matrix between 37 selected species or taxa (see Table 2) at the 50 stations was analysed by the principal component technique. Because of their frequent absence, 12 taxa were introduced into the analysis without interfering with the calculation of the factors. They were just plotted on the factorial planes according to their coordinates. These scarce species are indicated by letter codes on Table 2 and Figures.

The two first factors contributed 46% of the total variance (38% by the first component), which thus represented an important part of the total 'information'.

In order to identify these axes we made correlation between scores (factor loading) and selected ecological parameters. Correlation coefficients are shown on Table 3.

The first axis was positively correlated with the phyto- and zooplankton abundance and negatively with temperature, salinity, diversity of copepods and percentage of dinoflagellates. This factor, which explained 38% of the total variance, could indicate plankton biomass variations through the nearshore-offshore gradient.

The second component (8% of the variance) was linked to the high diversity of copepods, temperature and low zooplankton abundance. It could represent the degree of complexity of the zooplankton system.

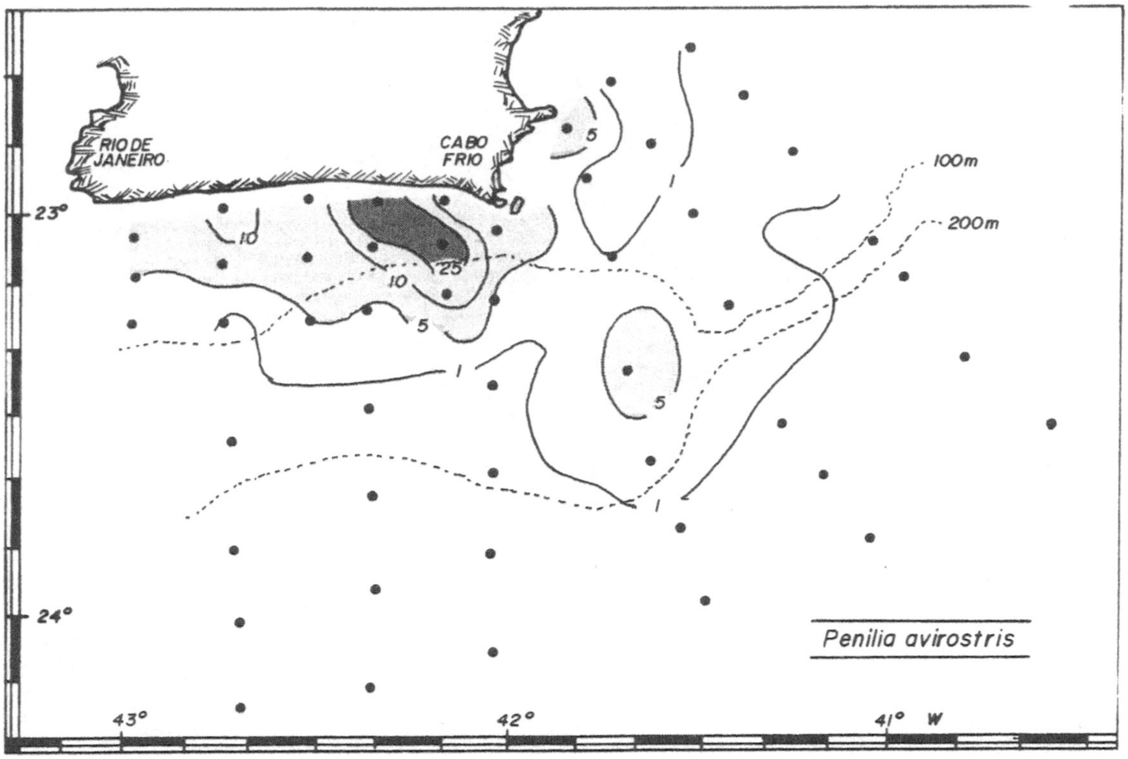

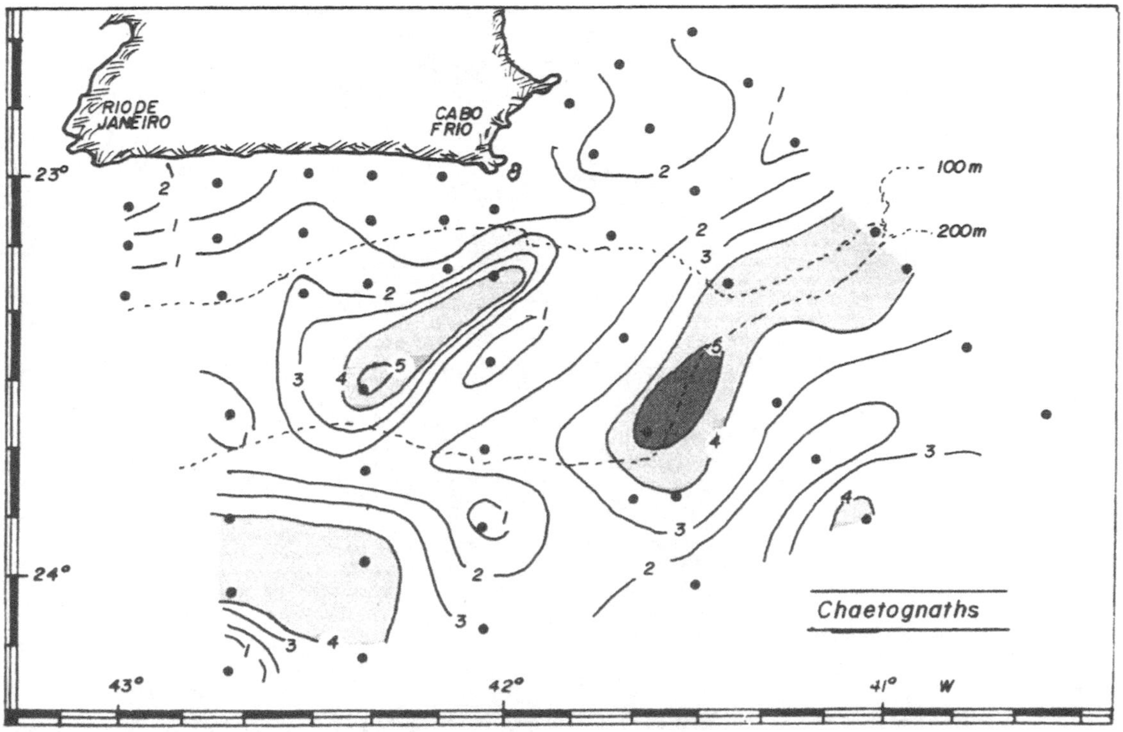

Fig. 9a. Cabo Frio region. Distribution of the most important taxa (% of total): *Penilia avirostris* and chaetognaths.

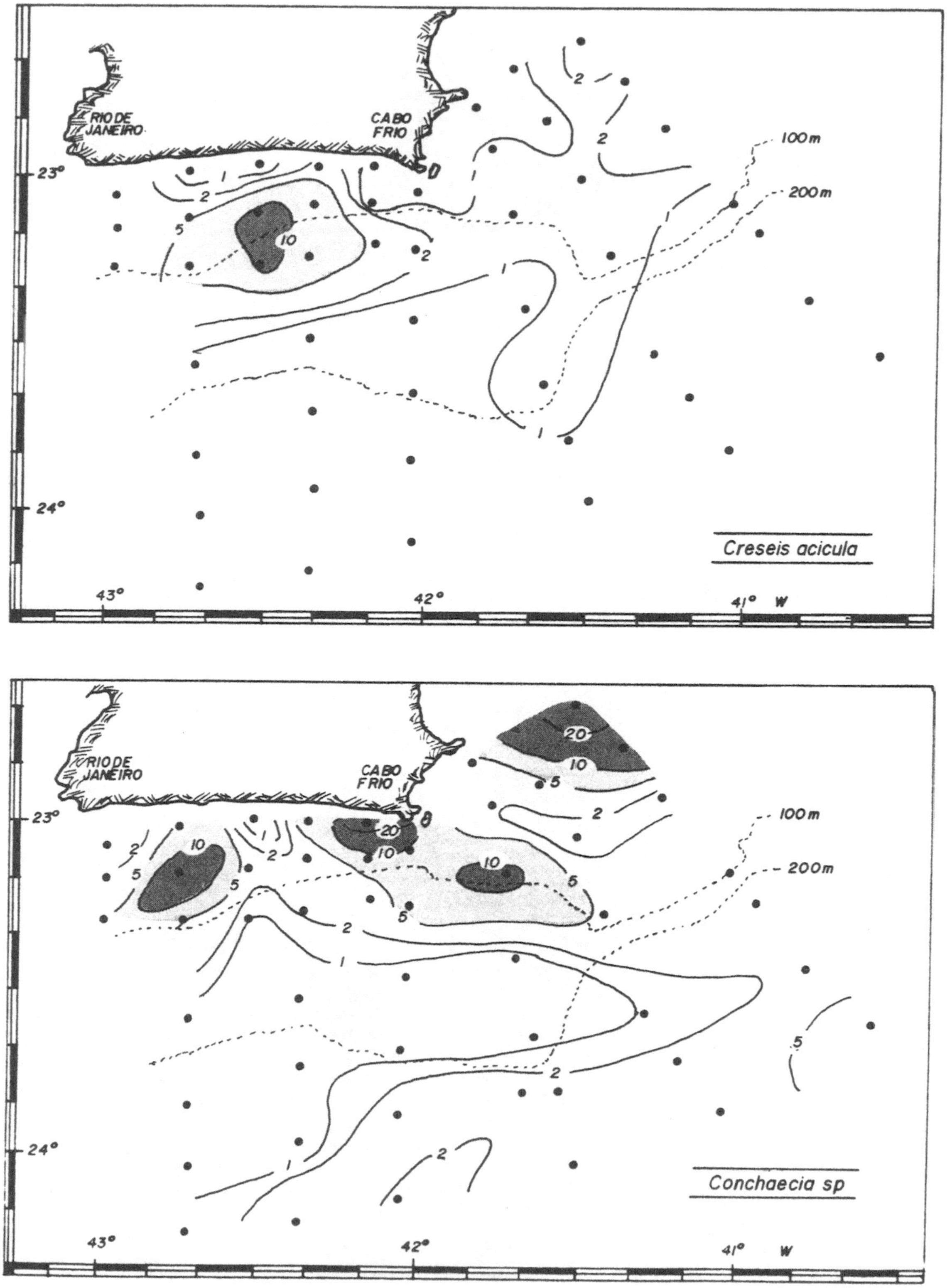

Fig. 9b. As for Fig. 9a: *Creseis acicula* and *Conchoecia* sp.

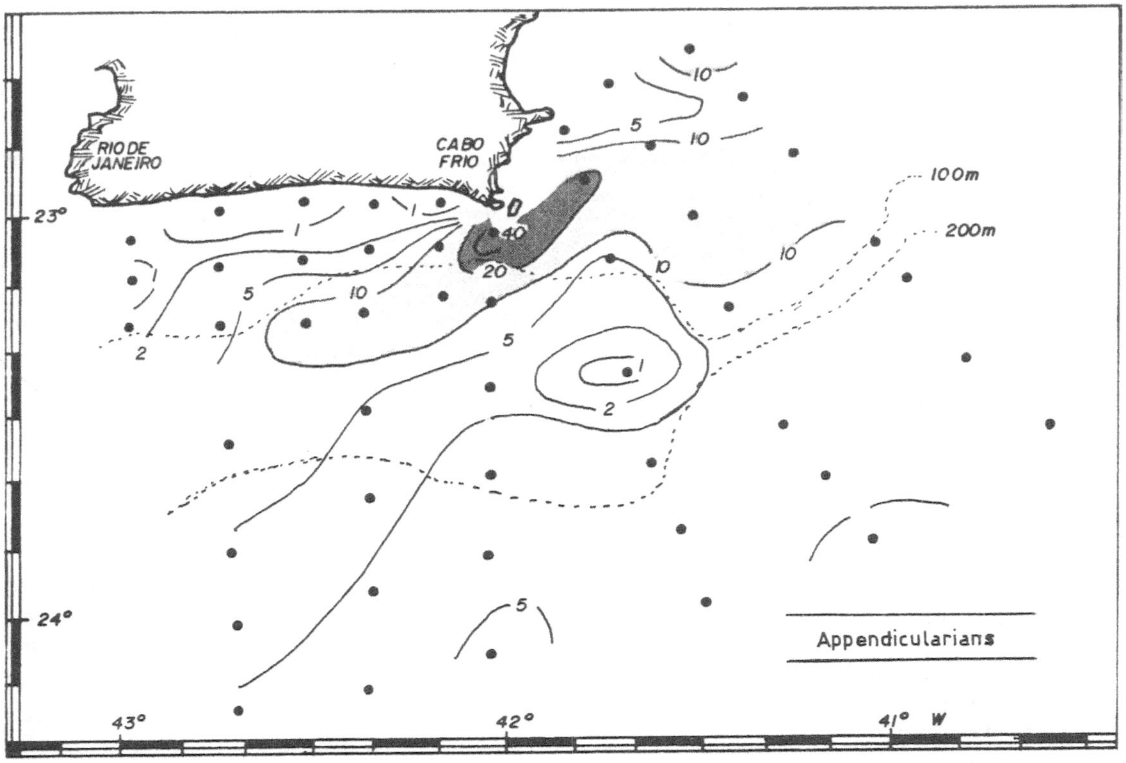

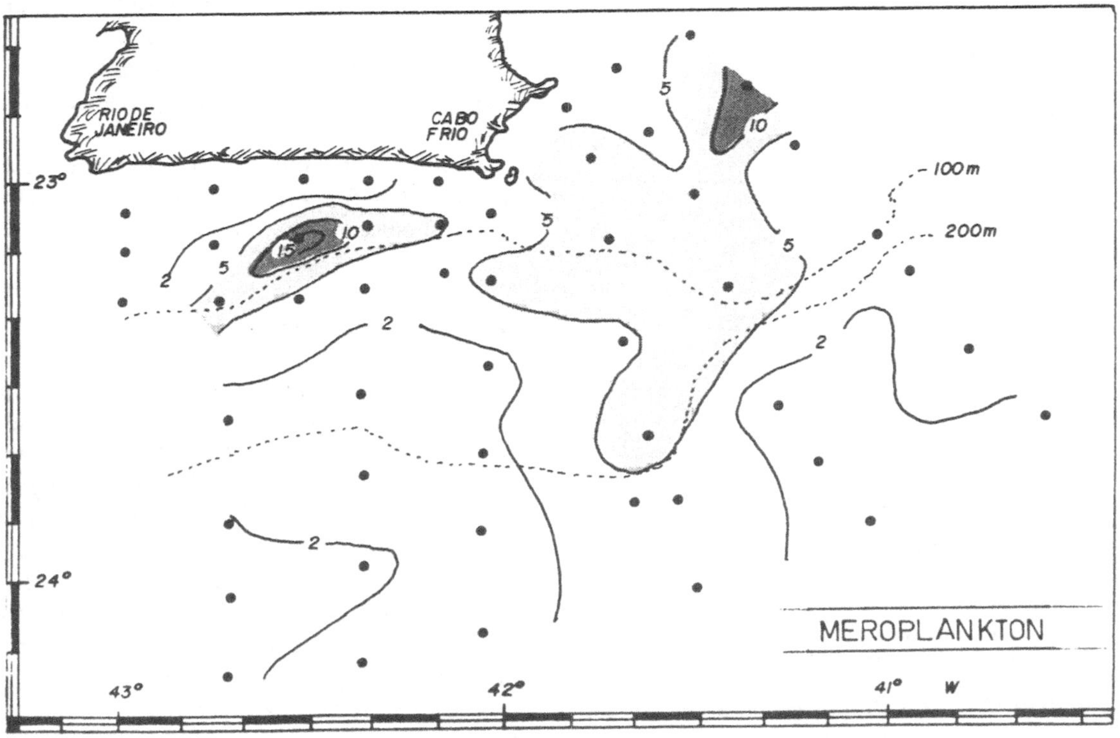

Fig. 9c. As for Fig. 9a: *appendicularians and meroplankton forms.*

Table 2. Zooplankton taxa selected for the statistical analysis. Taxa labelled by letter code (AM-PG) have not been introduced in the analysis because of their infrequent occurrence, but plotted on diagrams according to coordinate values.

1.	Medusa
2.	Siphonophora
3.	Creseis acicula
4.	Limacina sp.
5.	Sagitta enflata
6.	Sagitta friderici
7.	Conchoecia sp.
8.	Penilia avirostris
9.	Evadne tergestina
10.	Oikopleura longicauda
11.	Fritillaria pellucida
12.	Thalia democratica
13.	Doliolum nationalis
14.	Decapod larvae
15.	Polychaeta larvae
16.	Echinoderma larvae
17.	Ichtyoplankton
18.	Acartia lilljeborghi
19.	Calanoides carinatus
20.	Clausocalanus acuicornis
21.	Clausocalanus furcatus
22.	Coryceus amazonicus
23.	Coryceus giesbrechti
24.	Coryceus ovalis
25.	Coryceus typicus
26.	Corycella gracilis
27.	Ctenocalanus vanus
28.	Eucalanus pileatus
29.	Mecynocera clausi
30.	Oithona setigera
31.	Oncaea conifera
32.	Oncaea media
33.	Oncaea venusta
34.	Paracalanus parvus
35.	Temora stylifera
36.	Undinula vulgaris
37.	Copepod larvae
AM.	Amphipoda
CA.	Calanopia americana
CB.	Candacia bipinnata
CP.	Calocalanus pavo
FF.	Fritillaria formica
HL.	Haloptilus longicornis
HP.	Heterorhabdus papilliger
LC.	Cirriped larvae
LF.	Lucicutia flavicornis
MY.	Mysidacea
PA.	Paracalanus aculeatus
PG.	Pleuromamma gracilis

Zooplankton taxa were placed on a coordinate system of the above two components. Three communities could be separated on the first axis: A, B, and C (Fig. 10). This result was consistent with the cluster analysis (Fig. 11) which permitted the separation of the same associations according to an affinity for neritic (or oceanic) conditions.

Table 3. Principal component analysis: identification of factors 1 and 2 based on linear correlation between ecological parameters and score values.

Parameter	Factor 1	Factor 2
Temperature	-0.753[a]	0.349[b]
Salinity	-0.874[a]	0.288
Chlorophyll *a*	0.699[a]	-0.253
Phytoplankton	0.472[a]	-0.189
% Dinoflagellates	-0.344[b]	0.130
Total zooplankton	0.899[a]	-0.349[b]
Diversity	-0.486[a]	0.562[a]

[a] 0.99 level of probability.
[b] 0.95 level of probability.

A spatial structure analysis of the ecosystem based on zooplankton assemblages was performed by grouping stations according to the results of the analysis. Some homogeneous clusters appeared on the dendrogram (Fig. 12), labelled A–C. Areas were also identified by the 1-2 factorial plane (Fig. 13): coastal stations were lumped at the positive extremity of axis 1 (Area A), whereas all the oceanic ones (Area C) were grouped at the opposite side. In the center of the plane, the B cluster indicated an intermediate area characterized by transitory conditions.

Zooplankton communities and the ecological areas they define are shown on the map (Fig. 14) and will be commented upon below.

Discussion and conclusion

Forty percent of the zooplankton variability in the Cabo Frio region was explained by the near-shore-offshore gradient. Boundaries between coastal and oceanic zooplankton assemblages were mainly defined by the geographic shape of the upwelling region. These boundaries were close to the shore line on the western side of Cabo Frio, and far from the coast on the eastern side because the surface eutrophic water drifts from the northern center of the upwelling towards the south.

The greatest densities of zooplankton occurred at the near-coastal stations inside of the upwelling zones. Simultaneously, cold water reached the surface and thus the spatial successional pattern of the

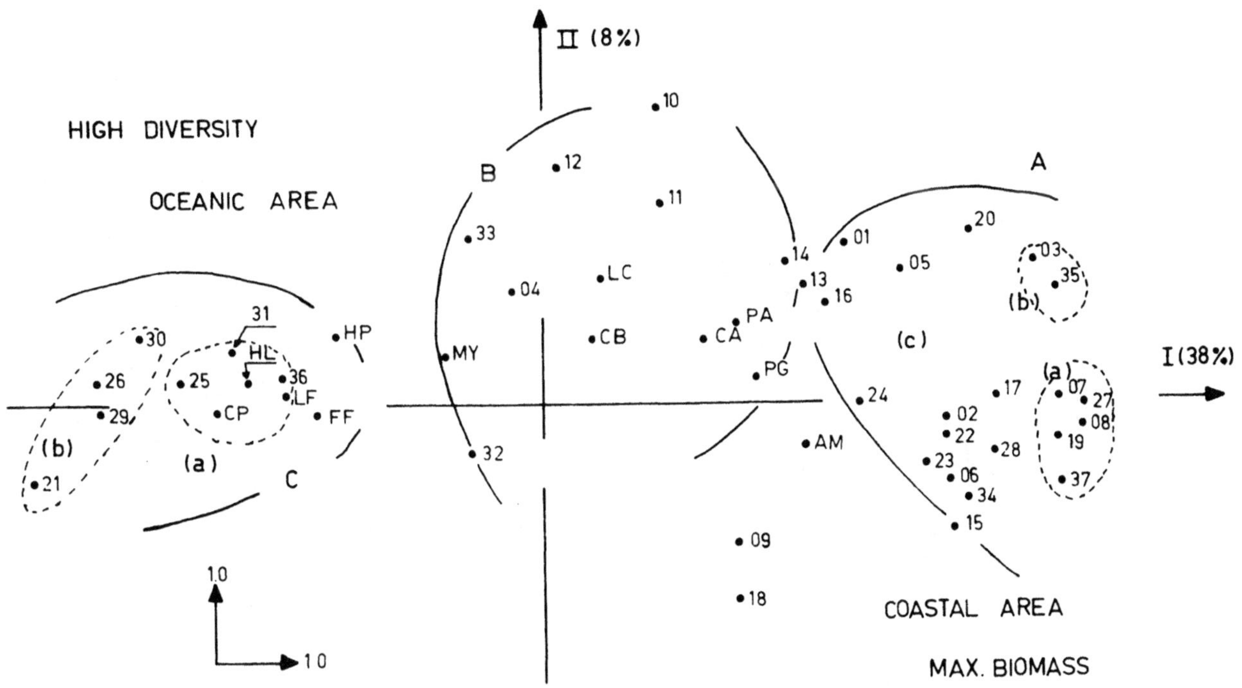

Fig. 10. Ordination model of zooplankton taxa in the plane of the first two principal components (for abbreviations see Table 2).

Table 4. Zooplankton communities according to their ecological areas.

Area	Designation	Characteristic taxa
A	Inner neritic	(a) *Calanoides carinatus, Ctenocalanus vanus, Conchoecia* sp. *Penilia avirostris,* Copepoda larvae
		(b) *Creseis acicula, Temora stylifera*
		(c) Hydromedusa, Siphonophora, Chaetognatha *(Sagitta enflata, S. friderici), Coryceus ovalis, C. amazonicus, Eucalanus pileatus, Clausocalanus acuicornis*
B	Outer neritic	*Oikopleura longicauda, Thalia democratica, Oncea venusta, Paracalanus aculeatus, Calanopia americana, Candacia bipinnata, Oncea media*
C	Oceanic (Brazil Current)	
	(a) neritic tolerant	*Coryceus typicus, Oncea conifera Undinula vulgaris, Calocalanus pavo, Haloptilus longicornis, Lucicutia flavicornis*
	(b) strictly oceanic	*Clausocalanus furcatus, Mecynocera clausi, Corycella gracilis, Oithona setigera*

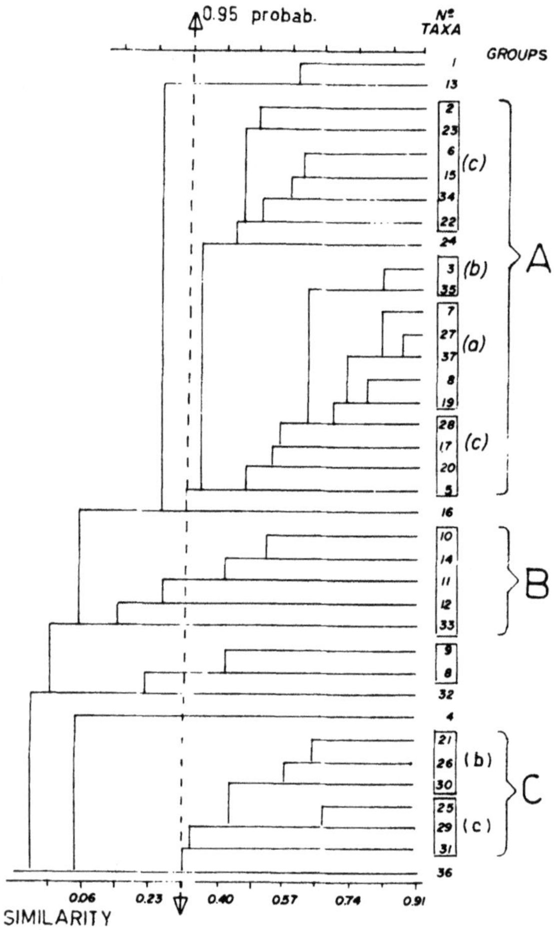

Fig. 11. Dendrogram for classification of taxa (for codes see Table 2). Clusters were separated at the 0.95 probability level.

food chain components, normally observed from eutrophic sources, was not found here. According to Vinogradov & Semenova (1975) this would be expected for intense upwelling from greater depths. In the case of Cabo Frio, upwelling is relatively weak and its characteristic community is in an advanced state of maturity.

The important occurrence of the herbivorous copepod *Calanoides carinatus* in the coastal area indicated deep water upwelling. This fact, often cited in the literature (cf. Bainbridge, 1960), was also recently reported at Cabo Frio (Valentin *et al*, 1976; Mureb *et al*, 1976). *Ctenocalanus vanus*, another herbivorous copepod traditionally cited as belonging to the upwelling community (Belfquih, 1980) is associated here with *Calanoides carinatus*.

The cladoceran *Penilia avirostris*, whose quanti-

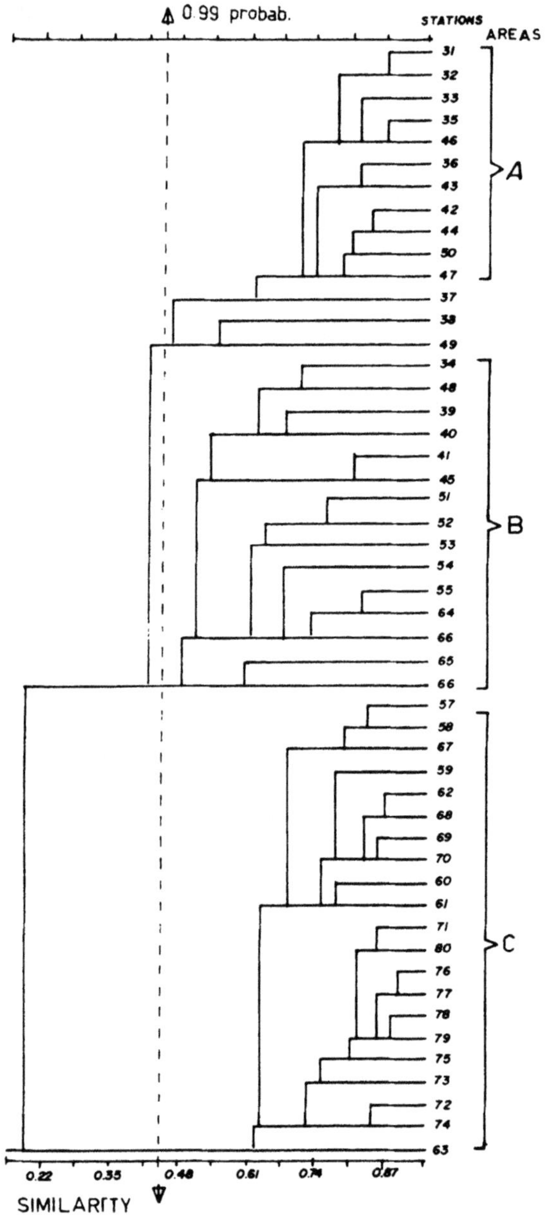

Fig. 12. Dendrograms for classification of stations with respect to zooplankton composition. Letter codes indicate ecological areas determined at 0.99 probability level and defined in text.

ty at some coastal stations exceeded that of copepods, was probably one of the key species in the ecosystem because of its high filtering capacity and high reproductive rate. *Penilia avirostris* is usually not an indicator of upwellings because it is scarce at water temperature below 18 °C, but it may profit from the specific biological conditions in up-

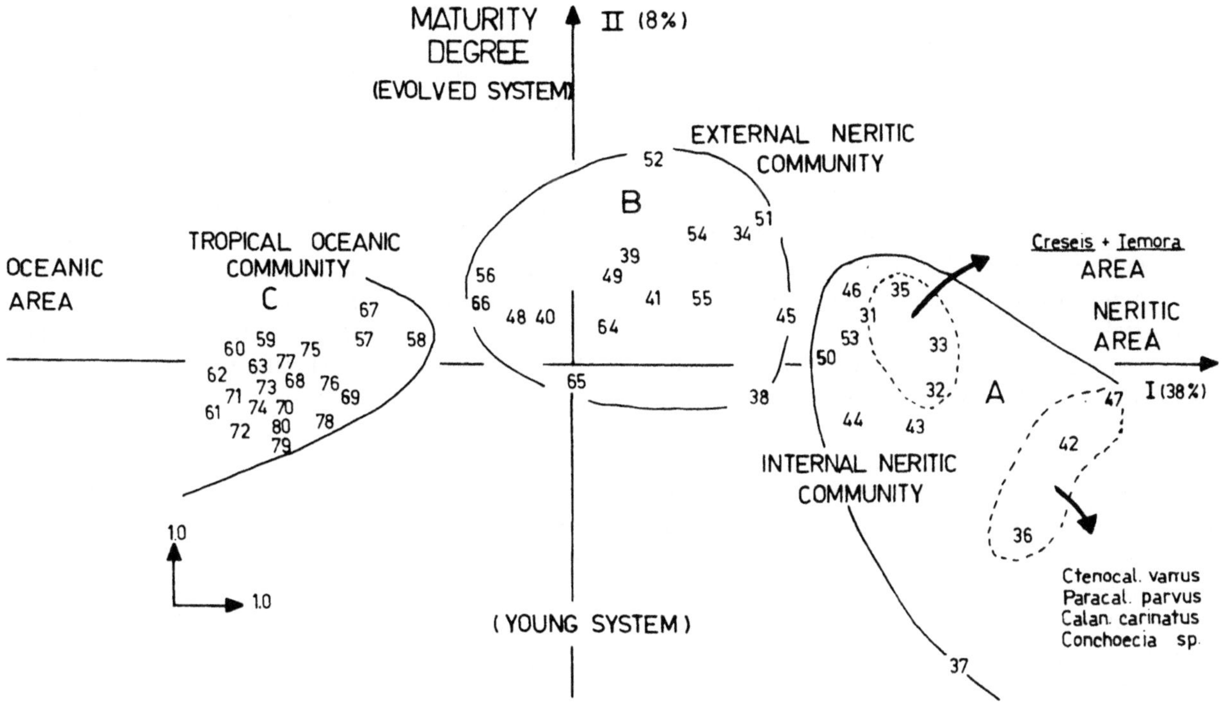

Fig. 13. Ordination model of stations in the plane of the first two principal components.

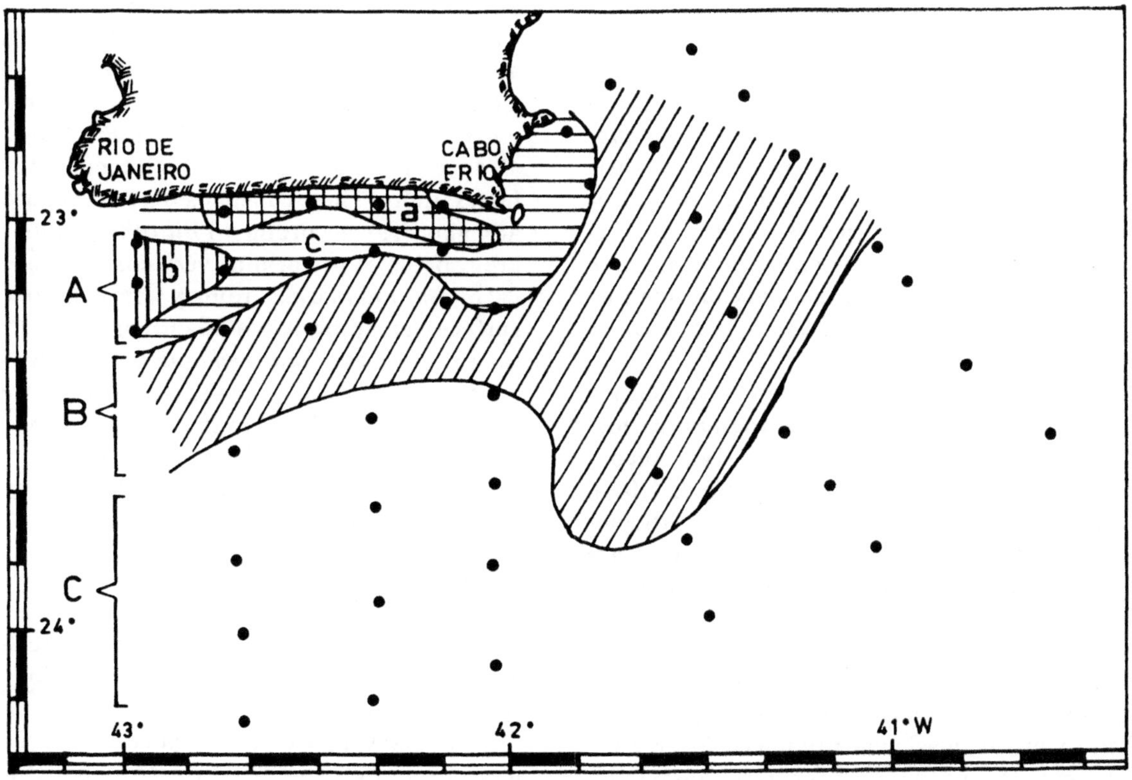

Fig. 14. Cabo Frio region. Ecological areas and zooplankton assemblages as defined in the text and Table 4.

welling regions due to its earlier mentioned characteristics. This could in part explain the frequent cladoceran maximum in strong upwelling periods (cf. Seguin, 1970; Frontier, 1975), and their great standing stock in the neritic community of the area A close to Cabo Frio.

An Ostracod (*Conchoecia* sp.) was also important in the close-coastal neritic community; its omnivorous feeding pattern made it perfectly adapted to the changing conditions of food items of the area: predatory in recently upwelled water, but herbivorous at high phytoplankton levels (Angel, 1970). According to Binet (1973) the continuous reproduction pattern of the Ostracoda would result in brief and strong quantitative variations.

It is interesting to note that taxa which comprised sub-group 'a' isolated by the two statistical analyses, *Calanoides carinatus, Ctenocalanus vanus, Conchoecia* sp. and *Paracalanus parvus,* were also reported by Ramirez (1977) at the subtropical convergence, especially at the level of the thermocline which separates the Brazil Current waters and the underlying Falkland Currents. According to this author, *Paracalanus parvus* was displaced by *Paracalanus aculeatus* further into the Brazil Current. This last species was in fact included in the outer neritic community (Group B).

The Pteropoda *Creseis acicula* and the Copepoda *Temora stylifera* which deliminated the sub-group 'b' in the coastal area were linked by their strong neritic affinity, but also by their similar trophic structure. These two species would show preference for a sestonic particle-enriched medium, in accordance with their mixed herbivorous diet (detritivorous and omnivorous) (Lebour, 1925 in Blanc & Leveau (1973)).

At the periphery of the 'herbivore' area defined by zooplankton from sub-group 'a', the main carnivorous taxa were found (sub-group c): Hydromedusa *Obelia* sp. (70%) and *Liriope tetraphylla* (10%), Chaetognatha *Sagitta enflata* with wide distribution and *Sagitta friderici* with coastal preference (Furnestin, 1957; Mostajo, 1973) and Siphonophora with *Muggiaea atlantica* as main species. Several species of the genus *Coryceus (C. ovalis, C. giesbrechti, C. amazonicus)* usually considered as omnivorous feeders completed the sub-group 'c' which indicated an advanced maturity stage and a certain trophic equilibrium in the inner neritic community.

The external neritic area was marked by a dominance of Appendicularia with *Oikopleura longicauda* as the most important species. This observation was in accordance with Bjornberg & Forneris (1956) and Forneris (1965) who report this species as common in mixing waters, e.g. at the boundary of coastal and oceanic areas. In association with Appendicularia, we observed the Salpa *Thalia democratica*, which, with a strictly vegetal diet, would cause an important decrease of all the other phytophages by food competition (cf. Hubbard & Pearcy, 1971). This would re-establish an equilibrium between the food chain levels and thus considerably alter the maturity of the system. In the outer neritic area there were besides the copepods *Oncea venusta, Paracalanus aculeatus, Calanopia americana, Candacia bipinnata* and *Oncea media,* also Pteropoda of the genus *Limacina* which were placed by Frontier (1975) among oceanic species with neritic tolerance.

Outside the 200 m isobathymetric line, the oceanic province was characterized by a low zooplankton abundance and 80% of copepods. The analysis seemed to discriminate two sub-groups in the offshore community: (a) oceanic with neritic tolerance: *Coryceus typicus* which differs from the other *Coryceus* species in group A by having a greater affinity to the region (cf. Binet & Dessier, 1971), *Calocalanus pavo, Oncea conifera, Undinula vulgaris, Lucicutia flavicornis, Haloptilus longicornis,* all species characteristic for high temperature and high salinity. These copepods were never numerous at Cabo Frio; according to Bjornberg (1963), they are capable of performing very considerable vertical migrations and accordingly were collected in the surface water especially at night; (b) strictly tropical oceanic species: *Clausocalanus furcatus* which, in contrast with *C. acuicornis*, is a tropical thermophilic copepod, absent in the upwelling zone. This species, together with *Mecynocera clausi, Corycella gracilis, Oithona setigera* therefore constituted a typical Brazil Current association.

Acknowledgements

I wish to thank all the researchers of the chemical and plankton laboratories of the Cabo Frio Project for their valuable help.

Financial support was provided by FINEP ('Financiadora de Estudos e Projetos') and Brazilian Naval Research Institute – Cabo Frio Project.

References

Angel, M. V., 1970. Observations on the behavior of *Conchoecia spinirostris*. J. mar. biol. Ass. U.K. 50: 731-736.

Bainbridge, V., 1960. Occurrence of *Calanoides carinatus* (Kroyer) in the plankton of the gulf of Guinea. Nature 188: 932-933.

Belfquih, M., 1980. Les copepodes du plateau continental Atlantomarocain: un cycle annuel dans les zones d'upwelling. These Doct., Univ. Aix-Marseille 1: 126 pp.

Binet, D., 1973. Note sur l'evolution des populations de copepodes pelagiques de l'upwelling mauritanien (Mars-avril 1972. Doc. Scient. Centre Rech. oceanogr., Abidjan 4: 77-90.

Binet, D. & A. Dessier, 1971. Premieres données sur les copépodes pélagiques de la région congolaise. 1. Liste des espèces et notes écologiques. Cah. O.R.S.T.O.M., sér. Océanogr. 9: 411-457.

Bjornberg, T. K. S. & L. Forneris, 1956. On the uneven distribution of the Copelata of the Fernando de Noronha area. Bol. Inst. oceanogr., S Paulo, Brazil 7: 105-111.

Bjornberg, T. K. S., 1963. On the marine free-living copepods off Brazil. Bol. Inst. oceanogr., S Paulo, Brazil 13: 3-142.

Blanc, F. & M. Leveau, 1973. Plancton et eutrophie. Aire d'épandage rhodanienne et Golfe de Fos (Traitement mathématique des données), thèse, Doc. Sci., Aix-Marseille, 681 pp.

Bougis, P., 1974. Ecologie du plancton marin. 2. Le zooplancton. Collection d'écologie, 3. Masson & Cie, Paris, 200 pp.

Davis, J. C., 1973. Statistics and Data Analysis in Geology. John Wiley & Sons, New York.

Forneris, L., 1965. Appendicularian species groups in Southern Brazil water masses. Bol. Inst. oceanogr., S Paulo, Brazil 14: 53-114.

Frontier, S., 1975. Contribution à la connaissance d'un écosystème néritique tropical: étude descriptive et statistique du peuplement zooplanctonique de la région de Nosy Be (Madagascar). O.R.S.T.O.M., Paris, 270 pp.

Furnestin, M. L., 1957. Chaetognathes et zooplancton du secteur atlantique marocain. Rev. Trav. Inst. Pêches marit. 21: 1-256.

Hamann, I., H. C. John & E. Mittelstaedt, 1981. Hydrography and its effect on fish larvae in the Mauritanian upwelling area. Deep Sea Res. 28A: 561-575.

Harmann, H. H., 1970. Modern factor analysis. University of Chicago Press, Chicago, 474 pp.

Hubbard, L. T. & W. G. Pearcy, 1971. Geographic distribution and relative abundance of Salpidae off Oregon coast. J. Fish. Res. Bd Can. 28: 1831-1836.

Moreira da Silva, P. de C., 1968. O fenômeno da ressurgência na costa meridional brasileira. Inst. Pesq. Mar., Rio de Janeiro O24: 31 pp.

Moreira da Silva, P. de C., 1973. A ressurgência em Cabo Frio. Inst. Pesq. Mar., Rio de Janeiro 78: 56 pp.

Mostajo, E., 1973. Quetognatos colectados en el Atlantico sud-occidental entre los 44°44'y 52°38'de latitud sur. Neotropica 19: 94-100.

Mureb, M. A., W. M. Monteiro, E. Pessotti & J. Valentin, 1976. O plancton na ressurgència de Cabo Frio (Brasil). 7. Os copépodos (periodo 21.06 a 28.02.1974). Inst. Pesq. Mar., Rio de Janeiro O96: 14 pp.

Ramirez, F. C., 1977. Planktonic indicators of Argentine shelf and adjacent areas. Proc. Symp. Warm Wat. Zooplancton spec. Publ.: 65-68.

Rodrigues, R. F., 1973. Upwelling at Cabo Frio (Brazil). MSci. Thesis, nav. post-Grad. School, Monterey, 89 pp.

Seguin, G., 1970. Zooplancton d'Abidjan (Côte d'Ivoire). Cycle annuel (1963-1964). Etude qualitative et quantitative. Bull. I.F.A.N. 32, 4: 607-663.

Utermöhl, H., 1958. Zur Vervollkommung der quantitativen Phytoplankton Methodik. Mitt. int. Ver. Limnol. 9: 1-38.

Valentin, J., 1974. O plancton na ressurgência de Cabo Frio (Brasil), 2. Primeiras observações sobre a estrutura física, química e biológica das águas da estação fixa (periodo 04.02 a 16.04.1973). Inst. Pesq. Mar., Rio de Janeiro, O83: 11 pp.

Valentin, J., 1980. L'écosystème d'upwelling à Cabo Frio (Brésil). Analyse en composantes principales du plancton à une station fixe. Oceanol. Acta 3: 43-50.

Valentin, J., M. A. Mureb, W. M. Monteiro & E. Pessotti, 1976. A origem das massas d'àgua na ressurgência de Cabo Frio (Brasil) vista através do estudo das comunidades de copépodos. Inst. Pesq. Mar., Rio de Janeiro 97: 6 pp.

Valentin, J., D. L. André, W. M. Monteiro-Ribas & D. R. Tenenbaum, 1978. Hidrologia e plancton da região costeira entre Cabo Frio e o estuário do Rio Paraíba (Brasil). Inst. Pesq. Mar., Rio de Janeiro 127: 24 pp.

Valentin, J. & A. P. Moreira, 1978. A materia orgánica de origem zooplanctónica nas águas da ressurgência de Cabo Frio (Brasil). An. Acad. Bras. Ciênc. 50: 103-112.

Vinogradov, M. E. & T. N. Semenova, 1975. Trophic characterization of pelagic communities of the equatorial upwelling. Tr. Inst. Okeanol. 102: 232-237 (in Russian).

Assessment of factors influencing the composition, body size and turnover rate of zooplankton in Parakrama Samudra, an irrigation reservoir in Sri Lanka

A. Duncan
Department of Zoology, Royal Holloway College, Englefield Green, Surrey TW20 9TY, U.K.

Keywords: tropical zooplankton, flushing, predation, food limitation, metabolism, rotifers

Abstract

This paper considers which of the following factors influenced the taxonomic composition, body size, fecundity and birth rates of the zooplankton inhabiting a tropical irrigation reservoir: (a) wash-out and/or dilution of the population which arose from the water regime to which the reservoir was subjected by management; (b) the size-selective or species-selective predation; (c) the effect of food limitation upon the body size, egg size, post-embryonic duration and fecundities of the planktonic species and (d) the metabolic consequences of high tropical temperatures. Many of these impinge upon the characteristics of tropical zooplankton and it is appropriate that this synthesis of published results on the zooplankton of Parakrama Samudra, Sri Lanka, was prepared for the Symposium on Tropical Zooplankton held at Sao Carlos, Brazil, in December 1982.

Daily losses of planktonic rotifers due to wash-out or by dilution were low and not significant compared with their capacity for recruitment but this may be a serious source of loss for the planktonic crustaceans which were virtually absent from this reservoir. Daily death rates (estimated by subtraction) were much higher and were mainly due to predation by a planktonic fish, *Ehirava fluviatilis*, and by *Asplanchnella brightwelli*. The main loss occurred during the day and on larger individuals of the brachionid species, *Trichocerca* spp. and *Filinia longiseta*, thus resulting in a small-sized rotifer community. This appeared to be due mainly to predation by the fish which was responsible also for the further reduction of rotifer body size in 1980 compared with 1979. Small rotifer body size was not caused by conditions of food deficiency because fecundity and birth rates were relatively high though not optimal. High tropical temperatures aggravate the cost of cumulative respiration resulting from food-limited prolongation of the juvenile phase and will raise the food threshold levels for growth and for reproduction above those adequate for temperate situations.

Introduction

This paper attempts to assess the importance of factors which may influence the taxonomic composition, body size and fecundity of the zooplankton inhabiting an irrigation reservoir called Parakrama Samudra in the Dry Zone of Sri Lanka. An ecosystem study of this tropical water body (7°55′N; 81°E; 58.5 m above sea level) was conducted by an international team of limnologists led by Dr. F. Schiemer of the Institute of Limnology, University of Vienna (Schiemer, 1981, 1983).

Parakrama Samudra is one of the largest reservoirs of an ancient irrigation system devised to compensate for the absence of a South-West Monsoon in the Dry Zone of the island and to allow rice cultivation. Its total area and volume vary throughout the year according to rainfall and the irrigational needs of the rice crops which are under the control of the local engineer. The area and volume of the reservoir and of each of its three constituent basins for the two periods of study are

Hydrobiologia 113, 201–215 (1984).

given in Table 1. Because of topographical differences in the three basins, the ratios between the basins' volumes at different water levels change and these differences affect the magnitude of the flow-through rates of water, calculated relative to the basin volumes. The magnitude of the daily relative rates of water flow may be one of the important factors influencing the densities of planktonic organisms. Table 1 gives the values during the periods of study for the northern basin (PSN) where most of the work was done. They were generally less than 10% d^{-1}, apart from early August 1979 when they were markedly higher and during September 1979 when there was no inflow. In future, it is likely that these flow-through rates will change as the planned expansion of the area of irrigated rice-fields under the Mahaweli Project is achieved. More recently, the reservoir has had a second function, that of providing a source of freshwater fish protein. In the 1950s, it was stocked with fry belonging to an exotic species of lacustrine fish, *Sarotherodon mossambicus* (Cichlidae), in order to start a fishery and this stocking has been continued up to the present day.

The aim of the ecosystem study, which is the first of its kind on Sri Lankan freshwater reservoirs, was to describe and to understand the ecological dynamics of an irrigation reservoir of this type in order to gain some insight into the possibilities of future management for supplies of both water and

Table 1. Area and volume of Parakrama Samudra and its basins, together with some values for the relative water inflow and outflow rates during 1979 and 1980.

	Year	PSS[a]	PSM[a]	PSN[a]	Total
Area (km^2)	1979	1.7	7.8	2.2	11.7
	1980	2.9	11.2	4.1	18.2
Volume ($10^6 \cdot m^3$)	1979	2.4	22.5	3.0	27.9
	1980	8.3	46.3	10.6	65.1

	1979: PSN(%·d^{-1}) In-flow	1979: PSN(%·d^{-1}) Out-flow		1980: PSN(%·d^{-1}) In-flow	1980: PSN(%·d^{-1}) Out-flow
Early August	17.7	33.8	Late February	2.9	6.7
Late August	7.2	4.4	March	3.9	1.0
September	0.0	5.4	Early April	2.5	3.5

[a] Parakrama Samudra South (PSS); North (PSN); Middle (PSM).

fish protein. There was additional practical and ecological interest in studying the impact of an exotic lacustrine fish on the mostly riverine endemic species of fish and any impact this could have upon the functioning of the reservoir ecosystem. The research on zooplankton was the responsibility of the present author together with Dr. R. Gulati (Duncan & Gulati, 1981, 1983a, b; Duncan, 1983a, b) and the present paper attempts to synthesise the published results for the purposes of the Symposium on Tropical Zooplankton at Sao Carlos but omits the details of methodology which can be found in the cited references.

Species composition and body size of the zooplankton

The open water invertebrate zooplankton of PSN consisted almost entirely of protozoans and rotifers together with a very few crustaceans. In general, the species or genera listed in Duncan & Gulati (1981, 1983a) and Duncan (1983a) belong to those widely recognised as tropical and planktonic, apart from some benthic rotifers and crustaceans brought into suspensions during windy periods. The virtual absence of planktonic crustaceans in the open water is striking and contrasts with earlier records for this reservoir (Fernando, 1980; Fernando & Rajapaksa, 1983). In 1979, only a few specimens of *Phyllodiaptomus annae* were caught in PSN in Clarke-Bumpus samples which filtered large volumes of water and, in 1980, this species, together with *Diaphanosoma excisum*, found refuge in the only patch of littoral macrophytes (*Ceratophyllum* sp) that ever managed to grow in the reservoir during the two visits. The protozoan zooplankton was mainly ciliate (*Vorticella* sp, *Lionotus* sp and some large species) and *Difflugia* sp and was more numerous than the rotifers in the 1980 samples (the only year when they were counted).

The percentage composition of the rotifer species present in the open water of PSN, given in Table 2, is influenced by the depth of water at which samples were collected. The weekly routine sampling in both 1979 and 1980 was done mostly during the morning from fixed stations at a depth of 0.5 m. In both years, by far the largest component were the *Brachionus* species, to which *B. angularis* contrib-

Table 2. Percentage composition of the rotifer species in the plankton of Parakrama Samudra North during 1979 and 1980, in order of abundance. (Duncan, 1983a; Duncan & Gulati, 1983a).

Routine sampling at 0.5 m depth at stations PSN3 and PSN8				Diurnal sampling at 4 depths and 7 times at PSN14 on 1.IX.79	
1979: 16 samples	%	1980: 14 samples	%	1979: 28 samples	%
Brachionus angularis	20	*B. caudatus*	21	*F. longiseta*	23
Polyarthra sp (small)	18	*Collotheca* sp	16	*Ptygura* sp	19
Unidentified spp	15	Unidentified spp	14	Unidentified spp	12
Filinia longiseta	10	*B. angularis*	14	*T. similis*	9
Trichocerca similis	8	*A. navicula*	13	*Polyarthra* sp (small)	9
Brachionus caudatus	6	*B. forficula*	6	*Collotheca* sp	6
Brachionus forficula	5	*Polyarthra* sp (large)	6	*Trichocerca* sp	6
Collotheca sp	5	*Lecane* sp	6	*A. navicula*	4
Ptygura sp	4	*Trichocerca* sp	2	*B. angularis*	4
Anuraeopsis navicula	3	*Polyarthra* sp (small)	2	*B. forficula*	3
Asplanchnella brightwelli	3	*T. similis*	0	*Polyarthra* sp (large)	2
Trichocerca sp	2	*F. longiseta*	0	*B. caudatus*	
Polyarthra sp (large)	1	*Ptygura* sp	0	*A. brightwelli*	1
Gastropus sp	*	*A. brightwelli*	0	*B. urceolaris*	*
Lecane sp	0	*Gastropus* sp	0	*Gastropus* sp	*
				Keratella tropica	*
Total	100		100		100
% of *Brachionus* spp	31		41		8

* Present at less than 1%.

uted most in 1979 and *B. caudatus* in 1980. The other striking difference is the high proportion of the small species of *Polyarthra, Filinia longiseta, Trichocerca similis* and the predator *Asplanchnella brightwelli* present in 1979 and their low numbers or complete absence in 1980. In contrast to the routine sampling, the whole water column was sampled during the diurnal study on 1st September 1979 and at different times of day from 06.00 to 17.45 but the same gear was used. This resulted in a different percentage composition (Table 2): *Filinia longiseta* and *Ptygura* sp both show higher proportions compared with the 1979 routine samples. This is probably due to both species' depth preferences and changes in density of species throughout the day.

A characteristic feature of the planktonic rotifers in this reservoir is their small body size (Duncan & Gulati, 1981, 1983a; Duncan, 1983a). Of the six species of rotifers for which a comparison in length could be made, there were four whose size was markedly smaller than those quoted in Koste (1978): *Brachionus angularis, B. caudatus, B. forficula* and *Filinia longiseta.* Table 3 lists the mean lengths of some of the commoner rotifer species of PSN and shows not only that many exhibited mean lengths of less than 100 μm in 1979 but also that

Table 3. Mean body length and 95% confidence limits of planktonic rotifers from Parakrama Samudra North during 1979 and 1980 (length in μm). (Duncan, 1983a).

Species present	in 1979	in 1980
Brachionus angularis	70.0 ± 3.7 (8)[a]	85.5 ± 6.7 (2)[a]
Brachionus caudatus	127.1 ± 8.1 (8)	70.6 ± 3.2 (7)
Anuraeopsis navicula	82.2 ± 3.2 (8)	75.2 ± 3.4 (5)
Trichocerca sp	78.1 ± 2.3 (30)	109.0 (1)
Polyarthra sp (large)	111.5 ± 8.4 (7)	92.0 ± 1.3 (2)
Polyarthra sp (small)	62.6 ± 1.3 (49)	31.0 (1)
Collotheca sp	64.9 ± 3.7 (39)	67.8 ± 10.3 (5)
Brachionus forficula	86.3 ± 2.0 (18)	–
Trichocerca similis	153.0 ± 5.8 (7)	–
Asplanchnella brightwelli	185.0 ± 7.6 (7)	–
Ptygura sp	75.9 ± 1.1 (136)	–
Filinia longiseta	84.2 ± 4.5 (8)	–
Total rotifer planktonic community		
Range of body length	20 to 185+	30 to 109
Mean length (μm)	83.2	75.4

[a] Number of specimens measured.

some species were even smaller in 1980. Because the linear dimensions were measured of all specimens counted, the mean length and range in size of the whole planktonic rotifer fauna can be calculated. Table 3 gives these values, which re-enforce the evidence of the overall smallness in body size in 1979 and the further reduction in size in 1980.

In addition to the invertebrates, there were also planktonic larvae, juveniles and adults of a fish species, *Ehirava fluviatilis* present in the reservoir. In 1979, larvae appeared in Clarke-Bumpus samples from PSN on September 2nd at a size of 4 mm and attained a juvenile size of 15 mm by the end of the month (Duncan, 1983a). It is not known whether the species was present in 1980 because the appropriate capture gear was lacking but, in 1982, adults as well as larvae and juveniles were found inhabiting the open water and they matured at a size between 16–17 mm (Newrkla & Duncan, in press). Examination of the ovaries showed that the fish spawned between July 24th and August 18th 1982. This was confirmed by the capture of planktonic larvae in mid August 1982. The spent adults probably move into the littoral regions, judging from the evidence of lift-net catches in this area (Duncan, 1983a; Newrkla & Duncan, in press).

Fecundity, egg size and duration of embryonic development

Eight of the fourteen species of rotifers in PSN during 1979 were carrying eggs compared with only six species in 1980 (Duncan & Gulati, 1981; Duncan, 1983a). However, there were many loose eggs identifiable as belonging to rotifers in the samples. If these are taken into account in order to estimate the fecundity of the whole rotifer community, the values given in Table 4 are obtained. The 1979 estimate of 0.402 eggs per female came from the diurnal study at PSN 14 for the depth of 40–80 cm and so is comparable with the 1980 estimates for the routine samples taken at 0.5 m. These were 0.500 and 0.798 eggs per female for PSN 8 and PSN 3, respectively. The fecundity of *Brachionus* spp was higher at PSN 3 compared with PSN 8 in 1979 (Duncan, 1983a). Table 4 gives the mean size of the attached and loose eggs combined, calculated as volume from two linear measurements, and shows that the eggs are small in size and that the mean

Table 4. Mean density, fecundity, daily birth rates and mean size of eggs of the rotifer community in Parakrama Samudra North in 1979 and 1980. (Duncan, 1983a; Duncan & Gulati, 1983a).

	1979 PSN 14[a]	1980 PSN 8	1980 PSN 3
Mean density of rotifers (ind · l^{-1})	3 881	544	657
Mean density of rotifer eggs (eggs · l^{-1})			
Attached eggs	274	70	106
Loose eggs	1 288	202	418
Total number of eggs	1 562	272	524
Fecundity (eggs per female)	0.402	0.500	0.798
Daily birth rates (eggs · $female^{-1}$ · d^{-1})			
Embryonic duration (D_e = 0.24 days)[b]	1.68	2.08	3.33
Embryonic duration (D_e = 0.44 days)[c]	0.91	1.14	1.81
Mean volume of egg (10^6 · μm^3 · egg^{-1})	0.061	0.018	0.025

[a] Calculated for the depth 40–80 cm for comparability with the sampled depth of 0.5 m at the other station.

[b] For small rotifer eggs, as existed in PSN.

[c] For larger, normal-sized eggs, as did not exist in PSN.

volume in 1980 was smaller than in 1979. The change in mean egg volume is quite large (about one third reduction) and may reflect the change in species composition in the two years (Table 2). However, some other factor was influencing the change in both mean body length and egg volume of *Brachionus caudatus* from PSN, which were 127 μm and 0.148 $\times 10^6$ μm^3 in September 1979 and 97 μm and 0.079 $\times 10^6$ μm^3 in April 1980, respectively. Comparable measurements for species of *Brachionus* from the Milk Factory Tank, a neighbouring reservoir near the town of Polonnaruwa, showed larger animals (for example, *B. caudatus* has a mean length of 184 μm) which carried larger eggs (for example, *B. calyciflorus* with a mean length of 295 μm had eggs of 0.641 $\times 10^6$ μm^3) (Duncan & Gulati, 1983b; Duncan, 1983b). The small size of rotifer eggs in PSN seems to be associated with the small size of the rotifers themselves and Duncan (1983b) was able to derive a significant linear relationship between these two parameters for *B. caudatus,* rather as did Pourriot (1973a) for temperate *B. calyciflorus*.

A consequence of small size in rotifer eggs was shown by Duncan (1983b) to be shorter embryonic duration than that of larger eggs. The duration of embryonic development of *B. caudatus* from PSN was determined at several consecutive temperatures and the subsequent regression equation predicted a duration of embryonic development of 5.67 hours at 28 °C, the prevailing temperature of the reservoir. For the larger *B. calyciflorus* from the Milk Factory Tank (MFT), with its larger eggs, the duration-temperature regression predicted 10.51 hours at 28 °C. The difference between these two regressions was significant (Duncan, 1983b). Comparison with durations of rotifer embryonic development in the literature showed that the results for *B. calyciflorus* from MFT are comparable but those for *B. caudatus* from PSN are shorter (Duncan, 1983b).

Density and turnover rate of rotifer populations

The rotifer populations in PSN were much denser during the 1979 research visit compared with that of 1980; the average density at 0.5 m depth of the two routine stations (PSN 8, PSN 3) for 8 sampling occasions in 1979 was 3 262 animals per litre but only 600 animals per litre for 7 sampling occasions in 1980. This density of rotifers at 0.5 m depth is representative of the whole water column during the frequent windy periods but not in calm weather when vertical differences in rotifer abundance appear and change throughout the day (Duncan & Gulati, 1983b). A comparison of areal abundance rather than column concentration, which takes into account the difference in reservoir depth (1.6 m in 1979 and 3.8 m in 1980), showed similar differences ($4.07 \times 10^6 \cdot m^{-2}$ in 1979 and $2.28 \times 10^6 \cdot m^{-2}$ in 1980). There was a consistent difference in the abundance of rotifers between the two bays of PSN. This appeared in both years of study but was most marked in 1979. In that year, the mean rotifer density was 1.5 times greater in the northern bay than in the southern one, probably due to both the dilution of southern bay populations with inflowing water from PSM which contained few rotifers and to the isolation of the northern bay populations from the main pathway of water flow. In 1980, when rotifer densities were less than 1 000 animals $\cdot$ l^{-1} throughout the reservoir, this horizontal difference was much smaller (1.2 times).

Another difference in PSN between the two years of study was that the density of the rotifer populations changed during the course of the two months' visit in 1979 whereas in 1980 it remained steady at less than 1 000 ind. l^{-1}. In 1979, the initial mean density at the two routine stations in mid August was also about 1 000 ind. l^{-1} but increased to 3 754 ind. l^{-1} by the end of August, after which there was a decline to less than 1 000 ind. l^{-1} by 7th September, and this was followed by another increase to 4 542 ind. l^{-1} by 19th September (Duncan & Gulati, 1981; Duncan, 1983a).

A more detailed analysis of changes in density of rotifer populations with depth and time of day was possible with a series of samples taken on 1st September 1979 (Duncan & Gulati, 1983a). These came from four depths at PSN 14 in a water column of 1.6 m depth and were sampled at 2 h intervals from 06.00 to 17.45. It was found that the mean concentration and areal abundance was 4 669 ind. l^{-1} and 7.61×10^6 ind. m^{-2}, respectively, and the number declined from an initial abundance of 8.381×10^6 ind. m^{-2} at 06.00 to a minimum of 3.784×10^6 ind. m^{-2} at 14.00 and rose again to 10.036×10^6 ind. m^{-2} at 17.45. Duncan & Gulati (1983a) provide more detailed information on these changes in the standing crop.

Knowing the fecundity of the rotifer community at each depth and time, the water temperature at each point and substituting into the regression relating duration of embryonic development to temperature for PSN *B. caudatus* (Duncan, 1983b), it proved possible to calculate hourly hatching rates (eggs per female per hour) and recruitment rates (numbers $\cdot l^{-1} \cdot h^{-1}$). By integration over depth and time, a daily recruitment rate of 14.107×10^6 ind. m^{-2} was obtained for PSN 14 on 1st September 1979. With a mean areal abundance of 7.61×10^6 ind. m^{-2} for this day, a turnover rate of 1.85 d^{-1} was derived. Despite the decrease in numbers which took place during the middle part of the day, the hourly recruitment rates were rather constant, ranging from 0.323×10^6 ind. $m^{-2} \cdot h^{-1}$ at 14.00 to 0.622×10^6 ind. m^{-2} h^{-1} at 06.00 but with a mean value of $0.54 \times 10^6 \pm 0.11$ (95% C.L.; d.f. 6). The order of magnitude of rotifer loss could be estimated by subtracting from the hourly recruitment rate the hourly changes in the standing crop. Details on how this was done are given in Duncan & Gulati (1983a); the calculated values should be positive, any negative values imply either a gain in numbers due to births or to an underestimation of the recruitment rate (e.g. by fecundity estimates too low). The hourly losses were relatively low (4–5% of the standing crop per hour) from 06.00 to 10.00 but were very high (25–32% per hour) between 10.00

and 15.00. From 15.00 to 17.45, the loss rates became negative, suggesting that some input of rotifers occurred from a source other than reproduction. This may be associated with the advent of a storm after 15.30, during which the water column was mixed down to 1.5 m depth (Duncan & Gulati, 1983a).

Factors potentially affecting the taxonomic composition, size structure and turnover rates of the zooplankton in PSN

Any discussion of why the zooplankton of PSN possesses the characteristics outlined earlier must consider, amongst others, the following environmental influences:

(a) wash-out and/or dilution of the populations which arise from the water flow regime to which the reservoir is subjected;
(b) size-selective or species-selective predation;
(c) the effect of food limitation upon the body size, egg size and fecundity of the planktonic species and
(d) the metabolic consequences of high tropical temperatures.

Many of these impinge upon the characteristics of tropical zooplankton in general and each will be discussed in turn.

Duncan & Gulati (1981) suggested that rotifer densities of less than 1 000 ind. l^{-1} in PSN were associated with either high flushing rates or with dilution with water containing few rotifers. With little or no through-put, as occurred in September 1979, the rotifer populations could exploit their high capacity for increase at tropical temperatures to attain densities of more than 5 000 ind. l^{-1}. There are similarities between the pattern of relative water flows (Table 1) and changes in the standing crop in 1979 but during the 1980 visit rotifer densities remained below 1 000 ind. l^{-1} for two months, despite the persistence of low relative flow rates and rather long retention periods. When rotifer populations can achieve turnover rates of 1.85 per day (Duncan & Gulati, 1983b), outflow losses of 0.1 d^{-1} are relatively insignificant and, even with outflows of 0.3 per day, populations could still increase in number. The possibility that the water management regime in PSN could cause wash-out of the planktonic crustaceans is more real but difficult to assess in their absence and with the paucity of data on the durations of tropical species' life cycles. Jayatunga (1982) has recorded four species of copepods and five species of cladocerans as well as rotifers and protozoans in Kalawewa reservoir (28.3 km^2; 123×10^6 m^3). The population densities of the rotifers were on an average about one order of magnitude lower than in PSN, although brief peaks of up to 1 000 ind. l^{-1} were recorded. The larger meshed netting used (80 μm) would not retain the smaller rotifer species found in PSN (where a 33 μm mesh net was used) where they formed 60–70% of the total. There were far more crustaceans than in PSN, the numbers ranging from 10 to 270 ind. l^{-1}. In this reservoir, the mean daily flow-through rates were about 0.02 d^{-1}, a little lower than those normally recorded for PSN (Table 1).

It was possible to estimate the daily death rates of the rotifer populations in PSN from the difference between daily changes in the standing crop and the sum of the daily gains (births plus immigration) and the daily losses (emigration) for periods in 1979 and 1980 when the reservoir was well-mixed both vertically and horizontally (Duncan, 1983a). These calculated death rates are high and could be nearly doubled solely by using the predicted embryonic duration from PSN of 5.67 hours instead of the more usual one of 10.51 hours from MFT to calculate the rotifer daily birth rates. The high death rates during September 1979 (0.48–1.66 per day) coincided with the appearance of planktonic larval stages of the clupeoid fish, *Ehirava fluviatilis*, on 2nd September and their gradual increase in numbers from 65 ind. m^{-3} to 163 ind. m^{-3} by the end of the month. The death rates in August 1979 (0.37–0.96 d^{-1}) are associated with the planktonic mature adults of this fish, not caught by the Clarke-Bumpus sampler but whose presence was subsequently demonstrated in August 1982 by use of a Bongo net (Newrkla & Duncan, in press). If it is assumed that all the calculated deaths for the period 14th–25th September were caused by immature fish predation, an estimated daily ingestion would be $54 \times 10^3 \pm 9.63$ (95% C.L.; d.f. 5) rotifers per fish. The hourly ingestion rate would be 2 250 rotifers per fish, if they fed throughout 24 hours, or 4 500, if they fed only during the hours of light. The diurnal study on 1st September showed that the greatest losses of rotifers took place between 10.00 and 15.00, a period of the day which coincides with the greatest penetration of light (Duncan & Gulati, 1983a), when a visual predator such as a fish may feed most effectively in the lighted part of the water column. The observed shift to smaller sized rotifers which occurred during the daytime (Duncan & Gulati, 1983a) and the changes in vertical distributions of different species may be accounted for by fish predation, both for the species being eaten and for those not taken. The species which showed the greatest decline in numbers were the brachionids, *Trichocerca* spp and *Filinia longiseta* but there was also a loss of the larger size classes.

Gut analysis of littoral *Ehirava fluviatilis* between 20 and 24 mm has revealed rotifers such as *Trichocerca similis* and *Brachionus caudatus* together with the protozoan *Difflugia* sp. The larger specimens of *E. fluviatilis* also contained crustacean remains, amongst which *Phyllodiaptomus annae* could be identified. It seems possible that sufficiently dense populations of this fish species could influence the taxonomic composition and/or size structure of the zooplankton in PSN by virtually eliminating the planktonic crustaceans and larger rotifer individuals which carried the larger eggs. The remaining unpreyed upon rotifer populations of smaller individuals carrying smaller-sized eggs would have accelerated turnover rates due to the shortened embryonic durations of small eggs. The densities of young fish during September 1979 may well have been high enough to produce these effects upon the zooplankton.

There are very few examples of large water bodies like Parakrama Samudra for which fish densities are known and which possess a zooplankton consisting of rotifers, protozoans and very few crustaceans. Table 5 lists some examples. The Japanese man-made estuarine Lake Kojima is one such water body. Like PSN, it has relatively high flushing rates of 0.06 per day but differs in being saline. As in PSN, its standing crop of zooplankton is quite dense and the mainly brachionid rotifers can contribute 91% of the dry weight biomass whereas the crustaceans are never more than 20% by dry weight. The main fish species is the silver crucian carp, *Carassius langsdorffi,* whose adults preferentially select open water rotifers and crustaceans whereas the juveniles feed within the littoral weed beds. It is suggested by the authors that the population density of this species together with its selective feeding may affect the species composition of the zooplankton. The rotifer species *Brachionus calyciflorus, B. angularis, Asplanchna sieboldi* and *Keratella valga* were the major species found in the gut whereas *Polyarthra trigla* and *Filinia* spp were not detected although present. In another Japanese water body, Lake Suwa (Table 5), in which the dominant species of zooplankton were rotifers, ciliates and *Bosmina* sp (in the summer) and where the densities

Table 5. Fish densities and/or biomass in various water bodies with a zooplankton consisting of rotifers, protozoans and few crustaceans.

Water body	Area (m^2) Volume (m^3)	Fish density or biomass	Zooplankton composition	Zooplankton density (ind. l^{-1})	Zooplankton biomass (gDW · m^{-3}	Authors
Lake Kojima	10.7×10^6 17.7×10^6	1.4 ind. m^{-2} 94.4 g · m^{-2}	*Brachionus* crustaceans	900 to 7 000	0.25 to 1.65	Mori & Yamamoto (1975)
Lake Suwa	14.5×10^6 64.3×10^6	1.81–3.73 g · m^{-2}	rotifers ciliates Bosmina	470 to 5 080	–	Mori & Yamamoto (1975)
Goldfish ponds	200 70	140–54 ind · m^{-2} 0.21–310 g · m^{-2}	rotifers ciliates Moina	58×10^3	–	Mori & Yamamoto (1975)
Lea Bridge	4 000 7 000	4–28 g · m^{-2}	cladocerans copepods	– –	1.73 1.73	O'Grady & Spillett (1981)
PSN 1979	2.2×10^6 3.0×10^6	up to 163 ind · m^{-3}	rotifers protozoans	3 262	–	this work
PSN 1982	4.1×10^6 10.6×10^6	up to 170 ind. m^{-3} about 8.0 g · m^{-2}	rotifers protozoans	about 2 000	–	this work

of zooplankton were also high, the young of the pond smelt (*Hypomesus alidus*), the most abundant fish species, favoured rotifers when it was less than 20 mm in length but turned mainly to *Bosmina* sp when it was larger. The total standing crop of fish in this lake doubled as the fish grew from being fry in May to attaining maturity in December. A more extreme example is the smaller fertilised Japanese goldfish culturing ponds (Table 5) in which the initial fry density was very high but had more than halved by harvest time a year later, when the biomass was 310 g m^{-2}. After 15 days during which the fish fry fed upon the very dense populations of rotifers and *Moina* sp that had developed prior to the release of the fry, *Moina* sp disappeared and the zooplankton consisted solely of rotifers, ciliates and a few nauplii. At Lea Bridge in the London area, (Table 5), carp (*Cyprinus carpio*) reared on dense populations of planktonic crustaceans yielded 28 g m^{-2} from an initial stocking of 4 g m^{-2} of fish 17.4 cm long. Such precise measurements of fish densities are not yet available for PSN but, assuming that the previously quoted densities for the planktonic *E. fluviatilis* apply to the open water of the whole basin, the standing crop would be ca. 8.0 g m^{-2}.

Another potential predator upon the rotifer populations of PSN is the carnivorous rotifer, *Asplanchnella brightwelli.* This species was present in fairly high densities (112 ind. l^{-1}) in 1979 but was rather sparse in 1980 (Duncan, 1983a). Although no gut analyses have been carried out on individuals from PSN, it is clear from Green & Oey Bianwlan (1974) that this species can ingest a variety of rotifer species. From their data on *A. brightwelli* from a Javanese sewage pond (where it was present in very high densities), it is possible to calculate Ivlev Electivity indices based upon the proportions of different species detected in the stomach or in the plankton. This index is high and positive for *B. caudatus* (+0.99), *B. angularis* (+0.91) and *B. rubens* (+0.33) which demonstrates high electivity for these species. Other species of rotifers are either strongly avoided or are taken indifferently. It cannot be ruled out that *A. brightwelli* may also affect the populations of *Brachionus* spp in PSN, but may take different sizes from the fish.

Another environmental factor which can influence size of body and possibly size of eggs is a limited food supply. The food effect is not simple because there are interactions between food quantity and temperature and between duration of development and body size. In rotifers kept under experimental conditions, this latter interaction is most easily detected in the adult reproducing for the first time (primipara). The period in the rotifer life cycle which is vulnerable to environmental influence is the juvenile phase from egg hatching up to first reproduction. Various authors divide this into an active feeding phase followed by a period of enlargement to adult size by stretching and incorporation of assimilated material (King, 1967; Pourriot, 1973a; Ruttner-Kolisko, 1974). What is in doubt is whether a juvenile rotifer with a limited food supply results in a small adult or in an adult of increased size due to prolongation of the juvenile phase, as suggested by Sterzynski (1979) and also whether the smaller females produce smaller eggs (or vice versa), as is implied by Pourriot's (1973a) results. King (1967, Fig. 2) plots the growth of individual *Euchlanis dilatata* at 22 °C and shows a doubling of the juvenile phase (from 0.88 days to 1.91 days) in animals grown in ten times less algal food (16.4 and 1.64 μgDW · ml^{-1}). The body length of the primipara female was smallest in the lowest food level and largest in 16.4 μgDW · ml^{-1} or more. Only one size is plotted for eggs and neonate juveniles.

Pilarska's (1977) eco-physiological studies on *Brachionus rubens* provide more detailed information. She cultured animals throughout their life cycle at 20 °C in ten concentrations of *Chlorella vulgaris* and, on the basis of the species' responses, was able to define levels of food which were deficient, optimal and excessive. If one accepts the very high cell weight for *C. vulgaris* (29×10^{-6} μgDW · $cell^{-1}$; Pilarska, 1977), these food levels come to 0.35, 29 and 290 μgDW · ml^{-1}; with more usual cell weights of 3–4 pgC · $cell^{-1}$ (Chalk, 1981) or $6\text{–}8 \times 10^{-6}$ μgDW · $cell^{-1}$, the food levels would be 0.084, 7 and 70 μgDW · ml^{-1}. In any case, the lowest food level was much lower than in King's experiments. The largest juveniles and mature females were those grown under optimal food conditions and these larger amictic females also produced the largest eggs. In food deficient conditions, the juveniles and females were about 30% smaller and the diploid eggs were smaller by 15–20%. As in King's experiment, there was a prolongation of the post-embryonic or juvenile period to 2.84 days in the lowest food compared with 1.26 days in optimal conditions. Pilarska also found a marked shortening of the period of egg production to 0.73 days in the low food compared with 6.8 days in the optimal food. The instantaneous rates of growth in dry weight usefully combine these two food effects on size and duration of the juvenile phase. These were 0.27 d^{-1} and 0.46 d^{-1}, respectively. This work confirms that food-deficient juveniles slowly grow into small adults which (if they can) produce small eggs but only for a very limited period; the food concentration at which this occurred in *B. rubens* was 0.35 μgDW · ml^{-1} (or 0.084 μgDW · ml^{-1}) at 20 °C. In apparent contrast to these results, Leimeroth (1980) found that the dry weight of female *B. calyciflorus* without eggs and their eggs increased with decreasing food (0.32 μg · $female^{-1}$ and 0.08 μg · egg^{-1} in 115 μgDW · ml^{-1} algal food; and 0.39 μg · $female^{-1}$ and 0.15 μg · egg^{-1} in 9.2 μgDW · ml^{-1}). Since he suggests that 9.2 μgDW · ml^{-1} of *Kirchneriella lunaris* was the best food level, these decreases in size occur in conditions of excessive food and so are consonant with Pilarska's results.

Amongst the many papers on mass culture of *B. plicatilis* as food for young fish, those of Scott & Baynes (1978) and Theilacker & McMaster (1971) provide further evidence for food effects on size and rates of increase. Scott & Baynes have shown that the body weight of individuals (measured as dry weight, total protein, carbohydrates and lipids) decreased markedly when a dense culture (31×10^3 ind. l^{-1}), kept at 25.6 °C, was allowed to graze down an initial algal food density of 125 μgDW · ml^{-1} for 7 days down to starvation levels. Rotifer numbers increased between days 2 and 5 at 0.38 d^{-1} (or a doubling time of 1.84 days). From day 2, the rotifers lost body weight at a rate between 0.14 d^{-1} to 0.26 d^{-1}, depending upon the variable being considered. At its maximum on day 1, the individual dry weight was 0.65 μg · ind^{-1} and this fell to 0.113 μg · ind^{-1} on day 6. Theilacker & McMaster explored the relationship between the concentration of algal food and increase rates or doubling time in *B. plicatilis* reared at 24 °C in large batch cultures. Their animals were smaller than those of Scott & Baynes: length 164–231 μm; dry weight 0.16 μg · ind^{-1}. At or above food levels of 1×10^6 algal cells · ml^{-1} (or 166 μgDW · ml^{-1}) the population increased at a maximum rate of 0.76 d^{-1} (or a doubling time of 0.9 days) whereas at or below a concentration of 2×10^5 cells · ml^{-1} (or 33 μgDW · ml^{-1}), the rate of increase was dependent upon food level and the following significant ($p = 0.005$) exponential relationship could be fitted: $Y = 0.194 \cdot e^{0.050X}$, where Y is the daily instantaneous rate of numerical increase at 24 °C and X is the food concentration in μgDW · ml^{-1}. The Brachionus species from PSN were much smaller than either *B. rubens* or *B. plicatilis* (*B. caudatus* PSN, 0.013 μgDW · ind^{-1}) but those from MFT were commensurate (*B. caudatus* MFT, 0.104 μgDW · ind^{-1}; *B. calyciflorus* MFT, 0.474 μgDW · ind^{-1}) (Duncan & Gulati, 1983b). A turnover rate of 1.85 d^{-1} was calculated for the rotifer populations of PSN during the diurnal study of 1st September 1979 using the rapid birth rates of PSN *B. caudatus* (that is, $D_e = 5.67$ h at 28 °C). This is equivalent to a rate of 1.69 d^{-1} at 24 °C (applying a Q_{10} of 1.25 from Scott & Baynes' data, 23 °C–28 °C). By substituting into the above equation, the predicted food concentration which would support this level of population increase can be calculated as 43.3 μgDW · ml^{-1} or 21.7 μgC · ml^{-1}, which is higher than the measured sestonic carbon concentrations in the Parakrama Samudra reservoir at this time (Duncan & Gulati, 1983a). These were 3–11 μgC · ml^{-1}. However, the short embryonic duration of PSN *B. caudatus* may not be applicable to all the species of rotifers in the reservoir. Using $D_e = 10.51$ h at 28 °C, the turnover rate reduces to 1.0 d^{-1} and the predicted food level required to support this to 30 μgDW · ml^{-1} or 15 μgC · ml^{-1}, a little closer to the measured sestonic values. These rates of increase suggest that PSN rotifers were not living under conditions of food deficiency and that small body size of some species was due to other causes.

Green (1960) records great variation in the size and spine length of the *B. caudatus* of the River Sokoto and the small mature form without spines in his Fig. 6 is very similar to the *B. angularis* of PSN. In several tropical crater lakes (Green, 1977), he found small forms of *B. caudatus* and *B. falcatus* which he termed dwarf rotifers, defined as mature individuals that are smaller than usual for the species. His *B. caudatus* had lorica lengths of about 100 μm, which is about the size of the smallest mature *B. caudatus* carrying eggs in PSN and individuals smaller than this, down to 54 μm, were probably juveniles. Green thought that the occurrence of dwarf rotifers may be associated with high temperatures, amongst several other possibilities. Pourriot (1973a) has found a well-correlated and inverse relationship between temperature and the size of bodies and of eggs in cultured well-fed *B. calyciflorus*. He presents a graph (Fig. 2) which illustrates that this is a consequence of loss of the larger size classes in high temperature cultures. This seems very relevant to a discussion on the body size of tropical zooplankton. Although his work provides evidence for a temperature effect on body size, Pourriot himself felt that it was the nutritive state of the juvenile phase which influenced the size of both the adult and the egg she produced. He was the first author to derive a significant linear regression between egg diameter and lorica length of animals from field samples and cultures and he found an interesting tendency for larger eggs to have a prolonged development, which is similar to the findings of the present work (Pourriot & Deluzarches, 1971).

If small body size indicates food limitation rather than the presence of size-selective predation, one might expect this to be accompanied by low fecundity, cessation of amictic reproduction or a switch-over to mictic reproduction. According to Pourriot (1973b), most planktonic rotifers can manage to produce on average more than two eggs per day over the reproductive period of their life cycle when cultured on an optimal diet at 20 °C. The mean value he obtained for amictic *B. calyciflorus* ranged from 2.93 to 3.59 eggs per female per day, when fed on an excess of *Phacus pyrum*. This contrasts with 0.83 and 0.62 eggs per female per day when fed on natural concentrations of phytoplankton. The former optimal values for amictic females approach the maximum possible birth rates at 20 °C, if the number of oocytes in the ovary is 32 rather than 64 and if the time interval between egg depositions is approximately $0.41 \times D_e$ (Ruttner-Kolisko, 1975). At 20 °C and with an embryonic duration (D_e) of 0.35 days, the maximum birth rate is 2.86 eggs per female per day. Wiktor (1968), who investigated the fecundity of field populations of *Brachionus* living in Szczecin Bay, Poland, recorded maximal fecundities of 2.66 and 1.55 eggs per female for *B. calyciflorus* and *B. angularis*, respectively, during May, June and July. In July, when the water temperature was 18 °C and D_e equals 1.03 days (Bottrell *et al.*, 1976), a fecundity of 2.66 eggs per female represents a birth rate of 2.59 eggs per female per day. In the light of the above comparative information, the level of fecundity (Table 4) and the daily birth rates (Duncan, 1983a) recorded for rotifer populations in PSN, though high compared with temperate values, are relatively low for the prevailing temperature conditions (28 °C). With a time interval between egg depositions of $0.41 D_e$ (Ruttner-Kolisko, 1975) and a calculated $D_e = 10.51$ hours (Bottrell *et al.*, 1976), or an observed $D_e = 5.67$ hours (Duncan, 1983a), rotifer populations in PSN should be able to produce between 5 and 10 eggs per female per day, given an optimal diet and assuming egg-laying throughout 24 hours.

During the visits to PSN in 1979 and 1980, the seston was visually dominated by blue-green algae and diatoms together with benthic detrital particles re-suspended during the windy periods. The main species in 1979 were *Anabaenopsis raciborski* (520 μm^3), *Melosira granulata* (4 700 μm^3) and Mougeotia spp (1 005 μm^3) (Dokulil *et al.*, 1983) but there was less *M. granulata* during 1980 (Rott, 1983). Net fractionation showed that most of the chlorophyll a biomass and the photosynthesis could be attributed to smaller algal species (Dokulil *et al.*, 1983). Rough estimates of the particulate food available to the planktonic rotifers were obtained by filtering sestonic samples through a mesh of 33 μm hole size. The carbon concentration of the filtrate was about 11 $mgC \cdot l^{-1}$ in 1979 and about 3 $mgC \cdot l^{-1}$ in 1980. Not all of this was algal carbon since there were detrital particles probably colonised by bacteria. These would be available as food to certain species of rotifers, although probably not to *Brachionus* spp (Starkweather & Bogdan, 1980). On 1 September 1979, when the total sestonic carbon was 11 $mgC \cdot l^{-1}$, the algal carbon was approximately 3.2 $mgC \cdot l^{-1}$ (obtained by converting algal volumes to their carbon equivalents) and the non-algal carbon by subtraction was about 8 $mgC \cdot l^{-1}$. By the same treatment, the algal carbon on 8 March 1980 was 2.1 $mgC \cdot l^{-1}$ and the non-algal fraction about 1 $mgC \cdot l^{-1}$, that is, much less than in 1979.

It was hoped that the feeding studies conducted by Duncan & Gulati (1983b) on natural seston (the fraction less than 33 μm) at field temperatures would indicate the food levels for rotifers in PSN by comparison with laboratory feeding rates at known food levels, reviewed by Starkweather (1980). The comparison is not simple since ingestion and clearance rates are influenced by body size of the consumer, by the size and shape of particles and by the biomass and numerical density of particles in relation to the level above which feeding is maximal. In many, otherwise excellent, quantitative studies on the feeding of rotifers, the size of food cell or consumer body is often not recorded. By calculating the daily food ration as the relative feeding rate ($\mu gC \cdot \mu gC^{-1} \cdot d^{-1}$; C/W – the daily consumption divided by body weight in common units), a comparison of the daily ingestion was possible between the PSN and MFT *B. caudatus* and the MFT *B. calyciflorus* (Duncan & Gulati, 1983b). These are presented in Table 6 together with values from the literature which are, however, available only for the larger species of *Brachionus* and in units of calories rather than carbon ($cal \cdot cal^{-1} \cdot d^{-1}$). The three-fold greater daily ration of *B. plicatilis* at 15 $\mu gDW \cdot ml^{-1}$ compared with that of similar-sized *B. rubens* at the same food level, known to be optimal for *rubens*, may be a function of the nature of the algal species provided as food, flagellate rather than chlorococcales. In both species, the daily ration decreased with decrease in food level and was very low for *B. rubens* in food deficient conditions. Bearing in mind the higher temperatures of the tropical measurements, the similar-sized MFT *B. caudatus* had lower daily rations than might have been expected. This may be related to the lower food supply available. Weight-specific effects are to be expected with the much large *B. calyciflorus* and the much smaller PSN *B. caudatus* but it seems likely that the measured values of both are too large for reasons which are not known.

In considering whether the concentration of edible particles (the fraction less than 33 μm) in PSN could represent optimal or limiting food levels for the rotifer populations or for the missing crustacean zooplankton, it is necessary to take into account what is known about food limitation in temperate species. For temperate cladocerans, food concentrations of between 2 and 8 $mgC \cdot l^{-1}$ are much higher than the known threshold values for growth and reproduction (Lampert & Schober, 1980; Rocha, 1983; Rocha & Duncan, unpubl. data). Chalk (1981) found that the critical food concentration for the feeding of *Daphnia magna* at 20 °C varied from 0.2 to 14.0 $mgC \cdot l^{-1}$, depending upon the size of food cell offered. She was able to derive a significant inverse log-linear relationship between the critical food concentration (in $cells \cdot ml^{-1}$) and the size of food cell (irrespective of how this was measured). A similar analysis of food cell size and critical food concentrations for species of *Brachionus*, obtained from Starkweather's (1980) summary, produced a similar inverse log-linear relationship (Duncan & Gulati, 1983b) but the *Brachionus* regression had a much steeper slope than the daphnid one. The two regressions overlapped at a food cell size of 1 000 μm^3 and a critical food concentration of 10^4 $cells \cdot ml^{-1}$. For cells less than 1 000 μm^3, the critical concentrations for rotifers were very much higher. The very high critical food

Table 6. Comparison of daily food rations of different species of *Brachionus*.

Species (and source)	Body weight ($\mu gDW \cdot ind^{-1}$)	Temperature (°C)	Nature of food	Concentration of food ($\mu gC \cdot ml^{-1}$)	Daily food ration (C/W) (d^{-1})	Authors
Brachionus calyciflorus (MFT)	0.474	26–28	PSN-Seston[c]	8.34	14.3–33.5[a]	Duncan & Gulati (1983b)
Brachionus caudatus (MFT)	0.104	26–28	MFT-Seston[d]	6.27	4.7[a]	Duncan & Gulati (1983b)
Brachionus caudatus (MFT)	0.104	26–28	PSN-Seston[d]	2.90	6.4[a]	Duncan & Gulati (1983b)
	0.104	26–28	PSN-Seston[c]	3.15	3.0[a]	Duncan & Gulati (1983b)
Brachionus caudatus (PSN)	0.013	26–28	PSN-Seston[c]	2.21	53.8[a]	Duncan & Gulati (1983b)
Brachionus plicatilis	0.158	20	Dunaliella	15–37	8.9–15.5[b]	Doohan (1973)
Brachionus rubens (adults)	0.13	20	Chlorella	0.18	0.86[b]	Pilarska (1977)
	0.16	20	Chlorella	15.0	2.74[b]	Pilarska (1977)
	0.13	20	Chlorella	150.0	4.68[b]	Pilarska (1977)

[a] C/W given as $\mu gC \cdot \mu gC^{-1} \cdot d^{-1}$; [b] C/W given as $cal \cdot cal^{-1} \cdot d^{-1}$; [c] less than 33 μm seston fraction; [d] less than 10 μm seston fraction.

levels of 100 μgDW · ml⁻¹ cited in Starkweather (1980) must therefore be associated with small food cells. Duncan & Gulati (1983b, Table 5) predict the critical food concentrations (in both numbers and carbon) required by *Brachionus* to attain maximum feeding rates on food cells between 1 μm^3 and 1 000 μm^3. For six of the seven abundant species of algae present in PSN, the critical food concentration would fall between 2 and 7 μgC · ml^{-1}. This is about the level of sestonic carbon measured in the reservoir during the feeding studies but assumes that the small PSN *B. caudatus* are capable of ingesting algae such as *Anabaenopsis raciborski* (520 μm^3), *Synedra acus* (405 μm^3), three species of *Scenedesmus* (about 860 μm^3) and *Mougeotia* sp (1 005 μm^3) (Rott, 1983; Dokulil *et al.*, 1983). For rotifers to attain maximum feeding rates on smaller algal cells, planktonic bacteria of small detrital particles, the food abundance would have to be greater than the sestonic carbon levels detected in the reservoir. This prediction confirms the impression that the somewhat low daily food rations measured for the MFT *B. caudatus* fed on PSN seston represent a food-limited but not food deficient rate.

Prolongation of the juvenile phase due to food limitation imposes a high cumulative respiratory cost and some species are able to overcome this at particular low food levels by maturing at a smaller body size. This has been shown to occur in nematodes (Schiemer *et al.*, 1980) and is likely to take place in daphnids (Duncan & Rocha, unpubl. data). It can be shown to occur in *B. rubens* by reworking the published results of Pilarska (1977) on the life cycle energy balance of this species reared at 20 °C in different food levels. Table 7 shows that the food-deficient juveniles are smaller throughout their post-embryohic phase which lasted 2.84 days. Their production, respiration and consumption for the whole juvenile period has been calculated from the published instantaneous energy budgets and used to calculate various efficiencies. Under conditions of food deficiency, respiration for the juvenile phase accounts for 75% of total assimilated energy (P + R) and 43% of consumed energy, both of which are very high when compared with the same efficiencies under optimal food conditions (40% and 9%, respectively). The food-deficient amictic female is small and is able to produce eggs only for 0.73 days compared with the much larger female reared under optimal food which produced eggs for 6.8 days. The reason for the shortness of the period of egg production under food deficiency is clear as respiration during this period accounts for 80% of assimilated and 54% of consumed energy, both of which are very high when compared with the optimal food efficiencies for the period of egg production (43% and 14%, respectively). All three life cycle phases were smaller under conditions of food defi-

Table 7. Energy budget of *Brachionus rubens* at 20 °C for the post-embryonic period and for the period of egg production and some resulting efficiencies, re-calculated from Pilarska (1977).

	Period of growth			Period of egg production		
Food concentration	Deficient	Optimal	Excessive	Deficient	Optimal	Excessive
10^6 cells · ml^{-1}	0.012	1.0	10.0	0.012	1.0	10.0
μgC · ml^{-1} (a)	0.18	15.0	150.0	0.18	15.0	150.0
Body size (μgDW · ind^{-1})						
Egg (2n)				0.04	0.06	0.06
Juvenile (young to old)	0.06–0.11	0.09–0.14	0.08–0.10			
Amictic female				0.13	0.17	0.14
Duration of period of body growth or egg production (days)	2.84	1.26	1.43	0.73	6.77	3.80
Energy parameters (cal · 10^{-3} · ind^{-1} · $period^{-1}$)						
Consumption	30.55	58.47	73.15	0.35	13.18	9.06
Production	4.26	8.00	3.96	0.05	2.34	0.32
Respiration	13.08	5.38	5.21	0.19	1.79	1.03
Efficiencies for the period						
Assimilation/consumption	0.57	0.23	0.13	0.68	0.31	0.15
Respiration/assimilation	0.75	0.40	0.57	0.80	0.43	0.77
Respiration/consumption	0.43	0.09	0.07	0.54	0.14	0.11

(a) Cell weight (29 × 10^{-6} μg DW · $cell^{-1}$; Rlarska, 1977) and carbon weight is 50% of dry weight.

ciency. It is not surprising that the food-deficient juveniles and adults also show higher assimilation efficiencies (57% and 68%, respectively) than the same stages grown in optimal food (23% and 31%).

Summary, conclusions and further studies

In Parakrama Samudra during 1979 and 1980, it appears that the effect of flushing rates of $0.1\ d^{-1}$ or more is small in rotifers whose population turnover rate can be $1.85\ d^{-1}$, although such rates may impose a serious daily loss on populations of planktonic crustaceans. Studies on the instar duration of tropical crustaceans at different food levels are required to assess whether this is likely. Much higher daily losses other than by wash-out were recorded for the rotifers during the periods of study when the planktonic stages of the planktivorous clupeoid fish were present. These losses took place mostly during the daylight hours when sunlight penetrated the water column deepest and the accompanying changes in the species composition and size-structure of the rotifer community point to the presence of species-selective and size-selective predation by a visual predator such as *Ehirava fluviatilis.* Gut analyses showed that the species and sizes of rotifers taken by the fish were those showing greatest decline in numbers. The effect is most marked with species of *Brachionus;* in *B. caudatus* from the reservoir, adults were small and bore small eggs. Further studies are required to confirm whether all the daily rotifer losses other than by wash-out can be accounted for by *Ehirava* predation alone: what needs to be known is the daily pattern of feeding and the number of rotifers that can be ingested hourly. Some part of the daily losses may be due to the other predator abundantly present, *Asplanchnella brightwelli,* which may be a competitor for the same species of rotifers. Further studies are needed to determine this. *Ehirava fluviatilis,* which was the most numerous fish in Parakrama Samudra North during the period of study, attained levels of biomass commensurate with those in other water bodies which possessed a zooplankton dominated by protozoans and rotifers and with a minimal crustacean contribution. The higher fecundity of the 1980 populations of rotifers compared with the 1979 indicated better food conditions and this suggests that the reduced densities and even small body sizes and eggs of the 1980 rotifers must be due to severe predatory stress by the planktonic fish whose density was unfortunately not determined. Selective predation by this fish species may be responsible for the species composition and size frequency distribution of the zooplankton in PSN and, particularly, for the absence of crustaceans and larger species of rotifers.

The small eggs borne by small adult *B. caudatus* from PSN had shorter embryonic durations than expected for the prevailing temperatures. When applied to diurnally sampled rotifer populations and taking into account the differences in temperature, fecundity and density at various depths, the predated-population turnover rate was $1.85 \cdot d^{-1}$. This is not particularly high, bearing in mind the tropical temperatures but it is almost twice that to be expected from rotifer populations with larger eggs. Whatever the factors influencing the appearance of small adults with small eggs, the resultant effect is to increase the rotifer populations' turnover rate.

There is evidence in the literature for temperate species that food deficiency prolongs the juvenile phase and results in small adults which lays small eggs for a limited period whereas at optimal food levels a brief juvenile phase ends in a large female which produces large eggs for a long period of time. The estimates of food available to PSN rotifers fell between the values quoted for these two food levels which suggests a condition of food limitation rather than food deficiency. The measured levels of fecundity, daily birth rates and daily weight-specific feeding rates are all lower than would be expected for an optimal diet at the prevailing temperatures and provide evidence for a food limited but not food deficient state in PSN rotifers. Judging from the available algal counts, algal sizes and sestonic biomasses, rotifers would have to capture and ingest relatively large algal species in order to attain maximal feeding rates at or above critical food concentrations and this may limit their food intake further. Further information is required both on this and on the densities and sizes of the smaller planktonic cells and particles available as food to the rotifers. There is as yet little published information on the food threshold levels for growth and reproduction or on the critical food concentrations for feeding of tropical species of planktonic crustaceans. It seems unlikely that the sparse presence of planktonic crustaceans in PSN is due to food deficiency since the available PSN food levels lie near to optimal levels for growth, reproduction and feeding in temperate daphnid species at 20 °C (Rocha, 1983; Chalk, 1981). However, evidence from the literature suggests that the high cost of cumulated respiration resulting from food-limited prolongation of the juvenile phase of planktonic animals is likely to be aggravated at tropical temperatures and may also increase the food threshold values for growth and for reproduction above those adequate for temperate situations. This assumes no respiratory acclimatization in the tropical animals, something which also requires further investigation.

Acknowledgements

I acknowledge with thanks the financial assistance received from The British Council and The Royal Society which enabled me to carry out this work and to attend the Symposium on Tropical Zooplankton at Sao Carlos, Brazil. I thank my colleagues, particularly Mr A. Wroot, for helpful criticism of the manuscript. This is paper no. 28 of the Parakrama Samudra Project.

References

Bottrell, H. H., A. Duncan, Z. M. Gliwicz, E. Grygierek, A. Herzig, A. Hilbricht-Ilkowska, H. Kurasawa, P. Larsson & T. Weglenska, 1976. A review of some problems in zooplankton production studies. Norw. J. Zool. 24: 419–456.

Chalk, E. A., 1981. Cladoceran filter feeding in a Thames Valley reservoir. Ph.D. Thesis CNAA Central London Polytechnic/Thames Water Authority, 266 pp.

Dokulil, M., N. Bauer & I. Silva, 1983. An assessment of the phytoplankton biomass and primary productivity of Parakrama Samudra, a shallow man-made lake in Sri Lanka. In F. Schiemer (ed.), Limnology of Parakrama Samudra – Sri Lanka: a case study of an ancient man-made lake in the tropics. Dr W. Junk, The Hague: 49–76.

Doohan, M., 1973. An energy budget for adult Brachionus plicatilis. Muller (Rotatoria). Oecologia 13: 351–362.

Duncan, A., 1983a. The composition, density and distribution of the zooplankton in Parakrama Samudra. In F. Schiemer (ed.), Limnology of Parakrama Samudra – Sri Lanka: a case study of an ancient man-made lake in the tropics. Dr W. Junk. The Hague: 85–94.

Duncan, A., 1983b. The influence of temperature upon the duration of embryonic development of tropical *Brachionus* species (Rotifera). In F. Schiemer (ed.), Limnology of Parakrama Samudra, Sri Lanka: a case study of an ancient man-made lake in the tropics. Dr W. Junk. The Hague: 107–115.

Duncan, A. & R. D. Gulati, 1981. Parakrama Samudra (Sri Lanka) Project – a study of a tropical lake ecosystem, 3. Composition, density and distribution of the zooplankton in 1979. Verh. int. Ver. Limnol. 21: 1001–1006.

Duncan, A. & R. D. Gulati, 1983a. A diurnal study of the planktonic rotifer population in Parakrama Samudra Reservoir, Sri Lanka. In F. Schiemer (ed.), Limnology of Parakrama Samudra – Sri Lanka: a case study of an ancient man-made lake in the tropics. Dr W. Junk, The Hague: 95–106.

Duncan, A. & R. D. Gulati, 1983b. Feeding studies with natural food particles on tropical species of planktonic rotifers. In F. Schiemer (ed.), Limnology of Parakrama Samudra – Sri Lanka: a case study of an ancient man-made lake in the tropics. Dr W. Junk, The Hague: 117–123.

Fernando, D. H., 1980. The freshwater zooplankton of Sri Lanka, with a discussion of tropical freshwater zooplankton composition. Int. Rev. ges. Hydrobiol. 65: 85–125.

Fernando, C. H. & R. Rajapaksa, 1983. Some remarks on long-term and seasonal changes in the zooplankton of Parakrama Samudra. In F. Schiemer (ed.), Limnology of Parakrama Samudra – Sri Lanka: a case study of an ancient man-made lake in the topics. Dr W. Junk, The Hague: 77–84.

Green, J., 1960. Zooplankton of the River Sokoto. The rotifera. Proc. zool. Soc. Lond. 135: 491–523.

Green, J., 1977. Dwarfing of rotifers in tropical crater lakes. Arch. Hydrobiol. Beih. Ergebn. Limnol. 8: 233–236.

Green, J. & Oey Biauw Lan, 1974. *Asplanchna* and the species of *Brachionus calyciflorus* in two Javanese sewage ponds. Freshw. Biol. 3: 223–226.

Jayatunga, Y. N. A., 1982. Studies on the hydrobiology and zooplankton in Kalawewa Area. M. Phil. Thesis, Univ. Colombo, 156 pp.

King, C. E., 1967. Food, age and the dynamics of a laboratory population of rotifers. Ecology 48: 111–128.

King, C. E., 1969. Experimental studies of ageing in rotifers. Expl. Gerontol. 4: 63–79.

Koste, W., 1978. Rotatoria. Gebr. Borntraegar. Berlin, Stuttegart, 673 pp.

Lampert, W. & U. Schober, 1980. The importance of 'Threshold' food concentrations. In W. C. Kerfoot (ed.), Evolution and Ecology of Zooplankton Communities. U.P. New England: 264–267.

Leimeroth, N., 1980. Respiration of different stages and energy budgets of juvenile *Brachionus calyciflorus*. Hydrobiologia 73: 195–198.

Mori, S. & G. Yamamoto, 1975. Productivity of Communities in Japanese Inland Waters. J.I.B.P. Synthesis 10: 436 pp.

Newrkla, P. & A. Duncan. The biology and density of *Ehirava fluviatilis* Deraniyagala (clupeoid), a zooplanktovorous species of fish planktonic as larva and adult in Parakrama Samudra, Sri Lanka. Verh. Internat. Verein. Limnol. 22: (in press).

O'Grady, K. T. & P. B. Spillett, 1981. The production of carp (*Cyprinus carpio*) in disussed filter beds. Proc. 2nd Br. F. W. Fish Conf.: 17–28.

Pilarska, J., 1977. Eco-physiological studies on *Brachionus rubens*. Ehrbg (Rotatoria). Pol. Arch. Hydrobiol. 24: 319–354.

Pourriot, R., 1973a. Rapports entre la température, la taille des adultes, la longeur des oeufs et le taux de développement embryonnaire chez *Brachionus calyciflorus* Pallas (Rotifère). Ann. Hydrobiol. 4: 103–115.

Pourriot, R., 1973b. Recherches sur la biologie des rotiferes, 3. Fécondité et durée de vie comparées chez les femelles amictiques et mictiques de quelques espèces. Ann. Limnol. 9: 241–258.

Pourriot, R. & M. Deluzarches, 1971. Recherches sur la biologie des rotifères, 2. Influence de la température sur la durée du développment embryonnaire et post-embryonnaire. Ann. Limnol. 7: 25–52.

Rott, E., 1983. A contribution to the phytoplankton species composition of Parakrama Samudra. In F. Schiemer (ed.), Limnology of Parakrama Samudra – Sri Lanka: a case study of an ancient man-made lake in the tropics. Dr W. Junk. The Hague: 209–226.

Ruttner-Kolisko, A., 1974. Plankton rotifers. Inld Wat. 26: 1–146.

Ruttner-Kolisko, A., 1975. The influence of fluctuating temperature on plankton rotifers: a graphical model based on life data of *Hexarthra fennica* from Neusiedlersee, Austria. In J. Salanki & J. E. Ponyi (ed.), Symp. Biol. Hung. 15: 197–204.

Schiemer, F., A. Duncan & R. Z. Klekowski, 1980. A bioenergetic study of a benthic nematode, *Plectus palustris* de Man 1880, throughout its life cycle. Oecologia (Berl.) 44: 205–212.

Schiemer, F., 1981. Parakrama Samudra (Sri Lanka) Project, a study of a tropical lake ecosystem. I. An Interim review. Verh. Internat. Verein. Limnol. 21: 987–993.

Schiemer, F., 1984. Limnology of Parakrama Samudra, Sri Lanka: a case study of an ancient man-made lake in the topics. Developments in Hydrobiology. Dr W. Junk, The Hague: 236 pp.

Scott, A. P. & S. M. Baynes, 1978. Effect of algal diet and temperature and the biochemical composition of the rotifer *Brachionus plicatilis*. Aquaculture 14: 247–260.

Starkweather, P. L., 1980. Aspects of the feeding behaviour and trophic ecology of suspension-feeding rotifers. Hydrobiologia 73: 63–72.

Sterzynski, W., 1979. Fecundity and body size of planktonic rotifers in 30 Polish lakes of various trophic state. Ekol. Pol. 27: 307–322.

Starkweather, P. L. & K. G. Bogdan, 1980. Detrital feeding in natural zooplankton communities: discrimination between live and dead algal foods. Hydrobiologia 73: 83–85.

Theilacker, G. H. & M. F. McMaster, 1971. Mass culture of the rotifer *Brachionus plicatilis* and its cultivation as a food for larval anchovies. Mar. Biol. 10: 183–188.

Wiktor, K., 1968. Seasonal modifications in fertility of some rotifer species. Pol. Arch. Hydrobiol. 15: 9–22.

Zooplankton feeding in the fish *Trichogaster pectoralis* Regan

Jiamjit Boonsom
National Inland Fisheries Institute, Bangkok, Thailand

Keywords: tropical zooplankton, *Trichogaster*, fish, fish culturing, planktivory, Thailand, rotifers, Crustacea

Abstract

The traditional culture method of *Trichogaster pectoralis*, using zooplankton produced from fermenting aquatic weeds as a source of food for the fry, leads to productions of ca 1 000 kg ha^{-1} a^{-1}. If chicken manure is applied at a rate of ca 450 kg ha^{-1} $month^{-1}$ instead, much more zooplankton is produced more rapidly, and fish production increases to ca 2 000 kg ha^{-1} a^{-1}. This results from a higher survival rate of the fry, which feeds on rotifers, ciliates, copepods and cladocera until a size of ca 15.0 mm, when plant material also becomes important in the diet.

Introduction and background

Trichogaster pectoralis Regan is a fish native to Thailand, common in swamps, rivers, lakes, and rice fields. It is a traditional food fish and its annual production is 42–48% of the total pond fish production of Thailand (Dept. Fish. 1981, 1982). The marketable fish is around 17–25 cm in size with a weight between 85 and 250 g.

The culture of this fish is centered on the irrigated coastal area of Samutprakran and Chachengsao Provinces, about 50 km east of Bangkok. This area was formerly an estuarine swamp, reclaimed for rice cultivation by building a dike along the sea shore in 1931. In 1957 the Department of Fisheries started a rice-cum-fish culture project in this area. *Trichogaster pectoralis* was found to be the best choice. Due to high salinity during the dry season and the low pH of the sulphate soil, rice production is non-profitable while fish production is very high. The rice field was consequently changed into a fish field since 1962 (Indrabarya, 1981).

Traditional culture system

The culture is extensive, and the average size of farms ranges between 3 and 20 ha. In the rice-cum-fish project the farmers raise the dike and build a peripheral ditch around the field. A ditch is 3 to 4 m wide and 75 cm deep (Fig. 1). The central platform area is overgrown by weeds.

Weeds are very important for *Trichogaster* farming. They serve as fertilizer, and protect the fish nests against strong winds, heavy rains, intense sunlight, and protect the fry and fingerlings against predators. Besides, when fermented, they raise the pH of the acid water within 7 to 10 days (Indrabarya, 1981).

Farmers stock their fields with spawners that are retained after the harvest. They do not select for sex or condition. The rate of stocking varies from 40 to 220 kg ha^{-1} (Sawaegwan, 1971; Varikul *et al.*, 1969).

The culture season starts in February when brood fish are released into the peripheral ditch for a while before permanently raising the water level to cover the platform area. A few days after flooding, the fish build nests and spawn among the weeds. Eggs hatch within 24 hours and the fry start to feed as

Hydrobiologia 113, 217–221 (1984).

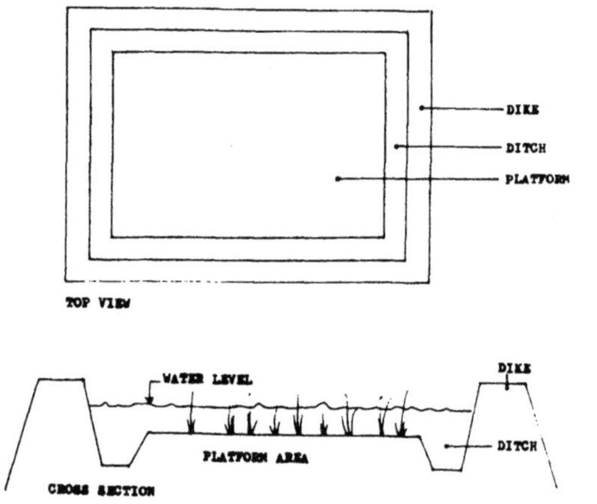

Fig. 1. Top view and cross section of *Trichogaster* field.

soon as the yolk sack is absorbed, around the sixth day and at ca. 5–6 mm in size (Pothongkam, 1968; Thongsa-nga, 1965). The only fertilizer used is green manure. The farmer cuts the weeds biweekly and piles or scatters them on the field to ferment.

The fish are usually harvested after 8–10 months. The yield of this type of culture ranges between 450 to 1 700 kg ha^{-1}.

Problems with the traditional culture system

Farmers always overstock their fields. Most are not aware that overstocking produces many small fish and a low price per kilogram. Fry production by farmers shows that the annual yield (Table 1) is only 6–23 offspring per pair of brood fish. This is quite low compared to a fecundity of 20 000–36 000 eggs per female (Pothongkam, 1968; Thongsa-nga, 1965; Hora & Pilay, 1962).

The high mortality rate of fry is due to many factors: low water quality, inadequate food at an early stage of life, and predation from predatory fish such as *Ophicephalus* (or *Channa*) *striatus* and *Anabas testudineus.* These two fishes are not easy to eradicate from the fields, especially during high water. Notonectidae and giant water bugs (Belostomatidae) are important insect predators of *Trichogaster* fry (Tong-u-thai, 1969).

Problems with water quality

Low dissolved oxygen results from fermenting water plants. Thirty-one species of weeds belonging to 15 families were found to be present. The emergent type is the most important group and is used as green manure in the field. It constitutes about 80% of the total area. Abundant species are *Eleocharis equisetoides, Paspalum conjugatum* and *Hymenachne myurus.*

Low pH in the early culture season is a result of a long exposure of the field to the sun during the dry season, but fry is reported to survive at pH values as low as 4.6 (Sitasit, 1969). Sea water seeps through the coastal embankment at low freshwater level. This is more serious for larvae and fry than for adults.

Food of the fry is very important too. A most suitable food for fish fry are Rotifera (Horvath, 1979). Fry needs more food than adults per unit body weight. As mentioned above, the source for producing natural food is the fermented weed. This green manure requires 7–10 days of fermentation before it releases sufficient nutrients to be effective. This is insufficient to feed the vast number of the fry produced by 650 kg of female fish per ha. Animal manure is a more effective fertilizer, producing a zooplankton bloom in a short time.

Table 1. Water quality in study pond from Jan. 29 to March 10, 1981. (Data of five stations averaged; after V. Srisuwanatach's unpublished data.)

Date	Z (cm)	WT (c°)		Sechi depth (cm)	DO (mg l^{-1})		CO_2 (mg l^{-1})		pH		Alkalinity (mg l^{-1})		gO_2 m^{-3} · d^{-1}		Sal. (‰)
		AM	PM		AM	PM	AM	PM	AM	PM	AM	PM	Prod	BOD	
29/01/81	36	26.0	33.0	17	0.0	3.0	5.5	4.0	6.9	6.7	169	138	2.8	14.6	0.1
10/02/81	42	26.5	31.2	31	0.0	2.3	2.5	3.0	7.3	7.4	226	225	2.4	3.0	0.6
24/02/81	46	27.7	32.1	43	0.2	6.7	2.7	1.0	7.0	7.3	224	233	2.7	2.0	0.6
10/03/81	51	27.8	33.7	29	0.0	2.0	5.0	4.0	7.2	7.2	337	230	2.5	9.3	0.6

Date of fertilization: Jan. 30, Feb. 11, 27, 1981.

A preliminary study of the stomach contents of fry showed that they were feeding on Cladocera as soon as the yolk sac was absorbed, and fry which fed on *Moina macrocopa* grew faster than fry fed rice bran (Bumrungsuk, 1970).

Material and methods

One way of increasing fish production is to increase the survival rate of the fry. This requires minimal management input and no drastic change of traditional culture systems. A pre-requisite for success, was collaborative work with a farmer.

A preliminary study on the application of chicken manure to stimulate production of microorganisms, showed that the maximum at which animal manure can be applied without polluting the fish pond was 31.25–62.50 kg $ha^{-1}d^{-1}$ (Srisuwanatach *et al.*, 1982). However, the experimental ponds used were new ponds without macrophytes or fish, while old *Trichogaster* fields are full of weeds and decaying matter. Therefore the rate of the chicken manure added to these fields reduced to 15.6 kg $ha^{-1}d^{-1}$ for 60 days during the fry rearing period. The total culturing time was 8 months.

The farmer who agreed to cooperate in this research has a 3.2 ha farm, with a water area of 2.9 ha. In 1980, the farmer stocked 400 kg of *Trichogaster*, about 4 000 fish, and after 8 months harvested 2 700 kg of *Trichogaster* and 200 kg of wild fish. The production was 1 000 kg ha^{-1}.

The experiment began in January 1981 by selecting 360 pairs of brood fish to stock in the field. The farmer added another 150 kg of *Trichogaster* and 50 kg of adult *Clarias macrocephalus*.

Every 10 days, a total of 450 kg fertilizer was placed at the four corners of the platform. This was continued for two months.

Water analysis, plankton, and stomach contents were collected at biweekly intervals. Dissolved oxygen, carbon dioxide, pH, alkalinity, transparency and temperature of air and water were recorded.

Plankton samples were collected using 80 μm nets. Samples were collected by 10 m surface hauls and preserved in 6% formalin. The samples were identified, and counted to calculate numbers per square meter of water area.

Results and discussion

Water analysis

The data obtained from 5 fixed stations were consistent, therefore the average value is used in Table 1. The transparency (secchi disc) and the alkalinity increased after fertilizing. Dissolved oxygen reached near zero in the early morning hours but no dead fish were found. Thus, dissolved oxygen is not a limiting factor for *Trichogaster*.

Plankton and stomach contents

Sixty-six species of plankters were found in the samples: 10 Protozoa, 44 Rotifera, 7 Cladocera, 3 Copepoda and one Ostracod (Table 2). Before fer-

Table 2. List of zooplankton found in *Trichogaster* pond.

Protozoa	*L. luna*
Arcella sp.	*L. lauterborni*
Centropyxis aculeata	*L. papuana*
Colpoda sp.	*L. sola*
Diffugia sp.	*L. (M) bulla*
Euglena oxyuris	*L. (M) harringi*
Euglena spp.	*L. (M) lamellata*
Glenodinium sp.	*L. (M) lunaris*
Phacus sp.	*L. (M) monostyla*
Trachelomonas lagenella	*L. (Hemimonostyla) inopinata*
Volvox sp.	*Mytilina bisulcata*
Rotifera	*M. ventralis*
Asplanchna sp.	*Testudinella patina*
Brachionus angularis	*Trichocerca bicristata*
B. calyciforus	*T. cylindrica*
B. caudatus	*T. tenuior*
B. leydigii	*Trochosphaera equatorialis*
B. plicatilis	Cladocera
B. quadridentata	*Alona davidi*
B. urceolaris	*Kurzia longirostris*
Keratella procurva	*Pleuroxus denticulatus*
K. tropica	*Ceriodaphnia cornuta*
Platyias quadricornis	*Macrothrix rosea*
Lepadella elliptica	*Moinodaphnia macleayi*
L. ovalis	*Moina macrocopa*
L. patella	*M. micrura*
Dicranophorus robustus	*Diaphanosoma excisum*
Euchlanis meneta	*D. sarsi*
Tripleuchlanis sp.	*Pseudosida* cf. *bidentata*
Polyarthra vulgaris	Copepoda
Filinia opoliensis	*Thermocyclops* sp.
Ascomcrpha saltans	*Mesocyclops thermocyclopoides*
Lecane aegana	Diaptomidae
L. aculeata	gen. sp. div.
L. closterocerca	Oatracoda
L. hastata	gen. sp. div.

Table 3. Quantity of zooplankton in fertilized *Trichogaster* farm Bangpli Samatprakan during Jan. 28 to March 10, 1981 (number m^{-2}). After S. Duangsawasdi, unpublished data.

Date	Day	Protozoa	Rotifera	Cladocera	Copepoda	Total	Notes
28/01/81	1	0	13 866	1 702	57 419	72 987	Stocking day
10/02/81	14	3 888	3 803	2 817	75 267	85 775	11 days after 1st manuring fry about 3-9 days old
24/02/81	28	28 410	7 845	2 170	14 178	52 603	13 days after 2nd manuring fry about 3-23 days old
10/03/81	42	15 966	6 270	39	17 115	39 390	13 days after 3rd manuring fry about 3-37 days old

Table 4. Frequency and percentage preference of food items in *Trichogaster* fry gut.

Size (mm)	Frequency	Phyto	Protozoa	Rotifera	Nauplii	Copepoda	Cladocera	Ostracoda	Insects
5.0- 5.9	1	-	-	1	1	-	-	-	1
6.0- 6.9	7	-	3	4	4	1	3	1	3
7.0- 7.9	2	1	2	2	1	-	-	-	-
8.0- 8.9	5	1	2	3	1	2	3	-	-
9.0- 9.9	5	1	2	2	2	3	4	-	-
10.0-10.9	1	-	-	1	1	-	1	-	-
11.0-11.9	2	-	1	1	1	-	2	-	1
12.0-12.9	3	-	-	3	1	3	2	-	-
15.0-37.0	7	7	7	5	3	1	-	-	-*
Total	33	10	17	22	15	10	15	1	5
Percentage preference		33.3	51.5	66.6	45.4	33.3	45.4	5.1	15.1

* 90% of the content is green mass.

Table 5. Items found in stomach content of *Trichogaster* larvae 5.6-12.0 mm in length.

I. **Zooplankton**	
Protozoa	Copepoda
Euglena sp.	Calanoid
Paramecium sp.	Cyclopoid
Phagus sp.	Nauplius
Rotifera	Cladocera
Brachionus leydigii	*Ceriodaphnia cornuta*
B. quadridentata	*Diaphanosoma* sp.
Platyias quadricornis	*Pseudosida bidentata*
Dicranophorus robustus	*Pleuroxus* sp.
Lecane bulla	Insect larvae
Polyarthra vulgaris	*Tendipes* sp.
II. **Phytoplankton**	
Oscillatoria sp.	
single cell algae	

tilizing, no Protozoa were found but after the second manuring the number of Protozoa increased tremendously to 28,410 org m^{-2} on Feb. 24. Rotifera were abundant at first and then dropped on Feb. 10 as they were grazed upon by the fry of age 3-9 days. The fish fry grazed not only on Rotifera but also on nauplii of Copepoda. The number of Copepoda decreased from 57,419 to 14,178 and 17,115 org m^{-2} on Feb. 24 and Mar. 10, respectively. Cladocera were the least abundant with only 39 org m^{-2} at a time when the fry was 3-37 days old (Table 3).

Gut content examination shows that larvae (5.0-37.0 mm in size) show a high preference to Rotifera, Protozoa, nauplii and Cladocera respectively, but it was noted that 90% of the gut content of the fry longer than 15.0 mm is a greenish mass (Table 4). It thus appears that the fry switch from a zooplanktivorus to a more herbivorous diet at this stage. The food items found in the gut content are listed in Table 5.

Fish yield

The total fish production was 6 382 kg, i.e. 2.000 kg ha^{-1}, or twice the yield of the preceding year. The survival of the fry was 58 fry per pair of brood fish.

Conclusions

In *Trichogaster* farming, fertilizing a pond with chicken manure at a rate of ca. 150 kg ha^{-1} 10 d^{-1} for two months during the fry rearing period increased natural food for the fry, increased fry survival rate, and doubled production (from 1 000 kg ha^{-1} to 2 200 kg ha^{-1} in 8 months).

Examination of the gut content of the fry showed that their food consisted of Rotifera (mainly *Brachionus* spp), which were the most important food item, while Protozoa, Cladocera and nauplii also play an important role. Fry switch from zooplankton feeding to algal or detritus feeding after reaching ca. 15.0 mm in length.

Acknowledgement

The author wishes to convey her thanks to Mr. Chen Foo Yan of FAO Network of Aquaculture Center in Asia for his guidance and for supplying the chicken manure for this experiment, to Mr. Prasert Sitasit of Fish Nutrition Section, NIFI, who contributed the supplementary food for the brood fish, to Mr. Narong Romerun who typed the manuscript, and to Dr. C. H. Fernando who kindly read and criticized the manuscript.

References

Fisheries Economics and Planning Sub-Division, Dept. Fish. 16: 1981. Freshwater fish farm production 1981.

Fisheries Economics and Planning Sub-Division, Dept. Fish. 11: 1982. Freshwater fish farm production 1980.

Bumrungsuk, S., 1970. Moina macrocopa for feeding Pla Salid fry Annu. Rep. Fish Cult. Sect., Freshwat. Fish. Div. Dept. Fish: 87-94.

Coleman, N. T. & G. W. Thomas, 1967. The basic chemistry of soil acidity. In R. W. Pearson & F. Adams (eds.), Soil Acidity and Liming: 1-41.

Gooch, B. C., 1966. Appraisal of North American fish culture fertilization studies. FAO Fish. Rep. 44 (4).

Hora, S. T. & T. V. R. Pillay, 1962. Handbook of fish culture in the Indo-pacific region. FAO Fish. Bio. Techn. Pap. 14: 10 pp.

Horvath, L., 1979. The rearing of warmwater fish larvae. Proc. World Symp. Finfish Nutrition & Fish Feed Technol., Hamburg 20-23 June 1978.

Indrabarya, B., 1981. Pla Salid (Trichogaster pectoralis) Thai Fish. Gazett. 34: 1945-1960. (in Thai, summary in English).

Pothongkam, K., 1968. Embryological and morphological development of Sepat Siam, *Trichogaster pectoralis* (Regan). Annu. Rep. Fish Cult. Sect. Freshwat. Fish. Div. Dept. Fish.: 36-75. (in Thai).

Rappaport, U., S. Sarig & Y. Bejerano, 1977. Observations on the use of organic fertilizers in intensive fish farming at the Ginosar Station in 1976. Bamidgeh 29: 57-70.

Schroeder, G. L., 1974. Use of fluid cowshed manure in fish ponds. Bamidgeh 26: 84-96.

Sitasit, P., 1969. Tolerance of fish fry in different pH water. Annu. Rep. Fish Cult. Sect., Freshwat. Fish Div. Dept. Fish.: 153-164 (in Thai).

Srisuwanatach, V., 1982. Impact of manure on the ecology of fish pond. Thai Fish Gazet 35: 101-108. (in Thai, abstract in English).

Thanchalanukit, W., 1966. The growth of Pla Salid Fry. In supranee Chinbutra 1974. Biology of Pla Salid. Annu. Rep. Fish. Cult. Sect. Freshwat. Fish. Div. Dept. Fish: 48-72. (in Thai).

Tong-U-Thai, K., 1969. Species and control of predatory insects on Sepat Siam *Trichogaster pectoralis* Regan. Annu. Rep. Fish Cult. Sect. Freshwat. Div. Dept. Fish: 165-177. (in Thai).

The freshwater zooplankton of Thailand (Rotifera and Crustacea)

Jiamjit Boonsom
National Inland Fisheries Institute, Bangkok, Thailand 10900

Keywords: tropical zooplankton, Rotifera, Cladocera, Copepoda, Thailand, biogeography

Abstract

A list is given of all zooplankton species hitherto reported from Thailand, based on the author's own observations and on the literature.

From a comparison with the fauna of some neighbour countries, it appears that important gaps remain to be filled-in.

Introduction

Thailand is a tropical country in South East Asia at a latitude of 5°32'N to 20°8'N and longitude 97°21'E to 105° 38'E. The average annual temperature is around 30 °C. The south-west monsoon brings rain each year during May–August. This creates a large numbers of natural lakes and swamps, especially on the central plain. Recently, a number of irrigation tanks, village ponds and large reservoirs were constructed in many parts of the country. Fish species and their composition in these reservoirs has been studied extensively, but the zooplankton has not.

Review of previous literature

The literature on Thai zooplankton is meagre. Boonsom (1970), in a preliminary study, reported and illustrated species from various parts of Thailand. Boonsom (1971), Boonsom *et al.* (1972), Pothipitak (1971), Bricker *et al.* (1978), and Kamolrat & Leelapatra (1973) also record some species, found in various rivers and reservoirs. Lai & Fernando (1981) made a detailed study of Calanoida found in the southern part of Thailand near the Malaysian border. De Ridder (1970) describes rotifers found in Bangkok, Chiangmai, Lampoon and Kanchanaburi.

Sangkhakul (1971) and Sangkhakul *et al.* (1971) record zooplankton species in Lum Takong and Sirindhorn reservoirs, respectively. Sangkhakul (1974) deals with the zooplankton of Bhumipal reservoir, the deepest reservoir in Thailand. Sangkhakul (1975) lists some species from Kangh Krachan reservoir, Sibke (1971) made observations on irrigation tanks in North East Thailand, and Sibke & Nujsombat (1974) worked on Chulabhorn reservoir, an altitudinal waterbody where air temperature may at times reach 4 °C. Srisuwanatach (1971) studied Lum Dom Noi reservoir, Waew-ngam (1973) worked on SIrikit reservoir, and, finally, the office of the national Economic and Social Development Board of the Kingdom of Thailand included the zooplankton in a survey of Bung Boraped and Nong Han, two natural lakes in Thailand.

Material and methods

Between 1966 and 1982, c. 500 plankton samples were collected from various types of habitats, using a 22 cm diameter net of 80 μm mesh size. The material was preserved in 6% formalin. Identification was done using a variety of sources listed in the references, and with the help of the experts listed in the acknowledgement.

Samples were collected from 7 rivers, 1 high altitude reservoir (+759 m · asl), 3 deep reservoirs (over 100 m deep), 7 moderately deep reservoirs (17–51 m in depth), 3 shallow reservoirs (1–6 m in depth), 4 irrigation tanks, 2 ponds and 2 fish fields. A detailed description of these 28 sites is available from the National Inland Fisheries Institute, Bangkok.

Results and discussion

The species listed in Table 1 include those found by the author and those listed in the literature. They amount to 80 species of Rotifera, 48 Cladocera, 21 Copepoda, 1 Ostracoda and 1 Branchiura. Among these, 102 are recorded in lentic environments, 34 in lotic environments and 21 were found in both. Only 14 species are confined to rivers. *Daphnia similis*, a species from the temperate zone (Brooks, 1957), was found in Chulabhorn Reservoir, at high altitude.

Table 1. List of freshwater zooplankton found in Thailand.

Species	Habitat							
	River	Large reservoirs				Tank	Pond	Field
		High	Deep	Mod.	Shallow			
I. **Rotifera**								
Asplanchnidae								
Asplanchna sp		+		+	+	+	+	+
Asplanchna brightwelli							+	+
Asplanchna priodonta				+				
Brachionidae								
Anuraeopsis fissa							+	+
Brachionus angularis				+				+
Brachionus calicyflorus			+			+		+
Brachionus caudatus								+
Brachionus diversicornis				+				
Brachionus donneri				+		+		
Brachionus falcatus		+	+	+	+	+	+	
Brachionus forficula							+	+
* *Brachionus leydigii*								+
Brachionus patulus			+	+	+			
Brachionus plicatilis				+			+	+
Brachionus quadridentata				+				+
Brachionus urceolaris								+
Keratells cochlearis			+	+		+		
Keratella cochlearis							+	
Keratella procurva							+	+
Keratella tropica			+	+	+	+		
Keratella valga	+		+	+	+	+		
Platyias sp.					+			
Platyias quadricornis								+
Colurellidae								
Colurella adriatica							+	+
Lepadella acuminata							+	+
* *Lepadella elliptica*								+
* *Lepadella ovalis*								+
* *Lepadella patella*								+
Conochilidae								
Conochilus sp.					+			
Conochilus unicornis					+			
Dicranophoridae								
* *Dicranophorus robustus*								+

Table 1. (cont.)

Species	Habitat							
	River	Large reservoirs				Tank	Pond	Field
		High	Deep	Mod.	Shallow			
Epiphanidae								
Epiphanes sp.	+							
Euchlanidae								
Dipleuchlanis propatula				+				
Euchlanis sp.				+				+
Euchlanis dilatata							+	+
* *Tripleuchlanis plicata*							+	+
Filinidae								
Filinia brachiata			+	+				
Filinia longiseta						+		
Filinia opoliensis			+					+
Filinia terminalis					+			+
Gastropidae								
Ascomorpha ovalis				+				
* *Ascomorpha saltans*								+
Gastropus sp.				+		+		
Hexarthridae								
Hexarthra sp.						+		
Lecanidae								
Lecane acus							+	+
* *Lecane aculeata*								+
Lecane aegana								+
Lecane arcuata							+	+
Lecane bulla				+	+			+
Lecane closterocerca					+			+
Lecane cornuta							+	+
Lecane curvicornis							+	+
Lecane hastata								+
* *Lecane harringi*								+
Lecane inopinata								+
* *Lecane lamellata*								+
Lecane lauterborni								+
Lecane luna			+	+	+		+	+
Lecane lunaris								+
Lecane monostyla								+
Lecane nana							+	+
Lecane papuana								+
Lecane quadridentata							+	+
* *Lecane sola*								+
Mytilinidae								
* *Mytilina bisulcata*								+
Mytilina ventralis					+			+
Synchaetidae								
Polyarthra sp.		+			+			
Polyarthra euryptera				+				
Polyarthra vulgaris				+				
Testudinellidae								
Pompholyx complanata							+	+
Testudinella sp.					+			
Testudinella patina			+	+		+		+
Trichoceridae								
Elosa sp.					+			

Table 1. (cont.)

Species	Habitat							
	River	Large reservoirs				Tank	Pond	Field
		High	Deep	Mod.	Shallow			
Trichocerca bicristata								+
Trichocerca cylindrica			+			+		+
Trichocerca pusilla							+	+
Trichocerca similis				+		+		
* *Trichocerca tenuior*								+
Trichotridae								
Macrochaetus sp.					+			
Trochosphaeridae								
* *Trochosphaera equatorialis*		+						+
II. **Cladocera**								
Bosminidae								
Bosmina sp.	+			+	+	+		
Bosmina longirostris			+	+				
Bosminopsis deitersi	+		+	+	+	+		
Chydoridae								
Alona sp.	+		+	+	+			
Alona affinis	+							
* *Alona davidi*	+							+
Alonella nana	+			+				
Alona karua						+		
Alona verrucosa			+					
Camptocercus rectirostris						+		
Ephemeroporus barroisi	+			+				
Chydorus eurynotus			+	+				
Chydorus sphaericus				+				
Chydorus ventricosus			+	+		+		
Dadaya macrops	+							
Dunhevedia crassa	+					+		
Euryalona orientalis	+	+		+				
Kurzia longirostris	+			+	+	+		+
Leydigia acanthocercoides	+							
* *Oxyurella singalensis*	+							
Pleuroxus aduncus	+							
* *Pleuroxus denticulatus*								+
Daphnidae								
Ceriodaphnia sp.					+			
Ceriodaphnia cornuta	+	+	+	+	+	+		+
Ceriodaphnia pulchella				+				
Ceriodaphnia reticulata				+				
* *Daphnia similis*		+						
Daphnia lumholtzi		+	+					
Scapholeberis kingi		+	+	+	+			
* *Simocephalus vetulus*				+				
Simocephalus latirostris	+	+		+				
Macrothricidae								
Ilyocryptus spinifer				+				
Macrothrix hirsuticornis	+							
Macrothricidae	+			+		+		
Ilyocryptus spinifer	+					+		
* *Macrothrix* cf. *triserialis*					+			
Steblocerus pygmaeus	+							
Moinidae								
Moinodaphnia macleayii				+				+

Table 1. (cont.)

Species	Habitat							
	River	Large reservoirs				Tank	Pond	Field
		High	Deep	Mod.	Shallow			
Moina macrocopa				+				+
Moina micrura	+	+		+	+	+		+
Sididae								
Diaphanosoma sp.				+	+			
Diaphanosoma excisum	+	+	+	+	+			+
* *Diaphanosoma* cf. *modigliani*				+	+	+	+	
Diaphanosoma sarsi	+	+		+				+
* *Diaphanosoma* cf. *volzi*				+				
Latonopsis australis				+			+	
* *Pseudosida* cf. *bidentata*								+
III. **Copepoda**								
CYCLOPOIDA								
Cyclopidae								
Acanthocyclops vernalis				+				
Cyclops nanus				+				+
Ectocyclops sp.	+							
* *Macrocyclops* cf. *distinctus*	+							
* *Mesocyclops* cf- *thermocyclopoides*		+		+	+	+		
Microcyclops bicolor			+	+				
Microcyclops varicans		+		+				
* *Paracyclops* cf. *fimbriatus*	+			+				
Thermocyclops cf. *crassus*	+	+	+	+	+	+	+	+
Tropocyclops prasinus	+			+				
Ergasilidae								
Ergasilus sp.				+	+			
CALANOIDA								
Diaptomidae								
Allodiaptomus raoi var. *membranigera*				+				
Diaptomus chaffonjoni	+	+						
Diaptomus javanicus		+						
Neodiaptomus botulifer			+	+	+			
Neodiaptomus malaysianus	+							
Neodiaptomus memphistopheles	+							
Neodiaptomus schmackeri			+	+	+	+	+	+
Phyllodiaptomus annae	+							
Temoridae								
Eurytemora sp.						+		
HARPACTICOIDA	+			+		+	+	
IV. **Ostracoda**								
Cypridae								
gen. sp. div.					+			+
V. **Brachiura**								
Argulidae								
Argulus sp.				+				
Total	35	18	25	65	35	30	26	68

High = High altitude reservoir (> 759 m a.s.l.)
Deep = Deep reservoirs
Mod. = Moderately deep reservoirs
Shallow = Shallow reservoirs
Tank = Irrigation Tanks
Field = Trichogaster fish fields
* = Not previously reported from Thailand

Fernando (1980a, 1980b) discussed the composition of the freshwater zooplankton from tropical Asia. The present records from Thailand include typical limnetic species such as *Ceriodaphnia cornuta, Moina micrura, Diaphanosoma excisum, Thermocyclops crassus* and *Mesocyclops thermocyclopides* and rotifers belonging to the genera *Brachionus, Hexarthra* and *Filinia.* However, it appears that many litoral species know from neighbouring regions have not been recorded yet from Thailand.

There are distinct circum tropical features in the zooplankton of Thailand. *Daphnia* is very rare (see also Dumont, 1980) and there are no carnivorous Cladocera of the genera *Polyphemus* and *Leptodora.*

In comparison with Malaysia, Sri Lanka and Singapore which have similar biotope and have been studied in detail, the 150 species reported here indicate that much remains to be uncovered.

A great additional effort will thus be required to get a comprehensive picture of the zooplankton of Thailand.

Acknowledgement

I thank Khun Waravudh Waravudh Jok-Ngoen, Khun Sumalee Dulyaanukit, Khun Nikom Poothamnong, Khun Rachadaporn Kittiworachet and Khun Tanasee Tubkratoke for their enthusiasm and willingness to help preparing this report. Khun Sutanya Ratanaphun I thank for typing and Dr. Thirapan Bhukaswan for his guidance. Prof. C. H. Fernando of Waterloo University and Prof. B. Dussart, Universite de Paris VI, for kindly reviewing the species listed in this paper. Prof. C. H. Fernando critically reviewed the manuscript.

References

Boonsom, J., 1970. Preliminary report of freshwater plankton found in Thailand. Fish. Res. Unit annu. Rep., Inld Fish. Div., Dep. Fish., Bangkok: 191-222. (in Thai).

Boonsom, J., 1971. Hydrobiological and fisheries survey in Nong Srang Kam Community fish pond. Fish. Res. Unit annu. Rep. Inld Fish. Div., Dep. Fish., Bangkok: 99-104. (in Thai).

Boonsom, J., S. Boonyaratpalin, P. Yama & W. Seekong, 1972. Hydrobiological and fisheries surveys in Pattani River. Fish. Res. Unit annu. Rep., Inld Fish. Div., Dep. Fish., Bangkok: 62-67. (in Thai).

Bricker, K. S., L. Wongrat & J. E. Gannon, 1978. Composition and distribution of crustacean plankton in twelve inland water bodies of Thailand. Kasetsart Univ. Fish. Res. Bull. 10. Bangkok, Thailand.

Brooks, J. L., 1957. The systematics of North American Daphnia Mem. Conn. Acad. Arts Sci. 13: 180 pp.

Brooks, J. L., 1963. Cladocera. In: W. T. Edmondson (ed.), Freshwater Biology, 2nd edn. John Wiley & Sons, N. Y.: 587-656.

Chengalath, R., C. H. Fernando & W. Koste, 1973. Rotifera from Sri Lanka (Ceylon) 1. The genus Lecane with descriptions of two new species. Bull. Fish. Res. Stn Sri Lanka 24: 13-27.

Chengalath, R., C. H. Fernando & W. Koste, 1973. Rotifera from Sri Lanka (Ceylon) 2. Further studies on the Eurotatoria including new records. Bull. Fish. Res. Stat. Sri Lanka 24: 29-62.

Chengalath, R., C. H. Fernando & W. Koste, 1974. Rotifera from Sri Lanka (Ceylon) 3. New species and records with a list of Rotifera recorded and their distribution in different habitats from Sri Lanka. Bull. Fish. Res. Stat. Sri Lanka 25: 83-96.

De Ridder, M., 1970. Raderdieren uit het verre Oosten. Biol. Jaarb.

Dumont, H. J., 1980. Zooplankton and the Science of Biogeography: the example of Africa. In: W. C. Kerfoot (ed.), Evolution and Ecology of Zooplankton Communities. University Press New England, Hanover (N.H.), Lond.

Edomondson, W. T., 1963. Rotifera. In W. T. Edmondson (ed.) Freshwater Biology, 2nd edn. John Wiley & Sons, N. Y.: 420-494.

Egat, 1981. Summary Report. EGAT Hydro-electric Project, Basic data (Project commisioned before 1982), Electricity Generating Authority of Thailand. Rep. 846-2401.

Fernando, C. H., 1969. A guide to the freshwater fauna of Ceylon. Suppl. 3, Bull. Fish. Res. Stat. Ceylon 20: 15-25.

Fernando, C. H., 1980a. The freshwater zooplankton of Sri Lanka, with a discussion of tropical freshwater zooplankton composition. Int. Revue ges. Hydrobiol. 65: 85-125.

Fernando, C. H., 1980b. The species and size composition of tropical freshwater zooplankton with special reference to the Oriental Region (South East Asia). Int. Revue ges. Hydrobiol. 65: 411-426.

Fernando, C. H. & N. P. Zankai, 1981. The Rotifera of Malaysia and Singapore with remarks on some species. Hydrobiologia 78: 205-218.

Fernando, C. H. & J. E. Ponyi, 1981. The freeliving Cyclopoid Copepod (Crustacea) of Malaysia and Singapore. Hydrobiologia 78: 113-123.

Harding, J. P. & W. A. Smith, 1974. A key to the British freshwater Cyclopoid and Calanoid Copepods. 2nd eds., sci. publ. 18. Freshwat. biol. Ass. Ambleside, Westmorland, 56 pp.

Idris, B. A. G. & C. H. Fernando, 1981. The Cladocera of Malaysia and Singapore with remarks on some species. Hydrobiologia 77: 233-256.

Kamolrat, P. & W. Leelapatra, 1973. Survey of plankton in Lam Pao irrigation tank, Kalasin Province. NE. agric. Center annu. Rep. Dep. Fish: 17-34 (in Thai).

Koste, W., 1978. Rotatoria. Die Rdertiere Mitteleuropas. Berlin, Stuttgart, 2, 673 pp.

Lai, H. C. & C. H. Fernando, 1978. Redescription of *Neodiaptomus botulifer*. Kiefer and one of its related species. (Calanoida: Copepoda). Hydrobiologia 59: 229-235.

Lai, H. C. & C. H. Fernando, 1981. The freshwater Calanoida Crustacea of Thailand. Hydrobiologia 76: 161-178.

Mamaril, A. C. & Fernando, C. H., 1978. Freshwater zooplankton of the Philippines, Rotifera, Cladocera and Copepoda. Nat. appl. Sci. Bull. Manilla 30: 109-221.

Mizuno, T., 1968. Illustration of freshwater plankton of Japan. Hoikusha, Osaka, 351 pp.

Office of the National Economic and Social Development Board, Kingdom of Thailand, 1981. Rehabilitation and Development of Bung Boraped and Nong Han, 1. Bung Boraped.

Office of the National Economic and Social Development Board, Kingdom of Thailand, 1982. Rehabilitation and Development of Bung Boraped and Nong Han, 2. Nong Han.

Pennak, R. W., 1953. Freshwater Invertebrates of the United State. The Ronal Press Company, N. Y.: 350-421.

Pothipitak, K., 1971. Hydrobiological and Fisheries Surveys of Meklong and Kwae River. Fish. Res. Unit annu. Rep. Inld Fish. Div., Dep. Fish., Bangkok, Thailand, 63-78 (in Thai).

Royal Irrigation Department, 1970. Lam Pra Plerng Project. Nakhon Rajasima Province. Planning Div. R. I. Dep. Bangkok (in Thai).

Royal Irrigation Department, 1970. Special paper on the dedication ceremony of the Lam Takong Dam in Nakhon Rajasima Province, 16 pp (in Thai).

Royal Irrigation Department, 1974. Nam Oon Project. Sakelnakorn Province. Planning Div. R. I. Dept, Bangkok (in Thai).

Royal Irrigation Department, 1974. Special paper on the dedication ceremony of the Vajiralongkorn Dam in Kanchanaburi Province, 11 pp. (in Thai).

Sangkhakul, S., 1971. Hydrobiological and Fisheries surveys in Lum Takong reservoir. Annu. Rep., Inld Fish. Div., Dep. Fish., Bangkok, Thailand: 51-62 (in Thai).

Sangkhakul, S., V. Srisuwannataj, W. Pokaew & W. Martjootaseranee, 1972. Hydrobiological and fisheries survey in Sirindhorn reservoir. Fish. Res. Unit annu. Rep., Inld Fish. Div., Dep. Fish., Bangkok: 12-29 (in Thai).

Sangkhakul, S., 1974. Zooplankton in Bhumibol Reservoir, Tak Province. M. Sci. Thesis, Kasetsart Univ., Bangkok, Thailand, 51 pp. (in Thai).

Sangkhakul, S., 1975. Plankton in Kang Krachan Reservoir. Fish. Res. Unit annu. Rep., Inld. Fish. Div., Dep. Fish., Bangkok: 40-56 (in Thai).

Scourfield, D. J. & J. P. Harding, 1966. A Key to the British Species of Freshwater Cladocera, 3rd edn. Scient. Publ. 5. Freshwat. biol. Ass., Ambleside, Westmorland, pp. 55.

Sibke, A., 1971. Hydrobiological and fisheries surveys of Rong Hua Chang reservoir. Fish. Res. Unit annu. rep., Inld Fish. Div., Dep. Fish., Bangkok, Thailand: 79-98 (in Thai).

Sibke, S. & A. Nujsombat, 1974. Hydrobiological and fisheries surveys in Chulabhorn reservoir. Fish. Res. Unit annu. Rep. Inld Fish. Div., Dep. Fish., Bangkok, Thailand: 50-60 (in Thai).

Srisuwanataj, V., 1971. Hydrobiological and Fisheries surveys in Lum Dom Noi reservoir. Fish. Res. Unit annu. Rep., Inld Fish. Div., Dep. Fish., Bangkok, Thailand: 37-50 (in Thai).

Tressler, W. L., 1963. Ostracoda. In: W. T. Edmondson (ed.), Freshwater Biology, 2nd edn. John Wiley & Sons, N. Y.: 657-734.

Victor, R. & C. H. Fernando, 1979. The freshwater ostracods (Crustacea: Ostracoda) of India. Rec. zool. Sci. India 74: 147-242.

Waew-Ngam, B., 1973. Limnological and Fisheries survey in Sirikit reservoir. Fish. Res. Unit annu. Rep., Inld Fish. Div., Dep. Fish., Bangkok, Thailand: 24-41 (in Thai).

Wilson, M. S. & H. C. Yeatman, 1963. Free-living copepods. In: W. T. Edmondson (ed.), Freshwater Biology, 2nd edn. John Wiley & Sons, N.Y.: 735-737.

Tropical lakes – functional ecology and future development: The need for a process-orientated approach

Jens Petter Nilssen
Zoological Institute, University of Oslo, Blindern, Oslo 3, Norway

Keywords: tropical zooplankton, lake typology, planktivorous fish, evolutionary ecology, life histories, photyplankton loss rates, environmental problems

Abstract

The major classes of tropical lakes include shallow, lowland lakes; deep, tertiary lakes; high altitudinal lakes; rainforests lakes; and man-made lakes at all latitudes and altitudes. Basic ecological processes are similar in temperate and tropical lakes, including grazing, competition, predation and abiotic adaptation. Small tropical lakes of intermediate age are probably not biotically more complicated than similar-sized temperate lakes. The structure of the areas of adaptative radiation and the dispersal ability of the species are important for the present distribution of taxa. Fish play a key role in the tropics since many species both consume zooplankton and compete with them for algal and pelagic sestonic food. This important co-evolution between fish and algae, leaving a fraction of the algal community with a predation refuge, may have decreased the ability of zooplankton to exploit algae. In addition, heavy predation from juvenile and adult fish may greatly simplify the zooplankton community, and have resulted in the scarcity of Cladocera, notably the efficient filter-feeder *Daphnia*. Little is known of possible physiological constraints to cladoceran distribution, however. Thus similar co-evolution as hypothesized between fish and algae seems not to have occurred to such a great extent between fish and zooplankton. Diurnal patterns in habitat selection of fish may also influence nutrient re-distribution in the tropics as in many temperate lakes. Serious environmental problems threaten tropical lakes, including eutrophication, clear-cutting of the rain forest, unwise introduction of new species not adapted to prevailing conditions, overfishing, extensive use of biocids, and probably acidic rain in areas with poorly buffered waters. Important processes in tropical lakes could be elucidated by concentrating research upon the fate of phytoplankton successional production, involving competition, grazing, sinking, fungi and bacterial attack. Co-evolution of fish and algae should be further investigated as it could in part explain the general scarcity and simplicity of the zooplankton community. Limnocorral experiments should also be used for further assessing processes in tropical lakes.

Introduction

Knowledge of the structural components of tropical lakes and reservoirs has expanded greatly over the past three decades, thanks to detailed studies concentrating on relatively few sites (Lakes Chad, George, Lanao, Valencia, Gatun, Castanho and several other African and South American lakes and reservoirs). As will be seen below, all major classes of tropical lakes have been investigated, except for smaller lakes situated at higher altitudes. Contrary to the common belief of many temperate limnologists, regular seasonal fluctuations in production and life histories are characteristic of tropical lakes (Schmidt, 1973; Beadle, 1974; Brandorff, 1977; Lévêque, 1978; Melack, 1979; Lewis, 1979; Rai & Hill, 1981). Of those lakes investigated so far, only Lake George is a notable exception due to special circumstances, including a regular diurnal wind pattern and the maintainance of a very stable

Hydrobiologia 113, 231–242 (1984).

water level (cf. Beadle, 1974). Since tropical, as temperate aquatic ecosystems, show seasonal patterns, the processes which bring about these changes are important to investigate in detail. I will argue in more detail later that processes in tropical lakes are not significantly different from those in the temperate region, the major difference being the greater speed of the processes in the warm tropics. Since aquatic sciences potentially are of greater economic and socio-cultural importance in tropical Africa and America than in North America or Europe, it is striking how little effort has been invested in studying processes in tropical aquatic ecosystems. Without a detailed knowledge of the processes which bring about seasonal changes in biota and water quality, sound management of lakes and reservoirs seems very difficult. This is much easier to say than to do, however. Tropical areas offer a great number of practical problems not encountered in the temperate region, where facilities are often close at hand. Many parameters and processes which could be used for comparisons have already been described from temperate lakes. The aim of this paper is to: 1) point out major similarities and differences of structures and processes between the two regions; 2) indicate where information from temperate areas already is available and could be used profitably by tropical aquatic ecologists; 3) provide examples of process-orientated approaches which may be possible also in the tropics. This is necessary because tropical aquatic ecosystems today are threatened by a number of serious environmental problems, which must be considered in the near future.

A structural typology of tropical lakes

No general agreement exists as to what is a tropical lake, and the different classes of tropical aquatic ecosystems presented here are based upon the structural design of the water bodies. Advances of knowledge about one type can thus be used to predict future changes in other water bodies within that specific type. Furthermore, I think that many findings from one class can be used only to a limited degree for predicting future events in other types. This division of tropical lakes can therefore be considered a practical one. The following classes were separated: flat-land lakes; old, tertiary lakes; high altitudinal lakes; lakes in tropical rain forests; man-made lakes at all altitudes and latitudes, and border zone lakes (sub-tropical lakes). The division is shown in Fig. 1. Comparisons between tropical and temperate lakes are indicated in Table 1.

Tropical flat-land lakes include lakes subject to considerable water level fluctuations, and many of them have therefore developed into salt-lakes with very high conductivity (see Beadle, 1974). They range from permanent lakes to temporary lakes and ponds. Processes could be compared with temperate ones with respect to loss of nutrients from the sediment, diurnal movements of fish and role of macrophytes. Tropical ponds show great similarities to temperate ones and life histories comparisons can easily be done (Rzóska, 1961; Champeau, 1972). A characteristic feature of tropical flat-land lakes and reservoirs is considerable water level fluctuations, resulting in a major part of the sediments being dry every year (lakes Chilwa, Chad). This

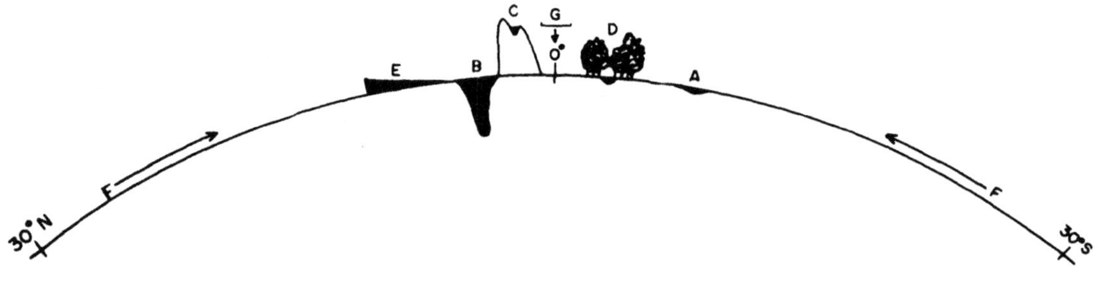

Fig. 1. Major classes of tropical lakes, subdivided according to functional units. *A*: Flat-lands lakes (Chad, Chilwa); *B*: Tertiary lakes (Tanganika, Malawi, Victoria, Rudolf, Albert); *C*: High altitudinal lakes (Titicaca, more examples given in Löffler, 1964); *D*: Lakes in tropical rain-forests (Amazonas, Kongo, South-East Asia); *E*: Man-made lakes; *F*: Border-zone lakes, lakes which often are considered sub-tropical lakes (Lake Kinneret); *G*: Indicate position of true equatorial lakes, they are of all types *A*–*E* above and characterized by nearly identical insolation all year round. Lake types *F* and *G* will not be treated further. Processes in *F* are similar to warm lakes in Southern range of the temperate region and *G* is treated in *A*–*E* (also see Melack, 1979).

Table 1. Tropical lake types and suggested areas and parameters for comparison from temperate lakes.

	Tropical lake type	Temperate type	Study of comparison
A.	Flatland lakes	Flatland lakes (Neusiedler See, Balaton) and other shallow lakes and ponds	Role of sediment, fish distribution, macrophyte ecology, timing of life histories (temperate ponds)
B.	Tertiary lakes	Tertiary lakes (Baikal, Ochrid)	Sympatric speciation, structural complexity, diversity and stability, role of planktivorous fish.
C.	High altitudinal lakes	Arctic, sub-arctic, alpine, sub-alpine lakes	All major processes: competition, predation, role of fish, life histories, fate of phytoplankton production.
D.	Rain forest lakes	Dystrophic lakes	All major processes, like C.
E.	Man-made lakes	Man-made lakes, depending upon class A-D	Major physical processes, depending upon geographical site
F.	Sub-tropical lakes	Border-zone, temperate lakes	All type of processes.

subsequent loss of nutrients from the lake basin probably has great chemical impact on the free waters. The major factor influencing circulation and also production in such lakes, as in many other tropical lakes, is wind rather than seasonal fluctuation of radiation and atmospheric temperature, as in the temperate zone. It should be noted that tropical lakes may be as stable as temperate lakes, provided they are deep enough, because of the density relation of the water (Denny, 1972) and the frequent occurrence of secondary thermoclines (cf. Lewis, 1979). Shallow lakes permanently affected by wind, which set up a diurnal pattern of stratification with breakdown of thermocline at night, show low nutrient loss from the sediments because they are permanently aerobic, thus restricting the number of microorganisms in the break-down processes (cf. Viner, 1975b). It has been suggested that the ecological time-unit in such lakes should be 24 h instead of one year (Viner & Smith, 1973). When considering gross chemical changes, including the circulation pattern, this time-unit could be justified. However, during such a limited time, processes involving adaptation of the biota through natural selection cannot take place as in seasonal succession of species in temperate lakes.

Much work in the tropics has been concerned with the old, tertiary lakes (cf. Beadle, 1974). Over geological time these lakes have produced many species with distribution often restricted to that specific lake, a speciation pattern which has greatly advanced the evolutionary theory (cf. Mayr, 1963; Fryer & Iles, 1972; Beadle, 1974). Since many of the generalizations of tropical lakes originate from the above lakes, it has made comparisons between tropical and temperate lakes difficult, because of an *a priori* assumption that tropical lakes were more complicated than temperate ones. It was seldom stated in which respect they were more complicated: in bacteria, protozoa, crustacea-zooplankton or fish structure, or energy flow, nutrient pattern and life histories. Old lakes in the temperate zone may also be structurally complicated as the African lakes, as are lake regions of similar sizes in Europe close to the Ponto-Caspian region. A further discussion on the ecology of the large African lakes is beyond the scope of this paper, but it is striking how little we know about the processes of these lakes after decades of work. This knowledge is urgently needed because some of the lakes are now seriously threatened by over-fishing which may disturb the delicate balance of ecosystems in these evolutionary and socio-culturally important lakes.

The class of tropical lakes where information is really meagre, is the high altitudinal lake. Very few have been studied in any detail, except for Lake Titicaca, which, because of its considerable age, forms a special case. Only the biogeography has been considered in any detail, and the characteristic cosmopolitan structure of the ecosystem has frequently been mentioned (Thomasson, 1956; Löffler, 1964). Population structure analysis and species delimitation are strong tools in modern population genetics, and the cosmopolitan nature of these populations will probably be strongly questioned (cf. Hebert, 1976). Except for some South American lakes, this lake class is probably of less economic importance than the others, but a thorough investigation of some high altitudinal lakes (although difficult from a practical standpoint) is strongly needed. It would be of great interest to study processes in such lakes, especially those associated with light in the temperate zone, and compare these with temperate, alpine and arctic lakes.

Rain forest lakes have mainly been studied during the last 20 years, and they behave in many cases strikingly similar to dystrophic lakes in the temperate zone. The major seasonal impact is, however, not temperature, but precipitation, flood and light regime in water (cf. Rai & Hill, 1981; Carvalho, 1982b). Phytoplankton and zooplankton in such lakes can easily be compared with temperate, dystrophic lakes both in structure and probably also in function. Phytoplankton is dominated by small, edible species (Unkovish & Schmidt, 1974), and zooplankton structure is similar to many temperate lakes (cf. Brandorff, 1977; Hardy, 1978). There is an urgent need for information on predation patterns in these lakes, but recent investigations show a strong impact from fish predation on zooplankton (Carvalho, 1982a).

There are most probably many more man-made lakes in the tropics than natural lakes (cf. Fernando, 1980), and information on these important aquatic ecosystems for man is still in its early phase. Man-made lakes in the various areas probably behave similar to lakes in that area, if lakes at all are present. In many reservoirs extensive fishing and aquaculture is taking place (cf. Ackermann *et al* 1973) and for proper management and multiuse of reservoirs a study of ecological processes seems urgent.

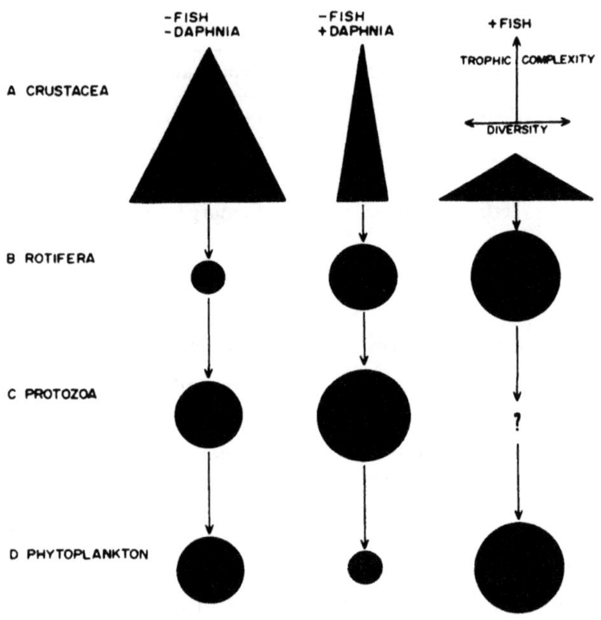

Fig. 2. General characteristics of major components with different combinations of planktivorous fish and *Daphnia* in temperate lakes. Information on *B–D* is scattered and must be considered a first approximation (compiled from literature, own unpublished data and Hessen, 1982). Size of circles indicate abundance of group.

Processes involving fish, competition and predation

Fish comprise a key group in most kinds of lakes, since they determine both the major pathways of energy-flux and the structure of zooplankton communities (Andersson *et al* 1978; Zaret, 1981). The role of planktivorous fish should be considered in context with other key-species, like the filter-feeder *Daphnia*, a major genus in temperate lakes, only affected by low and high pH and heavy fish predation (cf. Nilssen, 1978a; Potts & Fryer, 1979). The distribution of these two groups of animals may have important consequences for the rest of the community (Fig. 2). Heavy fish predation leads to simplified structure of the crustacean community, mainly due to a loss of large invertebrate predators. When fish are absent or found in inferior numbers, the presence of *Daphnia* generally determines the diversity of crustaceans (Nilssen *et al*, in prep.). More *Daphnia* results in lower numbers of other species of crustaceans. Rotifers compete with herbivorous crustaceans, which have been shown in numerous enclosures experiments, where they always exhibit inverse relationships (Nilssen *et al*, in prep.). Thus the condition under which highest abundance of rotifers is found, is when few invertebrates predators are present to suppress numbers of rotifers. The relations between the two groups and protozoans are far more complicated and depend on a number of other factors. It may tentatively be suggested that a combination of fish and *Daphnia* give the largest numbers of protozoans, since high abundance of *Daphnia* often results in high numbers of bacteria, which then in turn could lead to increased abundance of protozoans. The numbers of other microorganisms, including bacteria and phytoplankton, are even more difficult to predict, thus Fig. 2 is only a first approximation. One should remember, however, that Fig. 2 is only a tentative scheme showing the effects on the population structure and abundance of the other groups resulting from interactions between two key-groups in temperate lakes: planktivorous fish and *Daphnia*. In the tropics, the genus *Daphnia* is seldom reported (Fernando, 1980), the reason for which in a few cases is obscure (Lake Lanao), but in many cases it could be due to intensive fish predation. A lake in the Rio Doce Valley, Brazil, with low planktivorous fish predation, had high numbers of large-sized *Daphnia*, whereas other lakes with heavy fish predation had low numbers of small-sized *Daphnia* (dos Santos, 1980). Other less efficient filter-feeders, *Ceriodaphnia, Bosmina* and *Bosminopsis* dominated the Cladocera community. Enclosure experiments excluding fish have led to great increase in abundance of *Daphnia* (Arcifa, pers. commun.). Disappearance of *Daphnia* in the tropics has also been ascribed to increasing fish predation (Green *et al*, 1974). At present there is no reason to postulate any physiological contraint to *Daphnia* distribution in the tropics, but each case should be considered separately to reveal biotic processes that could account for their absence. The sampling of a waterbody once a year should not necessarily lead to the conclusion that *Daphnia* are absent from that locality. One should bear in mind that many species of *Daphnia* in the temperate zone are vulnerable to high and low pH, high accumulation of macrophyte and algae metabolites, filamentous algae, and intensive fish predation. These factors should be considered simultaneously when doing zoogeographical studies in the tropics. A lake in Rio Doce had high numbers of *Daphnia* outside the macrophyte belt, and lower numbers inside it. Another lake completely covered with macrophytes had no *Daphnia* at all (dos Santos, 1980). Both lakes had low fish predation.

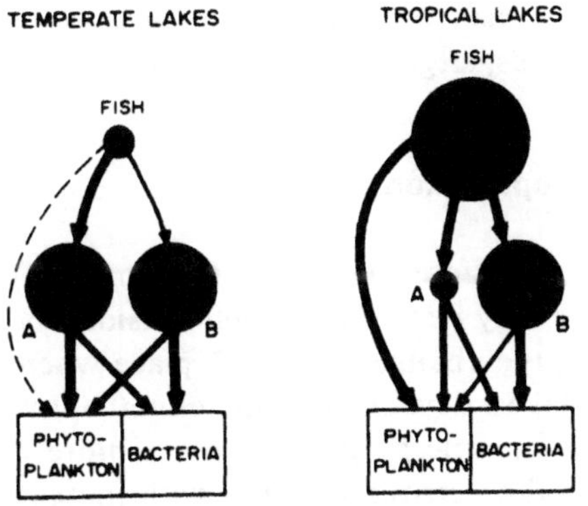

Fig. 3. Difference in role of fish in structuring temperate and tropical lakes at *lower altitudes*. At higher altitudes fish are faced with dispersal problems, fish community is simplified and probably function more like temperate lakes. Size of circle represents standing crop of the unit, and thickness of lines indicates importance of process. *A*: Crustacea; *B*: Rotifera and Protozoa.

The dual role of fish in the tropics

An important difference between the tropics and the temperate region is that phytoplankton-consuming fish are frequent in the tropics (cf. Moriarity *et al*, 1973; Lévêque, 1978). In this way fish both compete with zooplankton for sestonic food, and consume zooplankton (Fig. 3). Often different fish species have different food habits, and different age groups of fishes consume different kinds of food. When algae continuously are subject for heavy consumption by herbivorous fish, they probably have envolved adaptations to decrease this exploitation, like gelatinous sheats, toxic strains and large sizes. This important co-evolution between fish and algae may have decreased the ability of zooplankton to exploit algae, since by size-selective predation by fish they are in a size-range too small to consume large algae. Studies from Lakes Chad, George and Valencia suggest that some zooplankton are able to consume large algal food, however (Gras *et al*, 1971; Moriarity *et al*, 1973; Infante & Riehl, 1982). The role of fish is important to investigate further in tropical lakes and special emphasize should be given diurnal habitat selection. Studies by using quantitative echo sounder and nets in temperate lakes, have shown that fish do a net transport of nutrients into the pelagial of lakes from the sediments and that these nutrients immediately are available for algal growth (Brabrand *et al*, in prep.). The role of fish is therefore not only restricted to determine energy flow but also directly to influence pattern of nutrient paths, in addition to the indirect influence through selective predation on zooplankton. The processes above probably mainly work at lower altitudes, since at higher altitudes fish become faced with the same dispersal problems as in the temperate region, resulting in simplified fish distribution, often dominated by carnivorous, larger-sized fish species.

The eutrophication process

A major difference between temperate and tropical lakes may be realized when considering Fig. 4, showing the changes that take place when a temperate lakes developes from oligotrophy to eutrophy. The energy path becomes more directed towards the detritus pathway with increasing eutrophication (cf. Hillbricht-Ilkowska, 1977), whereas under oligotrophic conditions most energy flows through the grazing pathway, since most algae are edible and easily assimilated. In the tropics algae are often considered too large to be utilized by the small-sized zooplankton. This size structure of algae may not be a common phenomenon (cf. dos Santos, 1980). Even if smaller algae show inferior biomass, their higher turn-over give them a larger share in the production. The differences between tropical and temperate lakes may therefore be minor when all size groups of algae are considered.

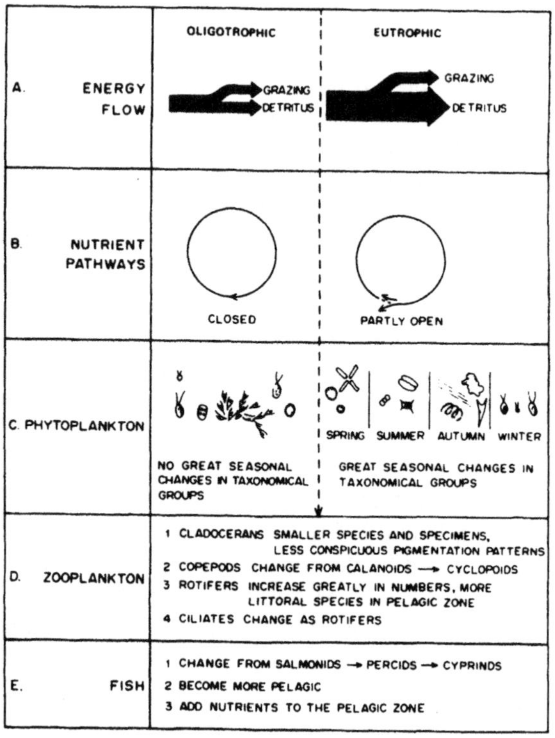

Fig. 4. Major patterns in development of a temperate lake from oligotrophy to eutrophy. Given the great diversity of temperate lakes and the very different biogeography this only aims at giving the major processes. *A*: Energy path changes towards detritus food chain. *B*: nutrient pathways are becoming partly open. *C*: Algae changes from edible to partly inedible species, mainly during late summer and autumn with bloom of large Cyanophyta and dinoflagellates. *D*: Zooplankton exhibit major changes in functional groups. *E*: Both fish behaviour and structure are altered.

Another important change during eutrophication is that the nutrient pathways generally become more open (Fig. 4). We have little information on nutrient processes in tropical lakes, and no general statement can at present be made. A full lake experiment will probably give insight into the changes of nutrient pathways during eutrophication in the tropics (cf. Zaret *et al*, 1981). The phytoplankton in temperate eutrophic lakes show a characteristic pattern with dominance of larger species late in the maximum growth season: late summer and early autumn. These larger species mainly belong to the cyanophyceae or dinophyceae groups. However, oligotrophic lakes seldom show any great seasonal changes in phytoplankton structure, although two smaller peaks may be detected during spring and autumn (cf. Nilssen, 1978b). Within the zooplankton community, major changes can be observed during the eutrophication development mainly due to changes in predation patterns and phytoplankton structure. Zooplankton frequently become smaller in size, more transparent, pigmentation of body and eyes become less conspicuous, calanoid copepods decrease while cyclopoids increase, rotifers increase greatly both in number and species diversity, the same may be seen with the protozoans. In tropical lakes a considerable predation pressure is present (mainly from fish, *Chaoborus* and water mites); thus such a system would probably behave more like a temperate eutrophic system. However, when tropical systems had low biomass of planktivorous fish, *Daphnia* dominated the system and was able to control phytoplankton (more than 90% nanno-algae). Phyto- and zooplankton fluctuated inversely throughout the year, suggesting that zooplankton grazing controlled phytoplankton abundance (dos Santos, 1980). Important changes take place in the fish community with increasing eutrophication in the temperate zone. Species tend to become more pelagic, and there are also qualitative changes in the fish community from salmonids to percids and finally to cyprinids. These comparisons are difficult to extend to the tropics where many of these fish are

Table 2. Important parameters in the zooplankton community and their implications for lake processes (compiled from literature and own unpublished data). In addition frequent sampling at many depth should be done to reveal life histories and species packing.

	Parameter	Taxa*	Give information on:
1.	Body-size	Clad., Cop., Rot.	Predation intensity: fish and invertebrate; food availability
2.	Colour (body, eggs)	Clad., Cop.	Intensity of fish predation; vulnerability to visual predators
3.	Pigmented are (eyes, body	Clad., Cop.	Intensity of fish predation; vulnerability to visual predators
4.	Distribution of morphs	Clad.	Predation intensity; predator habitat-selection
5.	Injuries on body	Clad.	Intensity of invertebrate predation
6.	Egg-size	Clad., Cop., Rot.	Intensity of predation, see Fig. 5 and text for further discussion
7.	Egg-number	Clad., Cop., Rot.	As for 6.
8.	Number of eggs per sample and female	All taxa with eggs	Reproductive strategy; food availability

* Cop.: copepods, Clad.: cladocerans, Rot.: rotifers.

not present. In addition, they often show a dense packing of fish species already. In this respect as well the tropics probably behaves more like a eutrophic lake in the temperate region, and comparison should therefore be made with fish processes in eutrophic lakes in the temperate zone. Moreover, whereas the driving force behind eutrophication in the temperate zone is phosphorus (Vollenweider, 1968), nitrogen seems of greater importance in the tropics (Talling & Talling, 1965; Farnworth & Golley, 1974; Viner, 1973, 1975a; Tundisi & Henry, in prep.). The different biogeochemistry of these two elements and their very different effect on algal physiology obscure a comparison of eutrophication in the two regions.

Life histories and evolutionary ecology

A major difference between publications in the tropical and temperate zone is that data generally are worked up to a much greater extent in the temperate zone; some notable exceptions being recorded however (Lakes George, Chad, Lanao, Valencia, Brazilian lakes and reservoirs). In addition to recording numbers and depth distribution one should also measure morphs of animals, size of individuals, areas of pigmentation, colour of animals, egg size, number of eggs per samples; all parameters which give additional information on predation pressure and life history strategies of the different populations (Table 2). Figure 5 shows how this information could be further used to elucidate the relationships between phytoplankton and zooplankton. The food-abundance curve (phytoplankton) is shown for the strata where the animals are living, but different strategies were recorded for the various zooplankton species. *Bosmina longispina* showed a complicated pattern. When food was low, it produced few, small eggs. During the first part of the spring bloom it produced few, large eggs, but a rapid shift took place in late spring and it produced many, small eggs. During the spring period there were both *K* and *r*-selected clones in the population and they reproduced according to the certainty of the environment to sustain the offspring. During summer the animals produced smaller numbers of medium sized eggs, a *K*-strategy. In early autumn they again produced few, small eggs. During the

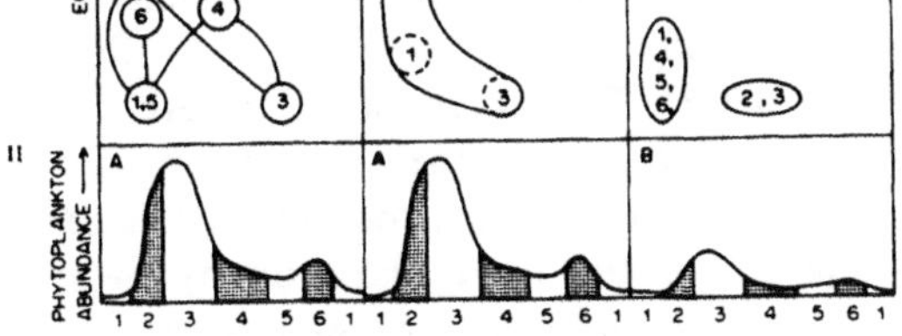

Fig. 5. Relationship between size and number of eggs in Cladocera exploiting richer (A) and marginal food abundance (B). I: Relationship between size and number of eggs, number refer to II. II: shape of food curve (phytoplankton), *1.* winter values, *2.* early spring, *3.* late spring, *4.* summer, *5.* late summer-early autumn, *6.* late autumn. Based on Nilssen (in prep.).

autumn bloom of algae they reproduced as in the early phase of the spring period: few, large eggs. The same patterns as observed for *B. longispina* were not recorded for *Daphnia* spp. or *Holopedium gibberum*, where a gradual trade-off between number and size of eggs was found (Fig. 5). Also here a greater tendency to produce few, larger eggs with food limitation was observed, but the animals produced many, small eggs when food was in excess. The above findings were supported by the reproductive strategy of the species *Daphnia longiremis*, which experienced food limitation by being distributed in the hypolimnion during the major part of the year. It produced mainly few, small eggs. Only when food showed greater abundance did it produce a larger amount of smaller eggs. A detailed analysis of egg size and number would be helpful in the tropics. Zooplankton in the tropics often carry few, small eggs, which may indicate that they are subject for fish selectively removing females with eggs, or that tropical zooplankton actually starve most of the year. The detailed analyses on egg numbers by Brandorff (1977) is interesting. He found decreasing numbers of eggs for many species during the low water phase in a várzea lake in the Amazonia. During this phase, many degradation products increase in the water (Rai, 1979; Rai & Hill, 1982), in addition to increased predation by planktivorous fish (Carvalho, 1982a).

A working strategy involving phytoplankton loss rates, enclosures and fish distribution

Detailed seasonal analyses of the fate of phytoplankton in lakes are scarce (Jassby & Goldman, 1974; Coveney *et al*, 1977; Kalff & Knoechel, 1978). Interesting studies were done by Polish hydrobiologists (e.g. Gliwicz, 1975; Gliwicz & Hillbricht-Ilkowska, 1975). Their major findings are shown in Fig. 6. They recorded the elimination of phytoplankton *in situ* and tried thereby to calculate the elimination of the phytoplankton by measuring other processes in the lake which influenced phytoplankton. This attempt was highly successful in that they were able to predict phytoplankton elimination with great certainty throughout most of the year (Fig. 6). This approach is simple to describe, but difficult to do. This could be one approach to understand the major processes in tropical lakes.

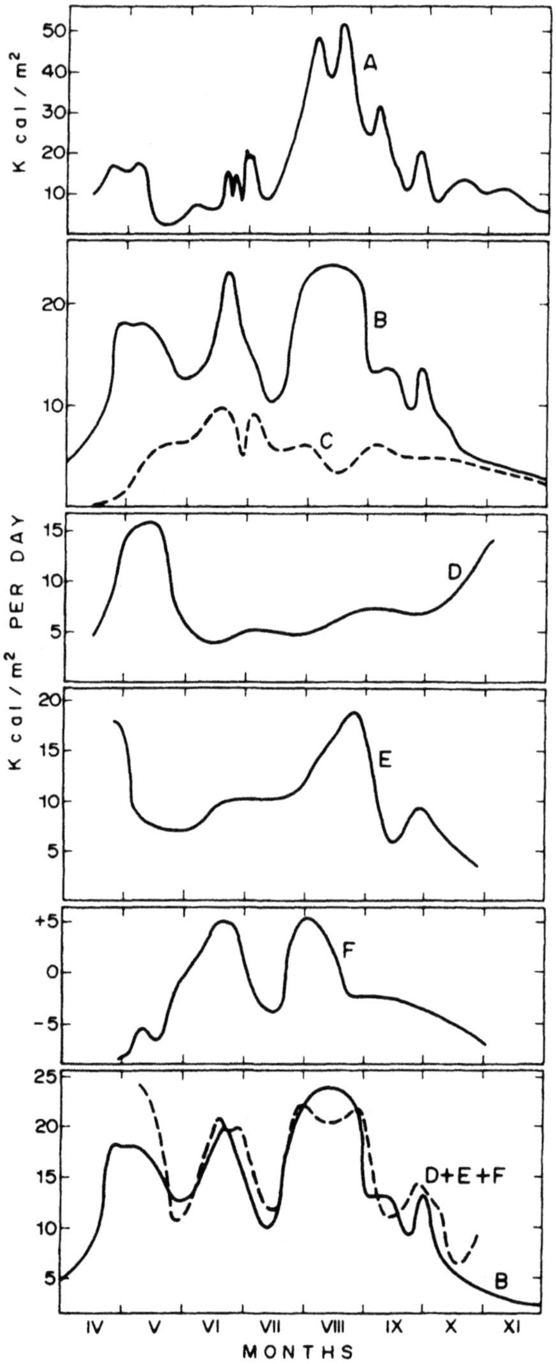

Fig. 6. Elimination of algal organic matter from lake Mikolajskie during the ice-free season. Biomass of algae (A); elimination (B); and the role of grazing (C) by filter feeders; elimination by sedimentation (D); (E): respiration by heterotrophs; and (F) accumulation of non-algal organic matter in the lake. D + E + F are compared with B. (After Gliwicz & Hillbricht-Ilkowska, 1975).

However, with lakes in the Congo and Amazon basin having bacterial production probably exceeding that of the phytoplankton, other working strategies must be used.

Experimental 'macrocosmos' or enclosures are now in common use to reveal lake processes (Lund & Reynolds, 1982; Brabrand *et al*, in prep.). Many of the processes include predation patterns from fish (Brabrand *et al*, in prep.) or elimination of phytoplankton production, as the elegant studies in England (Ferguson *et al*, 1982; Reynolds & Wiseman, 1982; Reynolds *et al*, 1982; Thompson *et al*, 1982). Such enclosures, both pelagic and with sediment contact, should be frequently used in tropics to reveal lake processes.

The keyrole of planktivorous fish, especially juveniles, has recently been assessed by using a quantitative echo-sounder (Borgström *et al*, in prep.). Although this approach mainly works in deeper lakes (⩾5 m), the great abundance of small juvenile and adult pelagic fishes in the tropics (cf. Duncan, 1982) and the great difficulties encountered when trying to catch this size group, imply the necessity of improvements in methods concerning assessment of fish stock, habitat selection and diurnal migration patterns.

Problems threatening tropical lakes

Many problems are threatening tropical lakes; some of the most important are shown in Table 3. Eutrophication is a major concern both for natural and man-made lakes, because it frequently increases the abundance of vectors which transmit serious illnesses. The eutrophication process in the tropics and the temperate region is probably different, since as earlier stated, lakes in the tropics are limited in nitrogen rather than phosphorous. The purification procedures are therefore greatly different.

Introduction of new species can also disturb the delicate food-web structure and balance of many tropical lakes. Management of these lakes is a compromise of many factors (see the approach of Gophen (1982)). Introduction of new species may also introduce new parasites which are harmful to humans. In many African lakes, overfishing or wrong fishing management have become a serious problem. This may be further transformed to zooplankton and phytoplankton structure, so that the whole lake balance is disturbed. With increasing use of biocids, this may have selective influence upon zooplankton and new, unwanted species associations may result. *Daphnia* especially are vulnerable to biocids (Shapiro, 1980). Tropical areas are also subject to another serious environmental problem, extensive damming up of wetlands. Although it is easily understandable that many of the man-made lakes constructed in the tropics were done so for the benefit of the local people or the country, great damage to large areas may result if this expensive damming of wetlands continues. One does not know much on the long-term effects of those extensive human impacts (cf. van der Heide *et al*, 1976; Goldman, 1979). With increasing industrialization of the developing countries, a greater amount of acid pollutants will be transported with the atmosphere to

Table 3. Problems threatening tropical water-bodies (compiled from literature).

	Environmental problem	Prospectus
1.	Eutrophication	Strong pollution (urban, industry); saprobiation (e.g. sugar cane industry); diminished water quality for drinking water and human use
2.	Clear-cutting of forests	Soil damage (a watershed problem); increased non-living seston, decreased primary production (?)
3.	Introduction of new species	Disturbance of ecological stability; introduction of parasites harmful to humans
4.	Over fishing	Declining fish protein for consumption; disturbance of ecological stability: fish-zooplankton-phytoplankton
5.	Use of biocids	Accumulation in organisms including fish used for consumption
6.	Acidification	Fish die-out; increased amounts of heavy metals in biota.
7.	Extensive damming	Destroy major terrestrial wetland biotopes; increase in water-dependent diseases.

remote places, as has already become a catastrophy for many lakes and rivers on the northern hemiphere (cf. Overrein *et al*, 1981). These pollutants may influence fish abundance in poorly buffered lakes as in a large part of Brazil. Simultaneously, they can also give higher productivity in more eutrophic lakes and reservoirs, since the rain transports large amounts of nutrients that are scarce in the tropics: nitrate and sulphate (Sternberg, 1968; de Moraes, 1978). It is therefore interesting to note a general increase in phytoplankton production after rainy periods and also after heavy rain episodes (cf. de Moraes, 1978). Oligotrophic lakes may be vulnerable, however, especially in the Amazonia region.

Conclusions

Research on tropical lakes has mainly been directed towards describing species; only a few studies are available where the structure of the lakes has been investigated in detail. Given the large numbers of environmental problems faced by tropical water-bodies, a process and evolutionary ecology-orientated approach is necessary for a sound management policy. Attention should be given to the fate of phytoplankton elimination during the year, since this would lead to an investigation of a large number of fundamental processes in the systems. Also, the zooplankton should be studied in more detail and in particular the life history strategies should be investigated, because they provide a link in the understanding of fish impact on the systems, and in addition give information on the food quality throughout the season. Fish distribution, quantity, diurnal patterns and nutrient regeneration should be given more attention than up to now.

Acknowledgements

I would like to thank the following people for discussing the ideas presented in this paper: Marlene S. Arcifa, Åge Brabrand, Miriam L. Carvalho, Annie Duncan, Henri Dumont, Bjørn Faafeng, Ernst J. Fittkau, Maciek Gliwicz, Alois Herzig, Dag Hessen, Takako Matsumura Tundisi, Heinz Löffler, Arnold Nauwerck, Fritz Schiemer, José G. Tundisi and S. Twombly. I would like to thank the Norwegian Council for Scientific and Industrial Research (NTNF) for financial support. This study forms a part of a research program on eutrophication partly supported by NTNF, Norwegian Institute for Water Research (NIVA) and the University of Oslo. Special thanks also go to M. Stella Leme for typing several versions of this manuscript, José G. Tundisi for using his library, José Ignácio Bertanha for drawing the figures, Arnold Nauwerck for access to unpublished material on African lakes, and the Brazilians for their hospitality and help during my stay in Brazil.

References

Andersson, G., H. Berggren, G. Cronberg & C. Gelin, 1978. Effects of planktivorous and benthivorous fish on organisms and water chemistry in eutrophic lakes. Hydrobiologia 59: 9-15.

Ackermann, W. C., G. F. White, E. B. Worthington & J. L. Ivens, (eds.), 1973. Man-made lakes: Their problems and environmental effects. Geophys. Monogr. Am. Geophys. Un. William Byrd Press, Richmond, Virginia 17: 1-847.

Beadle, L. C., 1974. The inland waters of tropical Africa, 2nd Edn. 1981. Longman, Lond., 475 pp.

Brandorff, G-O, 1977. Untersuchungen zur Populationsdynamik des Crustaceenplanktons in tropischen Lago Castanho (Amazonas, Brasilien). Thesis, Univ. Kiel, 108 pp.

Carvalho, M. L., 1982a. Influence of predation by fish and turbidity on *Daphnia gessneri* in an Amazonian flood-plain lake, Brazil. This volume.

Carvalho, M. L., 1982b. Efeitos de flutuação do nivel da água sobre a densidade e composiçâo do zooplâncton de várzea de Amazônia, Brasil. Acta Amazônica, in press.

Champeau, A., 1971. Recherches sur l'adaptation a la vie latente des Copépodes Cyclopoïdes et Harpacticoïdes des eaux temporaires Procençales. Bull. Soc. Ecol. 2: 151-167.

Coveney, M. F., G. Cronberg, M. Enell, K. Larsson & L. Olofsson, 1977. Phytoplankton, zooplankton and bacteria-standing crop and production relationship in a eutrophic lake. Oikos 29: 5-21.

De Moraes, E. de M., 1978. Ciclo sazonal, distribuição horizontal e vertical e interrelações ecológicas de nutrientes na Represa do Lobo (Brotas - Itirapina - SP). Thesis, Univ. São Paulo, 153 pp.

Denny, P., 1972. The significance of the pycnocline in tropical lakes. Afr. J. Trop. Hydrobiol. Fish. 2: 85-89.

Duncan, A., 1982. An assessment of factors influencing the composition, body size and turnover rate of the zooplankton in Parakrama Samudra, an irrigation reservoir in Sri Lanka. This volume.

Dos Santos, L. C., 1980. Estudo das populações de Cladocera em cinco lagos naturais (Parque Florestal do Rio Doce -

MG), que se encontram em diferentes estágios de evolução. Thesis, Univ. Fed. São Carlos, 260 pp.

Farnworth, E. G. & F. B. Golley (eds.), 1973. Fragile ecosystems. Evaluation of research and applications in the neotropics. Inst. Ecol. Springer-Verlag, Berlin, 258 pp.

Ferguson, A. J. D., J. M. Thompson & C. S. Reynolds, 1982. Structure and dynamics of zooplankton communities maintained in closed systems, with special reference to algal food supply. J. Plankton Res. 4: 523-543.

Fernando, C. H., 1980. Some important implications for tropical limnology. In Promotion of limnology in developing countries. First Workshop Kyoto, 1980: 103-107.

Fryer, G. & T. D. Iles, 1972. The cichlid fishes of the great lakes of Africa: their biology and evolution. Oliver & Boyd, Edingb., 641 pp.

Gliwicz, Z. M., 1975. Effect of zooplankton grazing on photosynthetic activity and composition of phytoplankton. Verh. int. Ver. Limnol. 19: 1490-1497.

Gliwicz, Z. M. & A. Hillbricht-Ilkowska, 1975. Ecosystem of the Mikolajskie Lake. Elimination of phytoplankton biomass and its subsequent fate through the year. Pol. Arch. Hydrobiol. 22: 39-52.

Goldman, C. R., 1979. Ecological aspects of water impoundment in the tropics. Unasylva, FAO 31(123): 2-11.

Gophen, M., 1982. The impact of zooplankton status on the management of Lake Kinneret (Israel). This volume.

Gras, R., A. Iltis & L. Saint-Jean, 1971. Biologie des Crustacés du lac Tchad, 2. Régime alimentaire des Entomostracés planctonique. Cah. O.R.S.T.O.M., Sér. Hydrobiol. 5: 285-296.

Green, J., S. A. Corbet & L. Betney, 1974. Ecological studies on crater lakes in West Cameroon. Debundsha Lake. J. zool. Soc. Lond. 173: 199-223.

Hardy, E. R., 1978. Composição do zooplancton em cinco lagos da Amazônia Central. Thesis, Univ. Fed. São Carlos, 143 pp.

Hebert, P., 1978. The population biology of Daphnia (Crustacea, Daphnidae). Biol. Rev. 53: 387-426.

Hessen, D., 1982. Eksperimentelle innhegningsforsøk i Gjersjøen med spesiell vekt på zooplanktonets relasjon til fytoplankton bakterier og planktivor fisk. Thesis, Univ. Oslo, (in Norwegian). 141 pp.

Hillbricht-Ilkowska, A., 1977. Trophic relations and energy flow in pelagic plankton. Pol. ecol. Stud. 3: 3-98.

Infante, A. & W. Riehl, 1982. The effect of Cyanophyta upon zooplankton in an eutrophic lake (Lake Valencia, Venezuela). This volume.

Jassby, A. D. & C. R. Goldman, 1974. Loss rates from a lake phytoplankton community. Limnol. Oceanogr. 19: 618-627.

Kalff, J. & R. Knoechel, 1978. Phytoplankton and their dynamics in oligotrophic and eutrophic lakes. Ann. Rev. Ecol. Syst. 9: 475-495.

Lévêque, C., 1978. Ecology of Lake Chad - A tropical shallow lake (Africa). Dr. W. Junk, The Hague.

Lewis, W. M., 1979. Zooplankton community analysis. Springer-Verlag, N.Y., 163 pp.

Löffler, H., 1964. The limnology of tropical high-mountain lakes. Verh. int. Ver. Limnol. 15: 176-193.

Lund, J. W. G. & C. S. Reynolds, 1982. The development and operation of large limnetic enclosures in Blelham Tarn, English Lake District, and their contribution to phytoplankton ecology. Prog. Phytoplankton Res. 1: 1-65.

Mayr, E., 1963. Animal, species and evolution. Harvard University Press, Cambridge, Mass., 797 pp.

Melack, J. M., 1979. Temporal variability of phytoplankton in tropical lakes. Oecologia 44: 1-7.

Moriarity, D. J. W., J. P. E. C. Darlington, I. G. Dunn, C. M. Moriarity & M. P. Tevlin, 1973. Feeding and grazing in Lake George, Uganda. Proc. r. Soc. Lond. B 184: 299-319.

Nilssen, J. P., 1978a. Selective vertebrate and invertebrate predation - some paleolimnological implications. Pol. Arch. Hydrobiol. 25: 307-320.

Nilssen, J. P., 1978b. Eutrophication, minute algae and inefficient grazers. Mem. Ist. ital. Idrobiol. 36: 121-138.

Overrein, L., H. M. Seip & A. Trollan, 1981. Acid precipitation - effects on forest and fish. Final Rep. SNSF-project 1972-1980. Res. Rep. Oslo-Ås, Norway, FR 19/80: 1-175.

Potts, W. T. W. & G. Fryer, 1979. The effect of pH and salt content of sodium balance in *Daphnia magna* and *Acantholeberis curvirostris* (Crustacea: Cladocera). J. comp. Physiol. 129: 289-294.

Rai, H., 1979. Microbiology of Central Amazon lakes. Amazoniana 6: 583-599.

Rai, H. & G. Hill, 1981. Bacterial biodynamics in Lago Tupé, a Central Amazonian water 'Ria Lake'. Arch. Hydrobiol., Suppl. 58: 420-468.

Rai, H. & G. Hill, 1982. Establishing the pattern of heterotrophic bacterial activity in three Central Amazonian lakes. Hydrobiologia 86: 121-126.

Reynolds, C. S. & S. W. Wiseman, 1982. Sinking losses of phytoplankton in closed limnetic systems. J. Plankton Res. 4: 489-522.

Reynolds, C. S., J. M. Thompson, A. J. D. Ferguson & S. W. Wiseman, 1982. Loss processes in the population dynamics of phytoplankton maintained in closed systems. J. Plankton Res. 4: 561-600.

Rzóska, J., 1961. Observations on tropical rainpools and general remarks on temporary waters. Hydrobiologia 17: 268-286.

Schmidt, G. W., 1973. Primary production of phytoplankton in three types of Amazonian waters, 3. Primary productivity of phytoplankton in a tropical flood-plain lake of Central Amazonia, Lago do Castanho, Amazonas, Brazil. Amazoniana 4: 379-404.

Shapiro, J., 1980. The importance of trophic-level interactions to the abundance and species composition of lakes in lakes. In J. Barica & L. R. Mur (eds.), Workshop on hypertrophic lakes, Växsjö, Sweden. Dev. Hydrobiol. 2: 105-116.

Sternberg, H. O., 1968. Man and environmental change in South America. In F. J. Fittkau, J. Illies, H. Klinge, G. H. Schwabe & H. Siolo (eds.), Biogeography and Ecology in South America. Dr. W. Junk, The Hague: 413-445.

Talling, J. F. & I. B. Talling, 1965. The chemical composition of African lake waters. Int. Revue ges. Hydrobiol. 50: 421-463.

Thomasson, K., 1956. Reflections on Arctic and Alpine lakes. Oikos 7: 117-143.

Thompson, J. M., A. J. D. Ferguson & C. S. Reynolds, 1982. Natural filtration rates of zooplankton in a closed system: the derivation of a community grazing index. J. Plankton Res. 4: 545-560.

Uherkovich, G. & G. W. Schmidt, 1974. Phytoplanktontaxa in dem zentralamazonischen Schwemmlandsee Lago do Castanho. Amazoniana 5: 243-283.

Van der Heide, J., P. Leentvaar & J. Meyer, 1976. Brokopondo Res. Rep. Suriname. Part II. Hydrobiology of the man-made Brokopondo Lake. Natuurw. Stud. Suriname Nederl. Antillen, Utrecht. 90: 1–95.

Viner, A. B., 1973. Responses of a mixed phytoplankton population to nutrient enrichment of ammonia and phosphate, and some associated ecological implications. Proc. r. Soc. Lond. B 183: 351–370.

Viner, A. B., 1975a. The supply of minerals to tropical rivers and lakes (Uganda). In Editor, Coupling of land water ecosystems. Ecol. Monogr.: 227–261.

Viner, A. B., 1975b. The sediment of Lake George (Uganda) I: Redox potentials, oxygen consumption and carbon dioxide output. Arch. Hydrobiol. 76: 181–197.

Viner, A. B. & I. R. Smith, 1973. Geographical, historical and physical aspects of Lake George. Proc. r. Soc. Lond. B 184: 235–270.

Vollenweider, R., 1968. Water management research. Scientific fundamentals of lakes and flowing waters, with particular reference to nitrogen and phosphorus as factor in eutrophication. O.E.C.D. DAS/CSI/68.27; 192 pp.

Żaret, T. M., 1981. Predation and freshwater communities. Yale University Press, New Haven and London, 187 pp.

Zaret, T. M., A. H. Devol & A. dos Santos, 1981. Nutrient addition experiments in Lago Jacaretinga, Central Amazon Basin. Verh. int. Ver. Limnol. 21: 689–692.

Influence of predation by fish and water turbidity on a *Daphnia gessneri* population in an Amazonian floodplain lake, Brazil

Mírian Leal Carvalho
Departamento de zoologia – área de ictiologia, Museu Paraense Emílio Goeldi, Caixa Postal, 399, 66000 – Belém, Pará, Brazil

Keywords: tropical zooplankton, predation by fish, *Daphnia gessneri*, Amazonian floodplain, turbidity

Abstract

The population behavior of *Daphnia gessneri* Herbst, 1967 in a floodplain lake (Lago Grande) of the lower Rio Solimões was investigated between April 1979 and March 1980 with regard to 1) predation by the fish called *tambaqui* (*Colossoma macropomum*, Characidae), 2) water level fluctuation and 3) water transparency. Zooplankton density samples were collected at two sites near mid-lake, where water depth and Secchi disc transparency were measured. In addition, qualitative samples of zooplankton and fish collections were taken at several sites in the adjacent floodplain areas. The author concludes that fluctuations in *Daphnia gessneri* populations correlate most with intense predation by fish and water turbidity.

Introduction

Daphnia species have received much attention throughout the world by limnologists and population ecologists. According to Herbert (1978), this interest is explained by the fact that *Daphnia* represents the major component of freshwater zooplankton, and furthermore, their short generation time makes culturing – and thus studying – them fairly easy. Most studies of *Daphnia* have been in temperate regions where environmental conditions are very different than in the lowland tropics. Tropical regions apparently have fewer *Daphnia* species than are found in the temperate latitudes and the mean size of individuals is only about 1.3 mm, which is equivalent to the average length of the smallest species in the latter regions (Fernando, 1980).

In Amazonia the genus *Daphnia* is monotypic and represented by *Daphnia gessneri*. To date little is known about the ecology of *D. gessneri*, though Brandorff (1977) observed that it is one of the most important species in the crustacean zooplankton of the floodplain lake of the lower Rio Solimões that he studied. Brandorff (1977) observed that *D. gessneri* disappears for a period of time each year, and furthermore, that ephippial females are present during the phases of highest population densities; he also found that there is a vertical separation of juvenile stages and adults.

D. gessneri appears to be missing from Amazonian blackwater rivers, though there are no experiments to indicate what physicai and/or biotic factors mitigate against its presence in these waterbodies. The low pH values of blackwater rivers, however, may be an important factor, but this hypothesis needs to be tested.

Zago (1978) studied the vertical migration of *D. gessneri* in the Americana reservoir in the State of São Paulo, Brazil, and concluded that the species presents no daily or seasonal vertical migration patterns. Furthermore, Zago (1976) stated that the structural variations of *D. gessneri* are not pronounced in comparison to temperate latitude species of *Daphnia*.

The principal objective of the present investiga-

Hydrobiologia 113, 243–247 (1984).

tion is to understand the ecology of *D. gessneri* in relation to fish predation and to water level fluctuation and water transparency.

Study area

Lago Grande is a floodplain lake of the right bank of the Rio Solimões (its coordinates are 3°22′ S and 60°35′ W, which places it about 60 km west of Manaus). During the dry season it encompasses about 11.2 km^2 and reaches a maximum length of 7.0 km; during the flood season its area increases to 33.2 km^2 and its length stretches to 11.0 km (measurements from Landsat satellite images of 09/12/76 and 31/07/77 respectively) (Fig. 1). Lago Grande belongs to a floodplain complex known as Manaquiri, but during the annual inundations it becomes connected with neighboring waterbodies to the west and east. The lake remains joined to the Rio Solimões during the entire year by a 35.0 km channel (Paraná do Manaquiri), through which it receives the major part of its water. Because of the length of the connecting channel, most of the suspended material that enters it in water invading from the Rio Solimões, is deposited before it reaches Lago Grande. Thus water reaching Lago Grande is 'decanted' and shows a higher transparency than the Rio Solimões from which it comes. During the low water period, however, the water depth of Lago Grande is reduced to a point where bottom sediments are mixed by wind action and thus the lake becomes turbid.

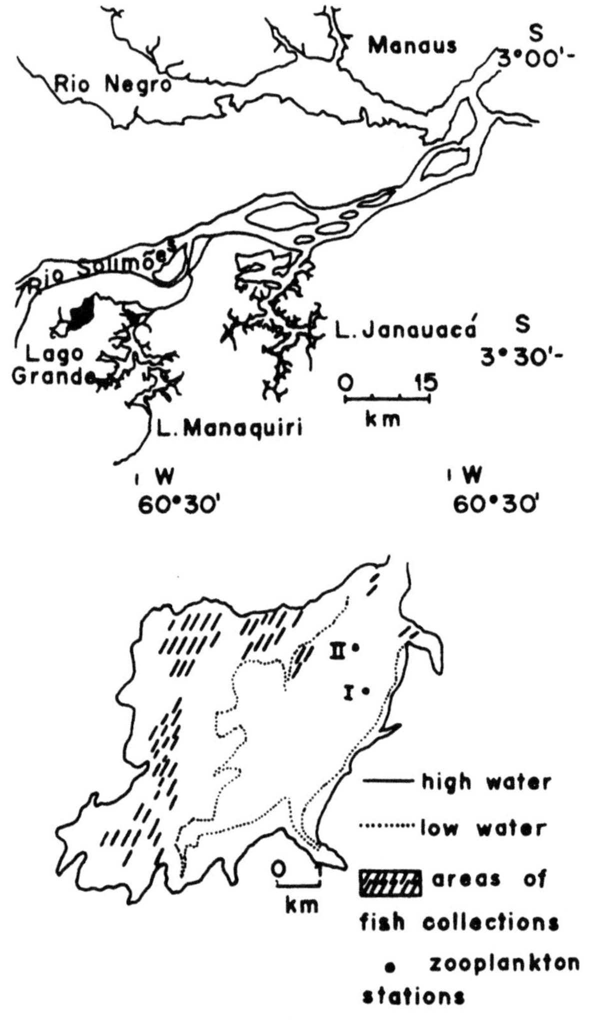

Fig. 1. Geographic position of Lago Grande, zooplankton stations, and areas of fish collections.

Methodology

Fish and zooplankton samples were collected between April 1979 and March 1980 on a monthly basis. The 97 specimens of young *tambaqui* (*Colossoma macropomum*, Characidae) analyzed were captured with gillnets at several sites within an area embracing approximately 5 km^2. Immediately after the capture of the fish specimens, their stomach contents were removed and preserved in 8% formaldehyde solution. Zooplankton identification and counts were realized from sub-samples taken with a 1 or 5 ml Stempel pipette (the choice of the sub-sample sizes was based on the quantity of organisms present in stomach contents).

Zooplankton density samples were collected at two stations located near mid-lake. These samples were taken with a water pump, and the entire water column was investigated. Additional zooplankton samples were collected by vertical hauls in several parts of the floodplain. In both cases a 70 μm plankton net was employed and the samples were preserved in 5% formaldehyde solution.

For the determination of the densities and relative abundances of the zooplankton species, total counts or sub-samples, using a 5 ml Stempel pipette, were made. *Daphnia gessneri* was identified according to Herbst (1967), whereas the other taxa were grouped indistinctly for the purpose of this study. Females with eggs, ephippial females, young (other than eggs), ephippial eggs and free embryos were counted. Water depth and Secchi disc transparency were measured at two of the stations where zooplankton was collected near midlake.

Results

Of the 97 *tambaqui* specimens analyzed, 71 individuals (74%) had eaten *Daphnia gessneri*. Figure 2 graphically represents the relative abundance of *D. gessneri* in the environment in relation to predation by young *C. macropomum*, water transparency and water level fluctuation. The relative abundance of *D. gessneri* was determined by using its monthly average density in relation to the mean densities of the other cladoceran taxa collected at the selected sites. The same was done in relation to stomach contents of fishes.

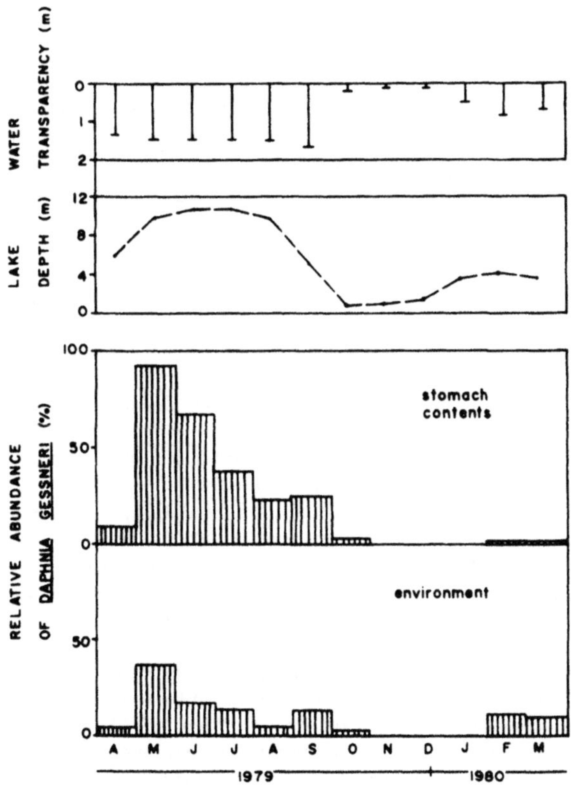

Fig. 2. The relative abundance of *Daphnia gessneri* in the environment in relation to predation by young *Colossoma macropomum*, water transparency and lake depth between April 1979 and March 1980 in Lago Grande.

The highest relative abundance of *D. gessneri* occurred in May (38%), and subsequently the population showed a decreasing relative abundance, until it disappeared completely in November, December, and January. A similar pattern was observed in the stomach contents of *C. macropomum*. During almost every month the relative abundance of *D. gessneri* in stomach contents showed higher values than those found in the environment.

A comparison between the relative abundance of *D. gessneri* in the environment and in stomach contents in relation to water transparency and fluctuation showed that the species disappears when the lake is at its lowest depth (0.80 m) and minimum transparency (0.10 m). As water level rises (4.20 m) and transparency increases (0.80 m) in February, *D. gessneri* appeared anew in the environment (12% of relative abundance) and in the stomach contents of *C. macropomum* (2% of relative abundance). Its relative abundance continued to increase as water transparency improved and water level rose, and attained its maximum in May. The maximum water depth was observed in June and July (11.0 m), whereas the highest transparency values were measured in September (1.70 m), which was the same time when *C. macropomum* consumed the most young *D. gessneri*. The presence of ephippial females and ephippial eggs was observed in May and November, respectively.

Discussion

There appear to be two factors responsible for the disappearance of *D. gessneri* during the low water period (November, December and January), namely, intense predation by fish and low water transparency due to bottom sediments that are mixed throughout the water column by wind action.

According to several authors, the majority of freshwater zooplanktivores are highly selective for *Daphnia* (Brooks, 1968; Galbraith, 1967). Although there were many other plantivores in Lago Grande, young *C. macropomum* appeared to be the most important. Moreover, *C. macropomum* showed a strong preference for *D. gessneri*, as was evidenced in stomach contents analyses. The fact that the relative abundance of *D. gessneri* in the environment was always less than in stomach contents of *C. macropomum*, strongly indicates that the fish ingests zooplankton by selective predation, i.e. rather than indiscriminate filtering of water, the fish actively selects prey.

Investigations of fishes that are selective zooplankton predators have shown that intense predation can cause the temporary disappearance or elimination of a prey species. This phenomenon has been most clearly shown after the introduction of planktivorous fishes into lakes where these taxa did not exist naturally (Brooks & Dodson, 1965; Galbraith, 1967).

Generally, planktivorous fishes capture the largest available prey. This is true both interspecifically and intraspecifically; in the latter case adults are consumed more than young (Brooks & Dodson, 1965; Galbraith, 1967; Green, 1967; Brooks, 1968). According to Zaret (1975), the largest zooplankton individuals possess more body pigments and are thus more easily seen by fishes, and this fact explains why predation on them is heavier than on smaller size-classes or species. Galbraith (1967) states that only the largest mature individuals are eaten, and hence a sufficient number of reproductive females remain to maintain the population. In Lago Grande in September, however, young *C. macropomum* consumed a proportionately high percentage of immature *D. gessneri*, and this may help explain why the zooplankton species disappeared from the environment. The high consumption of young *D. gessneri* may have been due to a scarcity of adults, which in turn disappeared because high water transparency improved visibility – and hence predation – of the zooplankton species by *C. macropomum*.

The decreased water transparency observed in Lago Grande during the low water period was a direct consequence of the mixing of bottom sediments by wind action. According to Moghraby (1977), large quantities of suspended particles in the water column act directly on zooplankton by obstructing their respiratory and natatorial organs, and these negative effects reduce the densities of these animals. Gliwicz & Rybak (1976) are even more specific, and state that a large quantity of suspended material interferes with filtration processes of *Daphnia*, though these particles appear to have no serious effect on species of *Bosmina*. In Lago Grande a high density of *Bosmina* was found during the period of lowest water transparency (low water season), whereas *D. gessneri* was not present (Carvalho, 1981). Brandorff (1977), however, found that in Lago Castanho of the lower Rio Solimões floodplain, the highest densities of *D. gessneri* appeared during the low water period. In Lago Castanho the minimum water depth was about 2.0 m, and transparencies did not reach the low points observed on Lago Grande.

In summary, it is hypothesized that intense predation by fishes, and especially by young *C. macropomum*, gradually reduced the *D. gessneri* population in Lago Grande, and finally, during the low water period, high turbidity subsequent to heavy predation, led to its disappearance in the water column. When water level arose anew and transparency improved, *D. gessneri* appeared again in stomach contents of *C. macropomum*.

Acknowledgments

I wish to thank Dr. Michael Goulding for the English translation and criticism of the manuscript. The research was financed by the Instituto Nacional de Pesquisas da Amazônia (INPA).

References

Brandorff, G. O., 1977. Untersuchungen zur Populationsdynamik des Crustaceenplanktons im tropischen Lago Castanho (Amazonas, Brasilien). Ph.D. Thesis. Univ. Kiel, 108 pp.

Brooks, J. L., 1968. The effects of prey size selection by lake planktivores. Syst. Zool. 17: 273–291.

Brooks, J. L. & S. L. Dodson, 1965. Predation, body size and composition of plankton. Science 150: 28–35.

Carvalho, M. L., 1981. Alimentação do tambaqui jovem (Colossoma macropomum Cuvier, 1818) e sua relação com a comunidade zooplanctônica do Lago Grande-Manaquiri, Solimões-AM. Master Thesis. Univ. Amazonas, 90 pp.

Fernando, C. H., 1980. The freshwater zooplankton of Sri Lanka, with a discussion of tropical freshwater zooplankton composition. Int. Revue ges. Hydrobiol. 65: 85–125.

Galbraith, M. G., 1967. Size selective predation on *Daphnia* by Rainbow trout and Yellow perch. Trans. am. Fish. Soc. 96 (1): 1–10.

Gliwicz, Z. M. & J. I. Rybak, 1976. Zooplankton. In: E. Pieczynska (ed.), Selected Problems of Lake Littoral Ecology. Widawnitwa uni ersytetu Warszaws Kiego, Warszawa, 238 pp.

Green, J., 1967. The distribution and variation of *Daphnia lumholtzi* (Crustacea, Cladocera) in relation to fish predation in Lake Albert, East Africa. J. Zool. Lond. 151: 181-197.

Herbert, P. D. N., 1978. The population biology of *Daphnia* (Crustacea, Daphnidae). Biol. Rev. 53: 387-426.

Herbst, H. V., 1967. Copepoda und Cladocera (Crustacea) aus Südamerika. Gewäss. Abwäss. 44/45 (5): 96-108.

Moghraby, A. J., 1977. A study on diapause of zooplankton in a tropical river - The Blue Nile. Freshwat. Biol. 7: 207-212.

Zago, M. A., 1976. A preliminary investigation on the cyclomorphosis of *Daphnia gessneri* Herbst, 1967, in a Brasilian reservoir. Bol. Zool. Univ. São Paulo, 1: 147-160.

Zago, M. A., 1978. Vertical migration of *Daphnia gessneri* Herbst, 1967, in Americana Reservoir, State of São Paulo, Brasil. Verh. int. Ver. Limnol. 20: 1720-1726.

Zaret, T. M., 1975. Strategies for existence of zooplankton prey in homogeneous environments. Verh. int. Ver. Limnol. 19: 1484-1489.

The impact of zooplankton status on the management of Lake Kinneret (Israel)

M. Gophen
Kinneret Limnological Laboratory, P.O. Box 345, Tiberias 14102, Israel

Keywords: tropical zooplankton, Lake Kinneret, zooplankton, fish management

Abstract

Monthly averages of standing stock wet biomass of zooplankton in Lake Kinneret (Israel) varied between 11 and 76 g m^{-2} during 1969–1981, with the exception of two months. Averaged contributions of different groups were: Cladocera 58%, Copepoda 35% and Rotifera 7%. Total standing crop wet biomass is highest during January–June, averages varied between 35 and 50 g m^{-2}, and decreases during summer–fall (23–36 g m^{-2}). The winter biomass of Cladocera fluctuated between 22 and 35 g m^{-2} and dropped to a range of 9–23 g m^{-2} in summer, whereas copepod biomass varied very little around an average of 18 g (ww) m^{-2} with the exception of low values from April to June. The stock biomass of Rotifera is relatively high during winter floods season (December–March) whilst in summer it is very low.

Young stages of fish in Lake Kinneret feed mostly on zooplankton and zoobenthic forms. The most abundant fish in the Kinneret ecosystem, *Mirogrex terraesanctae terraesanctae*, also feed on zooplankton at the adult stage throughout the year, and herbivorous fish consume zooplankton during the summer when lake plankton resources are limited.

The summer ecosystem of Lake Kinneret is characterised as a 'steady state' type, in which the impact of the zooplankton-chain is of great importance. Increase of predation pressure on zooplankton by fish can disequilibrate the balanced trophic relations existing between nannoplankton production and zooplankton grazing capacity. Such a situation can lead to organics accumulation as nannoplankton blooms, resulting in water quality deterioration. Management options aimed at preventing collapse of zooplankton populations are discussed.

Introduction

The role of zooplankton in freshwater ecosystems was intensively studied experimentally and *in situ*. Investigations on the trophic relations between filter feeding zooplankters, nannoplanktonic algae and fish were already undertaken 40 years ago (Hall *et al*, 1976). Moreover, eutrophication and oligotrophication processes due to changes in grazing capacity of zooplankton affected by fish, resulting in nutrients accumulation as algae, zooplankton or fish biomass were documented. Three main aspects were emphasized in these studies: 1) interrelationships between zooplankton communities and phytoplankton assemblages; 2) trophic relations between zooplankton and their predators, mostly fish; 3) direct and indirect impact of fish on phytoplankton biomass causing a push towards eutrophication in lakes. Consequently, it was concluded that fish population control may be an efficient tool for lakes management. The dependance of the trophic status of lakes on zooplankton grazing capacity were studied by Brooks & Dodson (1965), Nauwerck (1963), Gliwicz & Hillbricht-Ilkowska (1972), Patalas (1972), McNaught (1975), Nilssen (1975), Shapiro *et al* (1975), Lampert (1978), Serruya *et al*

Hydrobiologia 113, 249–258 (1984).

(1980), Geller & Müller (1981), Gophen (1981a), Wynne & Gophen (1981) and Edmondson & Litt (1982), among others.

Impact of fish populations on the quantity and quality of the species composition in zooplankton communities, i.e. elimination of large, visible components with low escapeability of a grazing assemblage, were presented by Hall *et al* (1970), Stenson (1972, 1976), Warshaw (1972), Drenner (1977), Gophen & Landau (1977), Gophen (1979), O'Brein (1979), Henrikson *et al* (1980), Zaret (1980), Drenner & McComas (1982), Drenner *et al* (1982) and others.

Feed-back interactions among three components of a system, namely fish, zooplankton and algae, and their implications on the trophic status were analysed by Hrbacek (1962), Brooks & Dodson (1965) Brooks (1969), Hall *et al* (1970), Stenson (1972), Warshaw (1972), Shapiro *et al* (1975), Anderson *et al* (1978), Stenson *et al* (1978), Henrikson *et al* (1980), Serruya *et al* (1980), Drenner *et al* (1982a, b), among others. It was concluded that the lake management option of regulating fish communities is important and efficient.

In the present paper I wish to analyse the status of zooplankton in Lake Kinneret which resulted in the optimal option for management.

Lake Kinneret – background

General background

Lake Kinneret, the only freshwater lake in Israel, as well as in the whole Middle-East region, is located in the northeastern part of the country. The lake currently supplies 30–35% of the national water supply. Lake water is utilized for agricultural irrigation, industry and drinking. The annual supply of lake water varies between 350 and 500 million cubic meters (MCM) from which 300–400 MCM are pumped and conveyed by the National Water Carrier (NWC) System and 50–100 MCM pumping by local consumers. Thus, lake water quality is an important national concern.

The maximal length of the lake is 21 km, north-south, and maximal width is 12 km directed east-west. The maximal depth is 42 m and average depth is 24 m. The total volume of the lake is 4 000 MCM when water level is 209 m below sea level (BSL). During the last 14 years the water level fluctuated between 208.31 BSL and 211.85 BSL.

Lake Kinneret is a warm monomictic lake which is stratified during 7–8 months. A very stable stratification every year from May to December is due to the high temperatures. During the stratification period the lake is separated into two layers with a very slight exchange between them (Serruya, 1975).

The watershed-area of Lake Kinneret (2 730 km^2) is located mostly north of the lake. External nutrient supply enriched the lake mostly during the winter–spring flood period. Consequently, nutrient concentrations in lake water are relatively high during winter–spring and diminish in summer and fall. So far, the summer fall epilimnetic ecosystem of Lake Kinneret is nutrient-limited. On the other hand the epilimnetic water temperatures are relatively high in summer (27–30 °C), generating high metabolic-energy requirements in secondary and tertiary consumers (Serruya *et al,* 1980).

Phytoplankton (Pollingher, 1978)

The phytoplankton assemblage is characterized by two main groups of algal species: 1) warm-lake species, common in African lakes, and 2) species typical to the flora of European freshwaters. The phytoplankton community structure in Lake Kinneret is dominated by *Pyrrhophyta (Peridinium)* and *Chlorophyta* whilst in other warm lakes the most common groups are *Bacillariophyta* (Diatoms) and *Cyanophyta.*

Under normal conditions, nannoplanktonic algae ($<20\ \mu$) are more numerous than netplanktonic forms. The ratio of nannoplankton to netplankton (nanno.:net.) in terms of cell number is less than 1 during winter when nutrient concentration is high. This ratio increases during summer when nutrients are depleted, *Peridinium* diminish, and phytoplankton biomass is dominated by nannoplanktonic species (Pollingher, 1978).

The Kinneret system is strongly phosphate-limited due to the high pH and high Ca^{+2} concentration (Serruya *et al,* 1974). In spite of this limitation, blooms of *Peridinium* developed during March–June (100–300 g (ww) m^{-2}). In terms of biomass, *Peridinium* contribute approximately 90% of the standing crop and nannoplankton less than 10% (5–20 g (ww) m^{-2}) (Pollingher, 1978). Primary productivity during *Peridinium* blooms in spring is 4 g

· $Cm^{-2}d^{-1}$, whilst in winter it varies between 0.8 and 1.0 g · $Cm^{-2}d^{-1}$ and in summer is 1.8 g · $Cm^{-2}d^{-1}$ (Pollingher & Berman, 1977).

Changes of P concentration may affect the relative contribution of nanno and netplankton. High concentrations of P, observed in 1973–74 resulted in relatively high biomass of nannoplankton (Pollingher & Berman, 1977).

Zooplankton

Zooplankton biomass is very stable, compared to that of phytoplankton. Monthly averages of zooplankton standing crop biomass during 1969–81 varied between 11 and 76 g(ww) m^{-2} with two exceptions: September 1970 (99 g(ww) m^{-2}) and September 1978 (8 g (ww) m^{-2}). During April–July the zooplankton communities are dominated by the cladoceran *Ceriodaphnia reticulata* whilst in other months the biomass of the predator cyclopoid *Mesocyclops thermocyclopoides (= leuckarti)* is similar to that of herbivorous cladocerans. The most abundant cladocerans, *Diaphanosoma brachyurum, Ceriodaphnia reticulata, C. rigaudii, Bosmina longirostris* and *B. longirostris var. cornuta* preferentially consume nannoplankton during most of the year. *Peridinium* is not grazed by cladocerans. This alga can be utilized by *Asplanchna* spp. which is rare in Lake Kinneret. The adult cyclopoids are predators which prey mostly on *Ceriodaphnia* spp. and *Diaphanosoma* sp. and consume detritus mostly in summer, whilst early stages are grazers of nannoplankton and detritus. The most abundant pelagic copepods are *Mesocyclops thermocyclopoides (= leuckarti)* (90%) and *Thermocyclops dybowskii* (10%). The Copepoda comprises 35%, Cladocera 58% and Rotifera 7% of the wet standing stock zooplankton biomass. In terms of biomass, rotifers are mostly abundant during winter-floods period.

Serruya *et al* (1980) pointed out that in the summer, the ecosystem of Lake Kinneret's epilimnion is characterised as a steady-state type. The main components of this ecosystem are nannoplanktonic algae, bacteria, herbivorous zooplankton and zooplanktivorous fish. The algal biomass is controlled by the zooplankton grazing activity, and the nutrient supply by bacterial degradation of zooplankton faeces. The biomass of zooplankton is regulated by fish predation.

Fishery

Nineteen fish species are present in the native ichthyofauna of Lake Kinneret (Ben-Tuvia, 1978). Since 1951, 12 exotic and two native fish species have been stocked into the lake (Ben-Tuvia, 1981). The lake currently supplies approximately 1 800 T (100 kg:ha) of commercial fish annualy (Golani, 1980). The commercial catch is composed of 12 species, whereas in terms of weight and marketing only 8 are important: three cichlids *(Sarotherodon galilaeus, S. aureus* and *Tristramella spp),* two barbels *(Tor canis* and *Barbus longiceps),* Kinneret bleak *(Mirogrex terraesanctae terraesanctae),* mugilids (4 species) and silver carp *(Hypophalmichthys molitrix).* From these 8 species, 4 are intensively introduced each year: mugilids, silver carp, Jordan St. Peter's fish *(S. aureus)* and Gallilee St. Peter's fish *(S. galilaeus).* Mugilids and silver carp are exotic fish and *S. aureus* was a very rare fish in Lake Kinneret prior to introduction (before 1958). *S. galilaeus* and mugilids have the highest commercial value in the catch, whilst the Kinneret bleak have the lowest value. Silver carp has a relatively low commercial value. Difficulties in marketing are mostly due to Kinneret bleak, a non-stocked fish.

Since 1950 there was a significant increase in the annual catch, due mostly to development and advances in fishing operations techniques and equipment and increasing effort: change from cotton to nylon gill nets in the late 1950's expanded fishing hours from the night to include daytime, introduction of purse seines, access to a new fishing area in the northeastern part of the lake (Beteicha lagoons), increasing average boat and gill net sizes, fishing in deeper waters, and introduction of about 30 sonar instruments on smaller boats in addition to the purse seiners (Reich, 1978; Lake Kinneret Fishermen's Organization, pers. commun.).

The impact of advances in techniques, introduction of new equipment, and increased effort on increasing total catch is evident when looking at annual catch values of the non-stocked fish, *Mirogrex* (bleak).

It was found (Drenner *et al,* 1982; Gophen, 1980b; Landau, 1979; Gophen & Scharf, 1981; Zismann, 1976; Ben-Tuvia, 1978) that fingerlings of most species feed mostly on zooplankton and zoobenthic organisms. Spataru (1976, 1978), Spataru & Zorn (1978), Gophen (1980b), Drenner *et al*

(1982), Landau (1979), Gophen & Landau (1977), Gophen & Scharf (1981) demonstrated the important role of zooplankton in the diet of adults of *S. aureus* throughout the year, *S. galilaeus* in summer, *Mirogrex* throughout the year and also silver carp in summer (Gophen, unpubl. data). Spataru (1976) showed the selective feeding on *Peridinium* by *S. galilaeus* as well as the efficient digestion of the algal cells (see also Gophen, 1980b).

Methods

Biomass assessment of phytoplankton (Pollingher, 1978) and zooplankton (Gophen, 1972, 1973, 1975–81, 1978a, b, 1980a, 1981a, b) are based on biweekly routine samples collected from 5 stations.

Information on fishery and fish introduction activities was taken from: Sarid, 1950–1979; Landau, 1974–1981; Reich, 1978; Ben-Tuvia, 1978, 1981; Landau, 1979; Golani, 1980–81; Bar-Ilan and Lake Kinneret Fishermen's organization (pers. commun.).

Food and feeding habits of Lake Kinneret fish were studied by the following:

1) *Sarotherodon (Tilapia) galilaeus (galilea, galilaeum):* Spataru (1976), Ben-Tuvia (1978), Landau (1979, 1974–1981), Gophen (1980b), Drenner *et al* (1982), Gophen *et al* (1982).
2) *Sarotherodon (Tilapia) aureus (aurea):* Ben-Tuvia (1978), Spataru & Zorn (1978), Gophen (1979).
3) *Mirogrex terraesanctae* (Bleak): Gophen & Landau (1977), Gophen & Scharf (1981).
4) Other species: Zismann (1976), Ben-Tuvia (1978), Spataru (1978), Gophen (1979 and unpubl. data).

Results and discussion

Zooplankton biomass (Fig. 1a, b; Table 1)

Results presented in Fig. 1 show the relatively low seasonal and multiannual fluctuations of Copepoda biomass. Conversely, the variations of Cladocera biomass are high seasonally as well as from one year to the next. Results in Fig. 1 indicate that the period of 1969–1981 can be divided into two different cycles: a) 1969–1974 with relatively high biomass of Cladocera and slightly higher biomass of Copepoda compared with b) 1975–1981 cycle with lower biomass for both groups. Thus, two representative years were chosen (1970 and 1981) together with multiannual averages of the monthly means and the data is presented in Table 1. It is noticeable that values of 1970 are mostly higher than the multiannual averages and those of 1981 vice versa. The average of annual averages of 1969–1974 cycle was found to be 42 g (ww) m^{-2} (± S.D.4) and that for the 1975–81 period - 32 g (ww) m^{-2} (± S.D.3). These differences are mostly due to the summer time. Two possible explanations can be given to the trend of decreasing zooplankton biomass which came to the lowest point in 1981: 1) food limitation and 2) increasing fish predation pressure. Nevertheless it was found that nutrient conditions, i.e. P concentrations, were favourable for nannoplankton during 1979–1981 and increase of nannoplankton was observed (Gophen, 1981c, 1982). Moreover it was pointed out (Gophen, 1981a) that the nutritional value of food sources for herbivorous zooplankton was higher during 1975–77 compared to 1972–74 cycle. So far, predation pressure of fish on zooplankton was found to be a more suitable explanation for the decrease of zooplankton biomass in the period of 1975–81. Consequently, such a decrease of zooplankton biomass accelerated the nannoplankton increase by lowering grazing pressure. The difference between Clad-

Table 1. Zooplankton standing stock wet biomass in g(ww) m^{-2}: monthly averages of two representative years, 1970, 1981 (see text) and mean of monthly averages for the period of 1969–1981 (± S.D.).

Month	1970	1981	Averages of monthly averages 1969–1981
1	45	53	47 (15)
2	61	26	46 (12)
3	32	37	50 (16)
4	27	30	42 (11)
5	67	27	42 (14)
6	60	20	35 (13)
7	31	13	23 (6)
8	46	27	27 (9)
9	99	19	32 (24)
10	32	15	25 (8)
11	23	25	31 (11)
12	26	35	36 (8)

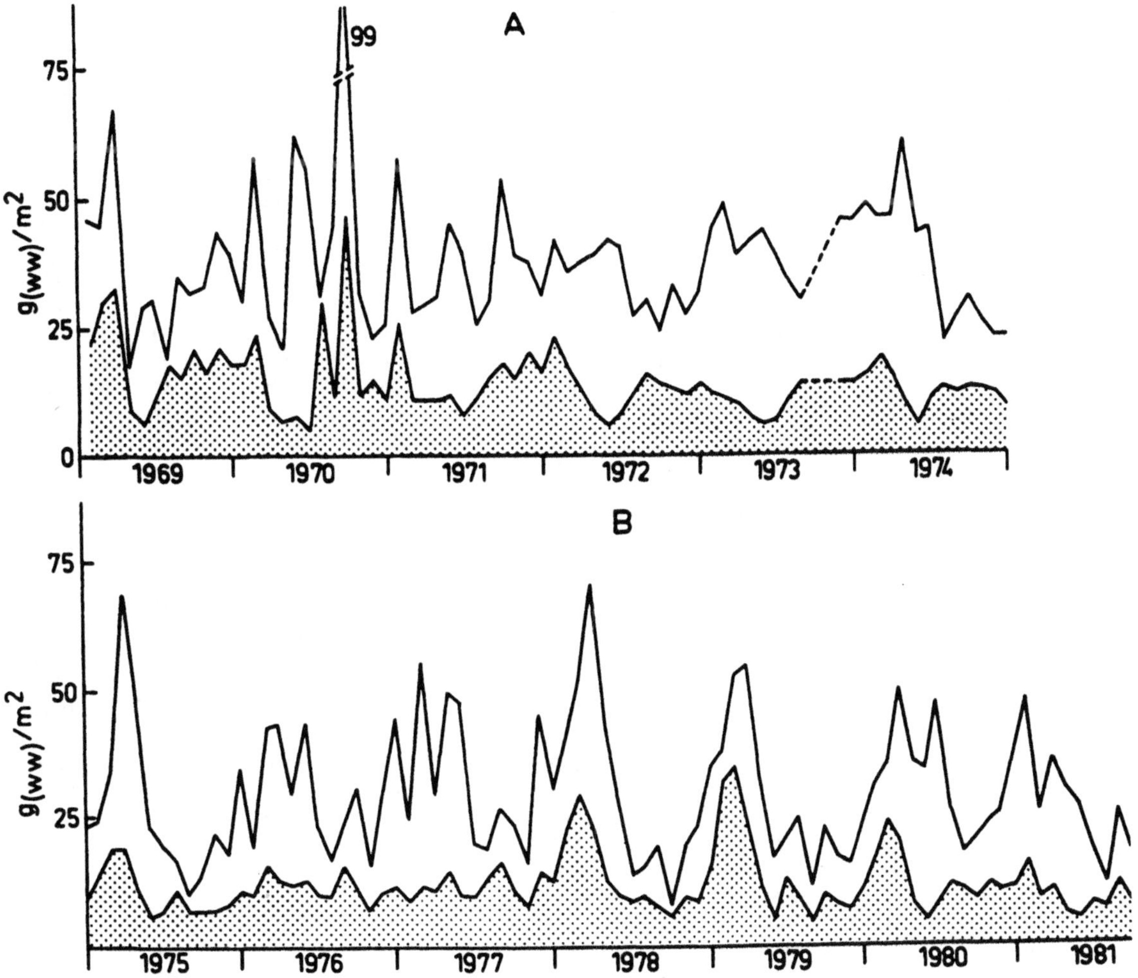

Fig. 1. Monthly averages of zooplankton standing stock wet biomass: a) 1969–1974; b) 1975–1981.

- Copepoda

- Cladocera

ocera and Copepoda biomass variations (Fig. 1a, b) are partly due to the fact that adult copepods are less affected by fish predation than Cladocera because of copepods' relatively high escapeability (Drenner *et al,* 1982).

Fishery and fish introduction (Tables 2, 3)

Information about fishery and stocking activities in Lake Kinneret is presented in Tables 2 and 3. These data indicate that the catch of *S. galilaeus* fluctuated highly, without clear correlation with the stocking activity of this fish (Reich, 1978). It is also noticeable that there was a long-term decrease in the *S. galilaeus* catch from the late 1950's to the middle 1970's. Landau (1979) showed a clear decrease of catch-per-unit-effort for *S. galilaeus.* On the contrary, a clear positive correlation existed between introduction and catch of *S. aureus,* which means that the reproduction of this fish in the lake is very poor (Reich, 1978). Total cichlids catch mostly *S. galilaeus, S. aureus* and small quantities of *Tristramella spp.* shows a high stability: the % of the cichlids catch from the total catch varied be-

Table 2. Lake Kinneret Fishery, 1950–1981: Annual catches in tons and % of the total Catch: *S. galilaeus*, *S. aureus*, total cichlids and total catch. (Sarid, 1950–1979; Golani, 1980–1981; Bar-Ilan, pers. commun.).

Year	*S. galilaeus* Tons	%	*S. aureus* Tons	%	Total Cichlids Tons	%	Mirogrex Tons	%	Total catch
1950	152	25	–	–	179	30	400	67	600
1951	83	12	–	–	96	14	600	86	699
1952	170	19	–	–	203	23	640	72	891
1953	72	11	–	–	89	14	506	81	627
1954	176	22	–	–	195	24	601	74	811
1955	79	12	–	–	93	14	570	87	656
1956	118	14	8	1	166	20	526	63	831
1957	307	30	–	0	343	34	560	55	1017
1958	291	31	–	0	332	36	525	57	925
1959	188	21	–	0	263	30	519	59	885
1960	159	11	3	0	208	15	1016	73	1400
1961	125	11	41	4	220	19	716	63	1142
1962	129	12	31	3	232	21	716	65	1110
1963	154	12	31	2	275	21	932	70	1329
1964	184	15	57	5	300	24	833	67	1249
1965	207	17	30	2	304	24	787	63	1241
1966	213	13	60	4	324	20	969	61	1594
1967	187	10	70	4	339	18	1157	63	1835
1968	230	13	116	6	411	23	1016	56	1820
1969	198	13	116	7	372	24	886	57	1551
1970	176	10	113	6	348	19	1152	63	1830
1971	153	7	159	7	403	19	1209	56	2159
1972	115	7	97	6	274	16	1016	58	1746
1973	101	7	79	5	229	15	969	63	1543
1974	73	5	51	3	193	13	949	65	1467
1975	121	8	84	5	267	17	1017	65	1571
1976	280	13	146	7	460	22	1248	59	2108
1977	182	10	98	6	308	18	1027	59	1735
1978	162	10	116	7	322	19	902	53	1692
1979	267	12	104	5	422	20	1313	61	2139
1980	309	16	83	4	556	28	852	43	1977
1981	224	12	64	10	397	21	965	51	1878

tween 13 and 28% with four exceptional years (1950, 1957–59) with values of 30–36%. The average % of cichlids catch from total catch during 1950–1981 is 21% (± S.D.6). The exceptionally high cichlids catch in 1980 (556 T) is due to an unusual catch of *Tristramella* spp. Moreover a very high correlation, $r^2 = 0.87$, was found between total catch and *Mirogrex* (bleak) catch, i.e. the increase of total catch is mostly due to an increase in the *Mirogrex* catch (Table 2).

The real contribution of the introduction of *S. aureus* to the fishery catch in Lake Kinneret is poor: *S. aureus* displaced *S. galilaeus* (Gophen *et al*, 1982). This actual policy of fishery-management in Lake Kinneret conducted by fisheries biologists was found unsuitable for the system by limnologists. *S. aureus* is a zooplankton feeder (Spataru & Zorn, 1978), and not a natural component of the system as present in current quantities. On the contrary, *S. galilaeus*, a natural component of the ecosystem which can utilize *Peridinium* efficiently (Spataru, 1976; Landau, 1979; Gophen, 1980b) should be, absolutely and relatively, more abundant. This hypothesis of inverse relations between *S. galilaeus* and *S. aureus* can also be supported by the data from 1978–81 (Table 2, 3) which show that no or very low quantities of *S. aureus* introduced during 1978–80 were followed by an increase in the *S. galilaeus* catch in 1979–81.

It is clear from data presented in Tables 2 and 3 that the only source for the catch of *S. aureus*, silver carp and mugilids was the stocked fingerlings. I

Table 3. Fish introduction activity in Lake Kinneret during 1951–1980. Fin = Number of stocked fingerlings in thousands. Catch = annual total catch of mugilids and silver carp. (Sarid, 1950–1979; Golani, 1980–1981; Bar Ilan, pers. commun.).

Year	*S. galilaeus*	*S. aureus*	Mugilids		Silver carp		Total
	Fin.	Fin.	Fin.	Catch	Fin.	Catch	Fin.
1951	243						243
1952	302						302
1953	480						480
1954	590						590
1955	795						795
1956	806						806
1957	650						650
1958	878	114	103				1100
1959	271	223	23				520
1960	366	531	30				960
1961	428	846	66				1370
1962	890	1043	357	2			2330
1963	1677	942	953	2			3570
1964	240	1448	2017	22			3710
1965	1140	1022	1629	56			3790
1966	1315	1035	1475	178			3830
1967	1385	925	5618	221			7930
1968	400	1165	1511	242			3080
1969	430	1925	1183	183	40		3540
1970	695	2449	3618	182	40		6800
1971	2150	1085	629	307	–		3860
1972	2100	1040	7	197	120		3270
1973	1550	250	33	201	634		2470
1974	1865	3185	–	195	1330	11	6380
1975	1270	3854	568	141	665	23	6360
1976	1750	3000	1400	188	800	29	6950
1977	1640	3216	850	202	1860	35	7570
1978	2250	–	770	214	1100	70	4120
1979	2000	–	1100	158	1400	108	4500
1980	1800	400	1200	181	1700	187	5100
1981				175		179	

think that stocking with *S. aureus*, as shown before, as well as with silver carp is not proper. Preliminary results (Gophen unpubl. data) indicated that silver carp consume *Peridinium* in winter–spring and zooplankton in summer–fall. Thus, in summer–fall the silver carp population will increase predation pressure on zooplankton, competing with natural species including *S. galilaeus*. I ask a simple question: Which is more suitable for proper management of Lake Kinneret: The exotic fish *S. aureus* with a relatively low commercial value or the natural species *S. galilaeus*, a very efficient *Peridinium* utilizer with the highest market demand? I suggest the latter. Moreover, unlike *S. galilaeus*, silver carp can be cultured successfully in fish ponds, so, why should it be done in the lake?

The greatest pressure on zooplankton is probably due to *Mirogrex* populations (Gophen & Landau, 1977; Gophen & Scharf, 1981). The fishery policy of this fish has to be included in the overall management program. It should be noted that the increase in the *S. galilaeus* population in the lake aimed at intensifying grazing pressure on *Peridinium* should be accompanied by increased fishing pressure on bleaks. This should initiate a partial channeling of zooplankton resources to *S. galilaeus* instead of *Mirogrex* in summer–fall.

The % of mugilids catch varied between 7 and 14% of the total catch (Table 3). This fish cannot reproduce in freshwater, and is known as an omnivorous feeder and has high commercial value. Nevertheless there is not enough information about its feeding habits in Lake Kinneret to make a correct decision on its status in the management program.

The importance of zooplankton in stratified ecosystems was pointed out by Nilssen (1978), Haney (1973), Porter (1977), and others. In such a system, algae, especially nannoplanktonic species, are strongly affected by zooplankton grazers, and storage of nutrients is in algae, zooplankton and planktivorous fish biomass. Intensive nutrient recycling characterises these systems (Peters, 1975; Serruya *et al*, 1980). Serruya *et al* (1980) and Landau (1979) pointed out the suboptimal conditions for vertebrates during summer conditions in Lake Kinneret. Food limitation is even more prominent in the shallow parts of the lake were fish are spawning, and these biotopes also serve as nursery grounds for the newborn fingerlings. So to say, zooplankton is the most available food source in the pelagic as well as the inshore zone of the lake during summer–fall. On the other hand, the zooplankton grazers prevent water deterioration by eliminating summer–fall nannoplankton blooms by their grazing activity.

If zooplankton food sources are the bottleneck for the growth of adult fish and recruitment of fingerlings in summer–fall, management should be very carefully designed, being aimed at increasing the most valuable (limnologically and commercially) zooplankton utilizing fish.

Serruya *et al* (1980) stated: 'The conditions of "zero net production" prevailing in summer are unfavourable to the development of large animals and most of the fish species are then in suboptimal conditions. In temperate zones, the reduction of algal production corresponds to the period of minimum feeding rate of large consumers; in warm stratified lakes the period of minimum net production corresponds to the period of maximum feeding of consumers.

Wet biomass of 5 000 tons of zooplankton feeding fish (Serruya *et al,* 1980) require approximately 250 T of wet zooplankton per day (5% of body weight) i.e. 1.5 g (ww) $m^{-2}d^{-1}$. Taking 1972 as a representative summer, it follows (Serruya *et al,* 1980): During July–October productivity of Cladocera was 4.2 g (ww) $m^{-2}d^{-1}$, of which 1.2 was consumed by fish and 3.0 by predator copepods. Food requirements of predator copepods were 5.4 g (ww) $m^{-2}d^{-1}$, of which 3.0 came from Cladocera-productivity and 2.4 as detritus from the organic pool; the productivity of copepods (0.3 g (ww) $m^{-2}d^{-1}$) was channeled to fish. The food requirements of grazing zooplankton were 2443 mg C m^2d^{-1}. In other words, the system is structured with 'close boxes' and increasing pressure on one may initiate a disequilibrated process.

For example: addition of fish might increase predation pressure on zooplankton, consequently, Cladocera will be preyed on more than copepods (Drenner *et al,* 1982b); the relative increase of predator copepods will add more pressure on Cladocera; the final result will be a lowering of the grazing capacity of herbivorous zooplankton and accumulation of organic matter as algae production. Management design should take into consideration the reduction of a certain zooplankton feeding fish in order to increase another one. So to say, in order to increase *S. galilaeus* populations, other fish biomass such as that of bleaks, *S. aureus* and silver carp has to be reduced.

During 1973–74 and 1979–81 an increase of P content in the Kinneret waters was observed. In 1973–74 a subsequent increase in nannoplankton and zooplankton was indicated (Pollingher & Berman, 1977; Gophen & Landau, 1977). In 1979–81 summer blooms of nannoplankton were observed but zooplankton biomass was reduced (Fig. 1a, b). It is suggested that in spite of favoured conditions for development of zooplankton, predation pressure by fish prevented this from occuring. The nannoplankton blooms in summer 1981 caused water quality deterioration. In recent years P concentrations in Lake Kinneret are increasing slightly and nannoplankton blooms in summer are predicted. Grazing activity of zooplankton can reduce nannoplankton production. Fish regulations should be aimed at preventing the diminution of zooplankton grazing capacity.

Conclusions

Recommendations for management of Lake Kinneret are composed of two parts: 1) reduction of the external nutrient load, which is not discussed in this paper; 2) regulation of fish populations by fishery policy and fish introduction.

Fish regulation was recommended as follows:

a) Intensive increase of *S. galilaeus* populations. This fish has a unique value in the market, can graze *Peridinium* efficiently and is a natural component of the ecosystem.

b) Eliminate introduction of *S. aureus* which is a zooplankton feeder and suppresses *S. galilaeus;* eliminate stocking of silver carp, an exotic fish with low commercial value, which feeds on zooplankton in summer and can be cultured more successfully in fish ponds.

c) Reduction of the bleaks population in the lake by increasing fishing pressure aimed at decreasing predation pressure on zooplankton.

References

Anderson, G., H. Berggren, G. Cronberg & C. Gelin, 1978. Effects of planktivorous and benthiovorous fish on organisms and water chemistry in eutrophic lakes. Hydrobiologia 59: 9–15.

Ben-Tuvia, A., 1978. Fishes, in Lake Kinneret, Monogr. biol., Dr. W. Junk, The Hague: 407–430.

Ben-Tuvia, A., 1981. Man induced changes in the freshwater fish fauna of Israel. Fish Management 12: 139–148.

Brooks, J. L., 1969. Eutrophication and changes in the composition of zooplankton, in eutrophication: Causes, consequences, correctives, 4. Natn. Acad. Sci., Wash. D.C.: 236–285.

Brooks, J. L. & S. I. Dodson, 1965. Predation body size and composition of plankton. Science 150: 28–35.

Drenner, R. W., 1977. The feeding mechanics of the gizzard shad *(Drosoma cepedianum).* Ph.D. Thesis, Univ. Kansas, 91 pp.

Drenner, R. W., F. deNoyelles & D. Kettle, 1982a. Selective impact of filter-feeding gizzard shad on zooplankton community structure. Limnol. Oceanogr. 27: 965–968.

Drenner, R. W., G. L. Vinyard, M. Gophen & S. R. McComas, 1982b. Feeding behaviour of the cichlid, *Sarotherodon galilaeus:* selective predation on Lake Kinneret zooplankton. Hydrobiologia 87: 17–20.

Edmondson, W. T. & A. H. Litt, 1982. Daphnia in Lake Washington. Limnol. Oceanogr. 27: 272-293.

Geller, W. & H. Müller, 1981. The filtration apparatus of Cladocera: Filter mesh size and their implications on food selectivity. Oecologia 49: 316-321.

Gliwicz, Z. M., 1977. Food size selection and seasonal succession of filter feeding zooplankton in a eutrophic lake. Ekol. pol. 25: 179-225.

Gliwicz, Z. M. & A. Hillbricht-Ilkowska, 1972. Efficiency of the utilization of nannoplankton production by communities of filter feeders measured in situ. Verh. int. Ver. Limnol. 18: 197-203.

Golani, D., 1980-1981. Lake Kinneret. In Israel Fisheries in Figures. State of Israel. Minist. Agric. Dep. Fish.

Gophen, M., 1969-1979. Zooplankton, in Lake Kinneret, C. Serruya (ed.), Rep. 1-9. Israel. oceanogr. limnol. Res. Co. Kinneret limnol. Lab.

Gophen, M., 1972. Zooplankton distribution in Lake Kinneret (Israel) 1969-1970. Israel J. Zool. 21: 17-27.

Gophen, M., 1973. Zooplankton, In T. Berman (ed.), Lake Kinneret Data Record Israel, Sci. Res. Conf.: 61-67.

Gophen, M., 1978a. Errors in the estimation of early stages of mesocyclops leuckarti (Claus). Caused by the diurnal periodicity of egg production. Hydrobiologia 5: 59-64.

Gophen, M., 1978b. The productivity of Mesocyclops leuckarti (Claus) in Lake Kinneret (Israel). Hydrobiologia 60: 17-22.

Gophen, M., 1979. Extinction of Daphnia lumholtzi (sars) in Lake Kinneret (Israel). Aquaculture 16: 67-71.

Gophen, M., 1980a. Artemia naupliias a food source for cyclopoids: Extrapolation of experimental measurements to the metabolic activities of copepods in Lake Kinneret, Israel. In G. Persoone, P. Sorgeloos, O. Roels & E. Jaspers (eds.), The Brine shrimp Artemia, 3. Universa Press, Wetteren, Belg.: 67-76.

Gophen, M., 1980b. Food sources, feeding behaviour and growth rates of Sarotherodon galilaeum (Linnaeus) Fingerlings. Aquaculture 20: 101-115.

Gophen, M., 1981a. Metabolic activity of Herbivorous zooplankton in Lake Kinneret (Israel) during 1972-1977. J. Plankton Res. 3: 15-24.

Gophen, M., 1981b. The metabolism of adult Mesocyclops leuckarti populations in Lake Kinneret (Israel) during 1969-1978. Verh. int. Ver. Limnol., 21.

Gophen, M., 1981c. Ed. Annual Report, 10 (1980). Israel oceanogr. limnol. Res. Co. Ltd. Kinneret limnol. Lab., 70 pp.

Gophen, M., 1982. Ed. Annual Report 11 (1981). Israel oceanogr. limnol. Res. Co. Ltd. Kinneret limnol. Lab., 62 pp.

Gophen, M., R. W. Drenner & G. L. Vinyard, 1982, in press. Fish introduction into Lake Kinneret: Call for concern. Fish. Management 4 (1).

Gophen, M. & R. Landau, 1977. Trophic interactions between zooplankton and Sardine Mirogrex terraesanctae populations in Lake Kinneret, Israel. Oikos 29: 166-174.

Gophen, M. & A. Scharf, 1981. Food and feeding habits of Mirogrex fingerlings in Lake Kinneret (Israel). Hydrobiologia 78: 3-9.

Hall, D. J., W. E. Cooper & E. E. Werner, 1970. An experimental approach to the production dynamics and structure of freshwater animal communities. Limnol. Oceanogr. 15: 839-928.

Hall, D. J., S. T. Threlkeld, C. Burns & P. H. Crowley, 1976. The size efficiency hypothesis and the size structure of zooplankton communities. Ann. Rev. Ecol. Syst. 7: 177-208.

Haney, J. F., 1973. An in-situ examination of the grazing activities of natural zooplankton communities. Arch. Hydrobiol. 72: 87-132.

Henrikson, L., H. G. Nyman, H. G. Oscarson & J. A. E. Stenson, 1980. Trophic changes without changes in the external nutrient loading. Hydrobiol. 68: 257-263.

Hrbacek, J., 1962. Species composition and the amount of zooplankton in relation to the fish stock. Rozpr. csl. Akad. Ved. Rada. M. P. V. 72 (10): 116 pp.

Lampert, W., 1978. Climatic conditions and planktonic interactions as factors controlling the regular succession of spring algal bloom and extremely clear water in Lake Constance. Int. Ver. Theor. Angew. Limnol. Verh. 21: 1436-1440.

Landau, R., 1974-1981. The fish population of Lake Kinneret. Internal Rep. Kinneret limnol. Lab., Israel Oceanogr. limnol. Res. Co. Haifa, Israel.

Landau, R., 1979. Growth and population studies on *Tilapia galilea* in Lake Kinneret. Freshwat. Biol. 9: 23-32.

McComas, S. R. & R. W. Drenner, 1982. Species replacement in a reservoir fish community: Silverside feeding mechanics and competition. Can. J. Fish. aquat. Sci. 39: 815-821.

McNaught, D. C., 1975. A hypothesis to explain the succession from calanoids to cladocerans during eutrophication. Verh. int. Ver. Limnol. 19: 724-731.

Nauwerck, A., 1963. Die Beziehungen zwischen Zooplankton und Phytoplankton im See Erken. Symb. bot. ups. 17 (5): 163 pp.

Nilssen, J. P., 1978. Eutrophication, minute algae and inefficient grazers. Mem. Ist. ital. Idrobiol. 36: 121-138.

O'Brein, W. J., 1979. The predator-prey interaction of planktivorous fish and zooplankton. Am. Scient. 67: 572-581.

Patalas, K., 1972. Crustacean plankton and the eutrophication of St. Lawrence Great Lakes. J. Fish. Res. Bd Can. 29: 1451-1462.

Peters, R. H., 1975. Orthophosphate turnover in Central European lakes. Mem. Ist. ital. Idrobiol. 32: 297-311.

Pollingher, U., 1978. The phytoplankton of Lake Kinneret. In C. Serruya (ed.), Lake Kinneret. Monogr. biol. Dr W. Junk, The Hague 32: 229-242.

Pollingher, U. & T. Berman, 1977. Quantitative and qualitative changes in phytoplankton of Lake Kinneret, Israel, 1972-1975. Oikos 29: 418-428.

Porter, K. G., 1977. The plant-animal interface in freshwater ecosystems. Am. Scient. 65: 159-170.

Reich, K., 1978. Lake Kinneret fishing in its development. Bamidgeh. 30: 37-64.

Sarid, Z., 1950-1979. Lake Kinneret. In Israel Fisheries in Figures. Minist. Agric., Dep. Fish.

Serruya, C., M. Edelstein, U. Pollingher & S. Serruya, 1974. Lake Kinneret sediments: nutrient composition of the pore water and mud-water exchange. Limnol. Oceanogr. 19: 489-508.

Serruya, C., M. Gophen & U. Pollingher, 1980. Lake Kinneret: Carbon flow patterns and ecosystem management. Arch. Hydrobiol. 88: 265-302.

Serruya, S., 1975. Wind, water temperature and motions in Lake Kinneret: general pattern. Verh. int. Ver. Limnol. 19: 73-87.

Shapiro, J. V., Lamavva & M. Lynch, 1975. Biomanipulation: An ecosystem approach to lake restoration. In Water quality management through biological control. Proc. Symp. Univ. Florida Dep. Euviron, Eng.: 85-96.

Spataru, P., 1976. The feeding habits of *Tilapia galilea* (Artedi) in Lake Kinneret (Israel). Aquaculture 9: 47-59.

Spataru, P. & M. Zorn. 1978. Food and feeding habits of (Gervais) (Cichlidae) in Lake Kinneret (Israel). Aquaculture 14: 327-338.

Spataru, P. & M. Zorn, 1978. Food and feeding habits of *Tilapia aurea* (Steindachner) (Cichlidae) in Lake Kinneret (Israel). Aquaculture 13: 67-69.

Stenson, J. A. E., 1972. Fish predation effects on the species composition of the zooplankton community in eight small forest lakes. Rep. Inst. Freshwat. Res. Drottingholm 52: 132-148.

Stenson, J. A. E., 1976. Significance of predator influence on composition of Bosmina spp. populations. Limnol. Oceanogr. 21: 814-823.

Stenson, J. A. E., T. Bohlin, L. Henrikson, B. I. Nilsson, H. G. Nyman, H. G. Oscarson & P. Larson, 1978. Effects of fish removal from a small lake. Verh. int. Ver. Limnol. 20: 794-801.

Warshaw, S. J., 1972. Effects of alewives (Alosa pseudoharengus) on the zooplankton of Lake Wononskopomuc, Connecticut. Limnol. Oceanogr. 17: 816-825.

Wynne, D. & M. Gophen, 1981. Phosphatase activity in freshwater zooplankton. Oikos 37: 369-376.

Zaret, T. M., 1980. Predation and freshwater communities. Yale Univ. Press.

Zismann, L., 1976. Food and feeding of young grey mullets. Prog. Rep. Israel oceanogr. limnol. Res. Co., Haifa, Israel: 1-5.

The free-living inland aquatic nematodes of Africa – a review

Ludo J. Jacobs
Laboratoria voor Morfologie & Systematiek, Rijksuniversiteit Gent, K. L. Ledeganckstraat 35, B-9000 Gent, Belgium

Keywords: tropical zooplankton, aquatic nematodes, zoogeography, Africa, checklist

Abstract

During the past three decades, much attention has been given to free-living nematodes and, in particular, to the species of the African continent. Despite the current absence of evidence for the existence of planktonic species, several conditions are suggested that could lead to intrusion of the planktonic habitat, e.g. turbulent shallow waters, high food densities and preadaptations such as swimming ability and negative geotaxis. A new and more practical ecological classification of the inland free-living nematodes is proposed to avoid further incoherency in this regard. All the free-living inland aquatic species described in Africa are presented here in a checklist, including notes on their ecology and distribution. Many of the cosmopolitan species, included in the checklist, are now increasingly being recognized as conglomerates of species, each species of which occupies a restricted area. Finally the distribution of free-living inland aquatic species in Africa is discussed and some preliminary notes are presented.

Introduction

Nematodes have been collected and identified from most habitats in almost all parts of the world. Especially in the last thirty years, the number of genera and species found has increased enormously, as has the need for taxonomical compilations and reviews of the body of information existing today. Mastering all these data is no longer possible for a single person. As a logical consequence, the few surveys that have been made as yet, always cover only a part of the field of nematology (e.g. Andrássy, 1978, 1981; Barker, 1962; De Coninck, 1965; Goodey, 1963; Heyns, 1971; Hopper & Cairns, 1959; Kir'yanova & Krall', 1968, 1971, 1980; Gerlach & Riemann, 1973, 1974; Mai & Lyon, 1975; Tarjan, 1980).

Studying aquatic nematodes of shallow African waters, I found that the decision made by nematologists on which nematodes are to be included within the definition 'aquatic' is to a large degree arbitrary. To avoid further incoherent use of this term, a definition of a free-living inland aquatic nematode is given.

Definition of free-living inland aquatic nematodes

In dealing with nematodes, two features stand out: the cosmopolitan distribution of most genera and species, and their presence in almost every environment where moisture is available. The occurrence of nematodes depends almost entirely on the presence of water, but food and pore size of the substrate (for benthic and terrestrial forms) are also significant.

Free-living nematodes are quite easy to distinguish from animal parasitic ones by their very different life-cycles, in spite of the usual existence of free-living stadia in the latter (Paramonov, 1962).

The distinction between marine and continental nematodes is a natural one, based on differences in

Hydrobiologia 113, 259–291 (1984).

Table 1. Ecological classification of freeliving nematodes.

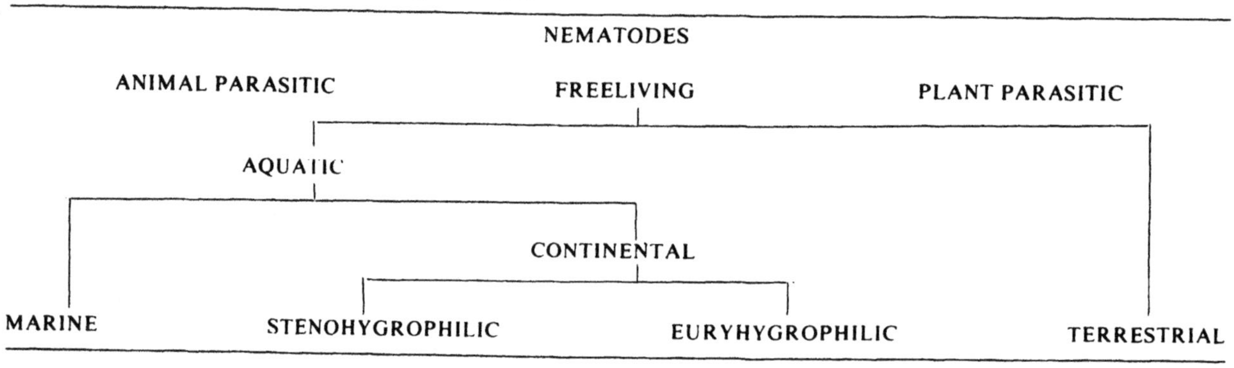

physiology and their ecological and taxonomical consequences (Nicholas, 1975; Zuckerman *et al*, 1971).

The distinction between limnicole and terricole forms, on the contrary, is an arbitrary one, because all nematodes are essentially aquatic. For instance, no sharp distinction can be drawn between nematodes that inhabit banks or shallows of e.g. rivers or lakes, and marshy soil. In spite of this, nematodes inhabiting water-saturated soils are nearly always considered as terrestrial (= soil inhabiting) and not as aquatic. Therefore, a more coherent ecological classification is proposed: stenohygrophilic = aquatic s.s. and euryhygrophilic = semi-aquatic forms, will both be regarded as aquatic, in contrast to soil-inhabiting or terrestrial nematodes (Table 1). This faunal subdivision, although still arbitrary today, may prove to be more practical than the rather elaborate classification proposed by Micoletzky (1922).

Free-living periphytic, epibenthic, benthic (especially in sediment with large pore sizes), bryophilic and planktonic (see, below) nematodes can be ranked as stenohygrophilic. A sizable part of this category is stenotopic, e.g. littoral species associated with algae, and submerged mosses. These two communities are the richest known from freshwater biotopes. Conversely, small rivers, streamlets and oligotrophic lakes have the poorest and often the most uniform nematode fauna (Gadea, 1962). Within the category of stenohygrophilic nematodes and especially in the periphytic and bryophilic species, nematodes are found with particularly long anterior and somatic setae, ocelli or simple pigmented 'eye-spots', well developed amphidial foveae and caudal glands, and filiform tails (Gerlach, 1953; Riemann, 1974; Wieser, 1959; Warwick, 1971). In addition, this ecological group can be split up into functional units according to water velocity (Hirschmann, 1952) and depth (Filipjev, 1929-1930).

Euryhygrophilic nematodes occur in water-saturated soils such as banks and shallows of waterbodies, marshy soil and swamps, phreaton, benthos (sediment with small pore- and/or grain-size) and caves. Faunal lists are given by Andrássy (1962), Micoletzky (1925) and Overgaard (1948). As a result of the variety of terrestrial (interstitial) aquatic habitats, this eurytopic group occupies a midposition in the classification. Most of its representatives have been reported as amphibious or terrestrial and they are characterised by a reputedly cosmopolitan distribution. Several among them are well known for their tolerance to dessication.

The range of soil-inhabiting species is determined by the distribution of terrestrial flora and flora with which they are associated, and which is, in turn, affected by the physical structure and physics of the soil. The nematodes' distribution also depends on their trophic specialization and on their resistance to dessication (Banage, 1962, 1963, 1964, 1966; Bassus, 1962; Deubert, 1959; Domurat, 1970; Johnson *et al*, 1972; Overgaard Nielsen, 1949; Wallace, 1958, 1959, 1971; Wasilewska, 1970; Winslow, 1960; a.o.). Soil nematodes include two groups: plant-parasitic (= phytohelminths) and free-living forms (eusaprobes, dyssaprobes and pararhizobes; Paramonov, 1962).

It is noteworthy that the important group of the dorylaimids, due to their omnivorous (eurytrophic) behaviour, is often not included in ecological classifications of soil nematodes. Those classifications, however, are nearly always centered on types of associations of nematodes with higher plants. For our present purpose, and as stated by Nicholas (1975), only nematodes that show close morphological and physiological associations with host plants should be considered as plant-parasitic, and all other ones as free-living.

Do planktonic nematodes exist?

To date no species of freshwater or marine nematode is known to be confined to the plankton and many nematologists are of the opinion that all planktonic finds must be considered as accidental (Filipjev, 1929). There are certainly arguments that bear out this view. Large numbers of free-living and plant parasitic nematodes have been noticed in streams after heavy rainfall and wherever large amounts of soil are transported by the water (solifluction in rainy seasons). In turbulent shallow waters, the benthic and periphytic nematodes are brought into suspension by force. In a case documented by Dumont & Van Vossel (1972), daytime navigation in a shallow river caused a distinct, unimodal, diurnal rhythm in the planktonic occurrence of nematodes. The planktonic occurrence of eelworms is therefore considered by many nematologists as a dispersing strategy, and species found in plankton samples are stated to represent only ubiquitous aquatic species from the benthos (Table 2; von Daday, 1910a, b; Micoletzky, 1915; Filipjev, 1929; Andrássy, 1964). However, such species as *Eumonhystera fuelleborni, Udonchus monhystera* and *Mactinolaimus hutchinsoni* have been found only in the plankton, although these species have not been recorded more than once or twice in all.

Assuming that nematodes are one of the most successful groups of animals and that they can be found in 'all' kinds of habitats, we may well ask why there are no planktonic species.

We can suggest three conditions that could lead to intrusion of the planktonic habitat: a) the existence of suitable niches; b) the occurrence of preadaptations; c) the presence of special ecological conditions that constitute starting points for invading the plankton.

One outstanding preadaptation is swimming ability. The most familiar example is that of the vinegar eelworm, *Anguillula aceti*. According to Peters (1952), the marked tendency to swim upwards (second preadaptation: negative geotaxis) is due to the rear half of the worm being heavier than the front, so that the head is directed upwards when the worm is at rest.

Borders of water bodies where organic material accumulates can, as stressed by Hischmann (1952), constitute special conditions leading to speciation. Food density is one of the important limiting factors for the occurrence of nematodes. Relatively high food densities (e.g. $6 \cdot 10^8$ *Acinetobacter* sp. cells ml^{-1} agar for *Plectus palustris*, a benthic nematode; Klekowski *et al*, 1979) are necessary to allow nematodes attain a certain age, size and stage of development to complete their life-cycle. There-

Table 2. Nematode species found in African plankton samples.

Nematodes found in African plankton samples		Locality	References
*Paradorylaimus filiformis**	n	Gizeh, Egypt	Daday, 1910a
*Eumonhystera fuelleborni****	1♂	Nyassa Lake, near Cambwe lagune, Tansania	Daday, 1910b
*Udonchus monohystera****	2♀♀	Tychopotamic plankton, Sambesi River, Simbabwe	Micoletzky, 1915
*Alaimus primitivus**	1j	Tychopotamic plankton, Sambesi river, Simbabwe	Micoletzky, 1915
*Eumonhystera vulgaris**	1♀	Baakens River, Port Elizabeth, South Africa	Micoletzky, 1915
*Neoactinolaimus africanus**	1♀	Matsume Lake, Mozambique	Filipjev, 1929
*Mactinolaimus hutchinsoni***	1♂, 1♀, 1♂j, 2jj	Freshwater pan, Weltevreden, South Africa	Filipjev, 1929
	2♂♂, 2♀♀	Rietfontein pan, Johannesburg, South Africa	Filipjev, 1929
*Ironus ignavus**	2♂♂, 2♀♀, 1j	Upper Kami Tarn, Mt. Kenya, Kenya	Andrássy, 1964
	1♂	Lower Kami Tarn, Mt. Kenya, Kenya	Andrássy, 1964
*Ischiodorylaimus gulliver***	1j	Lower Kami Tarn, Mt. Kenya, Kenya	Andrássy, 1964
*Mactionolaimus omercooperi***	1♂, 2♀♀, 6jj	Southern Hall Tarn, Mt. Kenya, Kenya	Andrássy, 1964
	2♂♂, 12jj	Southern Hall Tarn, Mt. Kenya, Kenya	Andrássy, 1964
	1♀	Lake, south of Enchanted Lake, Mt. Kenya, Kenya	Andrássy, 1964

*: cosmopolitan species; **: aquatic species; ***: possible planktonic species.

fore, starting points to invade the plankton will be situated where high food densities occur. These high food densities often occur in turbulent shallow waters. A large species diversity can be found at these ecological triplepoints for stenohygrophilic, euryhygrophilic and terrestrial forms. The food abundance leads to large saprobionthic populations (including rhabditids, diplogasterids, cephalobids) followed by epiphytic (plectids, monhysterids, chromadorids), omnivorous (dorylaims) and predaceous (tripylids and mononchs) forms. It is obvious that at least two dispersal strategies can be followed: the first by enlarging the available living-space by ubiquitous species; the second by speciation to stenotopic/stenovorous species.

Finally, the absence of more information on planktonic nematodes is probably due to the sampling methods. Most of the African samples were not taken by the nematologists who studied them. Probably the meshes of the plankton nets used were too large, allowing the smaller free-living species to slip through the meshes. A study specifically looking for possible planktonic nematodes is not available to date.

Notes on the distribution of free-living aquatic nematodes

Some general remarks on sampling free-living nematodes

In sampling nematodes, special methods are required. Although it is beyond the scope of this paper to present various sampling and fixation techniques, several important remarks deserve to be made.

As has already been noted, plankton samples have so far revealed only a few nematodes. Most of these nematodes are tylenchids or specimens of large-sized species like plant-parasitic dorylaims. It is obvious now that the smaller free-living aquatic forms can be found only accidentally unless meshes of plankton nets are smaller than 20 μm. Samples of mosses and especially submerged algae, usually contain a very rich nematode fauna because, being attached by their caudal glands, the eelworms do not slip through the meshes of the sampling device.

Usually, nematodes are killed in a cold fixative (4% formaldehyde) which often results in coiled specimens with distorted tissues that are difficult to study. Instead, hot fixative (60 °C, 4–6% formaldehyde, fixation on sieves) that relaxes the specimens first, should be used.

Sampled localities with transcribed names, for instance of Arabian or African origin, often present large problems in relocating them. The use of coordinates would avoid this problem.

The importance of biotic and abiotic data is often ignored in taxonomical papers and this information is only exceptionally given in descriptions of new taxa. On the other hand identifications in important ecological studies often remain at genus-level instaed of species-level due to problems of determination (Banage, 1964; a.o.).

Summarizing, well fixed specimens from precisely characterized localities will lead to a solid base for sound zoogeographical hypotheses.

Identification of nematodes

Nematodes are easily found and recognized as such by their very uniform shape and therefore will be familiar to anyone who has examined pond mud or freshwater algal slime. However, as a study subject, nematodes are reputedly 'difficult' or 'impossible' to identify by non-specialists because of the technical problems and the paucity of the type of illustrated keys that is widely used for other animal groups (Coomans *et al*, 1978). Even for specialists, identification to generic and especially to species-level is difficult.

Taxonomy of free-living nematodes started mainly in the second half of the nineteenth century (Dujardin, 1845; Bastian, 1865; Schneider, 1873, 1874; de Man, 1884; von Daday, 1893; Maupas, 1900). These early workers were unaware of the significance of characters that have since been recognized as diagnostic in taxonomic work (Clark, 1962; Coomans *et al*, 1978; Geraert, 1968; Goodey, 1959; Roggen & Asselberg, 1971). Older descriptions are therefore frequently quite incomplete (Andrássy, 1954, 1971a; Baqri & Coomans, 1973; Baqri *et al*, 1975; Coomans *et al*, 1978). The forms referred to in these older descriptions have become doubtful, and/or have been the subject of often unsound and incorrect interpretations, especially when type-material has been lost or is now in bad condition. In this respect, type-material when possible should be revised by (groups of) specialists and the exchange of identified material should be stimulated. As Steiner noted in Tarjan & Hopper (1974): "Synonymizing and desynonymizing thus have become a playground for an ever increasing confusion in nematological nomenclature'. Only during the past three decades has a breakthrough in taxonomic attention towards nematodes taken place and in this respect nematology is a young science.

It is of course quite natural that, following reexamination of the original material, most species described by these early workers, had to be transferred, to other genera, in some cases even to other families or to other suborders. Combining the latter with the taxonomical instability of a still expanding classification of the Nematoda, it is obvious that biogeographical hypotheses have to be formulated very cautiously.

Biogeography of free-living nematodes

It appears that a large number of free-living species and most of the free-living genera of nematodes are cosmopolitan. However, bearing in mind the insufficient collecting and the doubtful identifications in almost all cases (Table 3), it is at least today not possible to prove that the distribution of such species is more restricted than usually maintained.

Generally, a number of these cosmopolitans may in fact have acquired the same reproductive strategies as Rotifera (Dumont, 1980b, 1983): parthenogenesis, drought (and digestion?) resistant stages, rapid maturation upon hatching and short generation times. These strategies contribute to a maximal dispersal of the species, and therefore, cosmopolitanism seems to be a logical consequence of these strategies.

However, recapitulating our fragmentary information, arguments are found which indicate a continental distribution of many species, rather than a cosmopolitan one.

Free-living nematodes are usually the most abundant metazoans inhabiting benthic or soil ecosystems. Although extremely long life cycles (13 to 20 months) at low temperatures (7 °C) have been reported by Gerlach & Schrage (1972), most free-living nematodes maintain short life spans (2–30 days at 18–33 °C: Hopper *et al*, 1973). According to King (1980) and Stanley (1975), parthenogenetic nematodes can effectively generate as much genotypical variation by somatic mutation as can amphigenetic animals by sexual recombination, if they maintain short life spans and large populations (nematodes in the soil: up to 18.1 10^6 m^{-2}: Stockli, 1950; 8.3–20.0 10^6 m^{-2}: Overgaard Nielsen, 1949). Therefore, parthenogenetic nematodes have to be considered as highly adapted to survive changing environments. Because nematodes have a conservative structural organisation, we may well ask if the importance of somatic mutations has not been overemphasized. The organization plan of nematodes (strong cuticle, high turgor pressure system, eutely, etc.) does not seem to allow important departures from the basic form. However, contrary to the apparently inflexible system, the success of the Nematoda in many habitats has depended on minor modifications only. Many examples in which mutations caused different phenotypes have been described (Sturhan, 1970). Mutants of nematodes have been obtained under experimental conditions at increased temperatures (Dougherty & Nigon, 1950; Nigon, 1951) and with chemical mutagens (Nigon & Dougherty, 1954; Sturhan, 1967, 1970).

The strong ability for passive dispersal in parthenogenetic and autogame-hermaphroditic nematodes must have resulted in transportations over long distances in relatively short periods. This mobility must have brought about an intense geneflow between the large number of populations as well as the possibility of a constant emigration towards new and suitable habitats.

According to the vicariance model, biotic distribution is thought to result mainly from subdivision (vicariance) of ancestral biotas with secondary emphasis on pathways of dispersal and migration. In this model, and in accordance with Wegener's hypothesis, subdivision of mesozoic flora and fauna on the landmass of Pangaea was caused by movement of tectonic plates. The movement of tectonic plates induced important climatological changes. However, vicariance as seen by Croizat (1958, 1964) underestimated the significance of natural continental barriers for achieving dispersal patterns. Barriers of higher order (e.g. Pleistocene glaciations and Alpine orogenesis) must have had a large influence on the differential distribution of fauna and flora.

Once the continents drifted apart, the ability for passive dispersal of most of these so-called cosmopolitans must have resulted in a much longer period of faunal exchange between the land masses than

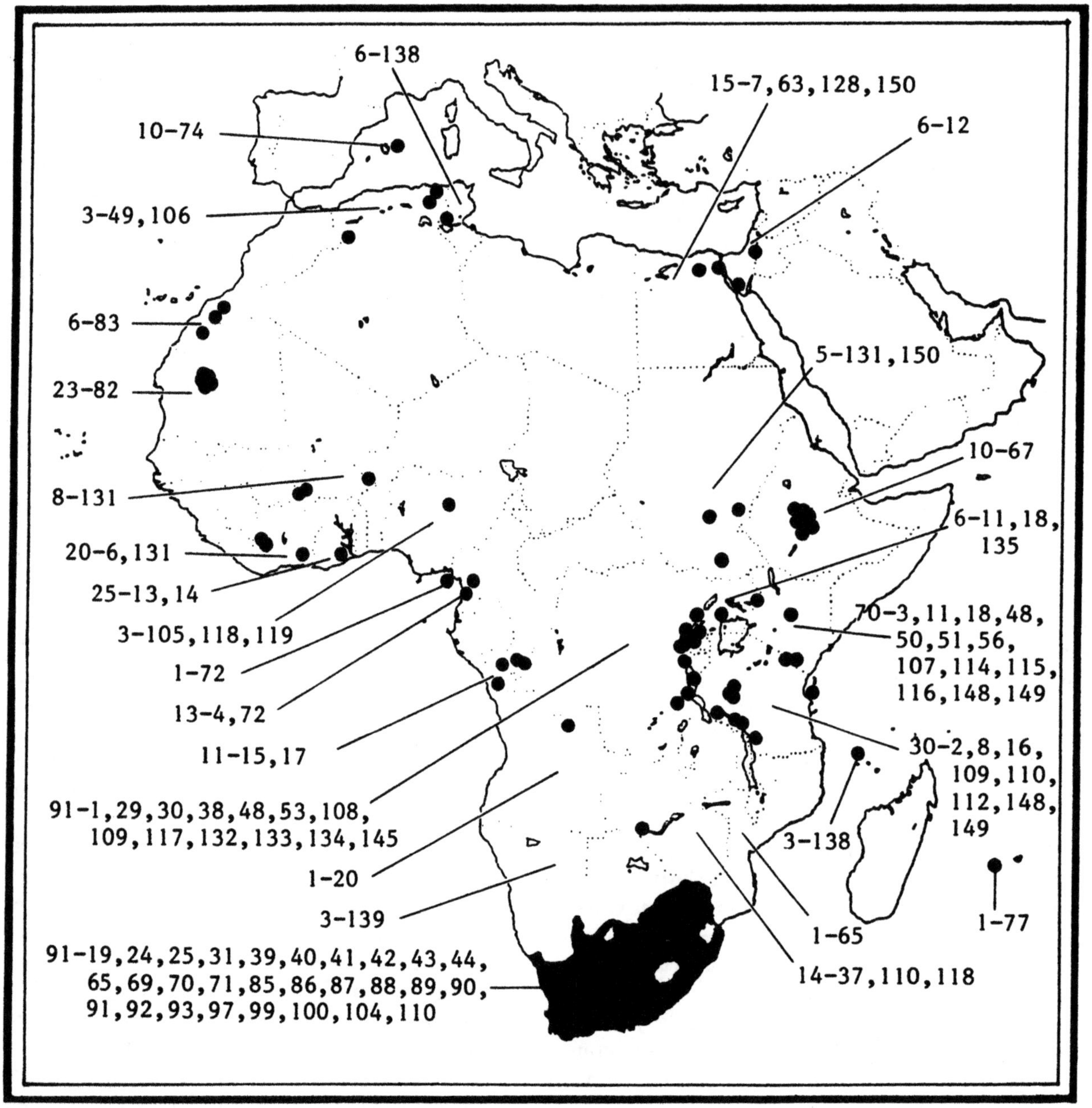

Fig. 1. Data points for free-living inland aquatic nematodes in Africa. First number: total number of different species found in the indicated state or island. Second and following numbers: reference numbers (see Table 3).

for other groups. Consequently, endemisation of parthenogenetic and autogame-hermaphroditic groups will have started much later than in amphigenetic species. In this respect, it is not known what will govern the evolution velocity when mutation velocity is constant and random. Therefore, only barriers of higher order and the presumed start for endemisation can be considered when compairing the radiation velocity of parthenogenetic taxa to amphigenetic taxa on one continent. The ability to avoid extinction by somatic mutation of parthenogenetic taxa in changing environments and the

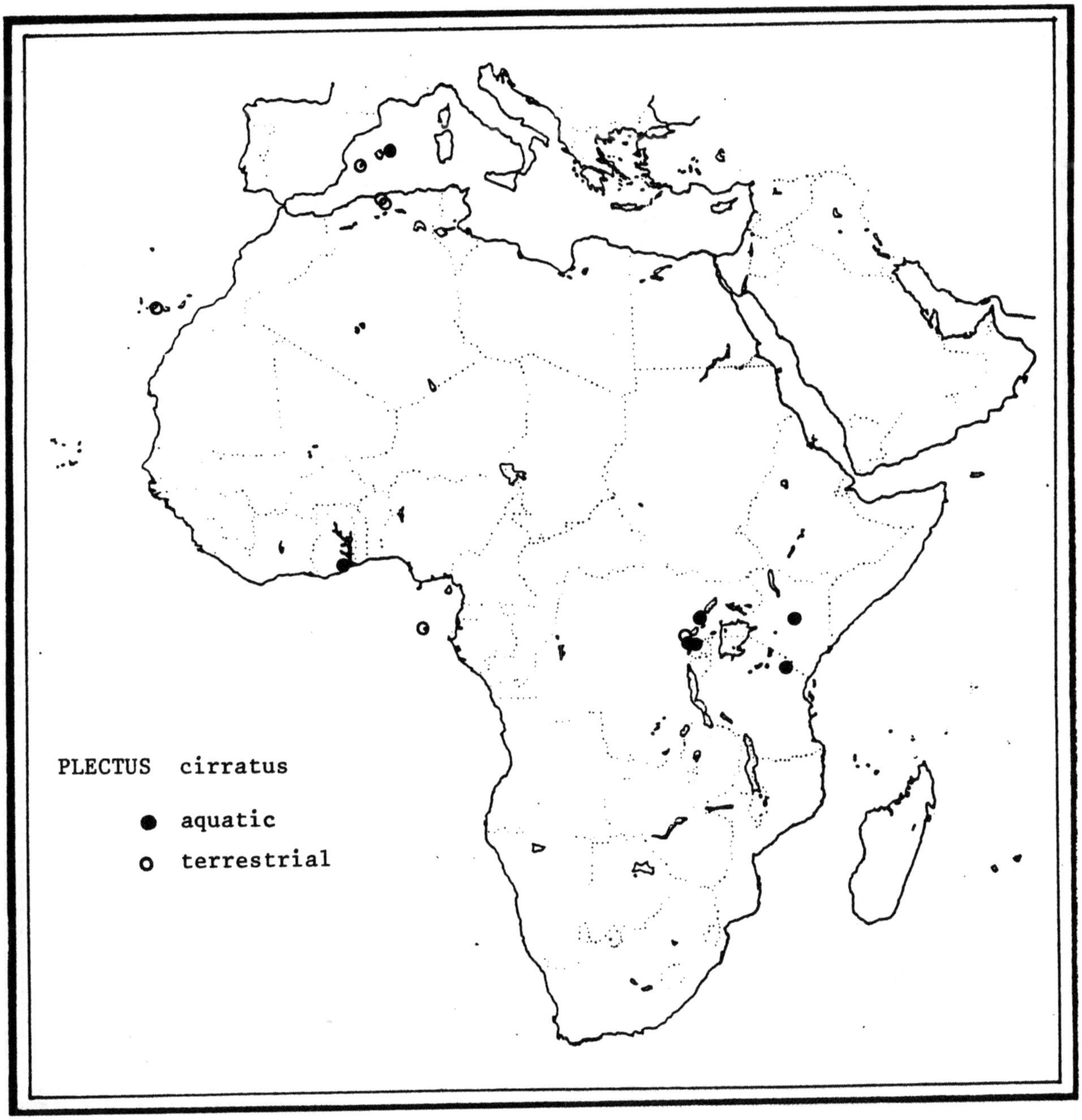

Fig. 2. Distribution of *Plectus cirratus* in Africa.

genetical isolation within one continent, have given rise to allopatric evolution. Many of the cosmopolitan species will have been able to survive, either as, at least, a relict population or as a mosaic of different populations of closely related species, each occupying a restricted area.

Several cosmopolitans (e.g. *Eumonhystera vulgaris, Monhystera wangi, Prismatolaimus dolichurus* and *Brevitobrilus graciloides*) are probably each a group of species and might be separated into a number of good taxa. We may establish a parallel with several other freshwater animal groups which, until recently, were considered to contain many cosmopolitan species. They are now being recog-

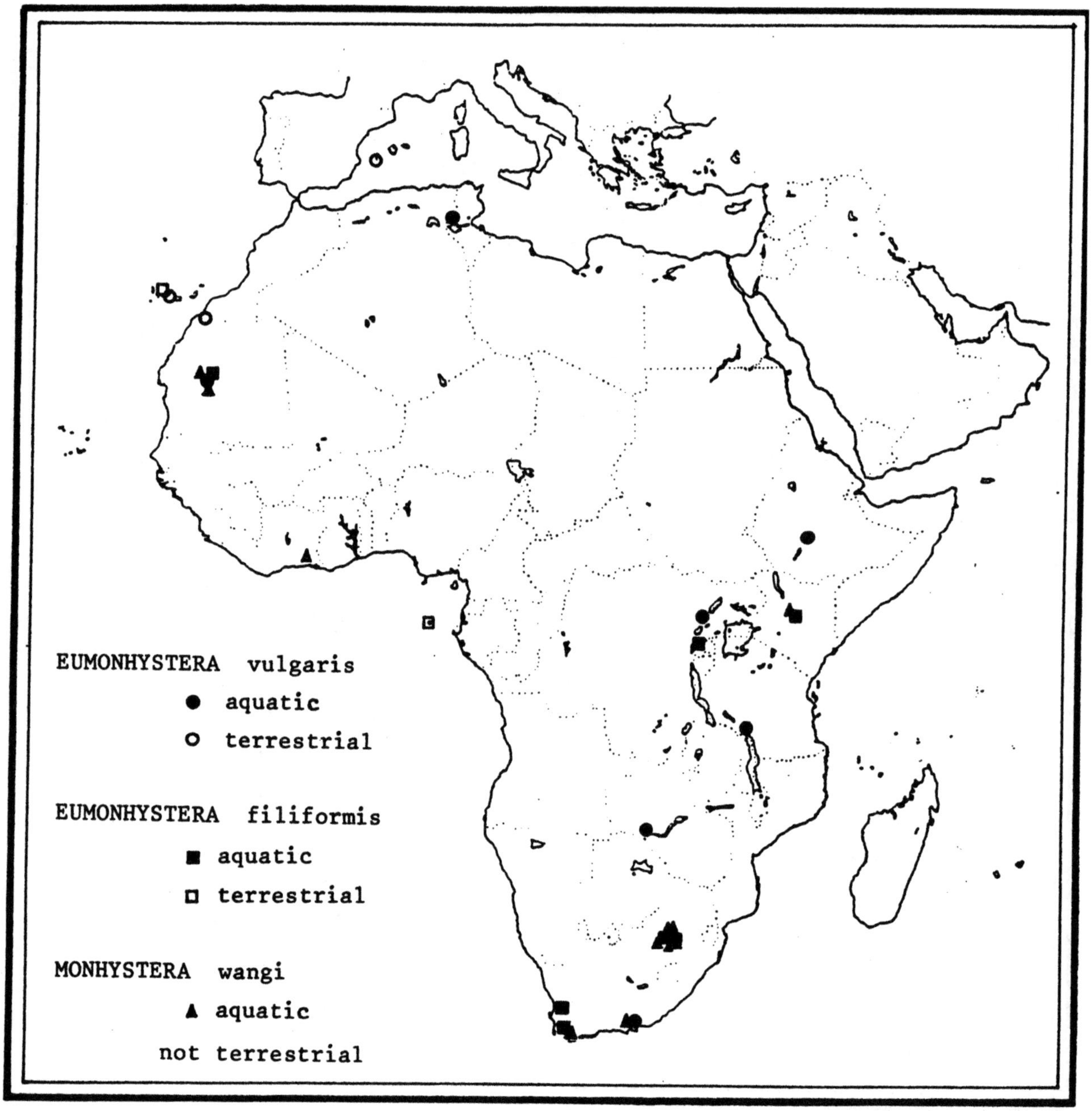

Fig. 3. Distribution of *Eumonhystera vulgaris, E. filiformis* and *Monhystera wangi* in Africa.

nized to a growing degree to have more species than previously anticipated, many of which occupy restricted areas (Rotifers: Pejler, 1977; Cladocera: Dumont, 1980a; Frey, 1982; Copepoda: Dumont, 1980a; Van de Velde, 1982).

As was mentioned before, the success of the Nematoda depended on minor (morphological) modifications. Therefore, minor morphological structures, often overlooked by early taxonomists might prove to be diagnostic (Coomans, 1971b), even when these structures are only visible by S.E.M. techniques. So, revisional work on cosmopolitan species is an urgent but tremendous challenge to taxonomists.

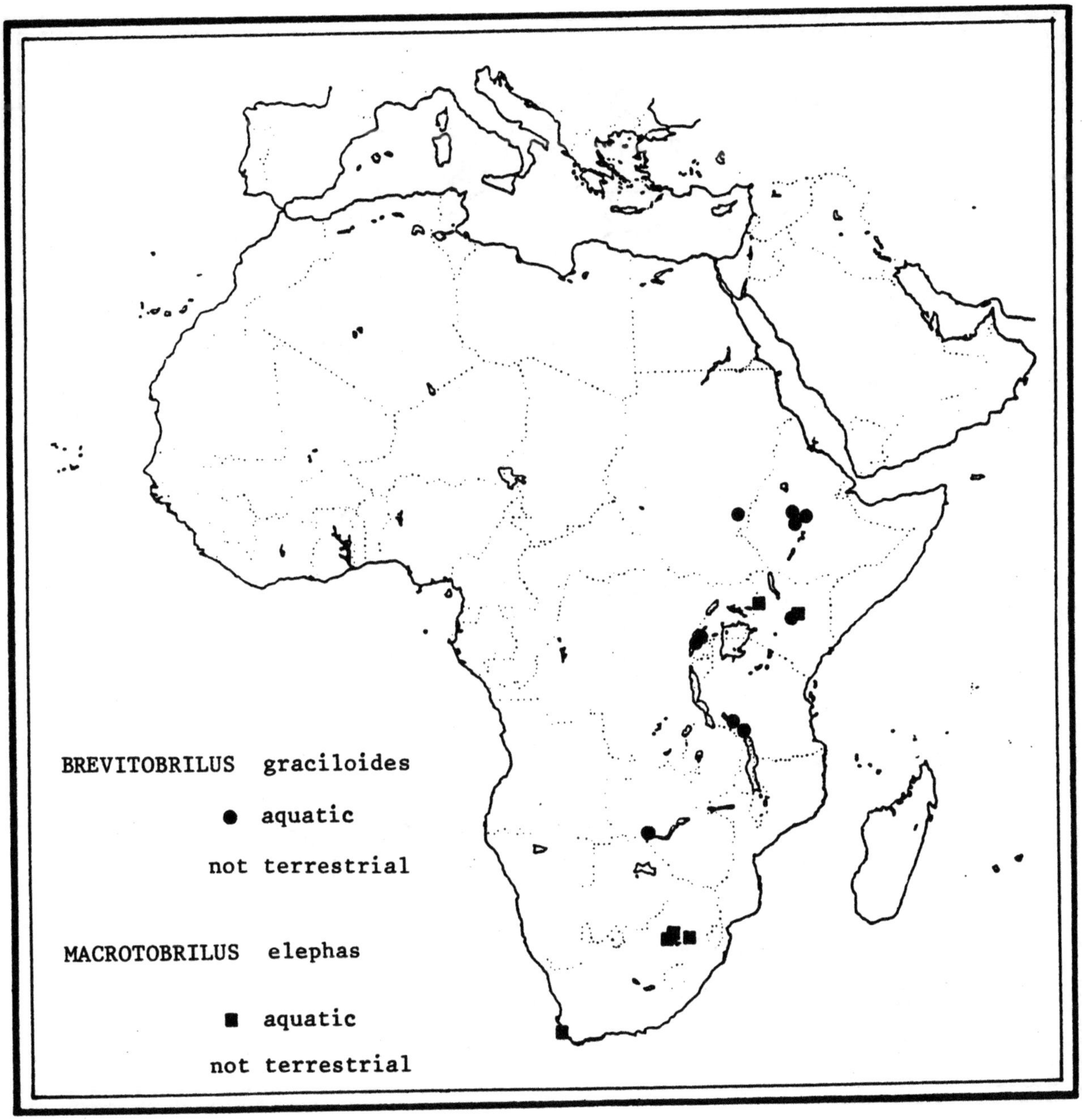

Fig. 4. Distribution of *Brevitobrilus graciloides* and *Macrotobrilus elephas* in Africa.

Preliminary discussion on the biogeography of the free-living inland aquatic nematodes of Africa

During the last thirty years, the number of new genera and species, the rearrangement of taxa and the total body of information on nematodes has grown to such an extend that mastering all these data has become impossible for any single student. Hence, there is a need for comprehensive reviews such as Goodey's 'Soil and Freshwater Nematodes' (1963), which is still the best general guide to genera, although sadly now out of date. Due to the increased number of publications on the taxonomy of free-living nematodes, and especially of species of

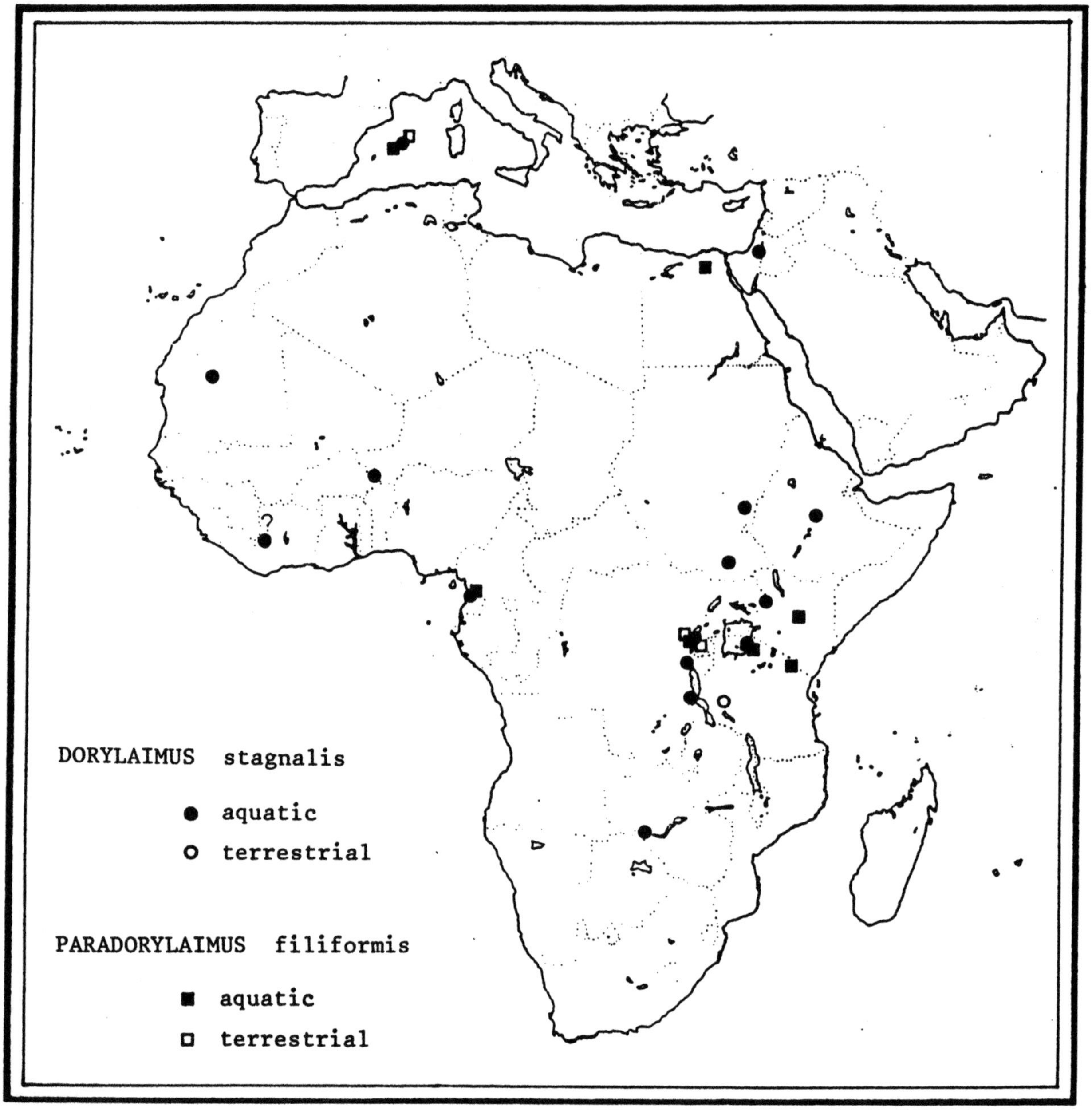

Fig. 5. Distribution of *Dorylaimus stagnalis* and *Paradorylaimus filiformis* in Africa.

the African continent, a taxonomical survey of the African species and a review of the literature is necessary (Heyns, 1971; Thomas & Taylor, 1968; Harris & Gooch, 1970). In Table 3, a checklist of free-living inland aquatic nematodes of Africa, with notes on their ditribution and ecology, is presented. The classification of the Areolaimida, Desmodorida, Chromadorida and Enoplida, is based on Gerlach & Riemann (1973, 1974), the Monhysterida on Andrássy (1981), the Tobrilidae on Tsalolikhin (1981), the Dorylaimida on Coomans (1971a) and the Secernentea on Goodey (1963) and Southey (1982). Nomenclatorial corrections have been made according to Tarjan &

Table 3. A checklist of the free-living inland aquatic nematodes of Africa, with notes on their distribution and ecology. For explanation of the abbreviations, used in this table, see the list at the end of this paper, pp. 286–287.

	Ecology		Number of localities (samples: n > 3)	Distribution in Africa and islands	Number of the references	Remarks
	aquatic	non-aquatic				
ADENOPHOREA (von Linstow, 1905) Chitwood, 1940						
Areolaimida De Coninck & S. Stekhoven, 1933						
TERATOCEPHALIDAE Andrássy, 1958						
Euteratocephalus capensis Heyns, 1977	$A_3A_4A_7$		1	SA	89	
Euteratocephalus crassidens (de Man, 1880) Andrássy, 1958						
syn. *Teratocephalus crassidens* de Man, 1880	$A_{10}A_{14}$	$B_1B_2T_2$	1	Z	53	
Metateratocephalus (de Man, 1880) Eroshenko, 1973	A_6A_7	T	n	K	56	
Euteratocephalus hirschmannae Heyns, 1977	A_4A_{13}		1	SA	89	
Euteratocephalus palustris (de Man, 1880) Andrássy, 1958	$A_6A_7A_{13}$		1	K	56	
Teratocephalus terrestris (Bütschli, 1873) de Man, 1876	$A_6A_7A_{10}A_{14}$	T_1T_2	3	Z, K	53, 56	
PLECTIDAE Örley, 1880						
Anaplectus granulosus (Bastian, 1865) De Coninck & S. Stekhoven, 1933	$A_2?A_{10}$	T_1T_4	n	Gh, Z	13, 134	dry well (Ghana, Winneba-Takaradi)
Plectus acuminatus Bastian, 1865	A_{15}	T	3	K	117	
Plectus armatus Bütschli, 1873	A_{15}	T	1	K	117	
Plectus cirratus Bastian, 1865	$A_4A_6A_{10}A_{14}$	$B_3T_1T_2$	n	Z, K, Bl	11, 149, 53, 74, 134, 150	
Plectus communsis Bütschli, 1873	A_4A_{14}	T_1	1	Bl	74	
Plectus globocephalus Mulk & Coomans, 1978	A_3A_6		1	K	117	
Plectus longicaudatus Bütschli, 1873	$A_4A_6A_{10}A_{14}$	B_1	n	Z, K	3, 53, 117	
Plectus palustris de Man, 1880	$A_3A_5A_6$	B_2	1	Bl, K	74, 117	
Plectus parientinus Bastian, 1865	A_5A_{10}	$B_2?T_1$	2	Bl, K	74, 117	
Plectus rhizophilus de Man, 1880	$A_2A_3A_5A_{10}$		1	Mau	82	
Plectus sambesii Micoletzky, 1916	$A_4A_5A_9$		n	S, Z	53, 109, 110	
Plectus tenuis Bastian, 1865	A_1		1	Tu	138	
Tylocephalus auriculatum (Bütschli, 1873) Anderson, 1966						
syn. *Wilsonema auriculatum* (Bütschli, 1873) Cobb, 1913	$A_2?$	$B_1T_1T_6$	1	Gh	13	dry well
RHABDOLAIMIDAE Chitwood, 1951						
Rhabdolaimus terrestris de Man, 1880	$A_6A_7A_9A_{10}A_{13}$	B_1	n	Z, Ta, Mau, K	8, 53, 56, 82	
syn. *Rhabdolaimus aquaticus* de Man, 1880	A			Gh	13	no specifications are given

Table 3. Continued.

	Ecology		Number of localities (samples: n > 3)	Distribution in Africa and islands	Number of the references	Remarks
	aquatic	non-aquatic				
AULOLAIMIDAE Jairajpuri & Hooper, 1969						
Aulolaimoides elegans Micoletzky, 1915	A_6		1	Mau	82	
Aulolaimus oxycephalus de Man, 1880	A_{10}		1	Gh	13	
Gymnolaimus exilis (Cobb, 1893) Cobb, 1913						
syn. *Cryptonchus abnormis* (Allgén, 1933) S. Stekhoven, 1951	A_3A_7	T_1	1	IC	6	
Cylindrolaimus abnormis Allgén, 1933	A_7?		1	Z	1	
LEPTOLAIMIDAE Örley, 1880						
Aphanolaimus aquaticus Daday, 1894	A_6A_{10}		1	Z	134	
Aphanolaimus longisetosus (Altherr, 1960) Andrássy, 1968						
syn. *Paraphanolaimus longisetosus* Altherr, 1960	A_{14}		1	Ca	4	
Aphanolaimus louisea Coomans & De Waele, 1979	$A_6A_7A_{14}$		n	K	48	
Aphanolaimus microlaimus De Coninck, 1935	$A_{10}A_{14}$	B_2	2	Z	48, 53	
Chronogaster sp. apud Andrássy, 1956	A_3A_7		1	IC	6	
Chronogaster africana Heyns & Coomans, 1980	$A_4A_7A_{13}$		n	SA	92	
Chronogaster glandifera Heyns & Coomans, 1980	A_3A_7		1	SA	92	
Chronogaster longicauda Heyns & Coomans, 1980	A_4		1	SA	92	
Chronogaster magnificus Andrássy, 1956	A_3A_7		1	IC	6	
Chronogaster multispinata Heyns & Coomans, 1980	A_3		n	SA	92	
Chronogaster typicus (de Man, 1921) De Coninck, 1935	A_1A_7		1	Mau	82	also in Arabian sea (Timm, 1963)
Paraphanolaimus tenuis (Daday, 1908) Gerlach & Riemann, 1973						
syn. *Mononchus tenuis* Daday, 1908	A_4		1	Ta	148, 149	
Paraplectonema pedunculata (Höfmänner, 1913) Strand, 1934						
syn. *Cylindrolaimus aberrans* Micoletzky, 1915	A_{10}		1	Z	134	
BASTIANIIDAE De Coninck, 1935						
Bastiania gracilis de Man, 1876	A_6A_7		1	K	56	
Bastiania parexilis De Coninck, 1935	A_{10}	B_3T_1	1	Z	53	
AXONOLAIMIDAE Filipjev, 1918						
Cylindrolaimus filicaudatus Allgén, 1933	A_4A_{13}		1	Z	1	
Monhysterida Filipjev, 1929						
MONHYSTERIDAE de Man, 1876						
Diplolaimelloides delyi Andrássy, 1958			?	Eg	7	publication not seen

Table 3. Continued.

	Ecology		Number of localities (samples: n > 3)	Distribution in Africa and islands	Number of the references	Remarks
	aquatic	non-aquatic				
Eumonhystera dispar (Bastian, 1865) Andrássy, 1981						
syn. *Monhystera dispar* Bastian, 1865	A_1		1	Tu	138	cosmopolite
Eumonhystera filiformis (Bastian, 1865) Andrássy, 1981						
syn. *Monhystera filiformis* Bastian, 1865	$A_2A_3A_4A_6A_9A_{10}$	$B_3T_1T_2$	n	Z, K, Mau, SA	11, 53, 82, 100	'cosmopolite' Andrássy (1981): sp.inq.
Eumonhystera fuelleborni (Daday, 1908) Andrássy, 1981						
syn. *Monhystera fuelleborni* Daday, 1908	A_6		1	Ta	148, 149	
Eumonhystera mwerazii (Meyl, 1957) Andrássy, 1981						also found in Rumania
syn. *Monhystera mwerazii* Meyl, 1957	A_4		1	Z	109	
Eumonhystera similis (Bütschli, 1873) Andrássy, 1981						Europe, USSR, Nepal
syn. *Monhystera similis* Bütschli, 1873	$A_2A_4A_5$	B_1	n	Ta, S, SA	110, 148, 149	
Eumonhystera vulgaris (de Man, 1880) Andrássy, 1981						cosmopolite
syn. *Monhystera vulgaris* de Man, 1880	$A_2A_3A_4A_5A_6A_9$ A_{10}	$B_1B_2B_3$	n	Ta, S, As, Tu, Z, Mau, Et, RdO	53, 67, 82, 110, 148, 149	
Monhystera africana Andrássy, 1964	$A_6A_7A_{14}$		2	K	11, 56	USSR + India
Monhystera bothriolaima Steiner, 1916			1	SWA	139	publication not seen
Monhystera elegantula S. Stekhoven, 1935	A		1	Eg	7	publication not seen
Monhystera gabaza Joubert & Heyns, 1980	$A_4A_7A_{14}$		1	SA	100	
Monhystera magnacephala Joubert & Heyns, 1980	$A_4A_7A_{14}$		2	SA	100	
Monhystera paludicola de Man, 1881	$A_2A_4A_7A_{10}$		n	K,SA,Mau,SWA	10, 17, 59, 102	cosmopolite
Monhystera somereni apud Argo & Heyns, 1973, nec Allgén, 1952	A_{11}		1	SA	25	soil on top of rock surrounded by the sea at high tide (Port St. John)
Monhystera taaiboschiensis Joubert & Heyns, 1980	$A_4A_6A_7$		2	SA	100	
Monhystera wangi Wu & Hoeppli, 1929						cosmopolite
syn. *Monhystera somereni* Allgén, 1952	$A_2A_3A_4A_5A_6A_7$ A_{13}			Eg, Ic, K, Mau, SA	3, 6, 7, 82, 100	
Monhystrella lepidura (Andrássy, 1963) Andrássy, 1968	A_1		1	Gh		also in Argentinia Canada, Europe, USSR
Monhystrella macrura (de Man, 1880) Andrássy, 1981						
syn. *Monhystera macrura* de Man, 1880	A_1		1	Tu	138	
Monhystrella paramacrura Meyl, 1954	$A_2A_5A_6A_{10}A_{12}$		2	Mau, RdO	82, 83	also in Europe & Mongolia
XYALIDAE Chitwood, 1951						
Theristus agilis (de Man, 1880) Filipjev, 1918			1		23	cosmopolite
syn. *Monhystera agilis* de Man, 1880			1	Z	133	
Theristus borosi Andrássy, 1958	A_{11} or A_{12}		1	Eg	7	publication not seen
Theristus flevensis S. Stekhoven, 1935			1	Eg	128	relict of old Bitter Lake fauna
Theristus cf. *flevensis* S. Stekhoven, 1935				Mad	77	resurgency of river on La Réunion island

Table 3. Continued.

	Ecology		Number of localities (samples: n > 3)	Distribution in Africa and islands	Number of the references	Remarks
	aquatic	non-aquatic				
Desmodorida De Coninck, 1965						
MONOPOSTHIIDAE Filipjev, 1934						
Nudora omercooperi Inglis, 1965	A_{11}?		1	SA	97	freshwater spring below high tide mark at Claytons Rocks
Chromadorida Filipjev, 1929						
CHROMADORIDAE Filipjev, 1917						
Ethmolaimus pratensis de Man, 1880	$A_6A_7A_{13}$	T	n	K	56	
Chromadorina germanica (Bütschli, 1874) Wieser, 1954			1	Eg	7	publication not seen
Chromadorita leuckarti (de Man, 1876) Filipjev, 1930	A_1		1	Tu	138	
CYATHOLAIMIDAE Filipjev, 1918						
Achromadora aetiopica (Allgén, 1933) Gerlach & Riemann, 1973						
syn. *Euchromadora aetiopica* Allgén, 1933	$A_4A_7A_{13}$		1	Z	1	
Achromadora ruricola (de Man, 1880) Micoletzky, 1925	$A_5A_6A_7A_9A_{13}A_{14}$		n	Ta, K, Z	8, 53, 56	
Achromadora semiarmata Altherr, 1952	$A_2A_6A_{12}$		2	Mau, RdO	82, 83	
Achromadora terricola (de Man, 1880) Micoletzky, 1925	A_6A_7		1	K	56	
Paracyatholaimus intermedius (de Man, 1880) Filipjev, 1930	A_6A_{12}		1	RdO	83	
Enoplida Filipjev, 1929						
TRIPYLIDAE de Man, 1876						
Paratripyla intermedia (Bütschli, 1873) Brzeski, 1964	A_{10}	T_1	n	Z	134	
Tripyla sp. apud Allgén, 1952	A_4		1	K	3	
Tripyla affinis de Man, 1880	A_{10}	B_3T_1	2	Z	53, 134	
Tripyla cornuta Skwarra, 1921	A_1A_{14}		n	Z, K, Ca	4, 11	
Tripyla glomerans Bastian, 1865	$A_4A_5A_6A_7A_{14}$	T_1		Et, SA, I, Su, Ca, K	4, 12, 19, 56 67, 131	
Tripyla setifera Bütschli, 1873	A_6	$B_3T_1T_2$	1	Z	132	
Trischistoma ursalae Argo & Heyns, 1973	A_{10}	B_6T_1	2	SA	25	
Tobrilia imberbis (Andrássy, 1953) Andrássy, 1967						
syn. *Tobrilus imberbis* Andrássy, 1953	A_{13}		1	Ta	8	
Brevitobrilus graciloides (Daday, 1910) Tsalolikhin, 1981						
syn. *Trilobus graciloides* Daday, 1910	$A_4A_5A_6A_{10}$		n	Ta, Et, Su, Z, R_w, K	3, 11, 67, 132, 134, 148, 149	
? *Trilobus gracilis* Bastian, 1865	A_4		1	S	101	

Table 3. Continued.

	Ecology		Number of localities (samples: n > 3)	Distribution in Africa and islands	Number of the references	Remarks
	aquatic	non-aquatic				
Brevitobrilus stefanskii (Micoletzky, 1925) Tsalolikhin, 1981						
syn. *Trobrilus stefanskii* (Micoletzky, 1925) Andrássy, 1954	$A_4A_6A_7$		2	SA	19, 99	
Epitobrilus meyli Tsalolikhin, 1981	A_6		1	Z	145	
syn. *Trilobus graciloides* aff. Daday, 1908	A_6		1	Z	109	
Eutobrilus heptapapillatus (Joubert & Heyns, 1979) Tsalolikhin, 1981						
syn. *Tobrilus heptapapillatus* Joubert & Heyns, 1979	$A_4A_6A_7A_{13}$		n	SA	99	
Macrotobrilus elephas (Andrássy, 1964) Tsalolikhin, 1981						
syn. *Tobrilus elephas* Andrássy, 1964	$A_4A_6A_7A_{13}$		n	SA, K	11, 19, 56	1 species genus, endemic
Neotobrilus ampieri (Joubert & Heyns, 1979) Tsalolikhin, 1981						
syn. *Tobrilus ampieri* Joubert & Heyns, 1979	A_4		n	SA	99	
Neotobrilus breviductus (Loof & Riemann, 1976) Tsalolikhin, 1981						
syn. *Tobrilus breviductus* Loof & Riemann, 1976	$A_4A_7A_{13}$		1	SA	19	
Neotobrilus diversipapillatus (Daday, 1905) Tsalolikhin, 1981						widespread benthic species, littoral cosmopolite
syn. *Tobrilus longus* (Leidy, 1851) Andrássy, 1959	A_4A_7sewage		n	SA	25	
Tobrilus diversipapillatus (Daday, 1905) Andrássy, 1959	A_4A_6		n	SA	99	
Neotobrilus floridensis (Joubert & Heyns, 1979) Tsalolikhin, 1981						
syn. *Tobrilus floridensis* Joubert & Heyns, 1979	$A_4A_6A_7$		2	SA	99	benthic
Neotobrilus telekiensis (Allgén, 1952) Tsalolikhin, 1981						
syn. *Trilobus telekiensis* Allgén, 1952	A_6		1	K	3	
Tobrilus telekiensis (Allgén, 1952) Andrássy, 1959	A_6A_7		n	K	11, 56	
Semitobrilus pellucidus (Bastian, 1865) Tsalolikhin, 1981						
syn. *Trilobus pellucidus* Bastian, 1865	A_4A_6		n	Ta, Et, K	3, 67, 139, 148, 149	Europe
Tobrilus annetteae Joubert & Heyns, 1979	$A_3A_4A_6A_7A_{13}$		n	SA	99	
Tobrilus gracilis (Bastian, 1865) Andrássy, 1959						
syn. *Trilobus gracilis* Bastian, 1865	$A_1A_3A_4A_5A_6A_7$		n	Tu, S, Eg?, K, Bl	3, 7, 74, 110, 138	
Tobrilus hoehnelensis (Allgén, 1952) Andrássy, 1959						
syn. *Trilobus hoehnelensis* Allgén, 1952	A_6		1	K	3	
Monochromadora monhysteroides Schneider, 1937	$A_2A_3A_5$		1	Mau	82	
syn. *Monachromadora monhysteroides* Schneider, 1937			1	Z	109	
Udonchus monohystera (Micoletzky, 1915) Riemann, 1970						
syn. *Chromadora monohystera* Micoletzky, 1915	A_4		1	S	110	tychopotamic plankton found only in Czechoslovakia in thermal spring
PRISMATOLAIMIDAE Micoletzky, 1922						
Onchulus oistospiculum (Allgén, 1963) Gerlach & Riemann, 1974						
syn. *Cyatholaimus oistospiculum* Allgén, 1933	$A_4A_7A_{13}$		1	Z	1	
Prismatolaimus andrassyanus Coomans & Mulk, 1980					51	

Table 3. Continued.

	Ecology		Number of localities (samples: n > 3)	Distribution in Africa and islands	Number of the references	Remarks
	aquatic	non-aquatic				
syn. *Prismatolaimus andrassyi* apud Mulk & Coomans, 1979, nec Khera & Chaturvedi, 1977	A_{15}	B_6	2	K	115	
Prismatolaimus dolichurus de Man, 1880	$A_6A_7A_{10}A_{15}$	B_3	n	Z, K	53, 56, 115	
Prismatolaimus intermedius (Bütschli, 1873) de Man, 1880	$A_4A_6A_7A_{10}$ $A_{13}A_{15}$	B_6T_1	n	Z, Ta, K	1, 8, 56, 115	
Prismatolaimus kenyensis Mulk & Coomans, 1979	A_{15}	B_3	n	K	115	
Prismatolaimus leptolaimus Andrássy, 1969	A_7		1	Co	17	phreatic
Prismatolaimus matoni Mulk & Coomans, 1979	A_{15}	B_6	n	K	115	
IRONIDAE de Man, 1876						
Ironus crassatus Argo & Heyns, 1972	A_{10}	T_1	n	SA	24	
Ironus ernsti Argo & Heyns, 1972	A_{10}	B_6T_1	2	SA	24	on an island in the Vaal-river
Ironus ignavus Bastian, 1865	$A_2A_3A_5$	T_1	n	Mau, K, Ug, Gh	11, 13, 82	
syn. *Ironus ignavus* var. *typicus* Bastian, 1865	A_1A_3	T_1	n	UV, IC, Z	131	
Ironus ignavus var. *brevicaudatus* Brakenhoff, 1914	A_{14}		1	Ta	112	
Ironus longicaudatus de Man, 1884	A_6A_{14}	B_3	n	Ug, Mau	11, 82	
Ironus tenuicaudatus de Man, 1876	$A_2A_4A_6A_7A_{10}A_{13}$		n	SA, K	19, 56	
Ironus luci Andrássy, 1956	A_3A_7		1	IC	6	
Ironus paludicola Schneider, 1937	A_5A_6		3	Mau	82	
syn. *Ironus macramphis* S. Stekhoven & Teunissen, 1938	A_5	T_1	1	Ta	109	
CRYPTONCHIDAE Chitwood, 1937						
Cylindrolaimus abnormis Allgén, 1933	$A_4A_7A_{13}$	T_1	2	Z	1	cosmopolite
ALAIMIDAE Micoletzky, 1922						
Alaimus asifkhalili Mulk & Coomans, 1979	$A_{10}(A_6)$	B_3	1	K	116	
Alaimus modestus S. Stekhoven & Teunissen, 1938	A_6		1	Z	134	
Alaimus primitivus de Man, 1880	A_4A_{10}	$B_3T_1T_2$	2	SA, Z, Eg?	110, 53	110: tychopotamic plankton; cosmopolite
Alaimus proximus Thorne, 1939	$A_6A_7A_{13}$		n	K	56	
Alaimus similis Thorne, 1939	$(A_5)A_{10}$	B_6	1	Gh	13	
Alaimus thompsoni Mulk & Coomans, 1979	A_6A_{10}	B_6	2	K	116	
OXYSTOMINIDAE Chitwood, 1935 (Filipjev, 1918)						
Halalaimus algeriensis Coomans & Jacobs, 1983	A_3		1	Al	49	2nd freshwater sp. of this marine genus, water: upper limit oligohaline
ONCHOLAIMIDAE Filipjev, 1916						
Oncholaimus sp. apud Andrássy, 1964	A_2A_{12}		1	I	12	4th continental sp. of this marine genus, slightly mineral spring south of Dead Sea
Oncholaimus deconincki Heyns & Coomans, 1977	$A_4A_7A_{13}$		3	SA	91	

Table 3. Continued.

	Ecology		Number of localities (samples: n > 3)	Distribution in Africa and islands	Number of the references	Remarks
	aquatic	non-aquatic				
Dorylaimida (de Man, 1876) Pearse, 1942						
MONONCHIDAE Chitwood, 1937						
Clarkus papillatus (Bastian, 1865) Jairajpuri, 1970						
syn. *Mononchus papillatus* Bastian, 1965	A_4A_{10}	T_1	2	Z, Gh	13, 134	
Clarkus pretoriensis (Coetzee, 1968) Jairajpuri, 1970						
syn. *Mononchus pretoriensis* Coetzee, 1968				SA	44	
Mononchus sp. apud Micoletzky, 1916	A_4		1	S	110	
Mononchus allgeni Meyl, 1957	A_5		1	Z	109	
syn. *Mononchus* sp. apud Allgén, 1933	$A_4A_7A_{13}$		1	Z	1	
Mononchus aquaticus Coetzee, 1968	$A_5A_6A_7A_{12}$		n	Mau, RdO, SA	44, 82, 83	cosmopolite
Mononchus cambelli Allgén, 1929						
syn. *Mononchus papillatus* apud S. Stekhoven & Teunissen, 1938, nec Bastian, 1865	A_{11}	T_1	1	Z	134	NZ + islands
Mononchus gerlachei de Man, 1904?	A?			Col	138	no specifications on the localities of samples are given; found also in Antarctica
Mononchus tenuis Daday, 1908	A_6		1	Ta	148, 149	Gerlach & Riemann, 1973: *Paraphanolaimus*; Cobb, 1917 not *Mononchus*; a *Prismatolaimus*?
Mononchus truncatus Bastian, 1865	$A_3A_4A_6A_7A_{10}$ A_{14}	T	n	K, SA	11, 19, 44, 50, 56	cosmopolite
syn. *Mononchus longicaudatus* Cobb, 1893	A_4A_6	T_1	n	Z, Rw	132, 134	
Mononchus macrostoma Bastian, 1865	$A_2A_3A_4A_5A_6A_{14}$	$B_3B_4T_1$	n	Ta, IC, UV	131, 148, 149	cosmopolite
Mononchus tunbridgensis Bastian, 1865	A_{10}		1	SA	43	
Prionchulus muscorum (Dujardin, 1845) Wu & Hoeppli, 1929	A_{10}	$B_3B_6T_2$	n	K, SA	44, 50	
syn. *Mononchulus muscorum* (Dujardin, 1845) Bastian, 1865	A_4A_{10}	T_1	n	Z	134	
MYLONCHULIDAE Jairajpuri, 1969						
Granonchulus subdecurrens Coetzee, 1966	$(A_4)A_{10}$	T_1	1	SA	40	mainly terrestrial
Mylonchulus agriculturae Coetzee, 1967	$(A_4)A_{10}$		1	SA	41	mainly terrestrial
Mylonchulus clavicaudatus (S. Stekhoven & Teunissen, 1938) Andrássy, 1958	A_4		1	Z	134	
Mylonchulus denticulatus (Cobb, 1917) Andrássy, 1958	A_4A_{14}		1	S	37, 118	described by Micoletzky (1916), named by Cobb (1917)
syn. *Mononchus* sp. apud Micoletzky, 1916			1	S	110	
Mylonchulus incurvus (Cobb, 1917) Andrássy, 1958	$(A_5)A_{10}$		1	SA	43	among grassroots near to Mac-Mac pools
Mylonchulus insolitus Andrássy, 1968	A_7		1	Co	15	phreatic

Table 3. Continued.

	Ecology		Number of localities (samples: n > 3)	Distribution in Africa and islands	Number of the references	Remarks
	aquatic	non-aquatic				
Mylonchulus lacustris (N.A. Cobb in M.V. Cobb, 1915) Andrássy, 1958						
syn. *Mylonchus montanus* (Thorne, 1924) Andrássy, 1958	$A_2A_4A_{10}$		2	SA	41	
Mylonchulus polonicus (Stefanski, 1915) Andrássy, 1958	$A_4A_{10}A_{13}$		2	SA, Eg?	7, 19	
Mylonchulus sigmaturus (Cobb, 1917) Altherr, 1953	$A_{10}A_{11}$	T_1	1	SA	41	mainly terrestrial, cosmopolite
COBBONCHIDAE Jairajpuri, 1969						
Cobbonchus abrubticaudatus (Altherr, 1960) Goodey, 1963						
syn. *Mononchus abrubticaudatus* Altherr, 1960	A_{14}	B_3	1	Ca	4	preferences for wet sites
Cobbonchus artemisiae Coetzee, 1968	$(A_5)A_{10}$		1	SA	43	
Cobbonchus coetzaeae Andrássy, 1970	A_2A_{13}		1	SA	19	soil among grassroots next to Mac-Mac pools
Cobbonchus dianae Coetzee, 1965	$(A_2)A_{10}$		+1	SA	39, 43	39: no locality is given
Cobbonchus megalus Coetzee, 1966	$(A_4)A_{10}$		1	SA	40	
Cobbonchus ockerti Coetzee, 1965	$(A_5)A_{10}$	T_1	2	SA	43	soil among grassroots next to Mac-Mac pools
ANATONCHIDAE Jairajpuri, 1969						
Miconchus sp. near *M. californicus* Mulvey, 1962	A_4A_7		1	Ni	119	
Miconchus dadayi (Micoletzky, 1914) Mulvey, 1962						
syn. *Mononchus dadayi* Micoletzky, 1914	$A_2A_4A_{13}A_{14}$	B_2B_3	n	UV, IC	131	
Miconchus rapax (Cobb, 1917) Andrássy, 1958						
syn? *Iotonchus dadayi* Micoletzky, 1914	A_4A_{14}		1	Ca	4	
Miconchus schneideri (Meyl, 1955) Andrássy, 1958						
syn. *Iotonchus schneideri* Meyl, 1955	A_6A_7		1	Z	108	psammon
IOTONCHIDAE Jairajpuri, 1969						
Iotonchus geminus Heyns & Lagerwey, 1965	$(A_4)A_{10}$	T_1	1	SA	93	
Iotonchus pauli Heyns & Lagerwey, 1965	$(A_4)A_{10}$		1	SA	93	at the mouth of the Sikambe river
Iotonchus riscoceiae (Carvalho, 1955) Andrássy, 1958	$(A_2)A_{10}$	T_1	n	SA	42	
Iotonchus transkeiensis Heyns & Lagerwey, 1965	$(A_4)A_{10}$		1	SA	93	at the mouth of the Sikambe river
Iotonchus vulvapapillatus Andrássy, 1964	A_6	T_1	n	K, Ug	11	
MONONCHULIDAE De Coninck, 1962						
Oionchus obtusicaudatus (Kreis, 1932) Andrássy, 1961, nec Thorne, 1935	$A_6A_7A_{13}$	T_1	1	Ta	8	

Table 3. Continued.

	Ecology		Number of localities (samples: n > 3)	Distribution in Africa and islands	Number of the references	Remarks
	aquatic	non-aquatic				
NYGOLAIMIDAE (Thorne, 1935) Meyl, 1960						
Nygolaimus gracillimus Andrássy, 1968	A_7		1	Co	15	phreatic
Nygolaimus intermedius (de Man, 1880) Loof, 1961 syn. *Dorylaimus intermedius* de Man, 1880	A_3A_{14}		n	Bl	74	
Nygolaimus trichodorus Andrássy, 1965	A_{10}		1	Gh	13	
DORYLAIMIDAE de Man, 1876						
Afrodorylaimus bwana Andrássy, 1964	$A_6A_7A_{14}$		n	K	11	
Chrysonema abyssinica Filipjev, 1931	A_6A_{13}		1	Et	67	
Chrysonema luettichaui Meyl, 1957	A_9A_{10}		2	Z	109	thermal spring
Chrysonema thornei Filipjev, 1931	A_6A_{13}		1	Et	67	
Discolaimoides arcuicaudatum (Furstenberg & Heyns, 1965) Das, Khan & Loof, 1969 syn. *Discolaimoides arcuicaudatum* Furstenberg & Heyns, 1965	A_{10}		1	SA	69	Island in the Vaal river
Discolaimus zicsii Andrássy, 1968	A_7		1	Co	15	phreatic
Dorylaimus sp. apud Allgén, 1933	$A_4A_7A_{13}$		1	Z	1	
Dorylaimus afghanicus Andrássy, 1969	A_4A_7		n	SA	19	104: at least two rivers are polluted, to some extent
Dorylaimus asymphydorus Andrássy, 1969	$A_5A_6A_7A_{10}A_{12}$		n	RdO, SA	83, 104	
Dorylaimus nyongi Altherr, 1960	A_4		1	Ca	4	
Dorylaimus pachydermis Daday, 1908	A_6		1	K	148, 149	
Dorylaimus pachys Andrássy, 1970	$A_4A_{10}A_{13}$		3	SA	19	
Dorylaimus stagnalis Dujardin, 1845	$A_2A_4A_5A_6A_7$ $A_9A_{10}A_{13}A_{14}$	B_3T_1	n	Et, S, Ta, Su, IC, UV, Z, K, I, Ca, Bl, Mau	11, 12, 67, 72, 74, 82, 108, 109, 110, 112, 131, 150	cosmopolite; freshwater, accidently brackish, rarely terrestrial 109: thermal spring 150: planktonic
Dorylaimus stekhoveni Baqri & Coomans, 1973	$(A_4)A_6A_{10}$		2	Z	29, 132	
Dorylaimus stenus Andrássy, 1970	A_6		1	SA	19	
Dorylaimus stephani Andrássy, 1969	A_3		2	K	18	
Dorylaimus tepidus Andrássy, 1959	A_6		1	SA	19	
Dorylaimus unicus Andrássy, 1970	$A_4A_6A_7A_{13}$		2	SA	19	
Drepanodorylaimus macramphidius Andrássy, 1971	A_2A_{10}		1	An	20	
Ecumenius monohystera (de Man, 1880) Thorne, 1974	A_5		1	Mau	82	
Eudorylaimus afer Andrássy, 1964	A_6A_7		2	K	11, 56	
Eudorylaimus brunettii (Meyl, 1953) Andrássy, 1959 syn. *Dorylaimus brunettii* Meyl, 1953	A_3A_7		1	IC	6	
Eudorylaimus carteri (Bastian, 1865) Andrássy, 1959	$A_6A_7A_{10}A_{14}$	B_3T_2	2	K	11	

Table 3. Continued.

	Ecology		Number of localities (samples: n > 3)	Distribution in Africa and islands	Number of the references	Remarks
	aquatic	non-aquatic				
Eudorylaimus consobrinus (de Man, 1918) Andrássy, 1959	A_6A_7		3	K	11, 56	
Eudorylaimus iners (Bastian, 1865) Andrássy, 1959		T_1		Z	29	
syn. *Dorylaimus iners* Bastian, 1865	A_4	T_1	1	Z	134	
Eudorylaimus leuckarti (Bütschli, 1873) Andrássy, 1959	$(A_5)A_{10}$	B_6	1	Gh	13	
Eudorylaimus metobtusicaudatus (S. Stekhoven & Teunissen, 1938) Andrássy, 1959						
syn. *Dorylaimus metobtusicaudatus* S. Stekhoven & Teunissen, 1938	A_{10}	T_1	1	Z	134	
Dorylaimus paraobtusicaudatus apud De Coninck, 1935, nec Micoletzky, 1922	A_9A_{14}		1	Z	53	
Eudorylaimus obtusicaudatus (Bastian, 1865) Andrássy, 1959						
syn. *Dorylaimus obtusicaudatus* Bastian, 1865	$A_2A_3A_4A_6A_7$ $A_{10}A_{14}$	$T_2T_3T_5$	n	IC, Z, Gh, BI	13, 74, 131, 132, 134	
Eudorylaimus paracentrocercus (De Coninck, 1935) Andrássy, 1959	A_6		1	K	11	
syn. *Dorylaimus paracentrocercus* De Coninck, 1935	$(A_2)A_6A_{10}A_{14}$		n	Z	53	
Eudorylaimus parvus (de Man, 1880) Andrássy, 1959						
syn. *Dorylaimus parvus* de Man, 1880	?	B_3T_2	1	Gl	138	no specifications on the samples are given
Ischiodorylaimus bathypyla Andrássy, 1970	A_6A_7		1	SA	19	
Ischiodorylaimus gulliver (Andrássy, 1964) Andrássy, 1969	$A_4A_6A_7A_{10}A_{13}$		2	K, SA	19, 56	
syn. *Dorylaimus gulliver* Andrássy, 1964	$A_5A_6A_7A_{13}A_{14}$		n	K	11	even in plankton
Ischiodorylaimus loeffleri (Andrássy, 1964) Andrássy, 1969	$A_6A_7A_{14}$		n	K	56	
syn. *Dorylaimus loeffleri* Andrássy, 1964	$A_6A_7A_{14}$		n	K	11	
syn. *Dorylaimus filiformis* apud Allgén, 1952 (partim), nec Bastian, 1865	A_6			K	3	
Ischiodorylaimus tessares Kleynhans, 1970	A_5		1	SA	104	
Ischiodorylaimus ugandanus Andrássy & Banage in Andrássy, 1969	$A_{10}A_{13}$		1	Ug	18	
Laimydorus sp. apud Baqri & Coomans, 1973	A_6A_{10}		1	Z	29	
syn. *Dorylaimus flavomaculatus* apud S. Stekhoven, 1944, nec v. Linstow, 1876	A_6A_{10}		1	Z	132	
Laimydorus acris (Thorne, 1938) Andrássy, 1969						
syn. *Dorylaimus acris* Thorne, 1938	$(A_6)A_{10}$		1	Z	109	
Laimydorus agilis (de Man, 1880) Siddiqi, 1969						
syn. *Mesodorylaimus agilis* (de Man, 1880) Goodey, 1963	A_7		1	Mau	82	phreatic
Laimydorus centrocercus (de man, 1880) Siddiqi, 1969						
syn. *Dorylaimus centrocercus* de Man, 1880	A_3A_{14}	B_3T_1	n	UV, IC	131	
Laimydorus conurus (Thorne, 1939) Siddiqi, 1969						
syn. *Dorylaimus conurus* Thorne, 1939	A_9		1	Z	109	thermal spring
Laimydorus dadayi (Thorne & Swanger, 1936) Andrássy, 1969	A_5A_6		n	Mau	82	Ethiopian relict
Laimydorus flavomaculatus (v. Linstow, 1876) Siddiqi, 1969						European
syn. *Dorylaimus flavomaculatus* v. Linstow, 1876	$A_4A_5A_6A_7$		n	S, Et, Su, Z	29, 67, 109, 110, 131, 132	

Table 3. Continued.

	Ecology		Number of localities (samples: n > 3)	Distribution in Africa and islands	Number of the references	Remarks
	aquatic	non-aquatic				
Laimydorus gazella Andrássy, 1970	A_5A_6		2	SA	19	
Laimydorus parhomalopapillatus (S. Stekhoven, 1944) Baqri & Coomans, 1973	A_7		1	Z	29	redescription
syn. *Dorylaimus parhomalopapillatus* S. Stekhoven, 1944	A_1		1	Z	132	national Parc Albert, no exact locality is given
Laimydorus pseudostagnalis (Micoletzky, 1927) Siddiqi, 1969						
syn. *Dorylaimus pseudostagnalis* Micoletzky, 1927	$A_6A_7A_{10}A_{13}$	T_1	n	K, Gh	11, 13, 56	
Laimydorus stenopygus (Andrássy, 1968) Siddiqi, 1969						
syn. *Dorylaimus stenopygus* Andrássy, 1968	A_7		1	Co	15	phreatic
Longidorella chappuisi (Schneider, 1935) Thorne, 1939	A_3A_{14}	B_3	1	Ca	4	
syn. *Dorylaimus chappuisi* Schneider, 1935	A_3A_{14}	B_3	1	UV	131	
Longidorella parva Thorne, 1939	$(A_5)A_{10}$		1	Gh	13	
Mesodorylaimus sp. apud Baqri & Coomans, 1973	A_{10}		1	Z	29	Baqri & Coomans, 1973: = *Paradorylaimus parafecundus* (De Coninck, 1935) Andrássy 1969 but the male is different as described by De Coninkc (1935)
syn. *Dorylaimus parafecundus* apud S. Stekhoven & Teunissen, 1938 (partim) nec De Coninck, 1935	A_4A_{10}	T_1		Z	134	
Mesodorylaimus aegypticus (Andrássy, 1958) Andrássy, 1959	A?		?	Eg?	7	publ. not seen
Mesodorylaimus africanus (Daday, 1908) Andrássy, 1959						
syn. *Dorylaimus africanus* Daday, 1908	A_{10}	B_3T_1	1	K	148, 149	
Mesodorylaimus augustus Andrássy, 1964	$A_6A_7A_{14}$		n	K	11, 56	
Mesodorylaimus arvensis (Cobb in Thorne & Swanger, 1936) Andrássy, 1959	A_4A_{10}	$B_3B_6T_1$	n	SA	31	
Mesodorylaimus bainsi Basson & Heyns, 1974	$(A_4)A_{10}$		1	SA	31	on a small island in a stream
Mesodorylaimus bastiani (Bütschli, 1873) Andrássy, 1959						
syn. *Dorylaimus bastiani* Bütschli, 1873	A_4A_6	$B_3T_1T_2$	2	Ta, S	110, 148, 149	Old World mainly terrestrial or bryo- & lichenedaphic
Mesodorylaimus effilatus (S. Stekhoven & Teunissen, 1938) Andrássy, 1959	A_6		1	Z	29	redescription
syn. *Dorylaimus effilatus* S. Stekhoven & Teunissen, 1938	A_6		2	Z	134	
Mesodorylaimus erdelyii Andrássy, 1965	A_{10}	B_6	1	Gh	13	
Mesodorylaimus exilis (Cobb, 1893) Andrássy, 1959						
syn. *Dorylaimus exilis* Cobb, 1893	A_3A_{14}		1	IC	131	
Mesodorylaimus ghanae Andrássy, 1965	$A_5A_7A_{10}$		2	Gh	13	
Mesodorylaimus importunus Basson & Heyns, 1974	$(A_4)A_{10}$		2	SA	31	
Mesodorylaimus intervallis (Thorne & Swanger, 1936) Andrássy, 1959						
syn. *Dorylaimus intervallis* Thorne & Swanger, 1936			?	Eg?	7	publ. not seen
Mesodorylaimus kamandeanus Baqri & Coomans, 1973	A_6		1	Z	29	redescription
syn. *Dorylaimus filiformis* apud S. Stekhoven, 1944, nec Bastian, 1865	A_6		2	Z	132	

Table 3. Continued.

	Ecology		Number of localities (samples: n > 3)	Distribution in Africa and islands	Number of the references	Remarks
	aquatic	non-aquatic				
Mesodorylaimus keilini (Lee, 1961) Andrássy, 1969						
syn. *Dorylaimus keilini* Lee, 1961	A_5		1	Ni	105	cryptobiotic
Mesodorylaimus kowyni Basson & Heyns, 1974	A_3A_{14}	T_1	1	SA	31	
Mesodorylaimus paralitoralis Basson & Heyns, 1974	A_6		1	SA	31	
Mesodorylaimus potus Heyns, 1963	A_6	T_1	n	SA	31	
Mesodorylaimus ruwenzorii (De Coninck, 1935) Andrássy, 1959	A_{10}	T_1	1	Z	29	redescription
syn. *Dorylaimus ruwenzorii* De Coninck, 1935	$(A_2)A_{10}$	B_3T_1	2	Z	53	
Actinolaimus ruwenzorii apud S. Stekhoven & Teunissen, 1938 (partim) nec De Coninck, 1935	A_{10}	T_1		Z	134	
Mesodorylaimus subtiliformis (Andrássy, 1959) Andrássy, 1959	$A_2A_3A_7$		1	Mau	82	benthic
Mesodorylaimus subtilis (Thorne & Swanger, 1936) Andrássy, 1959	A_3A_5		1	I	12	
Mesodorylaimus subtiloides (Paetzold, 1958) Andrássy, 1959	A_5		1	Mau	82	
Mesodorylaimus széchenyii Andrássy, 1961	A_6		1	Ta	8	
Mesodorylaimus szunyoghi Andrássy, 1968	$(A_4)A_{13}$		1	Ta	16	
Mesodorylaimus transkeiensis Basson & Heyns, 1974	A_{11}	B_6T_4	1	SA	31	
Mesodorylaimus usitatus Basson & Heyns, 1974	A_3A_{10}	B_6	2	SA	31	
Paradorylaimus filiformis (Bastian, 1865) Andrássy, 1969						cosmopolite
syn. *Dorylaimus filiformis* Bastian, 1865	$A_4A_5A_6A_9A_{10}$	T_1	n	K-Ta, K, Eg, Ta, Z, Ca, BI	3, 7(?), 53, 72, 74, 148, 149, 150	45: planktonic
Paradorylaimus parafecundus (De Coninck, 1935) Andrássy, 1969						Baqri & Coomans, 1973 syn. *P.p.* (D.C., 1935) A, 1969 = *Mesodorylaimus* spec. but differs from male De Coninck (1935) described
syn. *Dorylaimus parafecundus* De Coninck, 1935 (partim)	$A_4A_{10}A_{14}$		n	Z	53, 96	
Prodorylaimus dahli (Altherr, 1960) Andrássy, 1964					11	
syn. *Dorylaimus dahli* Altherr, 1960	A_3A_{14}	T_1	1	Ca	4	
Prodorylaimus ensis Kleynhans, 1970	$(A_4)A_{10}$		2	SA	104	
Prodorylaimus longicaudatus (Bütschli, 1874) Andrássy, 1959	$A_2A_3A_6A_7A_{14}$		3	K	11, 56	
Prodorylaimus paralongicaudatus (Micoletzky, 1925) Andrássy, 1959	A_{10}	B_3T_1	1	Z	29	
syn. *Dorylaimus longicaudatus* apud S. Stekhoven & Teunissen, 1938, nec Bütschli, 1874	A_{10}	T_1	1	Z	134	
Prodorylaimus rionensis (Gerlach, 1954) Andrássy, 1958	$(A_4)A_6A_7A_{10}$		2	SA, K	56, 104	
Thornenema africanum Andrássy, 1965	A_{10}		1	Gh	13	
Thornia juvenilis (De Coninck, 1935) Meyl, 1954						
syn. *Dorylaimus juvenilis* De Coninck, 1935	A_9		1	Z	53	
ACTINOLAIMIDAE (Thorne, 1939) Meyl, 1960						
Actinca gracillima Andrássy, 1964	$A_5A_6A_7A_{14}$	B_6	n	K	11, 56	
syn. *Actinolaimus papillatus* apud Altherr, 1960, nec Schneider, 1935	A_3A_{14}	B_3	1	Ca	4	
Actinca intermedia Andrássy, 1968	$A_4A_5A_{10}A_{13}$	B_6T_1	n	Ta, SA	16, 19	

Table 3. Continued.

	Ecology		Number of localities (samples: n > 3)	Distribution in Africa and islands	Number of the references	Remarks
	aquatic	non-aquatic				
Actinca papillatus (Schneider, 1935) Andrássy, 1964					11	
syn. *Actinolaimus papillatus* Schneider, 1935	$A_2A_3A_4A_5A_{13}$	B_2B_3	n	IC, UV	131	
Actinolaimus chappuisi Meyl, 1955	A_6A_7		1	Z	108	psammon
Actinolaimus mpoumensis Athlerr, 1960	A_3A_{14}		1	Ca	4	
Actinolaimus palmaeri Allgén, 1933	$(A_4)A_7A_{13}$		1	Z	1	
Actinolaimus perplexus Heyns & Argo, 1969	$(A_2)A_{10}$		1	SA	90	
Actinolaimus schuurmansstekhoveni Meyl, 1957	A_6A_7		1	Z	109	
syn. *Actinolaimus oomercooperi* apud S. Stekhoven & Teunissen, 1938, nec Filipjev, 1931	A_{10}	T_1	1	Z	134	
Actinolaimus straeleni Meyl, 1957	A_4A_5		3	Z	109	
Afractinolaimus noblei Andrássy, 1970	$A_4A_5A_7A_{10}A_{13}$		n	SA	19	
Brittonema taylori (Meyl, 1957) Thorne, 1967						
syn. *Actinolaimus taylori* Meyl, 1957	A_6		1	Z	109	
Mactinolaimus birketi (Altherr, 1960) Andrássy, 1970						
syn. *Actinolaimus birketi* Altherr, 1960	A_3A_{14}		1	Ca	4	
Mactinolaimus hutchinsoni (Filipjev, 1929) Andrássy, 1970						
syn. *Actinolaimus hutchinsoni* Filipjev, 1929	A_5		2	SA	65	
Mactinolaimus michaelseni (Steiner, 1916) Andrássy, 1970						
syn. *Actinolaimus michaelseni* Steiner, 1916	A_1?		?	SWA	139	publ. not seen
Mactinolaimus omercooperi (Filipjev, 1931) Andrássy, 1970						
syn. *Actinolaimus omercooperi* Filipjev, 1931	$A_2A_5A_6A_7A_{13}A_{14}$	B_2B_3	n	Et, IC	67, 131	
Mactinolaimus tenuis (Schneider, 1935) Andrássy, 1970						
syn. *Actinolaimus tenuis* Schneider, 1935	$A_3A_7A_{10}A_{14}$		n	UV, Su, Z	109, 131	
Mactinolaimus typicus Andrássy, 1970						
syn. *Actinolaimus neocyatholaimus* apud Andrássy, 1968, nec Kreis, 1936	A_7		1	Co	15	phreatic
Metactinolaimus sp. apud Baqri, Coomans & Van der Heiden, 1975		T_1		Z	30	redescription
syn. *Actinolaimus ruwenzorii* apud S. Stekhoven & Teunissen, 1938 (partim) nec De Coninck, 1935	A_4?	T_1		Z	53, 108	
Metactinolaimus leloupi Meyl, 1957	A_6		1	Z	109	
Neoactinolaimus africanus (Filipjev, 1929) Thorne, 1967						
syn. *Actinolaimus africanus* Filipjev, 1929	A_6		3	Moz, K, Z, Ta	2, 65, 132	65: plankton sample 2: 1400 found in zool. mus. Hannover; from Sansibar; 132: National Parc Albert, Zaire, without origin
Neoactinolaimus duplicidentatus (Andrássy, 1968) Andrássy, 1970						
syn. *Actinolaimus omercooperi* apud Andrássy, 1964 & 1965, nec Filipjev, 1931	$A_3A_5A_6A_7A_{13}A_{14}$	B_2B_3	n	K, Ug, Gh	11, 13	11: also in plankton
Neoactinolaimus elaboratus (Cobb, 1906) Heyns & Argo, 1969	A_2	T_1	1	SA	90	in a fountain; mainly terrestrial

Table 3. Continued.

	Ecology		Number of localities (samples: n > 3)	Distribution in Africa and islands	Number of the references	Remarks
	aquatic	non-aquatic				
Neoactinolaimus frostae (Allgen, 1952) Thorne, 1967						
syn. *Actinolaimus frostae* Alggén, 1952	A_6		1	K	3	
Neoactinolaimus intoni (Lee, 1961) Thorne, 1967						
syn. *Actinolaimus intoni* Lee, 1961	A_5		1	Ni	105	cryptobiotic
Neoactinolaimus transkeiensis Heyns & Argo, 1969	$(A_2)A_{10}$		1	SA	90	
Neoactinolaimus vaalensis Andrássy, 1970	$A_4A_{10}A_{13}$		n	SA	19	
Paractinolaimus capensis Heyns & Argo, 1969	$(A_4)A_{10}$	T_1	2	SA	90	
Paractinolaimus filipjewi (Schneider, 1935) Meyl, 1957						
syn. *Actinolaimus filipjewi* Schneider, 1935	$A_2A_3A_4A_5A_{14}$	B_2B_3	n	IC	131	
Paractinolaimus macrolaimus (de Man, 1880) Andrássy, 1964	$A_3A_6A_7A_{14}$	B_3	n	K	11	
syn. *Actinolaimus macrolaimus* (de Man, 1880) Steiner, 1916	$A_4A_5A_{10}$		n	Ta	110	
Dorylaimus macrolaimus de Man, 1880			n	S, Ta	84, 148, 149	
Paractinolaimus magaliesmontanus Heyns & Argo, 1969	$(A_4)A_{10}$		1	SA	90	
Paractinolaimus micoletzkyi (Schneider, 1935) Meyl, 1957						
syn. *Actinolaimus micoletzkyi* Schneider, 1935	$A_3(A_4)A_{14}$		2	IC	131	
Actinolaimus macrolaimus apud Micoletzky, 1925, nec (de Man, 1880) Steiner, 1916	A_1	B_3	1	Ta	112	
Paractinolaimus microdentatus (Thorne, 1939) Meyl, 1957	$A_3(A_4)A_5A_{10}$		2	I, SA	12, 90	
Paractinolaimus ruwenzorii (De Coninck, 1935) Andrássy, 1964	A_7		1	Z	108	
Paractinolaimus xosorum Heyns & Argo, 1969	$(A_2)A_{10}$		1	SA	90	
Paractinolaimus zairensis Baqri, Coomans & Van der Heiden, 1975	$A_5A_6A_{10}A_{14}$		n	Z	30	reexamination of the same samples
syn. *Actinalaimus omercooperi* apud S. Stekhoven, 1944, nec Filipjev, 1931	$A_4A_5A_6A_{14}$			Z	132	
Stomaglossa costatus (Schneider, 1935) Andrássy, 1968						
syn. *Actinolaimus costatus* Schneider, 1935	$A_2A_3A_4A_6A_{14}$	B_2B_3	n	IC	131	
Tachypleurosum conformis (Schneider, 1935) Andrássy, 1959						
syn. *Actinolaimus conformis* Schneider, 1935	A_4A_{14}	B_2B_3	1	IC	131	
Tachypleurosom labyrinthostoma (Cobb, 1893) Andrássy, 1959						
syn. *Actinolaimus labyrinthostoma* (Cobb, 1893) Cobb, 1913	A_1?		?	Col	138	no specifications of the locality ar given
BELONDIRIDAE Thorne, 1939						
Dorylaimellus afer Andrássy, 1969	A_{10}?		1	Co	17	soil & plantroots before the Adam-Hole, Kindamba
Dorylaimellus bambasea De Coninck, 1962	A_{10}	T_1	1	Gh	13	
Dorylaimellus striatus Cobb in Thorne, 1939	A_{10}		1	Gh	13	
Oxydiris oxycephalus (de Man, 1885) Thorne, 1939	$(A_5)A_{10}$	T_1	1	Gh	13	also Europe & N. America
Roquens heterurus (S. Stekhoven & Teunissen, 1938) Mulk, Coomans & Baqri, 1978		T_1			117	
syn. *Roquens africanus* Andrássy, 1970	A_4	B_6	1	SA	19	

Table 3. Continued.

	Ecology		Number of localities (samples: n > 3)	Distribution in Africa and islands	Number of the references	Remarks
	aquatic	non-aquatic				
LEPTONCHIDAE Thorne, 1935						
Discomyctus nagyi Andrássy, 1968	$(A_4)A_{10}$	B_6	1	Ta	16	
Dorylaimoides elegans (de Man, 1880) Thorne & Swanger, 1936						
syn. *Dorylaimus elegans* de Man, 1880	A_6?		1	K	148, 149	
Dorylaimoides limnophilum (de Man, 1880) Loof, 1964						
syn. *Dorylaimus limnophilum* de Man, 1880	$A_3A_4A_7A_{10}A_{14}$		n	K-Ta, BI	74, 148, 149	Europe, wide spread; terrestrial and pond sp.
Lenonchium longidens (Furstenberg & Heyns, 1966) Jairaipuri, 1967						
syn. *Dorylaimoides longidens* Furstenberg & Heyns, 1966	$(A_4)A_5$		1	SA	70	
Mumtazium mumtazae Siddiqi, 1968	A_{10}	B_6	1	Ug	135	
Proleptonchus saccatus (Clark, 1962) Andrássy, 1963	A_7	B_3	1	Co	15	phreatic
Tylencholaimus stecki Steiner, 1914	$(A_2)A_{10}$	B_3	1	Z	53	plantroots next to Adam-Hole, Kindamba
Tyleptus telyptus Andrássy, 1969	A_{10}?		1	Co	17	
DIPHTHEROPHORIDAE Thorne, 1935						
Diphtherophora vanoyei De Coninck, 1931	$(A_6)A_{10}A_{13}$	T_1	1?	K	107	
ISOLAIMIIDAE Timm, 1969						
Isolaimium incus Hogewind & Heyns, 1967	A_7		1	Co	15	phreatic
APORCELAIMIDAE Heyns, 1965						
Aporcelaimellus kikereensis Baqri & Coomans, 1973	A_{10}	T_1	1	Z	29	reexamination of the samples
syn. *Dorylaimus obtusicaudatus* apud S. Stekhoven & Teunissen, 1938 (partim) nec Bastian, 1865	A_{10}	T_1		Z	134	
Aporcelaimellus obtusicaudatus (Bastian, 1865) Altherr, 1968	A_4		2	SA	18	
syn. *Dorylaimus obtusicaudatus* apud S. Stekhoven & Teunissen, 1938 (partim), nec Bastian 1865	A_6A_{10}	B_6T_1	n	Z	134, 29	
Aporcelaimus sp. apud Andrássy, 1964	$A_3A_5A_{14}$		1	I	12	
SECERNENTEA von Linstow, 1905						
Rhabditida (Örley, 1880) Chitwood, 1933						
DIPLOGASTERIDAE (Micoletzky, 1922) Steiner, 1929						
Acrostichus angustilaimus (S. Stekhoven & Teunissen, 1938) Massay, 1966						
syn. *Diplogaster angustilaimus* S. Stekhoven & Teunissen, 1938	A_4A_{10}		2	Z	134	

Table 3. Continued.

	Ecology		Number of localities (samples: n > 3)	Distribution in Africa and islands	Number of the references	Remarks
	aquatic	non-aquatic				
Acrostichus microstoma (Goodey, 1929) Massey, 1966						
syn. *Diplogaster microstoma* Goodey, 1929	A_4		1	Z	134	
Diplogaster rivalis (Leydig, 1854) Bütschli, 1873	A_4A_{14}		1	Bl	74	
Diplogasteroides africana Micoletzky, 1916	A_4		1	S	110	
Mononchoides changi Goodrich, Heckler & Taylor, 1968	A_6		1	SA	19	Sasol-Dam
Paramonovnema filicaudatum Andrássy, 1968	$(A_5)A_{10}B_6$		1	Ta	16	
RHABDITIDAE Örley, 1880						
Ablechroiulus anchisporus Andrássy, 1966	A_5A_{10}		1	Gh	14	
Mesorhabditis paucipapillata (Paetzold, 1955) Paetzold, 1958	A_7		1	Mau	82	phreatic, heavily eutrophicated lakelet
Mesorhabditis spiculigera (Steiner, 1936) Dougherty, 1953			?	Eg	7	
Pelodera strongyloides (Schneider, 1866)						
syn. *Rhabditis strongyloides* (Schneider, 1860) Örley, 1880	A_4		1	Z	134	
Pelodera teres Schneider, 1866						
syn. *Rhabditis teres* (Schneider, 1866) Bütschli, 1873	$A_4A_6A_{10}$	T_1T_2	1	Z, Ru	134	
Rhabditis filiformis Bütschli, 1873	$A_4A_6A_{10}A_{14}$	$B_3T_1T_2$	n	Z, FPI	72, 134	
Rhabditis maupasi Seurat in Maupas, 1919	$(A_6)A_{10}$		1	Z	134	
Rhabditis producta (Schneider, 1866) Örley, 1880	A_4A_{10}	T_1T_2	2	Z	134	
Rhabditis viguieri Maupas, 1900	$(A_6)A_7A_{10}$		1	Al	106	
Rhabditioides incisocaudatus (De Coninck, 1935) Dougherty, 1955						
syn. *Rhabditis incisocaudatus* De Coninck, 1935	$(A_2)A_{10}$	B_3	1	Z	53	
PANAGROLAIMIDAE (Thorne, 1937) Paramonov, 1956						
Anguillula aceti (Müller, 1783) Ehrenberg, 1838						
syn. *Turbatrix aceti* (Müller, 1783) Peters, 1927				SA	87	
Macrolaimellus iucundus Andrássy, 1966	A_{10}		1	Gh	14	
Macrolaimus crucis Maupas, 1900	$(A_6)A_{10}$	T_1	1	Al	106	
Panagrolaimus rigidus (A. Schneider, 1866) Thorne, 1937	$A_5A_{10}A_{13}A_{15}$	T_1T_2	n	Gh	14	
CEPHALOBIDAE (Filipjev, 1934) Chitwood & Chitwood, 1934						
Acrobeles bushmanicus Heyns, 1969	A_4?	T_6	1	SA	86	dry river bed
Acrobeles ciliatus v. Linstow, 1877	A_2	T_5	1	Mau	82	
Cephalobus brachyuris Daday, 1908	A_6		1	K-Ta	148, 149	Thorne (1937): possible a *Plectus* sp.
Cephalobus palustris Daday, 1908	A_{10}		1	K	148, 149	idem
Cephalobus persegnis Bastian, 1865	A_9	T_1T_2	1	Z	53	
Cephalobus uncatus Daday, 1908	A_6		1	K-Ta	148, 149	Thorne (1937): sp. inq.; *Aphanolaimus* sp.?
Heterocephalus filiformis (de Man, 1880) Andrássy, 1967						
syn. *Cephalobus filiformis* de Man, 1880	A_6A_{10}	T_1	n	Z	134	

Table 3. Continued.

	Ecology		Number of localities (samples: n > 3)	Distribution in Africa and islands	Number of the references	Remarks
	aquatic	non-aquatic				
Paracrobeles laterellus Heyns, 1968	A_4?	T_1T_6	1	SA	85	dry sandy river bed
Tylenchida (Filipjev, 1934) Thorne, 1947						
APHELENCHIDAE (Fuchs, 1937) Steiner, 1949						
Aphelenchus kongoensis Allgén, 1933	$(A_4)A_7A_{13}$		1	Z	1	
APHELENCHOIDIDAE (Skarbilovich, 1947) Paramonov, 1953						
Aphelenchoides abyssinicus (Filipjev, 1931) Filipjev, 1934						
syn. *Aphelenchus abyssinicus* Filipjev, 1931	A_6A_{13}		1	E_t	67	
Aphelenchoides parietinus (Bastian, 1865) Steiner, 1932				Z	134	
syn. *Aphelenchoides parietinus* Bastian, 1865	A_2A_{10}	$B_3T_1T_2$	1	Z	53	
Seinura nodosa Andrássy, 1966	A_{10}		1	Gh	14	
TYLENCHIDAE Filipjev, 1934						
Pseudhalenchus anchilisposomus Tarjan, 1958	A_1?		1	Eg	142	barrage
Tylenchorhynchus africanus (Micoletzky, 1916) Micoletzky, 1922						
syn. *Tylenchus africanus* Micoletzky, 1916	A_4A_{14}	T_1	1	S	110	From Cinchona, no further speciations are given
Tylenchorhynchus alatus (Cobb, 1930) Filipjev, 1936	A_1?		1	Z	38	
Tylenchorhynchus brachycephalus Litainova, 1946	A_3A_7		1	IC	6	
Tylenchorhynchus quadrifer Andrássy, 1954	A_5A_7		1	Gh	14	
Tylenchorhynchus tarjani Andrássy, 1969	A_{10}	B_6	1	Co	16	Kindoumba, soil
Tylenchus davainei Bastian, 1865	$(A_2)A_{10}$	B_3	1	Z	53	
syn. *Anguillulina davainei* (Bastian, 1865) Goodey, 1932						
Tylenchus daveinei Bastian, 1865	$(A_2)A_{10}$	B_3T_2	1	Z	53	
syn. *Anguillulina daveinei* (Bastian, 1865) Goodey, 1932	$A_4A_6A_{10}A_{14}$	$B_3T_1T_2$	n	Ta, Ca	72, 112	
Tylenchus filiformis Bütschli, 1873						
Tylenchus leptosoma de Man, 1880	$(A_2)A_{10}$	B_3	1	Z	53	
syn. *Anguillulina leptosoma* (de Man, 1884) Goodey, 1932						
Insertae sedis						
Amblyura serpentulus (Müller, 1773) Ehrenberg, 1828	A_1?		1	Eg	63	Mount Sinai, aquatic?
syn. *Vibrio serpentulus* Müller, 1773						publ. not seen
Frostina pellucida (Allgén, 1952) Gerlach & Riemann, 1974						
syn. *Frostia pellucida* Allgén, 1952	A_4		1	K	3	

Hopper (1974, 1977, 1978).

Studies on free-living aquatic nematodes of Africa have not only been few but are also concentrated on the East-African Rift Valley, South Africa and the Niger-watershed (Fig. 1, Table 3). Contrary to the rest of the continent, large surveys on freshwater, soil and plant-parasitic nematodes have been conducted in southern Africa (Heyns, 1971, 1976; a.o.). No study has been made in about half the African countries and more than a quarter of the remainder has been sampled only once. Less than one third of the free-living aquatic species recorded from Africa has been reported more than once (Table 3), and most of them are known to be cosmopolitan or ubiquitous, e.g. *Euteratocephalus crassidens, Plectus cirratus* (Fig. 2), *Eumonhystera filiformis* (Fig. 3), *E. vulgaris* (Fig. 3), *Monhystera wangi* (Fig. 3), *Brevitobrilus graciloides* (Fig. 4), *Tripyla glomerans, Dorylaimus stagnalis* (Fig. 5), *Eudorylaimus carteri, E. obtusicaudatus, Paradorylaimus filiformis* (Fig. 5) and *Mesodorylaimus bastiani.* A relatively large number of African possible endemic species has been reported (Table 3). It is important to stress the scattered distribution of most aquatic nematode species (Figs. 2 & 3) or their occurrence as a single spot on the map. Given the often doubtful identification and the lack of a dense sample-grid, we can assume that revisional work on the species found in Africa will be marked by synonimisation of several so-called endemic species as well as by separation of conglomerated species (Andrássy, 1971a; Baqri & Coomans, 1973; Baqri *et al*, 1975).

Several typical aquatic cosmopolitans occupy a restricted area in Africa. *Brevitobrilus graciloides* for instance, is known only from the Rift Valley (Fig. 4). The areal of *Macrotobrilus elephas* on the other hand comprises at least the mountains of East Africa and the whole of southern Africa (Fig. 4). This species, endemic to Africa, is only found at Mid-African higher altitudes in Kenya and in several rivers of South Africa, which could indicate that the East African mountain ridge forms a corridor which allows cross-over of the lowlands of the Mid-African tropical region. Apparently ubiquitous nematodes like the freshwater species *Dorylaimus stagnalis* and the more eurytopic *Paradorylaimus filiformis* seem to be commonly distributed in Africa except in the south (Fig. 5), this in spite of a recent survey of freshwater nematodes in South Africa by Heyns (1976). The absence of these 'well-known' species can easily be understood since, conversely to earlier workers, the identification of such common species is today made very cautiously. As mentioned earlier, modern nematologists regard these species as groups of different species (Coomans, Heyns, pers. commun.).

An important feature for understanding the zoogeography of aquatic animals is the knowledge of their mechanisms of dispersal. Free-living aquatic nematodes yield a surprising variety of dispersing strategies. However, few of these are understood to date. Generally, it can be expected that we will gain a better insight into distributional patterns as our taxonomy improves.

Explanation of the abbreviations used in Table 3

Ecology

A_1: hygrophilic
A_2: wells & springs (crenon)
A_3: brooklets, brooks, rivulets (rhitron)
A_4: rivers & streams (potamon)
A_5: pond, pools, temporal waters
A_6: lakelets & lakes
A_7: phreatic water, psammon
A_8: caves
A_9: thermal water & hot springs
A_{10}: marshy, swampy soil & hygropetric zone
A_{11}: brackish & estuarine water
A_{12}: inland brackish saline water
A_{13}: between plants in water, on leafs; stems & leafs of water-plants
A_{14}: between algae & mosses in water or near water on humid places
A_{15}: humid places
B_1: associated with lower plants
B_2: associated with algae
B_3: associated with mosses
B_4: associated with lichenes
B_5: associated with fungi
B_6: associated with higher plants
T_1: terrestrial
T_2: bryoedaphic & lichenedaphic
T_3: dunes
T_4: beaches, shores
T_5: decomposed plants
T_6: dry riverbeds & wells
P_1: parasitic on (higher) plants
P_2: aquatic ectoparasitic
P_3: aquatic endoparasitic
P_4: terrestrial ectoparasitic
P_5: terrestrial endoparasitic

States and islands

Al	-Algeria	Mau	-Mauretania
An	-Angola	Moz	-Mozambique
Bl	-Balearic Islands	Ni	-Nigeria
Ca	-Cameroon	Rw	-Rwanda
Col	-Comores Islands	RdO	-Rio de Oro (=Spanish Sahara)
Co	-Congo-Brazzaville	S	-Simbabwe
Eg	-Egypt	SA	-South Africa
Et	-Ethiopia	SWA	-South West Africa (Namibia)
FPl	-Fernando Póo Island	Su	-Sudan
Gh	-Ghana	Ta	-Tansania
I	-Israel	Tu	-Tunesia
IC	-Ivory Coast	Ug	-Uganda
K	-Kenya	UV	-Upper-Volta
Mad	-Madagascar	Z	-Zaire

Acknowledgements

I am indebted to Prof. Dr A. Coomans of the Zoological Institute of Ghent, for the use of his library and for reading the manuscript. Prof. Dr E. Geraert is thanked for various suggestions and his help in solving some taxonomical problems. Dr H. J. Dumont is greatly acknowledged for his interest, valuable suggestions and editorial assistance. K. Martens and K. Roche are thanked for reading the manuscript and S. Wellekens is acknowledged for her encouragement and careful typing of the manuscript. This work was supported by the 'Instituut tot Aanmoediging van het Wetenschappelijk Onderzoek in Nijverheid en Landbouw' (Belgium).

References

1. Allgén, C., 1933. Über einige freilebende Nematoden aus dem Niederkongo. Zool. Anz. 103: 312–320.
2. Allgén, C., 1935. Das Männchen des Actinolaimus africanus Filipjev. Folia zool. hydrobiol. 8: 33–35.
3. Allgén, C., 1952. Über einige freilebende Süsswasser-Nematoden von Mt. Kenya (Brit. Ostafrika). Ein kleiner Beitrag zur Kenntnis tropischer Nematoden. Ark. Zool. Ser. 2, 3: 139–157.
4. Altherr, E., 1960. Results from the Danish Expedition to the French Cameroon (1949–1950). 28. Nématodes limnicoles. Bull. I.F.A.N. 22: 770–787.
5. Andrássy, I., 1954. Über einige von Daday beschiebene Nematoden-Arten. Zool. Anz. 152: 138–144.
6. Andrássy, I., 1956. Süsswasser-Nematoden aus Französisch-West-Afrika. Opusc. zool. Bpest 1: 3–18.
7. Andrássy, I., 1958. Nematoden aus ägyptischen Gewässern (Ergebnisse der zoologischen Aufsammlungen des Ungarischen Naturwissensch. Museums in Ägypten in Jahre 1957). Ann. hist. nat. Mus. natn. hung. (9) 50: 135–150.
8. Andrássy, I., 1961. Wissenschaftlichen Ergebnisse der ersten ungarischen zoologischen Expedition in Ostafrika. 2. Nematoda. Ann. hist. nat. Mus. natn. hung. 53: 281–297.
9. Andrássy, I., 1962a. Nematoden aus dem Psammon des Adige-Flusses. 2. Mem. Mus. Civ. Stor. Nat., Verona 10: 1–35.
10. Andrássy, I., 1962b. Nematoden aus dem Ufergrundwasser der Donau von Bratislava bis Budapest (Danubialia Hungarica, 17). Arch. Hydrobiol. Supp. 27: 91–117.
11. Andrássy, I., 1964. Süsswasser-Nematoden aus den grossen Gebirgsgegende Ostafrikas. Acta zool. hung. 10: 1–59.
12. Andrássy, I., 1964. Einige Nematoden aus der Umgebung des Toten Meeres. Israel J. Zool. 13: 89–97.
13. Andrássy, I., 1965. Erd- und Süsswasser-Nematoden aus Ghana. Klasse Adenophorea (Aphasmidia). Opusc. zool. Bpest 5: 127–151.
14. Andrássy, I., 1966. Erd- und Süsswasser-Nematoden aus Ghana. Klasse Secernentea (Phasmidia). Ann. Univ. scient. bpest. Rolando Eötvös, Sect. Biol. 8: 5–24.
15. Andrássy, I., (1966) 1968. The scientific results of the Hungarian soil zoological expedition to the Brazzaville-Congo. 31. Nematoden aus Grundwasser. Ann. Univ. scient. bpest. Rolando Eötvös, Sect. Biol. 9–10: 3–26.
16. Andrássy, I., 1968. Wissenschaftliche Ergebnisse der Ungarischen zoologischen Expedition nach Tansanien. 12. Bodemnematoden aus der III. Expedition. Acta zool. hung. 14: 239–257.
17. Andrássy, I., 1969. The scientific results of the Hungarian soil zoologica expedition to the Brazzaville-Congo 40. Vier neue Bodemnematoden-Arten. Opusc. zool. Bpest 9: 15–29.
18. Andrássy, I., 1969. Taxonomische Übersicht der Familien Prodorylaimidae n. fam. und Dorylaimidae de Man, 1876. Opusc. zool. Bpest 9: 187–233.
19. Andrássy, I., 1970. Nematoden aus einige Fluss-Systemen Südafrikas. Opusc. zool. Bpest 10: 179–219.
20. Andrássy, I., 1971a. Freilebende Nematoden aus Angola. 2. Über zwei Nematoden Arten. Publções cult. Co. Diam. Angola, Lisbon 84: 47–54.
21. Andrássy, I., 1971b. Überprüfung einiger von Micoletzky beschriebener Nematoden-Arten an Hand der Typenpräparate I. Mitt. zool. Mus. Berlin 47: 241–254.
22. Andrássy, I., 1978. Nematoda. In J. Illies (ed.), Limnofauna Europaea. 2nd ed. Gustav Fisher Verlag, N.Y.; Swerts & Zeitlinger B.V., Amst.: 98–117.
23. Andrássy, I., 1981. Revision of the order Monhysterida (Nematoda) inhabiting soil and inland waters. Opusc. zool. Bpest 17–18: 13–47.
24. Argo, A. D. & J. Heyns, 1972. Four new species of the genus Ironus Bastian, 1865 (Nematoda: Ironidae) from South Africa. Phytophylactica 4: 59–66.
25. Argo, A. D. & J. Heyns, 1973. New and little known species of the nematoda families Monhysteridae and Tripylidae from South Africa. Phytophylactica 5: 149–154.

25b. Baker, A. D., 1962. Checklist of the nematode superfamilies Dorylaimoidea, Rhabditoidea, Tylenchoidea, and Aphelenchoidea. E. J. Brill, Leiden, 261 pp.

26. Banage, W. B., 1962. Some nematodes from the Moor House National Nature Reserve, Westmorland. Nematologica 7: 32-36.
27. Banage, W. B., 1963. The ecological importance of free-living soil nematodes with special references to those of moorland soil. J. anim. Ecol. 32: 132-140.
28. Banage, W. B., 1964. Some aspects of the ecology of soil nematodes. Proc. e. afr. Acad. 2: 67-74.
28b. Banage, W. B., 1966. Nematode distribution in some British upland moor soils with a note on nematode parasitising fungi. J. anim. Ecol. 35: 349-361.
29. Baqri, Q. H. & A. Coomans, 1973. A taxonomic revision of the Nematode species described by S. Stekhoven and Teunissen (1938) and S. Stekhoven (1944) from National Virunga Park (Zaïre Republic). 1. Dorylaimidae, Aporcelaimidae and Longidoridae. Fondation pour favoriser les recherches scientifiques en Afrique, Bruxelles 102: 1-57.
30. Baqri, Q. H., A. Coomans & A. Van der Heiden, 1975. A taxonomic revision of the nematodes species described by S. Stekhoven & Teunissen (1938) and S. Stekhoven (1944) from National Virunga Park, Zaïre Republic. 2. Actinolaimidae. Revue zool. afr. 89: 567-586.
31. Basson, J. B. E. & J. Heyns, 1974. The genus Mesodorylaimus in South Africa (Nematoda: Dorylaimidea). Phytophylactica 6: 261-287.
32. Bassus, W., 1962. Über die Vertikalverteilung und den Massenwecksel der Nematoden in Waldböden Mitteldeutschlands. Nematologica 7: 281-293.
33. Bastian, C. H., 1865. Monograph on the Anguillulidae, or free Nematoids, marine, land, and fresh-water, with descriptions of 100 new species. Trans. linn. Soc. Lond. 25: 73-184.
34. Butschli, O., 1873. Beitrage zur Kenntnis der freilebenden Nematoden. Nova Acta Acad. Caesar. Leop. Carol. 36: 1-124.
35. Butschli, O., 1874. Zur Kenntnis der freilebende Nematoden, insbesondere den des Kieler Hafens. Abh. Senkenberg naturf. Ges. 9: 1-292.
36. Clark, W. C., 1962. Measurements as taxonomic criteria in Nematology. Nematologica 7: 10 (abstr.).
37. Cobb, N. A., 1917. The mononchs (Mononchus Bastian, 1866). A genus of free-living predatory nematodes. Contributions to a science of nematology (Cobb.). 4. Soil Sci. 3: 431-486.
38. Cobb, N. A., 1930. Tylenchus alatus n.sp. In R. P. Strong (ed.), The African republic of Liberia and the Belgian Congo. Harvard University Press, Cambridge: 487-489; Contr. Dep. trop. Med., Inst. trop. Biol. Med. 1: 487-489.
39. Coetzee, V., 1965. South African species of the genus Cobbonchus Andrássy, 1958 (Nematoda: Mononchidae). Nematologica 11: 281-290.
40. Coetzee, V., 1966. Species of the genera Granonchulus and Cobbonchus (Mononchidae) occurring in southern Africa. Nematologica 12: 302-312.
41. Coetzee, V., 1967a. Species of the genus Mylonchulus (Nematoda: Mononchidae) occurring in southern Africa. Nematologica 12: 557-567.
42. Coetzee, V., 1967b. Species of the genus Iotonchus (Nematoda: Mononchidae) occurring in southern Africa. Nematologica 13: 367-377.
43. Coetzee, V., 1968a. Monochidae (Nematoda) of southern Africa. S. Afr. J. agric. Sci. 11: 403-414.
44. Coetzee, V., 1968b. Southern African species of the genera Mononchus and Prionchulus (Mononchidae). Nematologica 14: 67-76.
45. Coomans, A., 1962. Morphological observations on Rotylenchus goodeyi Loof & Oostenbrink, 1958. 1. Redescription and variability. Nematologica 7: 203-215.
46. Coomans, A., 1971a. An outline classification of the nematoda. International Postgraduate Nematology Course, Wageningen, the Netherlands, 18 pp.
47. Coomans, A., 1971b. Morphology and nematode systematics. J. Parasitol. 57, Sect. 2, 2: 95-99.
48. Coomans, A. & D. De Waele, 1979. Species of Aphanolaimus (Nematoda, Araeolaimida) from Africa. Zool. Scripta 8: 171-180.
49. Coomans, A. & L. J. Jacobs, 1983. Halalaimus algeriensis n.sp. from the Sahara. Hydrobiologia 102: 39-44.
50. Coomans, A. & S. H. Khan, 1981. Mononchida from Mount Kenya. Biol. Jaarb. 49: 64-75.
51. Coomans, A. & M. M. Mulk, 1980. Prismatolaimus andrassyanus nomen novum for P. andrassyi apud Mulk & Coomans, 1979, nec P. andrassyi Khera & Chaturvedi, 1977. Nematologica 26: 283 (short commun.).
52. Coomans, A., L. De Coninck & C. Heip, 1978. Round table discussions. Ann. Soc. r. zool. Belg. 108: 109-122.
52b. Croizat, L., 1958. Panbiogeography, Vols 1, 2a, 2b. Published by the Author, Caracas, 1731 pp.
52c. Croizat, L., (1962) 1964. Space, time, form: the biological synthesis. Published by the Author, Caracas.
53. De Coninck, L. A., 1935. Contributions à la Connaissance des Nématodes libres du Congo Belge. 1. Les Nématodes libres des marais de la Nyamuamba (Ruwenzori) et des sources chaudes de Mont Banze (Lac Kivu). Revue zool. Bot. afr. 26: 211-232, 249-326.
54. de Man, J. G., 1884. Die Frei in der reinen Erde und in süssen Wasser lebenden Nematoden der niederlandischen Fauna. Eine systematische-faunistische Monographie. Leiden, 206 pp., 34 pls, 145 figs.
55. Deubert, K. H., 1959. Über die Bedeutung der Nematodenfauna ackerbaulich genutster Böden. Zertbl. Bakt. 112: 101-108.
56. De Waele, D., 1976. Morfologische, systematische en ekologische bijdrage tot de kennis van nematodenfauna van enkele hooggebergtemeren op Mount Kenya. M.Sc. Thesis, R. U. Gent, 205 pp.
57. Domurat, K., 1970. Nematode communities occurring in spring barley crops. Ekol. pol. 18: 681-740.
58. Dujardin, F., 1845. Histoire naturelle des helminthes ou vers intestinaux. Paris, 654 pp.
59. Dumont, H. J., 1980a. Zooplankton and the Science of Biogeography: the example of Africa. In W. C. Kerfoot (ed.), Evolution and Ecology of Zooplankton Communities. University Press New England, Hannover: p. 685-696.
60. Dumont, H. J., 1980b. Workshop on taxonomy and biogeography. In H. J. Dumont & J. Green (eds.), Rotatoria. Hydrobiologia 73: 205-206.
61. Dumont, H. J., 1983. Biogeography of Rotifers. Hydrobiologia 104: 19-30.
62. Dumont, H. & A. Van Vossel, 1972. Een diurnaal ritme in

het voorkomen van plankton in de Beneden-Samber (Mornimont). Modèle mathématique de pollution d'eaux intérieures. Rapport de synthèse. I Services du Premier Ministre, Commission intreministérielle de la politique scientifique: 274-288.
63. Ehrenberg, C. G., 1828. Symbolae physicae seu icones et descriptiones animalium evertebratorum sepositis insectis quae ex itinere Africam borealem et Asiam occidentalem Friderici Guilelmi Hemprich et Christiani Godofredi Ehrenberg. Berlin (Academie), no pagination.
64. Ferris, V. R., C. G. Goseco & J. M. Ferris, 1976. Biogeography of Free-living Soil Nematodes from the Perspective of Plate Tectonics. Science 193: 508-510.
65. Filipjev, I. N., 1929. Two new species of Actinolaimus from South Africa. Ann. Mag. nat. Hist. 10, 4: 433-439.
66. Filipjev, I. N., 1929-1930. Les nematodes libres de la baie de la Neva et de l'extremité orientale du Golfe de Finlande. Arch. Hydrobiol. 21: 1-64.
67. Filipjev, I., 1931. Report on freshwater Nematoda: Mr. Omer-Cooper's investigation of the Abyssinian freshwaters (Hugh Scott Expedition). Proc. zool. Soc. Lond. 2: 429-443.
68. Frey, D. G., 1982. Cosmopolitanism in Cladocera. Arch. Hydrobiol. 93: 484-502.
69. Furstenberg, J. P. & J. Heyns, 1965. Two new species of the genus Discolaimium Thorne, 1939 (Nematoda: Dorylaimoidea) from South Africa. S. Afr. J. agric. Sci. 8: 1155-1159.
70. Furstenberg, J. P. & J. Heyns, 1966a. Chitwoodia transvaalensis n.gen., n.sp. and Dorylaimoides longidens n.sp., two new nematodes from South Africa. Proc. helminth. Soc. Wash. 33: 26-30.
71. Furstenberg, J. P. & J. Heyns, 1966b. Chitwoodius nom.nov. from Chitwoodia Furstenberg and Heyns, 1966. Proc. helminth. Soc. Wash. 33: 165.
72. Gadea, E., 1960. Nota sobre algunos nematodos muscicolas de los territorios del Golfo de Guinea. Trab. Mus. Zool., Barcelona 2, 7 pp.
73. Gadea, E., 1962. The nematode communities inhabiting some freshwater and moss biotopes. Nematologica 7: 15 (abstract).
74. Gadea, E., 1963. Nota sobre nematodos dulciacuáticos de la isla de Menorca. P. Inst. Biol. apl. 34: 101-109.
75. Geraert, E., 1968. Morphometric relations in nematodes. Nematologica 14: 171-183.
76. Gerlach, S. A., 1953. Die biozönotische Gleiderung der Nematodenfauna an den deutschen Küsten. Z. Morph. Ökol. Tiere 41:
77. Gerlach, S. A., 1958. Deuxième contribution à la faune des Nématodes des eaux interstitielles littorales de Madagascar. Mém. Inst. scient. Madagascar F 2: 343-365.
78. Gerlach, S. A. & F. Riemann, 1973 & 1974. The Bremerhaven checklist of aquatic nematodes. A catalogue of Nematoda Adenophorea excluding the Dorylaimida. 1 & 2. Veröff. Inst. Meeresforsch. Bremerh. Suppl. 4, 736 pp.
78b. Gerlach, S. A. & M. Schrage, 1972. Life cycles at low temperatures in some free-living marine nematodes. Veröff. Inst. Meeresforsch. Bremerh. 14: 5-11.
79. Goodey, J. B., 1959. Data to be considered, observed and, where possible, reported upon when presenting descriptions of new species. Nematologica 4: 211-216.
80. Goodey, T., 1963. Soil and freshwater Nematodes, 2nd ed. revised by Goodey, J. B. Methuen & Co. Ltd., Lond., 544 pp.
81. Goodey, J. B. & D. J. Hooper, 1965. A neotype of Aphelenchus avenae Bastian, 1865 and the rejection of Metaphelenchus Steiner, 1943. Nematologica 11: 55-65.
82. Goossens, A., 1976. A study of the aquatic nematodes of the Adrar mountains in Mauretania. Biol. Jaarb. 44: 178-190.
83. Grootaert, P., (1976) 1977. A note on the aquatic nematodes from Rio de Oro (North-Western Sahara). Bull. I.F.A.N. 38 A: 832-835.
84. Harris, J. R. & P. S. Gooch, 1970. Plant nematology in Africa (Northern, 1932-69; Sub-Saharan, 1965-1969). Commonw. agric. Bur., Techn. Commun. 41, 32 pp.
85. Heyns, J., 1968. Paracrobeles laterellus n.gen.n.sp. from South Africa (Nematoda, Cephalobidae). Nematologica 14: 511-514.
86. Heyns, J., 1969. South African species of the genus Acrobeles von Linstow (Nematoda: Cephalobidae). Phytophylactica 1: 57-66.
87. Heyns, J., 1971. A guide to the plant and soil nematodes of South Africa. A. A. Balkema, Cape Town, 233 pp.
88. Heyns, J., 1976. Preliminary results of a survey of freshwater nematodes in South Africa. J. Limnol. Soc. s. Afr. 2: 43-45.
89. Heyns, J., 1977. Freshwater nematodes from South Africa. 1. Euteratocephalus Andrássy, 1958. Nematologica 23: 112-118.
90. Heyns, J. & A. D. Argo, 1969. Actinolaimidea of South Africa (Nematoda: Dorylaimida). Phytophylactica 1: 217-228.
91. Heyns, J. & A. Coomans, 1977. Freshwater nematodes from South Africa. 2. Oncholaimus deconincki n.sp. Revue zool. afr. 91: 906-912.
92. Heyns, J. & A. Coomans, 1980. Freshwater nematodes from South Africa. 5. Chronogaster Cobb, 1913. Nematologica 26: 187-208.
93. Heyns, J. & G. Lagerwey, 1965. South African species of the genus Iotonchus Cobb, 1916 (Nematoda: Mononchidae). S. Afr. J. agric. Sci. 8: 775-783.
94. Hirschmann, H., 1952. Die Nematoden der wassergrenze mittelfrankisher Gewasser. Zool. Jb. Syst. 81: 313-407.
95. Hopper, B. E. & E. J. Cairns, 1959. Taxonomic keys to plant, soil and aquatic nematodes. Alabama polytech. Inst. Auburn, 176 pp., 28 figs.
95b. Hopper, B. E. & S. P. Meyers, 1966. Aspects of the life cycle of marine nematodes. Helgoländer wiss. Meeresunters. 13: 444-449.
96. Hopper, B. E. & A. C. Tarjan, 1977 & 1978. Supplement for the year 1972 & 1973 to nomenclatorial compilation of plant and soil nematodes. Biosyst. Res. Inst., Res. Branch, Agric. Can., Ottawa, 17 & 21 pp.
96b. Hopper, B. E., J. W. Fell & R. C. Cefalu, 1973. Effect of temperature on life cycles of nematodes associated with the mangrove (Rhizophora mangle) detrital system. Mar. Biol. 23: 293-296.
97. Inglis, W. G., 1965. Nudora omercooperi sp. nov. (Nematoda: Free-Living) from a Unusual Locality in South

Africa. Hydrobiologia 25: 176–180.

98. Johnson, S. R., V. R. Ferris & J. M. Ferris, 1972. Nematode community structure of forest woodlots. 1. J. Nematol. 4: 175–182.

99. Joubert, A. P. & J. Heyns, 1979. Freshwater nematodes from South Africa. 3. Tobrilus Andrássy, 1959. J. Limnol. Soc. s. Afr. 5: 17–26.

100. Joubert, A. P. & J. Heyns, 1980. Freshwater nematodes from South Africa. 4. The genus Monhystera Bastian, 1865. S. Afr. Tijdskr. Dierk./S. afr. J. Zool. 15: 190–197.

101. Kir'yanova, E. S. & E'. L. Krall', 1969. Paraziticheskie Nematody Rastenii i Mery Bor' by s Nimi (Plantparasitic Nematodes and their control). 1. Nauka, Leningrad, 447 pp. (in Russian).

102. Kir'yanova, E. S. & E'. L. Krall', 1971. Paraziticheskie Nematody Rastenii i Mery Bor' by s Nimi (Plantparasitic nematodes and their control). 2. Nauka, Leningrad, 522 pp. (in Russian).

103. Kir'yanova, E. S. & E'. L. Krall', 1980. Paraziticheskie Nematody Rastenii i Mery Bor'by s Nimi (Plantparasitic nematodes and their control). 2. Amerind Publishing Co., Pvt. Ltd., New Delhi, 748 pp. (translated from Russian with index of volumes 1 & 2).

103b. Klekowski, R. Z., F. Schiemer & A. Duncan, 1979. A bioenergetic study of a benthic nematode Plectus palustris de Man, 1880, throughout its life cycle. 1. The respiratory metabolism at different densities of bacterial food. Oecologia 44: 119–124.

104. Kleynhans, K. P. N., 1970. New records and species of Dorylaimoidea (Nematoda) from South Africa. Phytophylactica 2: 251–261.

105. Lee, D. L., 1961. Two new species of cryptobiotic (anabiotic) freshwater nematodes, Actinolaimus hintoni sp. nov. and Dorylaimus keilini sp.nov. (Dorylaimidae). Parasitology 51: 237–240.

105b. Mai, W. F. & H. H. Lyon, 1960. Pictorial key to genera of plant parasitic nematodes. Art Carft of Ithaca, Inc., Ithaca, N.Y., 97 pp., 49 pls.

106. Maupas, E., 1900. Modes et formes de reproduction chez les nematodes. Arch. Zool. exp. gén. 8: 463–624.

107. Meyer, A. J. & A. Coomans, 1977. Freeliving nematodes from Mount Kenya. 1. Scientific report of the Belgian Mt. Kenya Bio-Expedition, 5. Revue Zool. afr. 91: 493–505.

108. Meyl, A. H., 1955. Nematodes aus dem Psammon des Tanganyika-Sees. Zool. Anz. 155: 135–138.

109. Meyl, A. H., 1957. Freeliving nematodes: Exploration Hydrobiologique du Lac Tanganyika (1946–1947). Mém. Inst. r. Sci. nat. Belg. 3: 26–51.

110. Micoletzky, H., 1916. Ergebnisse einer botanischen Forschungsreise nach Deutsch-Ostafrika und Sudafrika (Kapland, Natal und Rhodesien) Süsswasser-Nematoden aus Sud-Afrika. Denkschr. k. Akad. Wiss., Wien, Math. Naturw. Kl. 92: 149–171.

111. Micoletzky, H., 1922. Die freilebenden Erd-Nematoden. Arch. Naturgesch. A 87: 1–650.

112. Micoletzky, H., 1925a. Zur Kenntnis tropischer, freilebender Nematoden aus Suriname, Trinidad und Ostafrika. Zool. Anz. 64: 1–28.

113. Micoletzky, H., 1925b. Die freilebenden Süsswasser- und Moornematodes Dänemarks. K. danske Vidensk. Selsk. Skr., Ser. 8, 10: 58–310.

114. Mulk, M. M. & A. Coomans, 1978. Freeliving nematodes from Mount Kenya. 2. Scientific report of the Belgian Mt. Kenya Bioexpedition, 13. Revue Zool. afr. 92: 593–608.

115. Mulk, M. M. & A. Coomans, (1978) 1979. Freeliving nematodes from Mount Kenya. 3. Scientific report of the Belgian Mt. Kenya Bio-expedition, no. 15. Biol. Jaarb. 46: 145–157.

116. Mulk, M. M. & A. Coomans, 1979. Three new Alaimus-species (Nematodes: Alaimidae) from Mount Kenya. Nematologica 25: 447–457.

117. Mulk, M. M., A. Coomans & Q. H. Baqri, 1978. A taxonomic revision of the nematode species described by S. Stekhoven and Teunissen (1938) from National Virunga Park, Zaïre Republic. 3. Belondiridae. Revue Zool. afr. 92: 711–722.

118. Mulvey, R. H., 1961. The Mononchidae: a family of predaceous Nematodes. 1. Genus Mylonchulus (Enoplida: Mononchidae). Can. J. Zool. 39: 665–696.

119. Mulvey, R. H. & H. J. Jensen, 1967. The Mononchidae of Nigeria. Can. J. Zool. 45: 667–727.

120. Nicholas, W. L., 1975. The biology of free-living nematodes. Clarendon Press, Oxford, 219 pp.

120b. Nigon, V., 1951. Polyploidie expérimentale chez un nématode libre, Rhabditis elegans Maupas. Bull. biol. Fr. Belg. 85: 187–225.

120c. Nigon, V. & E. C. Dougherty, 1950. A dwarf mutation in a nematode. A morphological mutant of Rhabditis briggsae, a free-living soil nematode. J. Hered. 41: 103–109.

120d. Nigon, V. & E. C. Dougherty, 1954. Variations dans la vitesse de croissance chez un nématode Caenorhabditis elegans (Maupas). Caryologia, Suppl.: 796–797.

121. Omer-Cooper, J., 1930. Dr. Hugh Scott's Expedition to Abyssinia. A preliminary investigation of freshwater fauna of Abyssinia. Proc. zool. Soc. Lond. 1130, 1: 195–207.

122. Overgaard, C., 1948. Studies on the soil microfauna. 1. The moss inhabiting nematodes and rotifers. Publ. Sco. Sci. Lett. Aarhus 1: 3–98.

123. Overgaard Nielsen, C., 1949. Studies on the soil microfauna. 2. The soil inhabiting nematodes. Natura jutl. 2: 1–131.

124. Paramonov, A. A., 1962. Osnovy fitogel mintologii (plant-parasitic nematodes). 1. Izdatel'stvo Akademii Nauk SSSR, Moskva, 390 pp. (translated in 1968, I.P.S.T., Jerusalem).

125. Pejler, B., 1977. On the global distinction of the family Branchionidae (Rotatoria). Arch. Hydrobiol. Suppl. 53: 255–306.

126. Peters, B. G., 1952. Toxicity test with vinegar eelworm. 1. Counting and culturing. J. Helminth. 26: 97–110.

127. Riemann, F., 1974. On hemisessile nematoda with flagelliform tails living in marine soft bottoms and on microtubes fund in deep sediments. Mikrofauna Meeresboden 40: 1–15.

128. Riemann, F. & E. Rachor, (1972) 1973. Contributions to the knowledge of Suez Canal migration. Freilebende Nematoden aus dem Suezkanal. Israel J. Zool. 21: 167–187.

129. Roggen, D. R. & R. Asselberg, 1971. The use of ratios in nematology. Nematologica 17: 187–189.

130. Schneider, A., 1866. Monographie der Nematoden. Berlin, 357 pp.

131. Schneider, W., 1935. Voyage de Ch. Alluaud et P. A. Chappuis en Afrique Occidentale Française. (Déc. 1930-Avril 1931). 6. Freilebene Nematoden. Arch. Hydrobiol. Planktonk. 28: 1-20.
132. Schuurmans Stekhoven, J. H. Jr., 1944. Nématodes libres d'eau douce. Damas (1935-36). Explor. Parc natn. Albert 9: 1-31.
133. Schuurmans Stekhoven, J. H. Jr., 1951. Nématodes saprozoaires et libres du Congo Belge. Mém. Inst. r. Sci. nat. Belg., Sér. 2, 39: 3-79.
134. Schuurmans Stekhoven, J. H. Jr. & R. J. H. Teunissen, 1938. Nématodes libres terrestres. Mission de Witte G. F. Explor. Parc natn. Albert 22: 1-229.
135. Siddiqi, M. R., 1969. Mumtazium mumtazae n.gen., n.sp. (Nematoda: Tylencholaimidae) with the proposal of Laimydorus n.gen. (Thornematidae). Nematologica 15: 234-240.
136. Southey, J. F., 1982. Plant nematology. Her Majesty's Stationery Office, Minist. Agric. Fish. Food, Lond., 440 pp.
137. Stanley, S. M., 1975. Clades versus Clones in Evolution: Why We Have Sex. Science 190: 382-383.
138. Steiner, G., 1916a. Beiträge zur geographischen Verbreitung freilebender Nematoden. Zool. Anz. 46: 311-347.
139. Steiner, G., 1916b. Nématodes. Beitr. Kennt. Land- Süsswasserfauna Dt. Südw. Afr. 1: 377-411.
140. Stöckli, A., 1950. Über die quantitatieve Bestimmung der Bodennematoden. Z. PflErnähr. Düng 51: 1-22.
140b. Sturhan, D., 1967. Induktion von Mutationen bei Nematoden. Naturwissenschaften 24: 649-650.
141. Sturhan, D., 1970. Intraspezifische Variabilität und Mutationen bei Nematoden. In H. Sandner & J. Kozlowska (eds.), Proc. 11th int Nematol. Symp. Zeszyty Problemowe Postepow nauk Roln. 92: 209-220.
142. Tarjan, A. C., 1964. Plant-parasitic Nematodes in the United Arab Republic. F.A.O. Pl. Protect. Bull. 12: 1-8.
142b. Tarjan, A. C., 1980. An illustrated guide to the marine nematodes. Inst. Food agric. Sci., Univ. Florida, 135 pp., 49 pls.
143. Tarjan, A. C. & B. E. Hopper, 1974. Nomenclatorial compilation of plant and soil nematodes. Society of Nematologists Incorporated, U.S.A., 419 pp.
144a. Thomas, P. R. & C. E. Taylor, 1968. Plant nematology in Africa south of the Sahara. Commonw. agric. Bur., Techn. Commun. Bur. Helminthol. 39, 83 pp.
144b. Thorne, G., 1937. A revision of the nematode family Cephalobidae Chitwood and Chitwood 1934. Proc. helminth. Soc. Wash. 4: 1-16.
144c. Timm, R. W., 1963. Marine nematodes of the suborder Mohhysterina from the Arabian Sea at Karachi. Proc. helminth. Soc. Wash. 30: 34-49.
145. Tsalolikhin, S. Y., 1981. Revision of the genus Tobrilus (Nematoda Tobrilidae). Zool. Zh. 60: 1302-1313. (in Russian, with Englisch summary).
146. Van de Velde, I., 1982. Revisie van het genus Mesocyclops Sars, 1914 (Crustacea: Copepoda) in Afrika - met een bijdrage tot de kennis van Cladocera en Copepoda van de Sahara en randgebieden. Ph.D. Thesis, Rijksuniversiteit Gent, 361 pp.
147. von Daday, E., 1893. Beiträge zur Kenntnis der Microfauna der Natronwässer des Alföldes. Math. naturw. Ber. Ung. 11: 286-321.
148. von Daday, E., 1908. Adatok német-kelet-Afrika édesvízi mikrofaunajanak ismeretéhez. Math. és Természettud. Ertesitö Magyar Tudomány. Akad. Budap. 26: 1-42.
149. von Daday, E., 1910a. Die Süsswasser-Mikrofauna Deutsch-Ost-Afrikas. Zool. Stuttg. 23: 44-56.
150. von Daday, E., 1910b. Ergebnisse der mit Subvention aus der Erbschaft Treitl unternommen zoologischen Forsschungsreise Dr. Franz Werner's nach den ägyptischen Sudan und Nord-Uganda. 15. Beitrage zur Kenntnis der Mikrofauna des Nils. Sber. Akad. Wiss. Wien, Math.-naturw. Kl., Abt. 1, 116: 537-589.
151. Wallace, H. R., 1958 & 1959a. Movement of eelworms. 1, 2, 3, 4, 5. Ann. appl. Biol. 46: 74-85, 86-94, 662-668; 47: 131-139, 350-356.
152. Wallace, H. R., 1958 & 1959a. The movement of eelworms in water films. Ann. appl. Biol. 47: 366-370.
153. Wallace, H. R., 1971a. The movement of nematodes in the external environment. In A. M. Fallis (ed.), Ecology and physiology of parasites. University of Toronto Press, Toronto: 1-201.
154. Wallace, H. R., 1971b. Abiotic influences in the soil environment. In B. M. Zuckerman, W. F. Mai & R. A. Rohde (eds.), Plant parasitic nematodes. 1. Academic Press, N.Y.: 257-280.
155. Warwick, R. M., 1971. Nematode associations in the Exe estuary. J. mar. Biol. Ass. U.K. 51: 439-454.
156. Wasilewska, L., 1970. Nematodes of the sand dunes in the Kampinos forest. 1. Ekol. pol. 18: 429-443.
157. Wieser, W., 1959. Freeliving marine nematodes. 4. General part. Acta Univ. lund. N.F. II 55: 1-111.
158. Winslow, R. D., 1960. Nematode ecology. In J. N. Sasser & W. R. Jenkins (eds.), Nematology. University of North Carolina Press, Chapel Hill: 341-415.
159. Zuckerman, B. M., W. F. Mai & R. A. Rohde, 1971. Plant parasitic nematodes. 1. Academic Press, N.Y., 345 pp.

The effect of Cyanophyta upon zooplankton in a eutrophic tropical lake (Lake Valencia, Venezuela)

Aída Infante & Wendula Riehl
Escuela de Biología, Universidad Central, Apartado 47106, Los Chaguaramos, Caracas, 1041, Venezuela

Keywords: tropical zooplankton, disappearance of cladocerans, cyanophytes interfering, eutrophic lake

Abstract

The effect which Cyanophyta have upon the zooplankton varies according to the form of the alga (mucilaginous colonies or filaments) and its abundance. Periodical blooms of *Microcystis aeruginosa* were not detrimental for the zooplankton, in spite of the fact that copepods, cladocerans and rotifers consume small colonies. High concentrations of *Lyngbya limnetica* and *Oscillatoria limnetica* in Lake Valencia, Venezuela, proved to be inhibitory for cladocerans. A total absence of cladocerans was detected when filaments increased.

Introduction

The effects of Cyanophyta on zooplankton population dynamics have been a matter of controversy among limnologists, especially since eutrophication problems continue to intensify in many regions of the world. Much of the literature concerns the nutritional adequacy of this group of algae, with its reportedly toxic properties (Gorham, 1964; Stangenberg, 1968; Lampert, 1981) as potential food for zooplankton (Arnold, 1971; Porter & Orcutt, 1980; De Bernardi *et al.*, 1981).

In recent years, growing attention has focused on mechanical effects that filamentous Cyanophyta have upon the feeding of Cladocera (Hayward & Gallup, 1976; Webster & Peters, 1978; Gliwicz, 1980). Experimental and field results indicate that high concentrations of filamentous forms may affect the nutrition of cladocerans. The filaments clog the animal's filtering apparatus, and rejection movements of the post-abdomen become more frequent. Since filtration momentarily stops during the rejection movement (Burns, 1968), a reduced rate of food capture, combined with increased energy expenditure for rejection, may reduce the energy availability for growth and reproduction (Porter *et al.*, 1982).

In lakes with high production of bluegreens, such as Lake Valencia, cladocerans appear to be more affected than copepods or rotifers. In a previous paper (Infante, 1982) it was suggested that a long period of absence of all pelagic cladocerans species (*Ceriodaphnia cornuta, Diaphanosoma brachyurum* and *Moina micrura*) may have been caused primarily by the presence of high concentrations of *Oscillatoria limnetica* and *Lyngbya limnetica*.

In this paper, the relationship between the most common bluegreen algae in Lake Valencia and the zooplankton are analyzed. Emphasis is placed on changes noted in cladoceran populations that can be related to the filamentous Cyanophyta.

Material and methods

Lake Valencia

Lake Valencia is a large lake (area 350 km^2 and volume 6.3 km^3) in tropical South America (10° 10′N and 67°45′W). Large natural lakes are very scarce

Hydrobiologia 113, 293–298 (1984).

in that region, where fluvial systems predominate. There are other factors that make studies of this unusual lake very desirable. These are the progressive lowering of the lake level and the increase in salinity. High evaporation rates, a diminished income of underground water, the endorheic character of the lake and the discharge of untreated wastewater of about 2 million people, have been mentioned as the main causes of water reduction and increase in salt concentration (Böck, 1968; Tamers & Thielen, 1966; Lewis & Weibezahn, 1976).

The water temperature is high. At the surface, it varies between 26 °C and 29 °C. The lake has one annual period of water circulation from December to March (Infante, 1982). The depth of the mixing depends primarily on wind strength. During the rest of the year, the lake remains stratified and the deep waters (the deepest 10–15 m of the water column) become anoxic. Important amounts of toxic reduced substances accumulate and can produce mass mortality of fish and zooplankton when the lake starts circulating. However, recovery is very rapid (Infante *et al.*, 1979).

The phytoplankton assemblage and primary production of Lake Valencia reflect eutrophic condition. It is characterized by the continual dominance of cyanophytes, which at times produce blooms. The most numerous species are *Microcystis aeruginosa, Lyngbya limnetica* and *Oscillatoria limnetica.*

The major zooplankton species are *Brachionus calyciflorus, Keratella americana, Mesocyclops (Thermocyclops) crassus, Notodiaptomus venezolanus, Ceriodaphnia cornuta, Diaphanosoma brachyurum* and *Moina micrura*. Predators on zooplankton are an endemic planktophagic fish (*Xenomelaniris venezuelae*), larvae of several other fish species and *Chaoborus* sp.

Duplicate samples were taken every two meters vertically at two week intervals from October 1975 to April 1978. Twenty liters of water were pumped from the lake into a 60 μm plankton net to concentrate the zooplankton. Two aliquots of the concentrate were counted in Sedgwick-Rafter cells. Phytoplankton samples were taken concurrently with a bottle, identified and counted using Utermöhl chambers with an inverted microscope.

Results

Microcystis aeruginosa reaches very high concentrations in Lake Valencia (up to 3 000 colonies ml^{-1}). The size of the colonies varies from a few microns in diameter to big aggregates of some centimeters in width. However, most colonies have 100–300 cells. Although the colonies are found throughout the whole water column, the highest concentrations are observed in the upper 10 m. Quiet zones of the lake are often covered with a coat of *Microcystis* from January to April, when this alga produces blooms. At times, these blooms coincide with maxima of some zooplankton (Fig. 1), and small colonies (20–50 cells) are frequently found in the gut of most herbivores in Lake Valencia, including the three cladoceran species (Infante, 1978). This result does not indicate a toxic effect of *Microcystis* upon the zooplankton in Lake Valencia, when this alga blooms.

The other two dominant blue-green algae in Lake Valencia are filamentous forms. Interference by filaments on the abundance of copepods and rotifers was not observed (Fig. 1). During 1976 and 1977 when *Lyngbya limnetica* and *Oscillatoria limnetica* reached high concentrations in the lake, population densities of copepods and rotifers were also high. Previous studies on the food of zooplankton (Arnold, 1971; Moriarty *et al.*, 1973; Infante, 1978) showed that *Lyngbya* is frequently ingested by the copepods.

The situation was evidently different for the cladocerans (Fig. 2). A short period of cladoceran absence from September to December 1976 coincided with high concentrations of *L. limnetica* (200–500 filaments ml^{-1}) and the sustained absence from April 1977 to April 1978 corresponded with high densities of *Lyngbya* as well as *Oscillatoria.* During the peak of filament abundance in the lake (June and July 1977 and March 1978) they reached densities of 1300 filaments ml^{-1} in the upper 10 m of the water column, where zooplankton was concentrated.

Figure 3 shows the inhibitory concentrations of *Lyngbya* and *Oscillatoria* upon *Diaphanosoma, Ceriodaphnia* and *Moina*. At concentrations of *Lyngbya* and *Oscillatoria* higher than 9×10^7 filaments m^{-2} none of the cladocerans persisted in the lake.

Discussion

The results indicate that the effect which the cyanophytes have upon the zooplankton in Lake Valencia varies according to the form of the alga (mucilaginous colonies or filaments) and its abundance.

The fact that cladoceran populations develop well while ingesting *M. aeruginosa* seems to be in contradiction with experimental results showing a strong toxic effect of this alga upon *Daphnia* (Lampert, 1981).

According to Scott *et al.* (1980) there are two forms of *Microcystis aeruginosa* which have differing toxic effects on animals. The form *M. aeruginosa aeruginosa* always proved toxic, but never the form *M. aeruginosa flos-aquae*. The toxic strains are difficult to separate from the non-toxic by conventional methods. Contradictory results may possibly arise from the use of strains of different toxicity. However, this does not explain the results of Stangenberg (1968), who observed intense growth and reproduction of *Daphnia longispina* when fed fresh untriturated *Microcystis* colonies. Further, the death of the animals occurred when the cell contents of the same colonies were liberated.

De Bernardi *et al.* (1981) demonstrated experimentally that small colonies of *Microcystis* may be a suitable food for three species of *Daphnia*.

It is well known, that mucilaginous blue-green algal masses are very often associated with abundant bacteria and it is extremely difficult to remove the bacteria attached to the mucilage. Moreover, a mutualistic relationship between bacteria and planktonic blue-green algae has been suggested (Whitton, 1973). *Microcystis aeruginosa* seems to need associated bacteria in order to develop well, as Vance (1966) observed. He isolated a strain of this

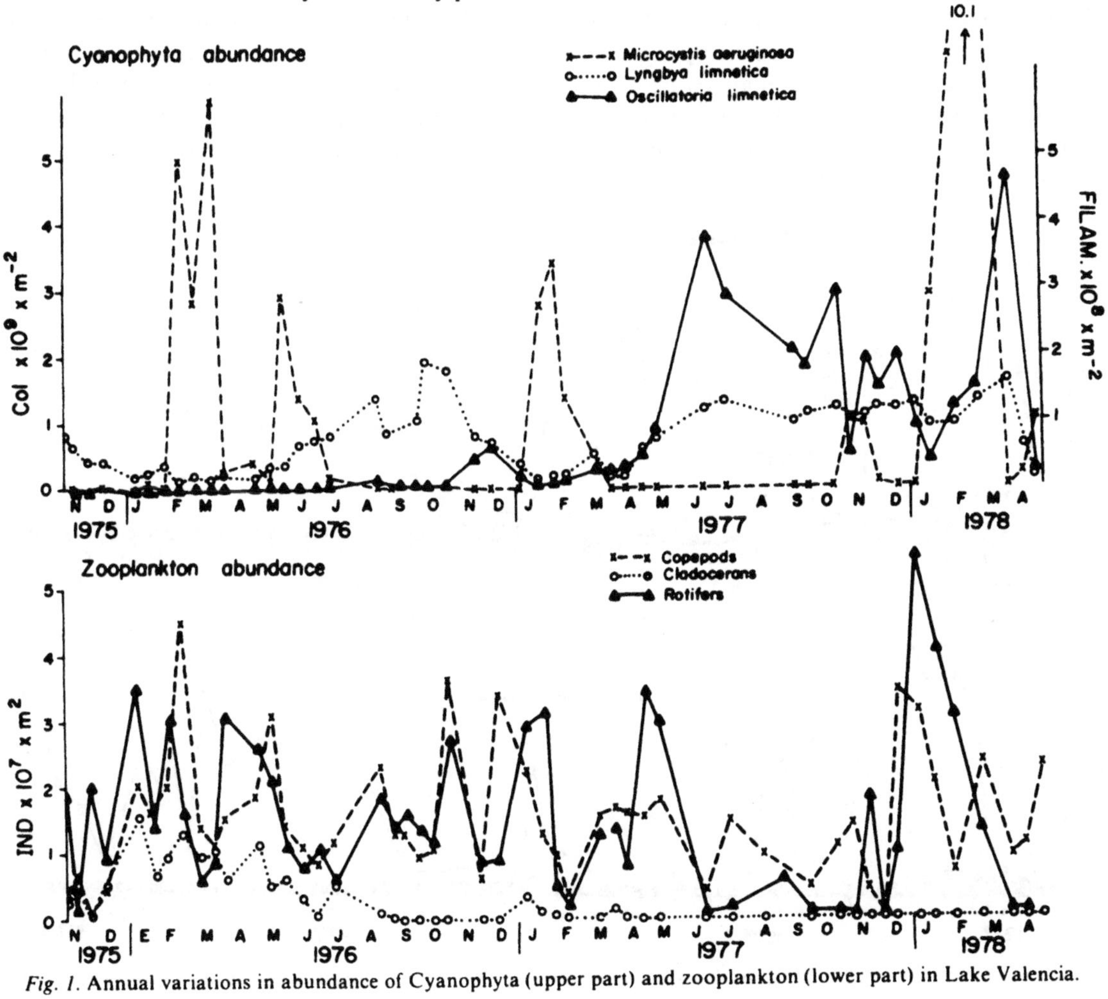

Fig. 1. Annual variations in abundance of Cyanophyta (upper part) and zooplankton (lower part) in Lake Valencia.

alga and was not able to subculture it in absence of associated bacteria. The unsuccessful attempts often reported in the literature to obtain colonies in the laboratory cultures, could be related to the absence of the right bacteria.

The fact that the toxic effect is detected only when single or paired cells (Lampert, 1981) as well as disrupted cells are provided (Stangenberg, 1968), but not when colonies are ingested, indicates that the cells in the colonies may be better protected against digestion, preventing the liberation of cellular endotoxin in the gut of the animals. The nutrition of the animals would then depend more on the mucilaginous part of the colony and on associated bacteria, rather than on *Microcystis* cells.

Another interesting result is the temporal absence of cladocerans in Lake Valencia. Similar cases of disappearance of cladocerans have been reported by Richards *et al.* (1975) in Lake Tahoe and by Gophen (1979) in Lake Kinneret. In both lakes, predation by introduced planktophagic species have been mentioned as the main cause for the elimination of the cladocerans. While not ignoring the influence that predation may have had on the temporary disappearance of the cladocerans in Lake Valencia, the increase of filamentous cyanophytes may have been more important. The reasons supporting this hypothesis can be summarized as follows:

1. The cladocerans developed and reproduced well during the period when filaments were in low densities. In the two instances where filaments increased, all three species disappeared and did not reappear until the filaments decreased.

2. During the first 18 months of the study no males or ephippia were observed. Shortly before the crash of the animals, some males were captured. This is an indication that unfavorable conditions were developing in the lake and this points to the dramatic increase in filaments, since food concentration, temperature, light and other factors remained at normal levels.

3. The number of eggs per adult female, as well as the percentage of females with embryos declined as filaments became abundant in the lake. This may be interpreted as an expression of a deficient nutrition of the animals. No accompanying decline in the amount of edible species of phytoplankton was detected; therefore, an interference in the intake of food is likely.

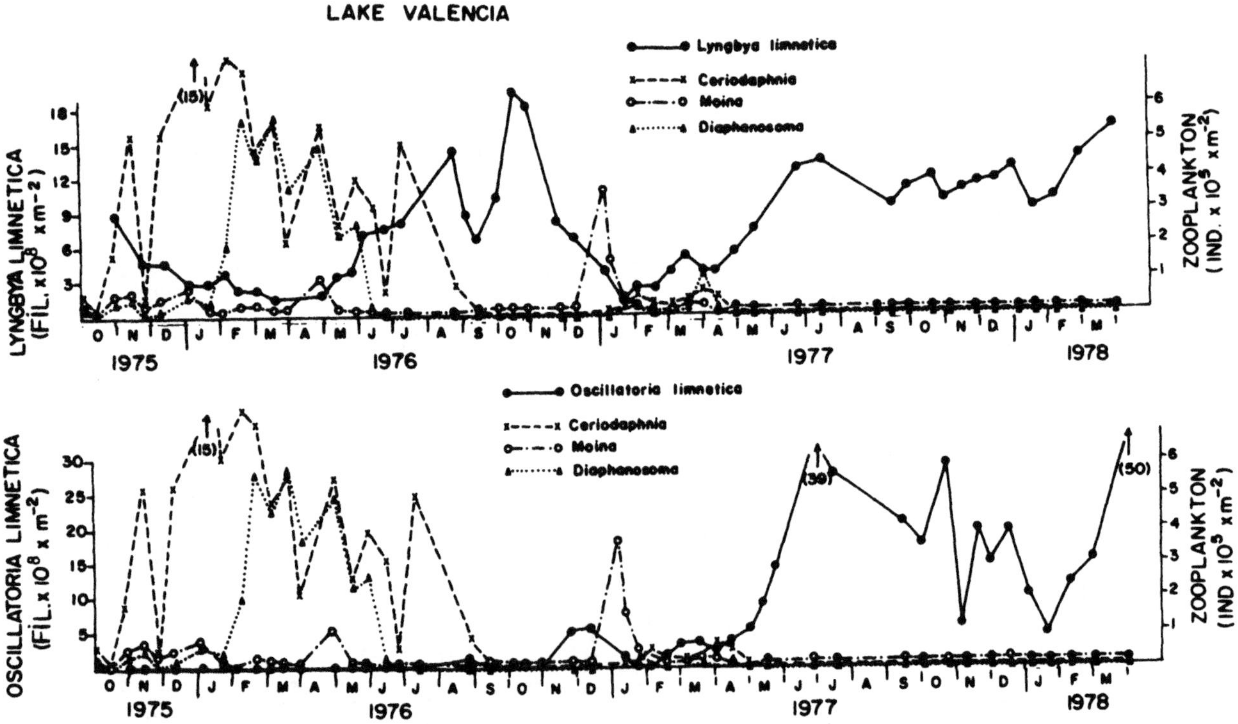

Fig. 2. Annual variations in abundance of *Lyngbya limnetica* and *Oscillatoria limnetica* related to the three cladoceran species

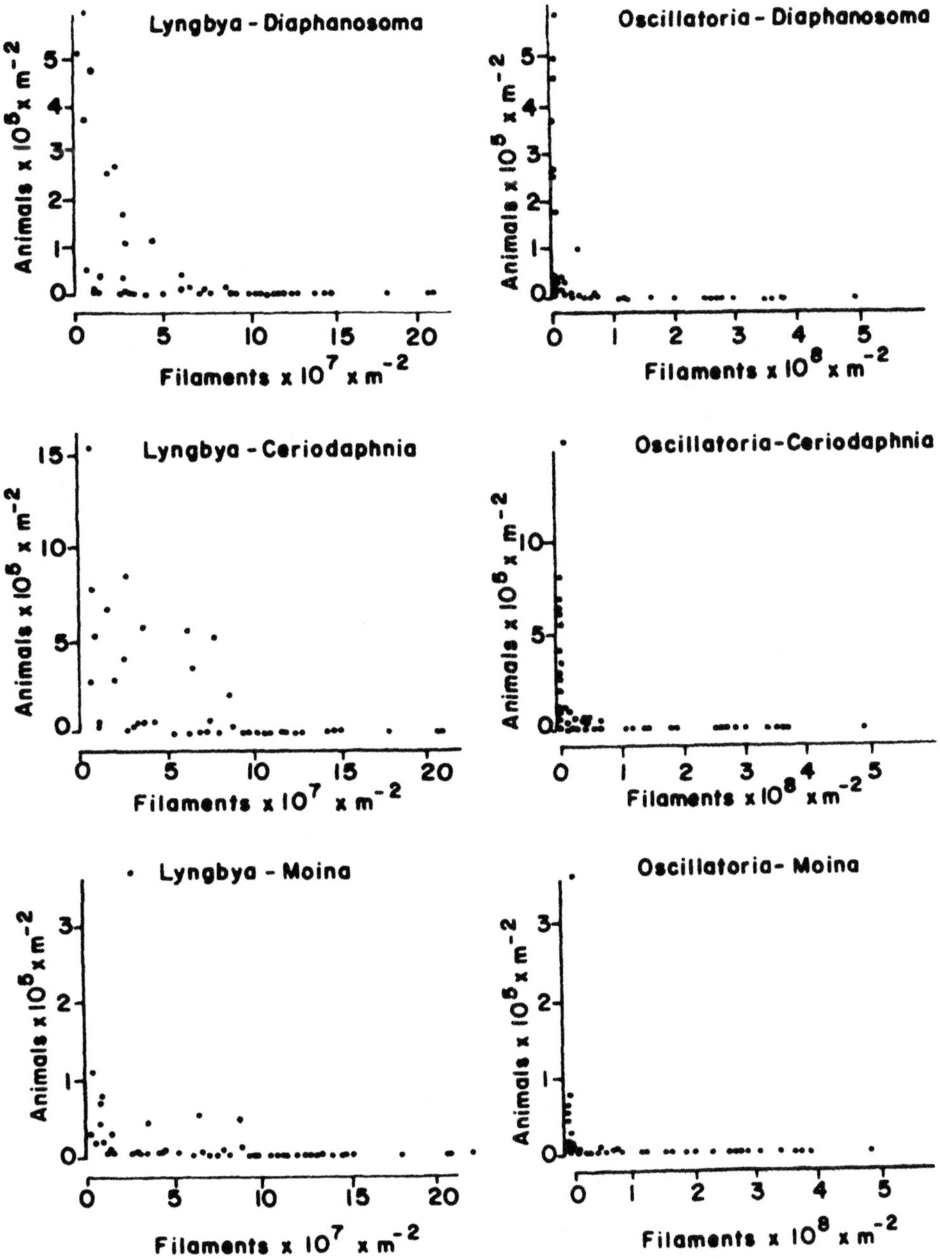

Fig. 3. Inhibitory concentrations of *Lyngbya limnetica* and *Oscillatoria limnetica* upon *Diaphanosoma brachyurum*, *Ceriodaphnia cornuta* and *Moina micrura*.

4. No unusual increases in the abundance of the main predators were noticed at the time of the decline.

Experimental evidence and field observations do not leave doubts about the detrimental effect that Cyanophyta filaments have upon cladocerans. Hayward & Gallup (1976) reported immobilization of the thoracic limbs by entangling filaments. Webster & Peters (1978) found a lowering in filtration rates of large cladocerans when filament concentrations increased. Gliwicz (1980) observed a decrease in food collection when *Anabaena* density was high. Edmondson & Litt (1982) suggested that *Oscillatoria* was probably an important factor in delaying the return of *Daphnia* after the decline of its main predator in Lake Washington.

In Lake Valencia, where conditions are highly favorable for the growth of blue-green algae, important alterations in community structure are occurring. The elimination of all cladocerans for prolonged periods of time may have significant and non-reversible effects on the food web of the lake.

Acknowledgements

This work was supported by CONICIT grant S1-0612. The authors are grateful to W. T. Edmondson, T. Zaret, A. Litt and S. Abella for reading the manuscript and for useful comments.

References

Arnold, D. E., 1971. Ingestion, assimilation, survival and reproduction by *Daphnia pulex* fed seven species of blue-green algae. Limnol. Oceanogr. 16: 906-920.

Bockh, A., 1968. Consecuencias de las actividades humanas sin control en el Lago de Valencia. Inst. Conserv. Lago de Valencia: 1-51.

Burns, C. W., 1968. Direct observation of mechanisms regulating feeding behavior of *Daphnia* in lake water. Int. Revue ges. Hydrobiol. Hydrogr. 53: 83-100.

De Bernardi, R., G. Giussani & E. Lasso Pedretti, 1981. The significance of blue-green algae as food for filterfeeding zooplankton: experimental studies on *Daphnia spp.* fed by *Microcystis aeruginosa*. Verh. int. Ver. Limnol. 21: 477-483.

Edmondson, W. T. & A. H. Litt, 1982. *Daphnia* in Lake Washington. Limnol. Oceanogr. 27: 272-293.

Gliwicz, Z. M., 1980. Filtering rates, food size selection and feeding rates in cladocerans. Another aspect of interspecific competition in filter-feeding zooplankton. In W. Ch. Kerfoot (ed.), Evolution and Ecology of Zooplankton Communities. University Press, New England, Hanover, New Hampshire & London: 282-291.

Gliwicz, Z. M. & E. Siedlar, 1980. Food size limitation and algae interfering with food collection in *Daphnia*. Arch. Hydrobiol. 88: 155-177.

Gophen, M., 1979. Extinction of *Daphnia lumholtzi* (Sars) in Lake Kinneret, Israel. Aquaculture 16: 67-71.

Gorham, P. R., 1964. Toxic algae. In: D. F. Jackson (ed.), Algae and Man. Plenum Press, New York: 307-336.

Hayward, R. S. & D. N. Gallup, 1976. Feeding, filtering and assimilation in *Daphnia schoedleri* Sars as affected by environmental conditions. Arch. Hydrobiol. 77: 139-163.

Infante, A., 1978. Natural food of herbivorous zooplankton of Lake Valencia (Venezuela). Arch. Hydrobiol. 82: 347-358.

Infante, A., 1982. Annual variations in abundance of zooplankton in Lake Valencia (Venezuela). Arch. Hydrobiol. 93: 194-208.

Infante, A., O. Infante, M. Márquez, W. Lewis & F. H. Weibezahn, 1979. Conditions leading to mass mortality of fish and zooplankton in Lake Valencia. Acta Cient. Venez. 30: 67-73.

Lampert, W., 1981. Toxicity of the blue-green *Microcystis aeruginosa*: effective defense mechanism against grazing pressure by *Daphnia*. Verh. int. Ver. Limnol. 21: 1436-1440.

Lewis, W. M. & F. H. Weibezahn, 1976. Chemistry, energy flow and community structure in some Venezuelan fresh waters. Arch. Hydrobiol., Suppl. 50: 145-207.

Moriarty, D. J., J. D. Darlington, E. G. Dunn, C. M. Moriarty & M. P. Tevlin, 1973. Feeding and grazing in Lake George, Uganda. Proc. r. Soc. Lond. 184: 299-319.

Porter, K. G. & G. D. Orcutt, 1980. Nutritional adequacy, manageability and toxicity as factors that determine the food quality of green and blue-green algae for *Daphnia*. In W. Ch. Kerfoot (ed.), Evolution and Ecology of Zooplankton Communities. University Press, New England, Hanover, New Hampshire: 282-291.

Porter, K. G., G. Gerritsen & G. D. Orcutt, 1982. The effect of food concentration on swimming behavior, ingestion, assimilation and respiration by *Daphnia*. Limnol. Oceanogr. 27: 935-949.

Richards, R. C., C. R. Goldman, T. C. Frantz & R. Wickwire, 1975. Where have all the *Daphnia* gone? The decline of a major cladoceran in Lake Tahoe, California-Nevada. Verh. int. Ver. Limnol. 19: 835-842.

Scott, W. R., D. J. Barlow & J. J. Hauman, 1980. Studies on the ecology, growth and physiology of toxic *Microcystis aeruginosa* in South Africa. In W. Carmichael (ed.), Environmental Science Research, 20. Plenum Press, New York & London: 49-70.

Stangenberg, M., 1968. Toxic effects of *Microcystis aeruginosa* Kg. extracts on *Daphnia longispina* O. F. Müller and *Eucypris virens* Jurine. Hydrobiologia 32: 81-87.

Tamers, M. A. & C. T. Thielen, 1966. Radiocarbon ages of ground water flowing into a dessicating lake. Acta cient. Venez. 17: 150-157.

Vance, B. D., 1966. Sensitivity of *Microcystis aeruginosa* and other blue-green algae and associated bacteria to selected antibiotics. J. Phycol. 2: 125-128.

Webster, K. E. & R. H. Peters, 1978. Some size-dependent inhibitions of larger cladoceran filterers in filamentous suspensions. Limnol. Oceanogr. 23: 1238-1245.

Whitton, B. A., 1973. Interaction with other organisms. In N. G. Carr & B. A. Whitton (eds.), The Biology of Blue-green Algae. University of California Press, Berkeley: 415-433.

Diel migration and vertical distribution of Cladocera in Lake D. Helvecio (Minas Gerais, Brazil)

T. Matsumura-Tundisi, J. G. Tundisi & L. S. Tavares
Universidade Federal de São Carlos, Departamento de Ciências Biológicas, Laboratório de Limnologia, Via Washington Luis, km 235, C.P. 676, 13.560, São Carlos, SP, Brazil

Keywords: tropical zooplankton, diel migration, thermocline, cladocerans, vertical distribution, summer, winter, tropical lake

Abstract

Six genera of Cladocera (*Diaphanosoma, Daphnia, Ceriodaphnia, Moina, Bosmina, Bosminopsis*), each of them usually with only one species were found in Lake D. Helvecio, a natural valley lake located in the eastern part of Brazil. Diurnal migratory movements of the organisms observed in this lake showed a different pattern in different species. Closely related species, which explore the same food source, live in different layers, thus avoiding interspecific competition. The migratory behaviour of the species was studied mainly in relation to temperature and oxygen distribution in the lake. Thus, analyses were made in the summer (January, 1978) when a strong stratification occurs with the establishment of a thermocline and an oxycline. Comparisons were made also with the data obtained in winter (July, 1978), when a complete mixing of water occurs.

Introduction

The vertical distribution of crustacean zooplankters is related to a species-specific migratory behaviour which depends on the influence of physical, chemical and biological factors. For many years light has been considered by several authors (Cushing, 1951; Banse, 1964; McNaught & Hasler, 1964; Hutchinson, 1967; Hart & Allanson, 1976; Buchanan & Haney, 1980) as the major factor responsible for the uneven distribution of the organisms in a vertical direction. However, biological factors such as phytoplankton distribution in the water column (Anderson *et al*, 1972), competition between species (Dumont, 1972) and predation (Zaret & Suffern, 1976), can also cause differences among species.

In environments where temperature and oxygen cause a discontinuity layer, such factors may also be considered responsible for the different migratory behaviour of the organisms.

There are few papers on fresh-water zooplankton vertical migration in tropical regions, most of them restricted to very shallow environments, ponds and reservoirs (Vaas & Sachlan, 1953; George, 1961; Khan *et al*, 1970). In Brazil, there is only one paper on this subject, dealing with *Daphnia gessneri* in Americana Reservoir (Arcifa-Zago, 1978).

The present paper examines the migratory behaviour of six cladoceran species (*Diaphanosoma brachyurum, Daphnia gessneri, Moina minuta, Ceriodaphnia cornuta, Bosmina coregoni* and *Bosminopsis deitersi*) in Lake D. Helvecio, eastern Brazil, in relation to the thermocline and chemocline (oxygen) present in this lake.

Study area

Lake D. Helvecio is one of hundreds of lakes in Rio Doce Valley, Minas Gerais State which was formed during the quaternary period. It has a dendrictic shape and covers a surface area of 6.8 m^2. Maximum depth is 31.0 m and mean depth is 12.0 m. Figure 1 shows a morphometric map of the lake

Hydrobiologia 113, 299–306 (1984).

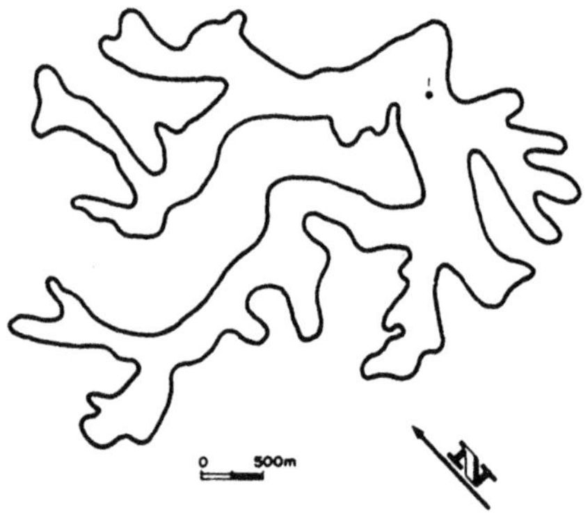

Fig. 1. Lake D. Helvecio (Lat. 19° 10'S; Long. 42° 1'W). Morphometry and position of sampling sites.

Morphometric	Data
Max. length (km)	9.1
Max. breadth (km)	1.0
Max. depth (m)	32.5
Mean depth (m)	12.0
Volume ($\times 10^5$ m^3)	831.495
Surface area ($\times 10^2$ m^2)	687.230
Shoreline (km)	45.0

and the positition of the sampling site. The lake has a typical thermal characteristic and can be considered as warm monomictic. It has a long period of permanent, strong stratification from May to September and a short period of vertical homothermy from June to August. The location of the lake in a valley surrounded by dense vegetation gives protection from wind. The winter circulation occurs by the progressive cooling that begins in May and is completed in July when the lake becomes isothermic.

During the period of stratification usually there is no measurable oxygen in the hypolimnion (below 15.0 m) and in the lower layers there is hydrogen sulphide.

The zooplankton community of Lake D. Helvecio consists mainly of crustaceans, comprising 90% of total zooplankton. Copepods are much more abundant than cladocerans. The distribution of these crustaceans in the water column is different during the period of stratification.

Material and methods

Samples for the study of the vertical migration of cladoceran populations were collected in January (period of stratification) and in July (period of mixing) at a single station in the deepest part of the lake at 5 m intervals from 0 to 25 m. A suction pump moved by gasolin was used taking from each depth 200 liters of material which concentrated through a plankton net of 68 μm mesh size. The samples were taken within 3-h intervals during a 24-h period.

For comparison with biological data temperature, dissolved oxygen and Secchi disc were taken simultaneously with plankton samples. The material was analyzed by counting all the cladocerans present in the samples and identifying the species.

Results

Marked differences have been found in vertical distribution and migratory behaviour of the six species of Cladocera present in Lake D. Helvecio. This distribution has been examined in relation to the vertical distribution of temperature and dissolved oxygen.

As can be observed on Fig. 2, the temperature of epilimnion (0–10 m) in January was 28.5 °C. The thermocline was well developed between 10 and 16 m depth showing a difference of 3.5 °C. The temperature of hypolimnion was 24.5 °C and no dissolved oxygen was present. The dissolved oxygen of the epilimnion was high with 8.0 mg l^{-1} but in the metalimnion it was reduced to 2.0 mg l^{-1}. In July the lake is vertically mixed with a homogeneous thermal and oxygen profile (25.0–25.2 °C and 6.5–7.0 mg l^{-1} respectively). High transparency was observed in this lake where the Secchi disc depth measured was 7.6 m in summer and 5.8 m in winter.

Figure 3 shows the vertical movement of cladoceran species in the summer when the lake is stratified. *Daphnia gessneri* is more abundant in the metalimnion and shows diurnal vertical movements between 10 and 15 m. The species begins at 15 m at dusk and ascents to a depth of 10 m at sunset (18.00 hours). *Ceriodaphnia cornuta* is also an inhabitant of metalimnion, remaining during the day in the deepest layers, but is shows a broader movement than *Daphnia gessneri*, going up to 5 m at night. The other four species – *Diaphanosoma brachyurum, Moina minuta, Bosmina coregoni* and *Bosminopsis deitersi* – are epilimnetic species showing a vertical distribution between 0 and 10 m. *Diaphanosoma*

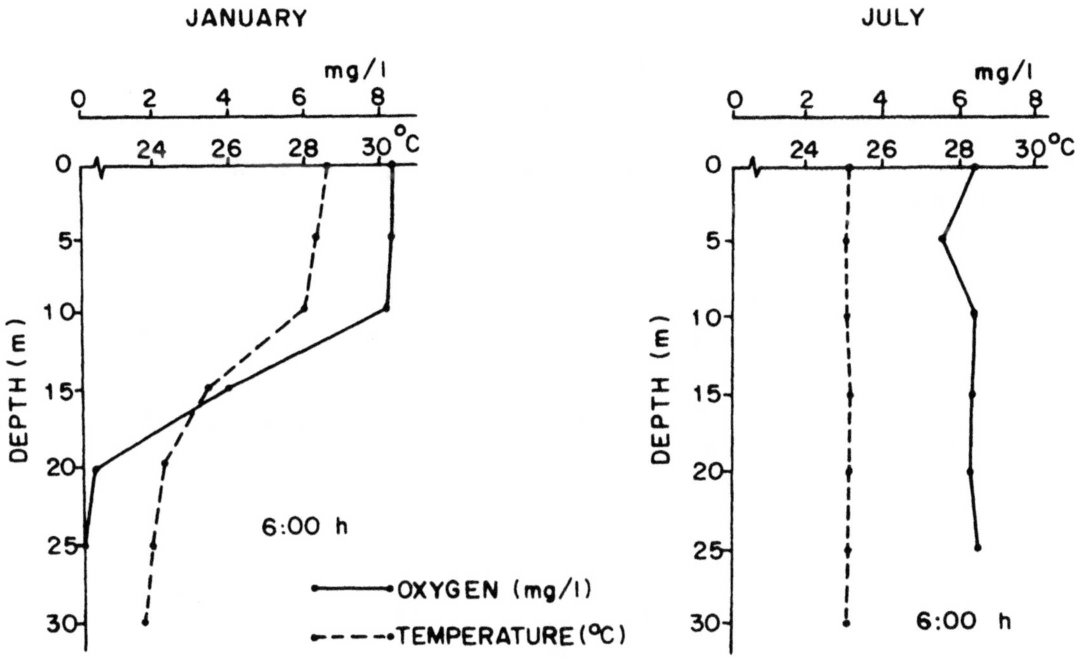

Fig. 2. Vertical distribution of temperature and oxygen in Lake D. Helvecio in January and July at 6.00 a.m.

brachyurum and *Bosmina coregoni* showed a vertical migration at dusk, rising toward the surface at 18.00 hours and descending at 6.00 hours. Most *Bosminopsis deitersi* were found at 10 m, and only for a short period during predawn and dawn (3.00–6.00 hours) was an ascent towards 5 m observed. *Moina minuta* was also distributed between 5 and 10 m for most of the diel cycle and only from 24.00–3.00 hours did a rise of part of this population toward the surface occur.

In the winter (July) when the lake showed homogeneous conditions of temperature and dissolved oxygen from surface to bottom, all species were distributed in the whole column until 25 m as can be observed on Fig. 4. At this time it was not possible to observe a clear vertical migration in any species. *Moina minuta, Bosmina coregoni* and *Bosminopsis deitersi* seemed to have a weak vertical movement lacking, however, a defined pattern.

The results of the organism countings are listed as appendixes I and II.

Discussion

Although light intensity is considered the main factor controlling the migratory movement of zooplankton, the real causes of the vertical migration of these organisms are still speculative. According to Ringelberg (1980) the abiotic environmental variables such as day length, oxygen content, temperature and the biotic factors such as the presence of food, crowding and predation can influence the physiological state of the organisms determining a sort of migratory behaviour. However, it is not possible to point out a unique pattern of migratory behaviour shown by animals. A complex operation of variables is probably responsible for the vertical movements. As pointed out by Ringelberg (*op. cit.*) one factor can cancel the effect of the other and it can be observed that in the presence of a thermocline the effect of light intensity is canceled. Thus, some animals can migrate through the thermocline while others are hindered.

The cladoceran species present in Lake D. Helvecio probably have different adaptations to light intensity, temperature and oxygen content showing a distribution with vertical segregation. During the period of stratification the Daphnidae (*Daphnia gessneri* and *Ceriodaphnia cornuta*) live in the metalimnion. These species probably tolerate low oxygen content and large ranges of temperature variation. The other species *Diaphanosoma brachyurum,*

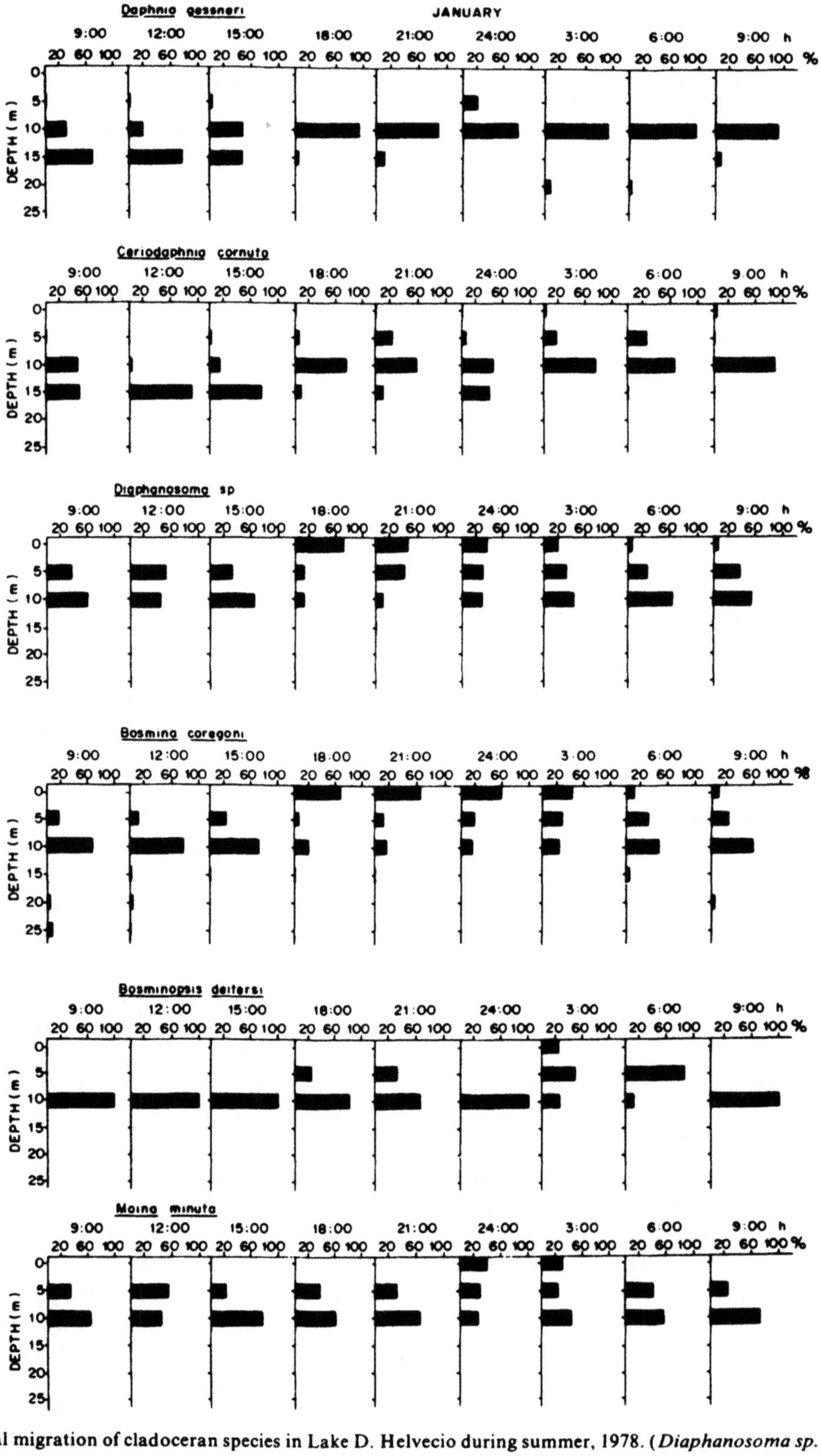

Fig. 3. Diel vertical migration of cladoceran species in Lake D. Helvecio during summer, 1978. (*Diaphanosoma sp.* = *Diaphanosoma brachyurum*).

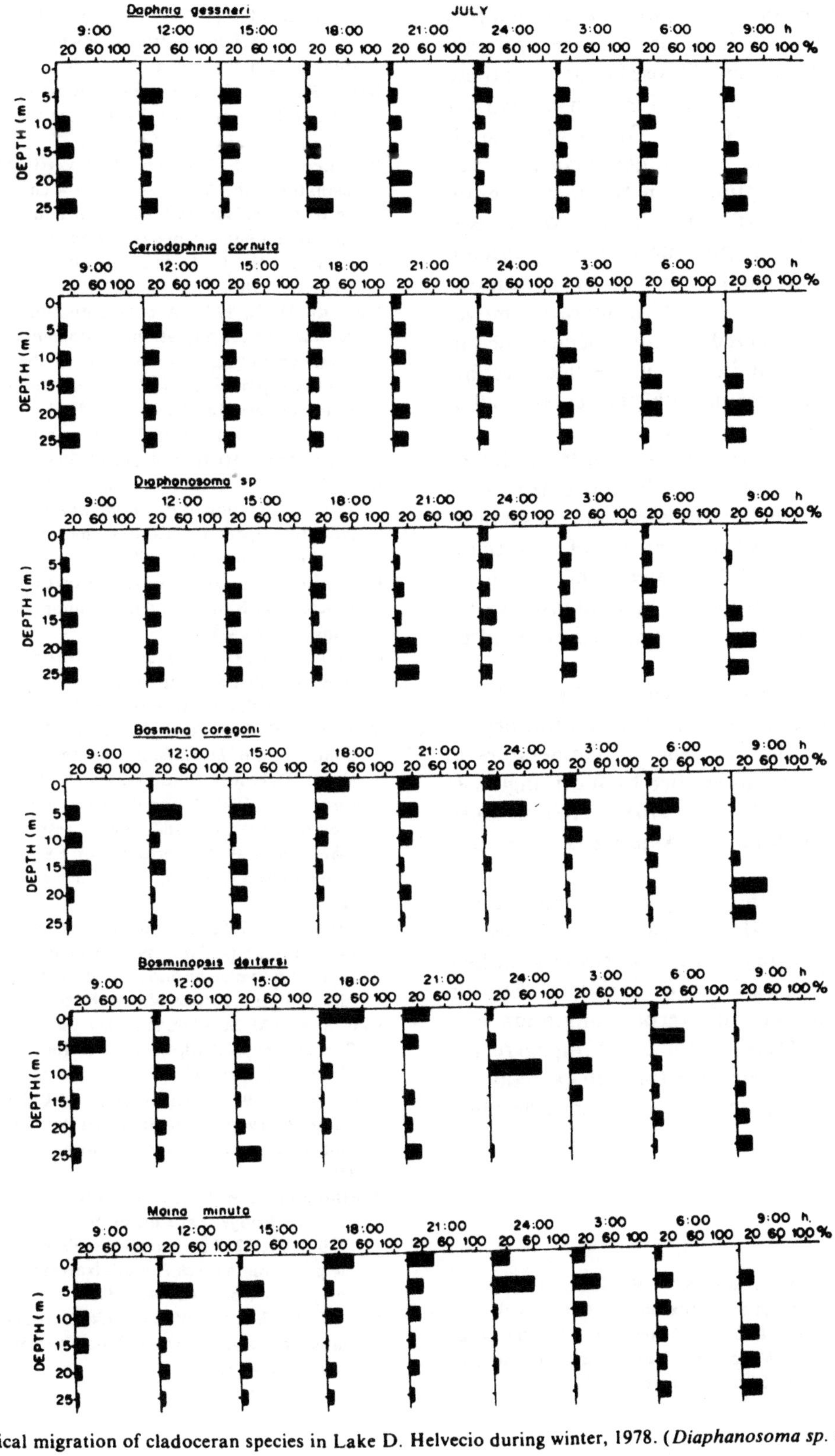

Fig. 4. Diel vertical migration of cladoceran species in Lake D. Helvecio during winter, 1978. (*Diaphanosoma sp. = Diaphanosoma brachyurum*).

Moina minuta, Bosmina coregoni and *Bosminopsis deitersi* live in the epilimnion and are probably sensitive to the variation of those factors. For these species the thermocline could function as a barrier to their migratory movements. On the other hand, *Daphnia gessneri* and *Ceriodaphnia cornuta* may have temperature optimum in the metalimnion; this fact, associated with the tolerance to low oxygen content makes it possible to have a specific niche segregated from the other species. The influence of light intensity, controlling migratory movements was not observed for any species since in winter when the conditions of the lake become completely homogeneous, no species showed a defined migratory movement.

The responses of the organisms to the different external stimuli which cause the movement reactions must be related to the evolutionary adaptations occurring during the life history of the species. As verified by Meyer (1980), some limnetic cladoceran species rapidly respond to light intensity variations while littoral cladoceran species respond to oxygen variations.

However, the thermocline seems to be a real barrier to the vertical displacement of several organisms even when the difference in temperature is about 1 °C or 0.5 °C as verified by Worthington & Ricardo (1936) and Viner (1970) in African lakes and by Arcifa (1978) in the Americana Reservoir, Brazil.

Thus, the vertical migratory movements of these cladocera in Lake D. Helvecio, are probably mechanisms of species segregation in a vertically stratified environment such as Lake D. Helvecio. Also, it must be emphasized that vertical migration is a possible system of fertilization of the impoverished epilimnion of stratified lakes through excretion of zooplankton during the vertical ascent or descent.

References

Anderson, G. C., B. W. Frost & W. K. Peterson, 1972. On the vertical distribution of zooplankton in relation to chlorophyll concentration. In A. Y. Takenouti (ed.), Biological Oceanography of the Northern North Pacific Ocean and Bering Sea, Idemitsu Shoten, Tokyo: 341–345.

Arcifa-Zago, M. S., 1978. Vertical migration of *Daphnia gessneri* Herbst 1876 in Americana reservoir, State of São Paulo, Brazil. Verh. int. Ver. Limnol., 20: 1720–1726.

Banse, K., 1964. On the vertical migration of plankton in the Sea. In M. Sears (ed.), Progress in Oceanography, 2. Pergamon Press, Oxford: 53–125.

Buchanan, C. & F. J. Haney, 1980. Vertical migrations of zooplankton in the Arctic: A test of the environmental controls. In W. C. Kerfoot (ed.), Evolution and Ecology of zooplankton communities. University Press of New England: 69–79.

Cushing, D. H., 1951. The vertical migration of planktonic Crustacea. Biol. Rev. 26: 158–192.

Dumont, H. J., 1972. A competition-based approach of the reverse vertical migration in zooplankton and its implications, chiefly based on a study of the interactions of the Rotifer *Asplanchna priodonta* (Gosse) with several Crustacea Entomostraca. Int. Revue Ges. Hydrobiol. 57: 1–38.

George, M. G., 1961. Diurnal variations in two shallow ponds in Delhi, India. Hydrobiologia 18: 265–274.

Hart, R. C. & B. R. Allanson, 1976. The distribution and diel vertical migration of *Pseudodiaptomus hessei* (Mrázek) (Calanoida: Copepoda) in a subtropical lake in southern Africa. Freshwat. Biol. 6: 183–198.

Hutchinson, G. E., 1967. A Treatise on Limnology, 2. Introduction to Lake Biology and the Limnoplankton. John Wiley & Sons, Inc. New York.

Khan, A. A., A. Q. Siddiqui & M. Nazir, 1970. Diurnal variations in a shallow tropical freshwater fish pond in Shahjahampur, V.P. (India). Hydrobiologia 35: 297–304.

McNaught, D. C. & A. D. Hasler, 1964. Rate of movement of populations of *Daphnia* in relation to changes in light intensity. J. Fish. Res. Bd Can. 21: 291–318.

Meyers, D. G., 1980. Diurnal vertical migration in aquatic microcrustacea: Ligh and oxygen responses of littoral zooplankton. In W. C. Kerfoot (ed.), Evolution and Ecology of Zooplankton Communities. University Press of New England: 80–90.

Ringelberg, J., 1980. Introductory remarks: Causal and teleological aspects of diurnal vertical migration. In W. C. Kerfoot (ed.), Evolution and Ecology of Zooplankton Communities. University Press of New England: 65–78.

Vaas, K. F. & M. Sachlan, 1953. Limnological studies on diurnal fluctuations in shallow ponds in Indonesia. Verh. int. Ver. Limnol. 12: 309–319.

Viner, A. B., 1970. Hydrobiology of Lake Volta, Ghana. II. Some observations on biological features associated with the morphology and water stratification. Hydrobiologia 35: 230–248.

Worthington, E. B. & C. K. Ricardo, 1936. Scientific results of the Cambridge expedition to the East African lakes, 1930–1931, 17. The verical distribution and movements of the plankton in lakes Rudolf, Naivasha, Edward and Bunyoni. J. linn. Soc. Zool. 40: 33–69.

Zaret, T. M. & J. S. Suffern, 1976. Vertical migration in zooplankton as a predator avoidance mechanism. Limnol. Oceanogr. 21: 804–813.

Appendix I

Density of organisms (in January) expressed in no./200 l.

Time	Depth (m)	*Diaphanosoma* spp	*Moina minuta*	*Daphnia gessneri*	*Ceriodaphnia cornuta*	*Bosmina coregoni*	*Bosminopsis deitersi*
9:00	0.0	3	-	-	-	-	-
	5.0	425	6	-	3	15	-
	10.0	664	11	23	133	58	1
	15.0	4	-	52	135	-	-
	20.0	-	-	-	1	5	-
	25.0	-	-	-	-	7	-
12:00	0.0	-	-	-	-	-	-
	5.0	578	11	1	27	10	-
	10.0	521	9	12	81	62	1
	15.0	2	-	45	1267	2	-
	20.0	1	-	-	12	3	-
	25.0	-	-	-	1	2	-
15:00	0.0	-	-	-	1	-	-
	5.0	498	6	2	27	28	-
	10.0	956	20	24	114	76	7
	15.0	6	-	24	540	1	-
	20.0	3	-	-	4	-	-
	25.0	-	-	-	-	-	-
18:00	0.0	4231	-	-	2	91	-
	5.0	766	13	-	19	9	1
	10.0	814	22	19	177	30	5
	15.0	11	-	1	25	3	-
	20.0	7	-	-	1	-	-
	25.0	3	-	-	-	-	-
21:00	0.0	1503	-	-	1	128	-
	5.0	1373	5	-	33	27	1
	10.0	337	10	8	75	37	2
	15.0	37	-	1	16	4	-
	20.0	-	-	-	1	1	-
	25.0	1	-	-	-	-	-
24:00	0.0	1336	14	1	1	64	-
	5.0	1147	11	1	10	22	-
	10.0	1126	10	4	68	20	4
	15.0	10	1	-	60	1	-
	20.0	-	-	-	-	-	-
	25.0	-	-	-	-	-	-
3:00	0.0	789	9	-	6	50	3
	5.0	1323	7	-	32	34	6
	10.0	1704	13	12	129	28	3
	15.0	2	-	-	1	-	-
	20.0	-	-	1	-	-	-
	25.0	-	-	-	-	-	-
6:00	0.0	143	-	-	-	6	-
	5.0	677	9	-	45	19	7
	10.0	1566	12	22	109	27	1
	15.0	-	-	-	4	3	-
	20.0	2	-	1	-	-	-
	25.0	1	-	-	-	-	-
9:00	0.0	56	-	-	4	7	-
	5.0	441	5	-	2	7	-
	10.0	615	14	26	79	44	1
	15.0	4	-	2	1	-	-
	20.0	2	-	-	-	3	-
	25.0	2	-	-	-	-	-

Appendix II

Density of organisms (in July) expressed in no./200 l.

Time	Depth (m)	*Diaphanosoma* spp	*Moina minuta*	*Daphnia gessneri*	*Ceriodaphnia cornuta*	*Bosmina coregoni*	*Bosminopsis deitersi*
9:00	0.0	135	5	2	6	–	3
	5.0	318	43	14	46	13	40
	10.0	412	23	86	65	14	13
	15.0	597	22	111	82	22	9
	20.0	551	11	97	83	6	3
	25.0	559	4	121	111	4	9
12:00	0.0	74	4	5	1	4	2
	5.0	433	37	235	105	36	5
	10.0	477	14	136	92	11	6
	15.0	490	7	115	80	18	4
	20.0	364	10	106	64	4	3
	25.0	559	5	167	66	6	2
15:00	0.0	49	3	4	–	–	–
	5.0	286	21	201	133	15	6
	10.0	464	13	160	90	3	7
	15.0	406	6	176	111	10	2
	20.0	427	9	105	116	10	3
	25.0	421	5	57	72	5	9
18:00	0.0	654	50	60	53	50	28
	5.0	449	13	69	159	17	3
	10.0	581	28	169	83	17	7
	15.0	258	2	221	66	9	1
	20.0	544	15	256	68	10	5
	25.0	370	10	410	90	–	–
21:00	0.0	171	37	51	56	13	9
	5.0	226	22	137	86	12	5
	10.0	407	16	194	90	8	–
	15.0	242	7	129	39	3	3
	20.0	927	13	368	106	7	2
	25.0	985	5	346	99	2	5
24:00	0.0	429	32	184	23	27	11
	5.0	653	81	303	104	67	17
	10.0	492	7	163	91	2	122
	15.0	773	4	215	107	10	2
	20.0	513	8	129	81	–	1
	25.0	503	–	243	62	3	5
3:00	0.0	210	39	54	29	17	10
	5.0	445	90	213	72	39	9
	10.0	432	41	222	122	25	13
	15.0	596	21	163	85	10	7
	20.0	660	16	286	94	5	–
	25.0	610	3	160	79	7	–
6:00	0.0	235	15	21	41	14	18
	5.0	451	46	83	83	92	94
	10.0	616	38	184	92	34	27
	15.0	670	27	189	166	28	20
	20.0	672	22	200	162	21	27
	25.0	384	36	111	42	8	8
9:00	0.0	1	–	2	–	–	–
	5.0	149	14	57	48	2	54
	10.0	*	*	*	*	*	*
	15.0	452	18	78	114	5	17
	20.0	810	18	125	183	22	23
	25.0	558	20	129	121	14	26

Biomass and production of *Argyrodiaptomus furcatus*, a tropical calanoid copepod in Broa Reservoir, southern Brazil

Odete Rocha & Takako Matsumura-Tundisi
Universidade Federal de São Carlos, Departamento de Ciências Biológicas, Laboratório de Limnologia, Via Washington Luis, Km. 235 Caixa Postal: 676, 13.560, São Carlos, SP, Brazil

Keywords: tropical zooplankton, biomass, production, copepod, development time, tropical region

Abstract

The biomass and the production of *Argyrodiaptomus furcatus* (Sars), the most abundant copepod in Broa Reservoir (São Carlos, São Paulo State), were estimated, determining in the laboratory the development time and the quantity of organic carbon and establishing the relationship between these two parameters. The daily production was calculated from $P = B(1 - e^{gt})$ and the annual production was obtained by integrating daily production against time. The maximum production of *Argyrodiaptomus furcatus* in the reservoir depends on the region considered and on the period of the year. The maximum production was 45.15 mg C $m^{-3}d^{-1}$ in March, 1976 at station II, region of macrophytes and 6.74 mg C $m^{-3}d^{-1}$ at station IV, near the dam. The mean production for the year is 6.26 mg C $m^{-3}d^{-1}$ at station II and 1.43 mg C $m^{-3}d^{-1}$ at station IV.

Introduction

The knowledge on the sources of organic matter and factors controlling productivity in tropical areas is still in its preliminary stages, although a big effort has been made in the past ten years in order to catch up with some lines of limnological research already fully developed in temperate regions.

Studies on population dynamics and production of freshwater zooplankton are extremely scarce. However since much of the production of higher trophic levels is dependent upon invertebrate production, particularly that of zooplankton, more investigations need to be directed towards the role of zooplankton communities in the biological productivity of freshwater basins.

In this study, the biomass and production of *Argyrodiaptomus furcatus*, a copepod commonly found in natural and man-made lakes in the southeast of Brazil, was estimated for a one year period, between 1975 and 1976 at Broa reservoir.

The species is the dominant and the largest herbivore in that zooplankton community and may therefore account for a significant proportion of the total secondary production. Detailed information about the spatial distribution, population dynamics and reproduction performance were obtained and have been partially reported elsewhere (Rocha *et al.*, 1982). Further laboratory measurements of development times and carbon–length relationship provided the time and weight elements required to the biomass and production calculations.

It is hoped that this preliminary assessment of *A. furcatus* production might indicate vital shortcomings to which future research will be directed.

General characteristics of Broa Reservoir

The water body is a shallow and relatively small man-made reservoir, initially constructed for hydro-electric power supply. It has a maximum length of 7.5 km, mean depth of 3.0 m and 6.8 km surface area. The climate is characterized by well-defined dry and rainy seasons with annual precipitation of about 1 500 mm. The rainy season coincides with the

Hydrobiologia 113, 307–311 (1984).

warm spring and summer, lasting from September to March. Water level changes considerably (1.5–2.0 m) and water temperature also varies seasonally between 15 and 28 °C. Many of its distinctive features, such as low pH (5.0–6.5), low conductivity (5.0–34.0 μmho cm²), low nutrient levels and low primary production (155.0 mg °C m²d⁻¹, according to Tundisi, 1977), are related to its localization in the tropical savanna and being surrounded by poor, acid and sandy soils.

The reservoir supports a high diversity in both phytoplankton and zooplankton communities.

Material and methods

Quantitative zooplankton samples were collected with a 6 ℓ Van Dorn sampler at weekly intervals in four fixed stations (Fig. 1). A total of 60 ℓ for each site was sieved through a 68 μm mesh and preserved in formalin. At station IV, in the deepest part of the reservoir, samples were collected at four depths, 0, 2, 4 and 6 m.

Numerical density was determined from counts of the total sample or subsamples depending on the abundance. Different instars (egg, total nauplii, four copepodite and male and female adults were counted separately.

Length–carbon relationship was obtained from a preserved sample. Length measurements of 40–80 individuals of the same instar were taken and total organic carbon was determined by wet dichromate oxidation. This relationship was used to convert numbers to biomass for each development instar.

Duration of development was determined in the laboratory underconstant temperature (25 ± 1 °C), but uncontrolled food conditions. Females with developed ovaries were selected from the reservoir and maintained in large test tubes with 50 ml of lake water. Experimental tubes were kept inside a thermostatically controlled water bath and the medium was renewed daily.

Egg duration and post embryonic development were followed from egg to egg.

Mean carbon weights for each instar were plotted against durations and from an exponential fit, the instantaneous growth rate 'g' was obtained.

Daily production in terms of body growth was calculated from: $P = B\ (1 - e^{gt})$. Since the only growth rate available was for 25 °C, a Q_{10} of 2.25 was applied to calculate production at other temperatures. Adult reproductive production was computed as egg production, and with basis on egg durations and field temperature, one or two broods per week were considered.

The annual production was obtained by integrating the daily production against time.

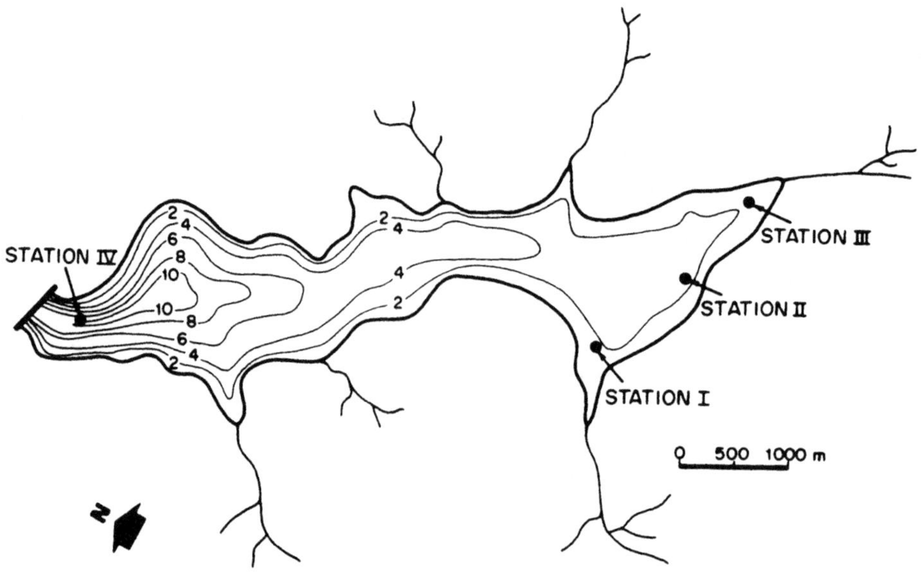

Fig. 1. Bathymetric map of Broa reservoir showing the sampling stations.

Results and discussion

Figure 2 shows the relationship between body length and organic carbon including all developmental stages of *A. furcatus*. A power function gives the best fit to the data.

The duration of embryonic and post embryonic development are presented in Table 1. The value of two days observed for the egg development is very close to the predicted duration of 1.97 days calculated from the curvilinear regression given by Bottrell *et al.* (1976), for calanoids. Similar egg durations have also been reported for other tropical diaptomids. *Pseudodiaptomus hessei* has an egg duration of 44 h at 27 °C (Hart *et al.*, 1975) and *Neodiaptomus schmackeri* has an egg duration of 2 days at 28–30 °C (Dunn, 1970).

A total duration of 29 days for the post embryonic development of *A. furcatus* is also very close to that of 30 days reported for *N. schmackeri* at

Table 1. Duration of development of *A. furcatus* at 25 °C.

Stage	Duration (days)	Mean duration
Egg	2	2
Nauplii (I–VI)	7–9	8
Copepodite (I–VI)	19–23	21
Egg to egg	27–35	31

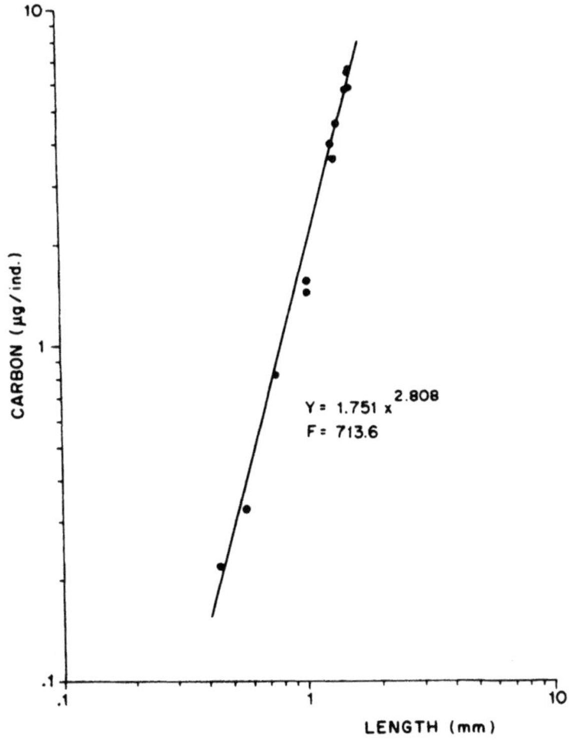

Fig. 2. Relationship between body length and organic carbon of *A. furcatus.*

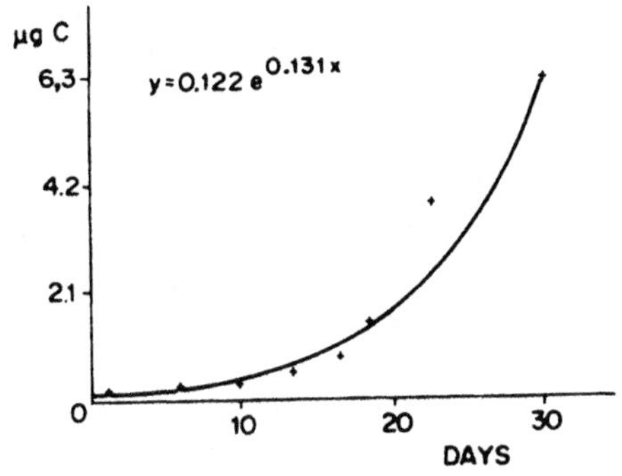

Fig. 3. Relationship between body carbon and duration of development of *A. furcatus.*

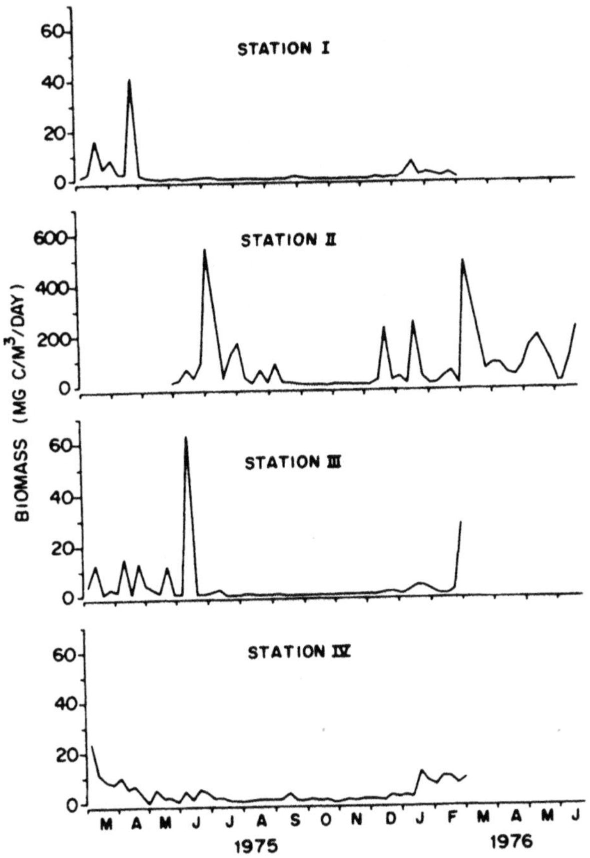

Fig. 4. Seasonal changes in the biomass of *A. furcatus* at different stations.

28–30 °C, but much longer than 21.5 days observed for *P. hessei* at 23 °C (Hart *et al.*, 1975). However, since important factors influencing the duration of development such as the size of eggs, food concentration and other environmental factors have not been controlled, comparisons cannot be made.

Figure 3 plots the relationship between carbon and duration of development. The data is best fitted by an exponential and the value of the slope, 0.131, corresponds to the instantaneous growth rate. The only comparable data is that for *Thermocyclops hyalinus* from Lake George, Uganda, which an instantaneous growth rate of 0.145 (Burgis, unpublished results, in P.E.G. 2nd contribution).

Figure 4 shows the seasonal changes in the daily biomass of the population in each sampling station, separately. Total biomass is similar in stations I, III, and IV, but one order of magnitude higher at station II. Throughout the year, the population was heterogeneously distributed in the lake being always more concentrated at station II, situated in the shallowest part of the reservoir, and a region densely covered by macrophytes.

The population has a marked seasonal pattern with two main peaks of abundance, one in the end of the autumn, beginning of winter. The biomass at station IV represents the mean values for the water column from 0 to 6 m, and this is possibly the reason for the somewhat lower values in relation to stations I and III where only surface was sampled.

Production estimates for stations I and IV and also temperature are shown in Fig. 5. Production reflects the same strong seasonality observed for the biomass and also the production is much higher at station II. It is possible that the patchiness in *A. furcatus* distribution can be related to higher food availability and perhaps shelter from predators, provided by the macrophytes.

There are two distinct peaks of production, one in the summer and another in the winter, showing no relation to the seasonal variation of temperature.

An important factor determining the seasonal occurrence of many species is food availability. No quantitative analysis of food supply has yet been made.

The maximum of production was 45.15 mg C $m^{-3}d^{-1}$ in March at station II, and 6.74 mg C $m^{-3}d^{-1}$ at station IV.

The mean production for the year is 6.26 mg C $m^{-3}d^{-1}$ at station II and 1.43 C $m^{-3}d^{-1}$ at station IV.

A daily P:B coefficient of 0.10 was obtained at station II, which falls inside the range of P:B coefficients reported for other tropical copepods, such as 0.11 for *Pseudodiaptomus hessei* (Hart *et al.*, 1975) and 0.078 for *Thermocyclops hyalinus* (Burgis, 1974).

Uncertainties about the present production estimates remain considerable and studies about growth, duration of development and feeding habits of the species need to be extended with improved techniques.

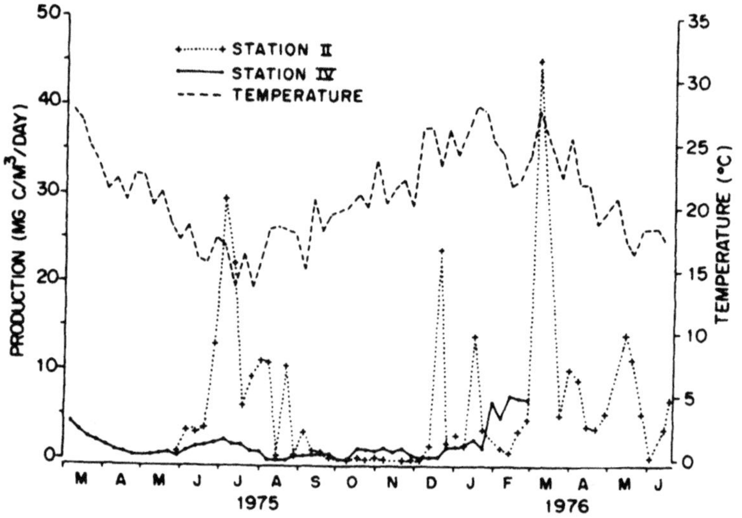

Fig. 5. Seasonal changes in the daily production of *A. furcatus* at station II and station IV and seasonal fluctuation in water temperature.

The underlying causes of the temporal and spatial fluctuations in biomass and production, and also the part played by other species and trophic levels are important questions to be investigated in the near future.

Acknowledgements

The authors express their thanks to Dr. A. Duncan for discussion and suggestions on production calculations, and the CNPq (National Research Council of Brazil) for financial support.

References

Bottrell, H. H. *et al.*, 1976. A review of some problems in zooplankton production studies. Norw. J. Zool. 24: 419–456.

Burgis, M. J., 1974. Revised estimates for the biomass and production of zooplankton in Lake George, Uganda. Freshwat. Biol. 4: 535–541.

Dunn, I. J., 1970. Recovery of a tropical pond zooplankton community after destruction by algal bloom. Limnol. Oceanogr. 15: 373–379.

Edmonson, W. T. & G. G. Winberg, 1971. A manual of methods for the assessment of secondary productivity in fresh waters. IBP Handbook 17, Blackwell Scientific Publications, Oxford.

Plankton Ecology Group - 2nd Contr. The evaluation of zooplankton production calculations. (in prep.).

Rocha, O., T. Matsumura-Tundisi & J. G. Tundisi, 1982. Seasonal fluctuations of *A. furcatus* population in Lobo Reservoir (São Carlos, SP - Brazil). Trop. Ecol. 23: 134–150.

Tundisi, J. G., 1977. Produção primária, standing-stock, fracionamento do fitoplancton e fatores ecológicos em ecossistema lacustre artificial (Represa do Broa, São Carlos). Tese de Livre Docência, USP, 410 pp.

The nature and origin of the crustacean zooplankton of Sahelian Africa, with a note on the Limnomedusa

Henri J. Dumont & Hans M. Verheye
Rijksuniversiteit Gent, Instituut voor dierkunde, Gent, Belgium

Keywords: tropical zooplankton, Cladocera, Calanoida, Cyclopoida, Limnomedusa, Sahel, Pleistocene, climate change, biogeography

Abstract

The zooplankton of the major Sahel river basins Nile, Shari (Chad), Niger, and Senegal, is different from that found in the Sahara and in Equatorial Africa. Similarities and differences between the individual basins are numerous as well. Many species are shared by the Nile and Lake Chad, by Lake Chad and the Niger (plus Senegal), or occur in all four basins, or are restricted (endemic) to only one basin.

These patterns are identical to patterns found in fish, molluscs, and macrophytes and show that crustacean zooplankton obeys the same laws of dispersal as these groups, in spite of its apparent preadaptation to passive dispersal.

The patterns can be explained by the climatic fluctuations of the Upper Pleistocene and Holocene. Following a general dry period over Africa between 20 000 and 13 000 BP, high river and lake levels prevailed between 12 500 and 8 400 BP. This was the period of maximum faunal interchange between all basins, and even with the Zaire basin. After a regression (8 000–7 000 BP), wetter conditions returned around 6 000 BP, but the Sahel itself remained dry, although its rivers and lakes, fed by waters of southern origin, showed higher levels than today. They flooded large areas of the southern Sahara, permitting aquatic animals and plants to reach the Adrar of Mauretania, the Tibesti, and the Ennedi mountains.

Since 3 000 BP, present day conditions developed. This last period is characterized by species extinctions, as exemplified by the droughts in Lake Chad in historical times, and in spite of the tremendous diversity still extant here today. Between 6 000 BP and the present, however, very little speciation took place, and faunal exchange between basins was very limited.

Introduction

In recent years, considerable progress has been made in the acquisition of data, and in the taxonomy of the zooplankton of Africa. A tentative biogeographical synthesis was presented by Dumont (1980). It appeared that only the Congo (Zaire) basin was still almost unstudied, and that situation has not changed since. Other river and lake systems north of the equator that require more study are the Volta basin, Lake Kyoga, Lake Albert, and Lake Turkana.

In the present paper, we discuss the nature of the zooplankton found in the following river systems: the Nile (including its lake sources and the smaller rift valley lakes), the Lake Chad basin, the Niger basin, and the Senegal basin (including the Gambia River, and some other short coastal rivers that spring in the Fouta Djalon mountains). The Cladocera, the Cyclopoid and Calanoid Copepods, and the genus Limnocnida (Coelenterata: Limnomedusa) will be considered. Rotifers are not included, since much ongoing but still unpublished work is likely to become available in the near future.

Our sources of information, basinwise, are:

Hydrobiologia 113, 313–325 (1984).

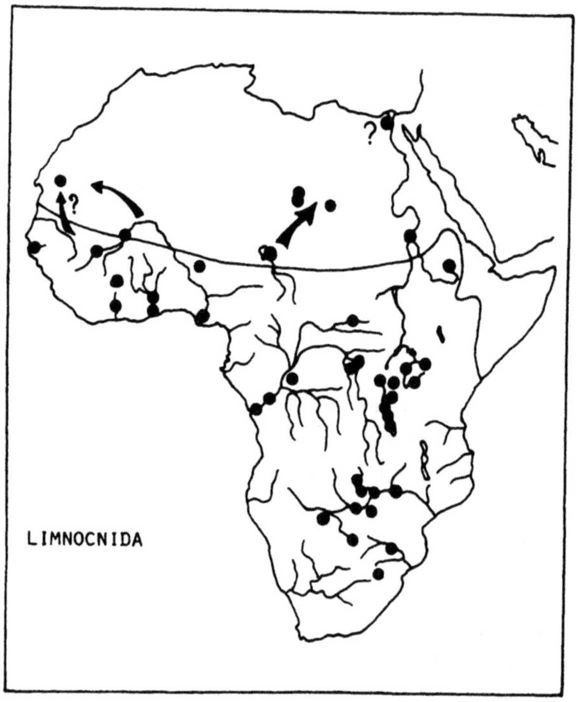

Fig. 1. Distribution of the genus *Limnocnida* in Africa. Arrows indicate most probable pathways of dispersal into the Sahara. The Senegal river is an improbable pathway, because of its mid-Holocene salinity.

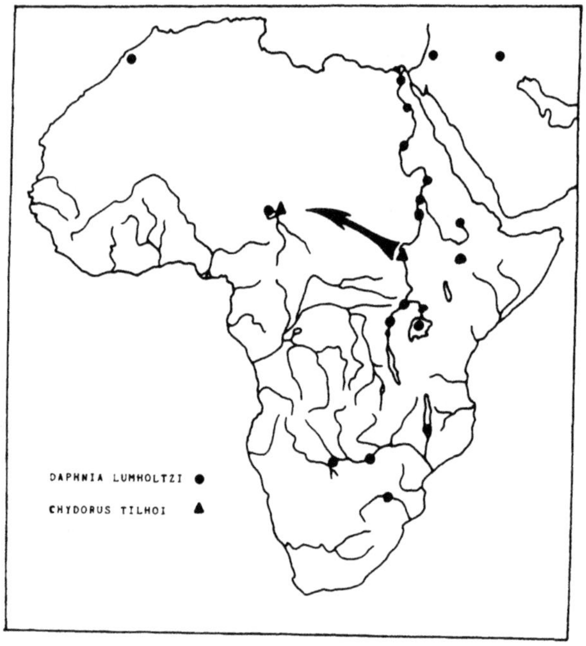

Fig. 2. Distribution of *Daphnia lumholtzi* Sars and of *Chydorus tilhoi* Rey & St Jean.

Gauthier (1951), Van de Velde (1978), Van de Velde & Dumont (1982) for the Senegal and Gambia basins; Brehm (1933), Brehm & Kiefer (1958), Dumont & Van de Velde (1977), Green (1962), Humes (1960), Kiefer (1933), Dumont, Pensaert & Van de Velde (1981), Lindberg (1958), Dussart (1980) for the Niger basin; Gauthier (1939), Dussart & Gras (1966), Rey & St Jean (1968, 1969) for Lake Chad; Rzoska (1976, which includes many older references), Kiefer (1978), Verheye & Dumont (1984), for the Nile system.

Earlier attempts to synthesise this body of knowledge (Dumont 1979, 1980), have shown that the plankton communities of the desert and semi-desert belt are very different from those of the Sahel belt, and the latter again differ strongly from those of the equatorial rainforest belt. Great similarities were encountered in the communities that live in the different Sahel river basins, but differences also became obvious, and these differences are just as interesting as the similarities. We will, hereafter, present a large number of individual cases, compare them with patterns observed in other groups of biota, and attempt to analyse their causes.

Limnomedusa

1. The genus Limnocnida *(Fig. 1)*

The number of species in this genus is still debated. It may possibly be not more than one or two (Goy, 1977), or up to five (Bouillon, 1957). However, the important point about these medusae is that they spread by way of rivers, but develop in numbers only in lakes or ponds. Consequently, few observations are available from the many rivers of Africa, but spectacular outbursts occur in man-made lakes shortly after damming, like in lakes Kariba and Volta. The relict occurrence of *Limnocnida* in desert mountain lakelets is therefore particularly significant. Populations are known from gueltas (lakelets) in the Adrar of Mauritania, from the south slopes of Tibesti, and from the Ennedi mountains. They must have reached these sites by riverine transport (see further). It should be noted that *Limnocnida* has been found in most African river systems, but not in the Senegal (and its adjacent lakes Guiers and Rkiz). Rzoska's (1976) statement that it is absent from the Nile is incorrect:

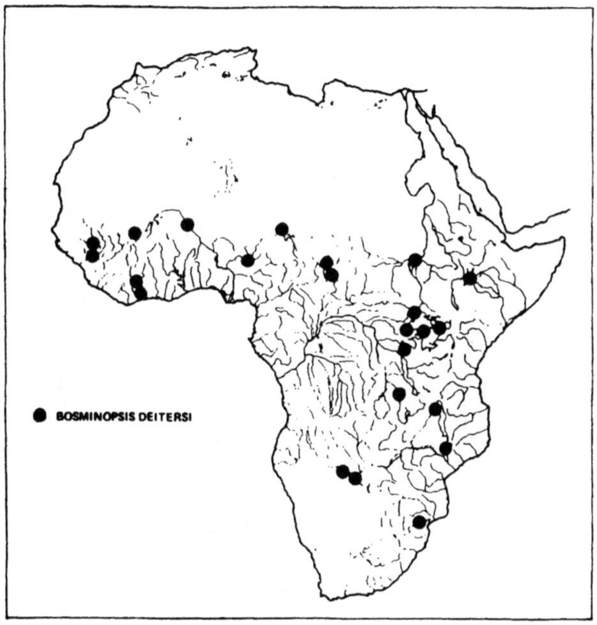

Fig. 3. Distribution of *Bosminopsis deitersi* Richard in Africa.

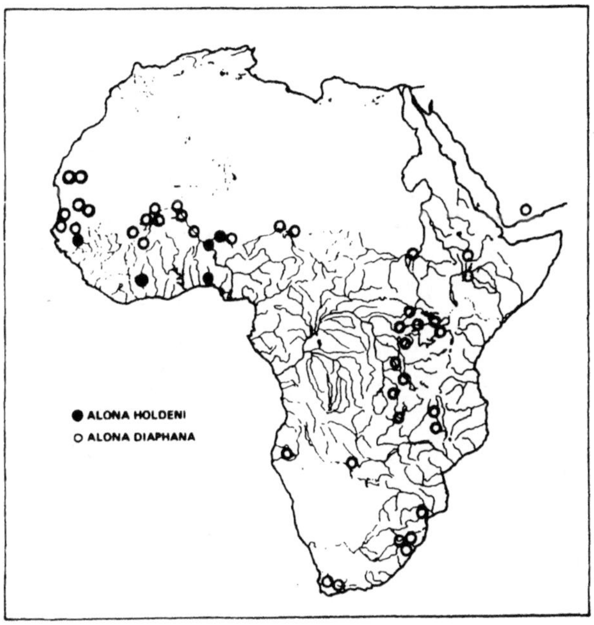

Fig. 4. Distribution of *Alona diaphana* (Sars) and of *Alona holdeni* Green in Africa.

there is a record from Lake Tana (Thiel, 1973), and specimens have been seen at Khartoum (el Moghraby, pers. commun.), while there has also been a short bloom of an unidentified medusa in Cairo (Elster & Vollenweider, 1961).

Cladocera

2. The genus Daphnia

Dumont (1980) showed that three limnetic species (*D. longispina, D. barbata, and D. lumholtzi*) extend from south Africa, across the raised East African plateau, to the Nile. *D. longispina* extends into Europe and temperate Asia. *D. lumholtzi* even reaches Australia, but *D. barbata* is restricted to Africa. It occurs in the Nile, Chad, Niger, and Senegal systems. A fourth limnetic species, *D.cf laevis* is also African, but even more local. It has been found only in the eastern part of South Africa and on the East African plateau. It is, however, at least as curious that *D. lumholtzi* does not extend to the Niger basin (Fig. 2). Its limit of westward extent is Lake Chad.

3. Other Daphniids and non-Chydorid families

Scapholeberis kingi Sars, and three species of *Simocephalus* (*S. vetulus* et ssp., *S. latirostris* Stingelin, *S. serrulatus* Koch) have a pattern of distribution similar to that of *D. barbata,* but *S. serrulatus* also extends into the equatorial rainforest, where the genus *Daphnia* is at all absent. The genus *Ceriodaphnia* with 4, but possibly quite a few more species in Africa, is in need of revision and cannot, at present, be included. Among the Bosminidae, *B. longirostris* (O.F.M.) and *Bosminopsis deitersi* (Fig. 3) show a pattern related to that of *D. barbata* again, while, in the Moinidae, *Moinodaphnia macleayi* behaves more like *S. serrulatus.* Like *D. lumholtzi,* it is a wide-ranging tropical form that extends to Australia. The Sididae *Diaphanosoma excisum* Sars, *D. sarsi* (Richard), and *Pseudosida szalayi* are more similar cases.

4. Chydoridae

Most chydoridae are littoral, some even benthic or phreatic in nature. Few species occur, and rather erratically, in the pelagial of large lakes. The genus *Chydorus,* and in particular *C. sphaericus* used to be considered, until very recently, as cosmopolitan and ubiquitous. However, for unknown reasons, the genus *Chydorus* is absent from the Western, Central and North-Eastern Sahara (Dumont, 1980), while at the species level a recent redescription of *C. sphaericus* (Frey, 1980) has opened possibilities for re-examining the numerous records of

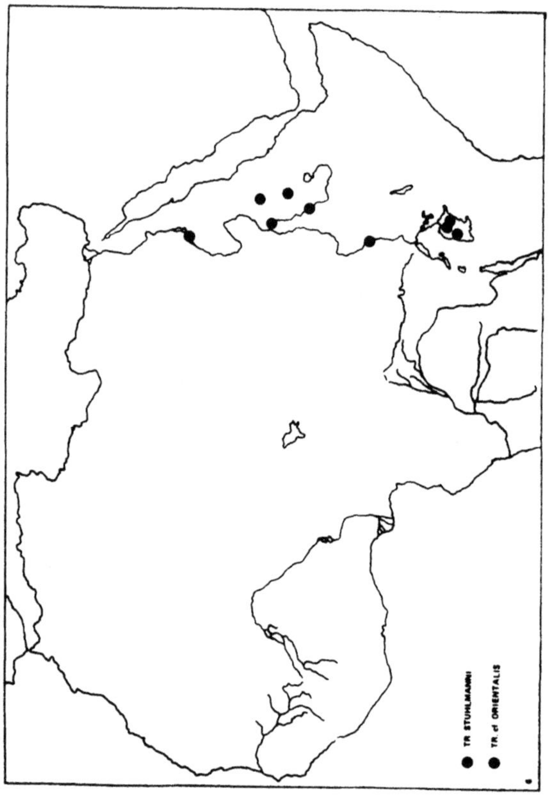
TR STUHLMANNI
TR cf ORIENTALIS
6

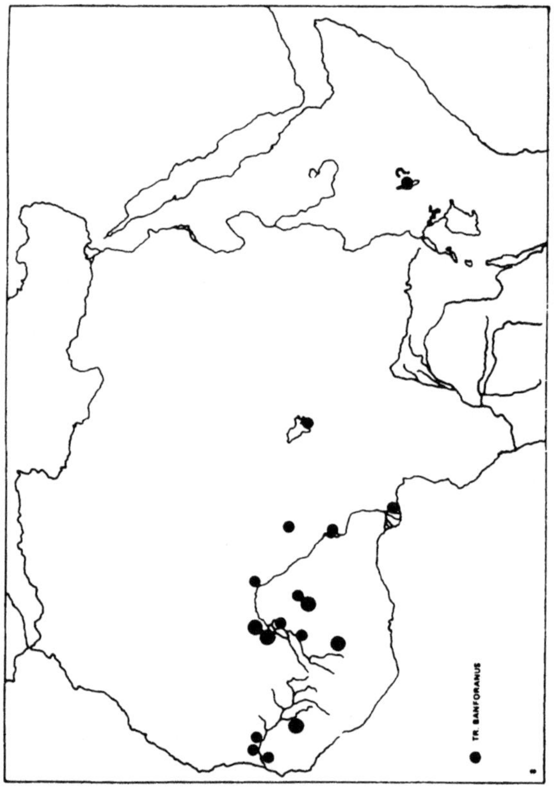
TR. BANFORANUS
8

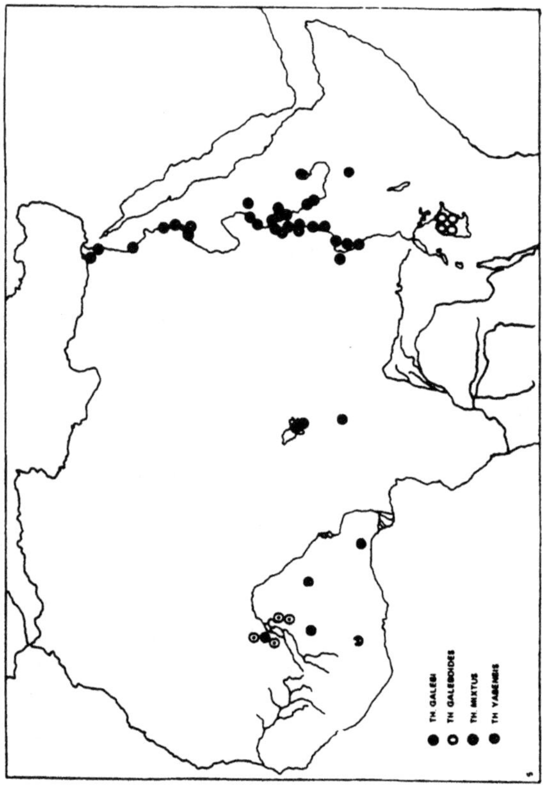
TH. GALEBI
TH GALEBOIDES
TH. MIXTUS
TH YABENSIS
5

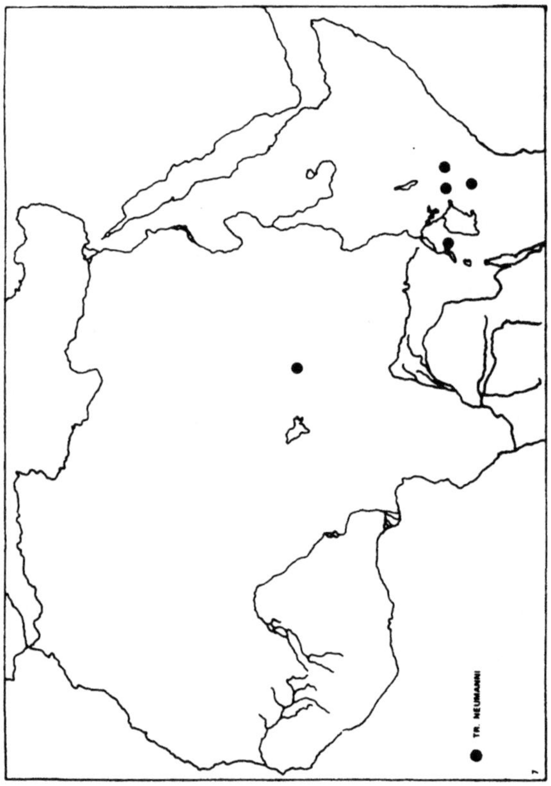
TR. NEUMANNI
7

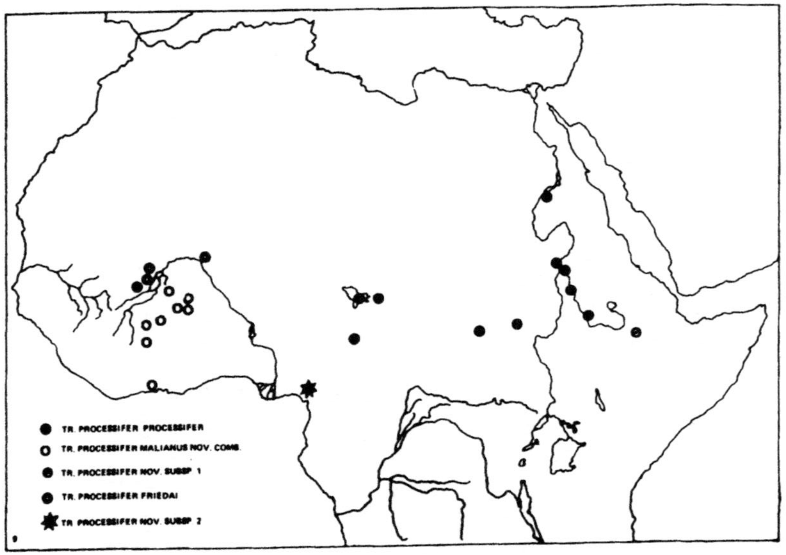

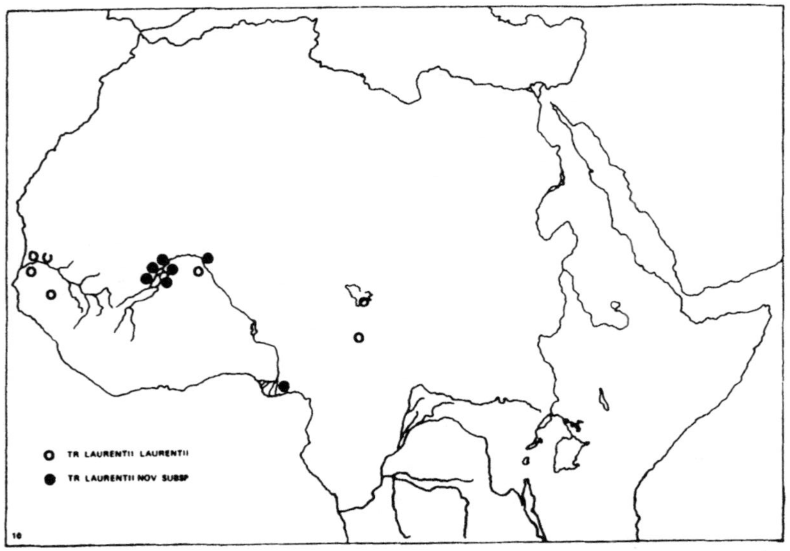

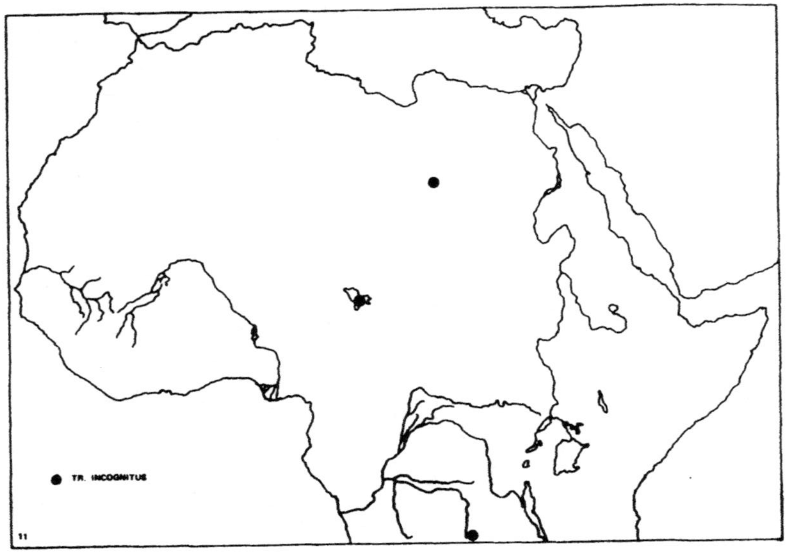

Figs. 5-11. 5: Distribution of the northern African species of *Thermodiaptomus;* 6: Distribution of *Tropodiaptomus* cf. *orientalis* and *Tropodiaptomus stuhlmanni* (Mrazek) in northern Africa; 7: Distribution of *Tropodiaptomus neumanni* (Van Douwe) (= T. magnus Kiefer) in northern Africa; 8: Distribution of *Tropodiaptomus banforanus* (Kiefer). The record from Lake Turkana is doubtful; 9: Distribution of *T. processifer* (Kiefer) and subspecies; 10: Distribution of *T. laurentii* Gauthier. A distinct subspecies, of probable mid-Holocene age, inhabits the internal delta of the Niger; 11: Distribution of *T. incognitus* Dussart & Gras (assuming *T. symoensi* Einsle to be a junior synonym).

that species from elsewhere in Africa. One very characteristic Chydorus is *C. tilhoi* Rey & St Jean. It was originally described from Lake Chad, but also occurs in the White Nile (Frey, 1982) (Fig. 2).

Among the many species of *Alona* found in Africa, *A. diaphana* Sars shows the familiar pattern seen before: South Africa, East Africa, the Nile, Chad, Niger, and Senegal, with relicts in the Western Sahara. It seems to avoid the wet rainforest zones (Fig. 4), where it is replaced by a number of endemic species (e.g. *Alona cannelata* Brehm, *Alona simonei* Dumont). Some of these may restrict to the coastal zones or to the Fouta Djalon mountains very narrowly, but others, such as *Alona holdeni* Green, have spread into the middle and lower Niger (Fig. 4), without, however, crossing into the Chad system.

5. *Calanoida: Thermodiaptomus (Fig. 5)*

This genus is exclusively African and is composed of 6 species and a possible subspecies. *Th. mixtus* (Sars) extends throughout Eastern Africa, reaching the Nile valley. *T. galebi* is the dominant species of the Nile River. It is extremely common in Lake Tana, but it also occurs in Lake Chad, and in some small tributaries of the Logone River draining towards Lake Chad (e.g. the Uham River: Kiefer, 1929). Very recently, its presence in Lake Ziway, Ethiopian rift valley, was established (Dumon, 1983). *T. galeboides* is endemic to Lake Victoria, and *T. yabensis* (Wright & Tressler) is confined to the Niger catchment.

6. *Calanoida: Tropodiaptomus*

This large genus, widespread in Australasia and Africa, is represented by possibly as many as 35 species and subspecies in Africa alone (Dussart & Defaye, 1983), in spite of several unsettled synonymies (Verheye & Dumont, 1983).

Tropodiaptomus cf. orientalis (Brady) is known from South Africa to the Nile (Fig. 6), but is replaced in Lake Victoria by the very closely related, possibly racially connected *T. stuhlmanni* (Mrazek). This taxon has not been found west of the Nile in Africa to date. *T. neumanni* (Van Douwe) occurs in lakes on the East African Plateau and Rift Valley, including Lake George, and it has also been reported once from the Chad basin (Gauthier, 1939) (Fig. 7). It does not, however, extend west of Lake Chad. *T. banforanus* Kiefer, conversely, occurs commonly in the Senegal valley and Fouta Djalon mountains, and it is also widespread in the Niger valley, reaching Lake Chad in the east (Fig. 8). A record from Lake Turkana by Lowndes (1936) is extremely doubtful. Species that are entirely confined to the Niger valley are *T. lateralis* (Kiefer) and *T. digitatus* Dussart.

While no evidence of subspeciation could be found in *T. banforanus* in the area discussed here, such a phenomenon is clearly present in the next two species. *T. processifer* (Kiefer 1926) occurs over the entire width of the African continent, and is present in its typical form in the Nile and Chad basins (Fig. 9). However, not less than two subspecies occur in the Niger basin. *T. p. malianus* Humes occurs south of the Niger River and is particularly common in the upper Volta valley. An undescribed subspecies (Verheye, in prep.) inhabits the Middle Niger, especially the marshy lakes of the internal delta. There exist also two isolated, altitudinal subspecies, one in Ethiopia (*T. p. friedai* Verheye & Dumont), and a second, as yet undescribed subspecies in the Cameroun Mountains. Subspeciation has also taken place in *T. laurentii* Gauthier, a West African species that extends as far east as the Lake Chad basin (Fig. 10). The nominal subspecies is found in the Senegal valley, in Lake Chad and in Lac de Lere, and also in some lakes to the southeast of the internal delta of the Niger, and poorly connected to that river (lakes Gorarou and Do). In the internal delta and the river itself, a group of populations is found which consistently differs in morphology, and should be considered as a distinct subspecies (Verheye, in prep.).

T. incognitus Dussart & Gras (Fig. 11), finally, is the dominant calanoid of Lake Chad when this lake is close to its 'normal' level (corresponding to a surface area of ca. 20 000 km^2). It is, in all probability, the same species as *T. symoensi*, described by Einsle (1971) from Lake Bangweolo in the Zaire catchment. Importantly, this species was found in at least two samples from Kufra oasis, East Libya. It is therewith the first Tropodiaptomus to be found within the limits of the Sahara, and north of the Tropic of Cancer.

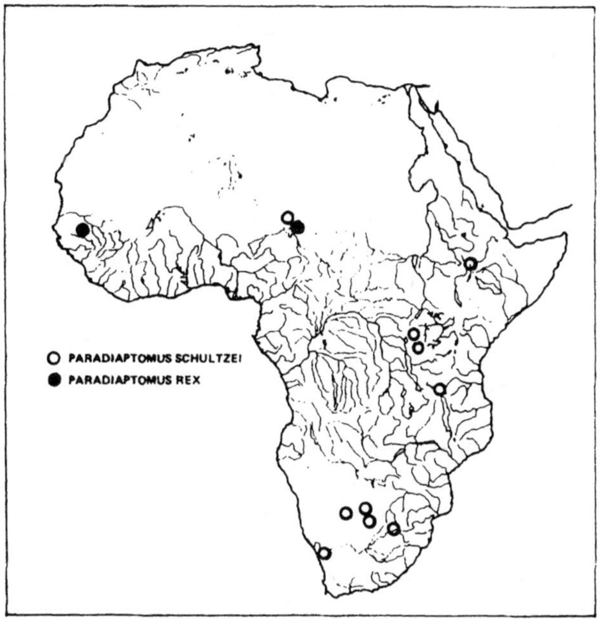

Fig. 12. Distribution of *Paradiaptomus schultzei* (Van Douwe) and of *P. rex* Gauthier in Africa. The range of both species meets in the Chad basin.

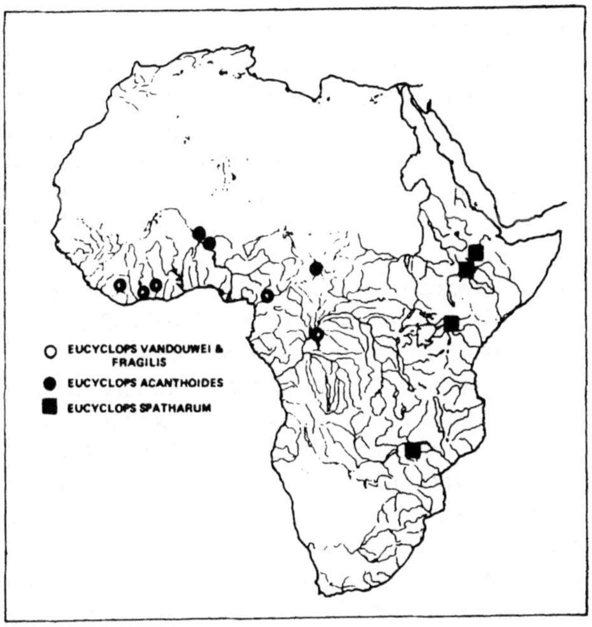

Fig. 13. Distribution of *Eucyclops spatharum* Harding, of *E. vandouwei* Brehm and *fragilis* Kiefer, and of *E. acanthoides* (Van Douwe) in Africa.

7. Calanoida: Paradiaptomus

Representatives of any of the four Paradiaptomid genera rarely co-occur with *Tropodiaptomus* or *Thermodiaptomus* species. In lowland areas, they are found predominantly in arid and semi-arid regions, where they are most common in semi-permanent and ephemeral biotopes, while at higher altitudes, they are found in permanent lakes as well. Paradiaptomids are consequently not found in the wet equatorial lowland areas of Africa (see also Dumont, 1980). There is circumstantial evidence that this pattern does not reflect true adaptation. Rather does it seem that Paradiaptomids have no access to permanent water in the well-watered lowlands of Africa because of the competitive superiority of the true Diaptomids in these latter conditions. The case of *Paradiaptomus schultzei* (Van Douwe) illustrates this point well. It is a species that was originally described from temporary pans in the Kalahari semi-desert, later found back in rainpools on the Ussangi steppe, East Africa, and in numerous pools (at least part of which seem to be permanent) in the Kagera park, Rwanda (Dussart, 1977). Recently, this animal was also discovered on the Ethiopian plateau and in Lake Chad area (Verheye & Dumont, 1984) (Fig. 12). The tendency to occur in permanent biotopes at higher altitudes (where Diaptomids are scarce) is obvious. *Paradiaptomus rex* Gauthier was originally described from a semi-permanent swamp in Senegal, and identified from the Chad basin on unpublished figures produced by R. Gras and communicated to us by C. Lévêque (Fig. 12). This little known species, which is clearly to be expected in the Niger valley as well, is thus living in a narrow zone on the border area of the Paradiaptomid and Diaptomid ranges. For the distribution of the other regional Paradiaptomids (*Metadiaptomus mauretanicus* Kiefer & Roy, *Paradiaptomus greeni* (Gurney) and *Neolovenula alluaudi* (De Guerne & Richard)), see Dumont, 1979, 1980 and Dumont *et al.*, 1981. These species typically occur in the Sahara and Northern Sahel. At about 16 °N, replacement by true Diaptomids takes place.

8. Cyclopoida

A meaningful discussion of this group is handicapped by the fact that many genera are in need of revision. The genus *Mesocyclops*, however, was recently revised by Van de Velde (1982). Among the conclusions of her study, it is worth stating that the reputedly cosmopolitan *M. leuckarti* (Claus), cited more than a hundred times from Africa, does not occur on that continent, but is replaced here by some 12 other taxa, all restricted to Africa. Some of them extend over most of the continent (*M. major* Sars, *M. aequatorialis* Kiefer et ssp., *M. ogunnus* Onabamiro, and *M. salinus* Onabamiro), while others have extremely limited distributions (for example, *M. tenuisaccus* is restricted to the Cape area only). *M. rarus* Kiefer occurs in South-East Africa and on the East African plateau, and has reached the Chad basin (a relict population is still present in the Tibesti mountains: Van de Velde, 1982). *M. aspericornis* (Daday) is found on the East African plateau, in the Nile valley, and in the Niger and Volta basins. Its absence from Lake Chad is, in all probability, due to confusion with *M. leuckarti*. *M. paludosus* Lindberg is only known from the East African plateau. The same holds true for *Thermocyclops schmeili* (Poppe & Mrazek), a representative of a related, species-rich genus. A revision of the African *Thermocyclops*-species will be needed before their individual ranges on that continent can be fully discussed (Maas, 1982). Probably, *T. neglectus* (Sars) is widespread, but 5–6 other species might show restricted ranges. Species with narrow ranges should be expected before all in littoral genera, representatives of which are less likely to be carried away by rivers. The large number of *Eucyclops* species endemic to the littoral of Lake Tanganyika (Sars, 1909) illustrates this point, even though excessive splitting may have occurred. An example of a species restricted to the East African plateau is *E. spatharum* Harding (Fig. 13), while *E. acanthoides* (Van Douwe) is limited to the Niger–Chad Zone. A species (or two related, vicariant species) typical of the rainforest zone are *E. vandouwei* Brehm and *E. acanthoides* (Van Douwe) (Fig. 13). It is almost certain that in many other genera, similar patterns will be discovered. In *Tropocyclops*, for example, the West African mountain and coastal zone has been shown to be an important centre of speciation (Dumont, 1981).

Discussion

In order to understand the origin of the patterns observed, it is necessary to dwell upon the relief, the climate, and the late-Pleistocene palaeoclimates of Northern Africa. First, however, we will compare the patterns found in zooplankton with those known in other groups of aquatic animals, and in higher aquatic plants.

Distribution patterns of fish, molluscs, and higher aquatic plants

The great similarity between the fish-faunas of the Nile (excluding its source lakes), the Chad basin, the Middle and Lower Niger, and the Senegal are so well-known (Greenwood, 1976; Beadle, 1981) that they need little comment. The term 'Soudanian fauna' has been coined specifically to identify that fauna. However, beside similarities, the faunas of these juxtaposed basins also show differences, and one wonders which of both are more interesting. Table 1 summarizes the situation, brought up to date for the Chad basin (Carmouze *et*

Table 1. Fish faunas of the different Sahel basins, and of two adjacent Saharien mountain ranges and their relationships. The central position of Lake Chad is emphasized.

	Total species	Shared with Chad basin	Endemic	Endemicity shared with Chad basin
Nile basin (without lakes of Upper White Nile)	c. 115	89	c. 20	7
Jebel Marra	16	16	–	–
Tibesti mountains	12	10	2 ssp.	–
Chad basin	c. 140	140	1*	1
Niger basin	c. 130	106	c. 20	8
Zaire basin	min. 700; max. 1 000	47	560	13

* Possibly up to 20 more in the upper Shari and Logone rivers, not occurring in the lake itself. *Alestes dageti* is the only endemic of the lake and the lower reaches of the Shari.

al., 1983). Note that the fauna of Jebel Marra is a pure extract of the Nile-Chad fauna. The Zaire basin stands out for three reasons: 1) its total number of species is at least 5 times that of either Soudanian system; 2) its level of endemicity is at least 70%; 3) it shares only few species with the Soudanian rivers (ca. 7%). Even when considering the size of the Zaire basin as a factor contributing significantly to its faunal richness, this difference of almost an order of magnitude indicates that 1) the basin as an entity must have had considerable stability in time; 2) it has had some, but no major connections with its northern and eastern neighbour basins. Obviously, the reverse is true for the Soudanian basins. Lake Victoria and the West African mountain waters tend to behave like the Zaire basin. Lake Victoria alone has more than 200 fish species, 80-90% of which are endemic. The Fouta Djalon mountains, where the rivers Senegal, Gambia, and Niger spring, in addition to a number of short coastal rivers, are still imperfectly known (Leveque, pers. comm.), but a single collection (Daget, 1962) yielded close to 100 species, 60-70% of which were restricted to that area (the *amphiguinean* fauna of Daget). These could be split even further into *atlantico-guinean* species (coastal side of the mountains only), and *soudano-guinean* species (landward side of the mountains only). The reader will easily equate all these types with types found in the zooplankton. Where there is endemicity in fish, there is endemicity in zooplankton as well. The relative proportions (percentages) may be higher in fish, but the difference is of a quantitative, not a qualitative nature.

Strikingly congruent patterns are found among the aquatic mollusca. Brown (1980) and the Danish Bilharziasis Laboratory (1978) amply illustrate the fact that the Guinean zone of West Africa harbours many endemic gastropod species, in contrast with the Middle and Lower Niger, while the Zaire basin is again rich in species and with a high degree of endemicity. Van Damme (1984) discusses the distribution of all the Mollusca of Northern Africa. The types of distribution encountered before are again apparent, even within single genera. Thus, *Lanistes carinatus* (Olivier) is restricted to the Nile system, *Lanistes ovum* Peters is shared by the Nile and Chad (but is not found in the Niger), and *Lanistes varicus* (Muller) occurs in the Senegal and Niger valleys, but not in Chad. Other species are shared by the Senegal, Niger, and Chad basins, but do not reach the Nile (e.g. *Aspatharia chaiziana* Rang). Endemics to individual basins are known for each basin as well, although the Middle and Lower Niger are remarkably poor. Even Lake Chad has its share of endemics among Gastropods (*Gabbiella tchadiensis* Mandahl-barth, *G. neothaumaeformis* (Germain), and *Biomophalaria tchadiensis* Mandahl-Barth), but not among bivalves. Thus, Molluscs stand out in the fauna of Lake Chad, for among fish only one species, *Alestes dageti* is endemic to that lake, while in the zooplankton the cyclopoid *Thermocyclops incisus tchadensis* Dussart & Gras is the only probable case.

Even the higher aquatic plants conform to the same patterns (Raynal-Roques, 1980). Some species extend from West Africa eastwards as far as Lake Chad, but do not reach the Nile (e.g. *Aponogeton subconjugatus* Schum. & Thonm.); others, occurring in South Africa, the East African plateau, the Nile, and sometimes even extending into Asia, do not occur west of the Chad basin. Examples are *Centrostachys aquatica* (R.Br.) Well, and, of course, *Cyperus papyrus* L. Besides, numerous species span the entire Sahel from East to West. *Nymphoides ezannoi* Berh. may be cited here, but the congeneric *N. guineensis* Raym. is restricted to the Fouta Djalon only.

Origin and age of the different distribution patterns

Northern Africa (i.e. Africa north of the Zaire basin and west of the Ethiopian highland) is nowadays composed of a desert (the Sahara) and its northern (the Maghreb) and southern (the Sahel) fringing steppes. The Maghreb will not be included in the present discussion. It has a Mediterranean climate, mitigated in the west by the influence of the Atlantic Ocean. South of the Sahara, there is a progressive increase of the vegetation cover as one moves south of 15 °N. Equatorial rainforest used to extend in Africa as far north as 8 °N, but has now regressed considerably by overexploitation.

Most of this enormous area is a low plateau, with few isolated mountain islands, such as the Fouta Djalon in the west, the Cameroun highlands and the Adamoua plateau in the centre, and the Darfur mountains and Bongo plateau in the east. None of these is very high, and they are morphologically closer to plateaus, than to true mountain chains. An immediate implication is that the rivers that spring from them are not well isolated from one another: overflows from one basin into another are not infrequent. It is precisely from these plateaus that the Senegal, Niger, and Logone-Chari spring. These three major rivers have another common particularity as well: taking their origin in well-watered, southern areas, they subsequently flow north into semi-desert and desert country. The Senegal reaches the Atlantic Ocean after a broad bend to the west. The Niger has a complex history: it may have reached the Atlantic not far from the mouth of the Senegal until shortly before 20 000 BP, when a general dry spell, synchronous with the Wurm III glaciation in Europe made it cease to flow. Its bed eventually became blocked by the sands of the erg Ouagadou, and when the river resumed its course, it formed a series of large, shallow endorheic lakes ('Lake Arouan') north of its present Central Delta, in the western desert (Monod, 1958). The lower course of the present Niger was then still part of the Oued Tilemsi, draining the Adrar des Iforhas. Around 6 000 BP, the Niger cut through the Tosaye sandstone, separating it from the Tilemsi, and thereby captured that river. A southern remnant of the former endorheic situation is formed by the so-called internal delta, a huge complex of marshes and shallow lakes. Through these events, the Niger's course became questionmark-shaped.

The Logone-Chari rivers drain into an enormous flat basin from which they do not, nowadays, find an outlet. Lake Chad collects their water and was, in every respect, an analogue to the Niger until some 6 000 years ago.

The Nile, finally, has a very long and complex history as well. While the Blue Nile has contributed to the River throughout the Pleistocene, and probably much longer (Said, 1981), it is now well established that the White Nile connected to its lake sources around 12 500 BP only (Livingstone, 1980). Before that time, much of the water supply to the White Nile must have come from the Bahr el Ghazal, perhaps even from Jebel Marra. Like the other major rivers of Northern Africa, the Nile springs in the south and runs north. Unlike these, however, it crosses the entire Sahara and empties into the Mediterranean Sea. It is the longest river in the world.

Palaeoclimates

Knowledge on Late Pleistocene climates and on climatic change has greatly expanded in recent years. An intense dry spell occurred over most of Africa between 20 000 and 18 000 BP. It was contemporaneous with the height of glaciation in Europe and the Atlas Mountains and caused a temporary interruption in the flow of the Niger, but also of that of the Nile (Livingstone, 1980), while Lake Chad dried up completely (Servant & Servant, 1983). Later, the climate progressively became wetter, and around 12 500 BP Lake Albert began to overflow into the White Nile, while shortly after 13 000 BP numerous lakes appeared in the Lake Chad basin. In their initial phase, these lakes fluctuated strongly, indicating that climatic conditions may still have been unstable. Shortly thereafter (11 000–8 000 BP), the Atlantic reached a level of ca. 5 m above today's, corresponding to maximum deglaciation (Michel, 1980). Seawater penetrated deep inland, and this was felt very strongly in the Senegal valley. A later and weaker transgression, the Nouakchottien, around 6 000–4 000 BP was indeed still sufficient to cause seawater incursions as far as Bogue (230 km inland), and invaded lakes Guiers and Rkiz (Michel, 1980).

Precipitation was higher than today in all climate zones now hyperarid, arid, or semi-arid. Lake Turkana was open to the Nile at 9 500 BP and may have intermittently remained so until shortly before 8 000 BP (Cohen, 1981). The Nile in the Sudan was also high between 12 500 and 8 100 BP (Williams & Adamson, 1980), and the major Saharian rivers (Tafassasset, Azouak, Tilemsi) were active. Rivers were draining the northern flanks of the Tibesti mountains and reached the Mediterranean (Jakel, 1979). Humidity peaked at 10 000–8 500 BP. No specific data for the Upper and Middle Niger are available for that period, but is cannot have behaved very differently from its neighbour systems. Considerable expanses of the present Western Sahara, possibly as far north as El Djouf, must have been covered by the waters of Lake Arouan.

Gasse (1980), finally, showed that high water levels were present in all East African plateau and rift lakes after 12 000 BP, with a peak between 9 500 and 8 400 BP. At that time, the Galla lakes in the Ethiopian rift were united, and supposedly had a functional outlet in the north, towards the river

Awash (see also Beadle, 1981).

A general and drastic recession followed between 8 000 and 7 000 BP.

At 6 000–5 000 BP, however, Lake Chad reached its Holocene peak level (Servant & Servant, 1983), the Atlantic rose again to +2 m and invaded the Senegal valley, and the Nile in the Sudan rose to fairly high levels, although not to the same extent as during the previous humid spell. The southern Sahara and Sahel remained or became hyperarid (Dumont, 1978) and, while the Niger rose to peak levels high enough to break through the Tosaye sandstone in the East, it probably captured only the preexisting but then dry bed of the Tilemsi. Lake Arouan emptied into the new river course and disappeared.

All these observations are consistent with an almost exclusive riverine supply of water of southern origin to these Sahel lakes, where local precipitation had become insignificant. This riverine supply, however, may have been considerably above today's, as if the former Sahel precipitation had been added to that of the Guinean zone. Very high lake levels recorded at Tichitt, Central Mauretania, before 4 500 PB (Munson, 1981) were thus probably due to transgressions of the Senegal, and not to local precipitation. This situation, with fluctuations, continued until ca. 3 000 BP, when a final minor wet spell was felt simultaneously on the Senegal, Chad, and Nile basins.

Biogeographical interpretation

It is clear that the generalized arid phase of the terminal Pleistocene (ca. 20 000–13 000 BP) must have been destructive to the floras and faunas of the great Sahel rivers. Probably, relicts survived in the upper reaches of the rivers Niger, Chari, and of the White Nile, explaining the remanent endemism in their respective basins today.

The humid period at 9 500–8 000 BP is the only likely candidate for the numerous basin-crossings that are evident from the mixture of biota in all basins. The outlet of Lake Turkana, which was functional during this period *only*, must have allowed the calanoid *Thermodiaptomus galebi* to reach the Ethiopian rift valley. Evidently, the Nile–Chad connection functioned too: *T. galebi, P. schultzei, Daphnia lumholtzi, Chydorus tilhoi*, and many other species crossed from East to West or *vice versa*. The crossing between Lake Chad and the Niger had to go through the stepping stone of the still active Oued Tilemsi. More species are, consequently, shared by the Nile and Lake Chad than by the Niger and Lake Chad. The Senegal contributed least to the faunal exchanges, because of its high mid-Holocene salinity. Exchanges between the Zaire and Lake Chad were also made possible, and introduced *T. incognitus* to the Chad basin.

Since the maximum extent of Lake Chad (and of the Arouan Lakes) was around 6 000 BP, it is extremely likely that the river-dispersed medusa, *Limnocnida*, reached the Adrar of Mauretania, the Tibesti, and the Ennedi at that point in time. At that same time, *T. incognitus* must have crossed between the Tibesti and Ennedi mountains and reached Kufra oasis. The relict fish faunas found on the southern slopes of Tibesti, Ennedi, and Jebel Marra (see review in Dumont, 1981; and Bailey-Watts & Rogers, 1970) should have the same isolation-time.

This lapse of time (5–6 000 y) is clearly insufficient to produce much allopatric speciation, although some signs of it are seen in the Tibesti fish. An endemic subspecies (a melanotic race) of *Sarotherodon galilaeus*, ssp. *borkuanus* Pellegrin, and another possible melanotic subspecies, *Labeo annectens tibestii* (Pell.), are found here. Both are derived of widely distributed Soudanian stock. Incipient speciation was also encountered in the Niger subspecies of the calanoids *T. laurentii* and *T. processifer*.

However, in both these species, two distinct subspecies were found to inhabit the general area at the middle valley of the river. The nominal subspecies of *T. laurentii* is found in the Senegal valley, in a few lakes to the south-east of the internal delta of the Middle Niger, outside of the floodable area, and very loosely connected to the river itself (lakes Goràrou and Do), and, finally, in the Chad basin. An undescribed subspecies inhabits the delta lakes and swamps, and was also recorded close to the mouth of the River (leg. A. Egborge). Likewise, a subspecies of *T. processifer*, distinct from *T. p. malianus* which is found south of the Niger, occurs in the floodable zone. There is little or no alternative but to consider both taxa as having locally arisen from pre-existing stock-populations. One, (*T. laurentii spp.*), has now expanded downstreams.

Finally, a word needs to be said about the strongly fluctuating conditions to which the Chad and Niger delta lakes are subject today. The droughts of the Sahel during the 1970s were accompanied by a considerable shrinking of Lake Chad, which reduced from its average size of 25 000 km^2 to 6 000 km^2, and its northern half dried up completely. Major changes occurred among the biota: *Tropodiaptomus incognitus* gradually became rare, and finally (temporarily?) disappeared. *Thermodiaptomus galebi* took over, but eventually declined also. Before the dry spell, other calanoids were found as well. In samples collected by P. Compere (Brussels) in Jan.-March 1968, *T. laurentii* and *T. banforanus* occurred. In collections studied by R. Gras (ORSTOM), probably dating from the later, 'lesser' Chad period, *T. processifer*, and the paradiaptomids *P. rex* and *P. schultzei* were found. With the decline of the lake level, *Daphnia* species decreased in abundance and disappeared, and many fish species reacted in the same way. There were also strong modifications in the macrophytes, and, notably, the *Papyrus*-stands dwindled to near-extinction (Carmouze *et al.*, 1983). Similar droughts happened earlier in this century as well, and those of 1904–1915 were possibly more severe than those of the 1970s (Servant & Servant, 1983). These short-term unfavourable periods illustrate well how devastating the long dry spell of 5 500-3 000 BP may have been to the Sahel faunas. Thus, the late Pleistocene–Holocene has been characterized by two opposite trends. On the one hand, considerable faunal and floral contact between adjacent basins produced a Soudanian type of aquatic biota, through which Lake Chad, by virtue of its central position, acquired a tremendous diversity. On the other hand, numerous extinctions, due to environmental change and biotic interactions, also took place, and, in all probability continue to take place. In contrast, speciation, the production of novel biota adapted to their particular environment, has been of very subordinate importance since ca. 12 500 BP.

Acknowledgements

This study was carried out as part of grants 2.0009.75 and 2.0033.81 of the Fonds voor Kollektief Fundamenteel Onderzoek, Belgium. We thank P. Compere, C. Lévêque, and A. Egborge for unpublished data and plankton collections.

References

Bailey-Watts, A. F. & J. F. Rogers, 1970. The fish collection of Jebel Marra. Annu. Rep. hydrobiol. Res. Unit Khartoum 12: 7–9.

Beadle, L. C., 1981. The Inland Waters of Tropical Africa. 2nd ed. Longman, London, 475 pp.

Blache, J., 1964. Les poissons du Bassin du Tchad et du bassin adjacent du Majo Kebbi. Mem. ORSTOM 4: 483 pp.

Bouillon, J., 1957. Etude monographique du genre Limnocnida (Limnomeduse). Ann. Soc. r. zool. Belg. 87: 252–500.

Brehm, V., 1933. Voyage de Ch. Alluaud et P. A. Chappuis en Afrique Occidentale Francaise. II. Cladoceren. Arch. Hydrobiol. 26: 50–90.

Brehm, V., 1958. Mission M. Blanc-F. d'Aubenton (1954). VI. Cladoceren, Euphyllopoden und Calanoide Kopepoden. Bull. IFAN A 20: 100–114.

Brehm, V. & F. Kiefer, 1958. Cladocères, Copépodes et Rotifères du Sudan (A.O.F.). Bull. IFAN A 20: 95–99.

Brown, D. S., 1980. Freshwater Snails of Africa and Their Medical Importance. Taylor & Francis, London, 433 pp.

Burke, K., A. B. Durotoye & A. J. Whiteman, 1971. A dry phase south of the Sahara 20 000 years ago. W. afr. J. Archaeol. 1: 1–8.

Carmouze, J.-P., J.-R. Durand & C. Lévêque (eds.), 1983. Lake Chad. Monographiae Biologicae 53. Junk, The Hague, 575 pp.

Cohen, A., 1981. Palaeolimnological research at lake Turkana, Kenya. Palaeoecology of Africa 13: 61–82.

Daget, J., 1954. Les poissons du Niger superieur. Mem. IFAN 36: 391 pp.

Daget, J., 1962. Les poissons du Fouta Dialon et de la basse Guinee. Mem. IFAN 65: 207 pp.

Danish Bilharziasis Laboratory, 1978. Distribution of African Freshwater Snails of Medical or Veterinary Importance. Copenhagen, 20 pp.

Dumon, D., 1983. Morfologische studie van Copepoda en Cladocera van Ethiopie. M.Sc. Thesis, University of Gent, 102 pp.

Dumont, H. J., 1978. Neolithic hyperarid period preceded the present climate of the Central Sahel. Nature 274: 356–358.

Dumont, H. J., 1979. Limnologie van Sahara en Sahel. Unpublished D.Sc. Thesis, University of Gent, 557 pp.

Dumont, H. J., 1980. Zooplankton and the science of biogeography: the example of Africa. In: W. C. Kerfoot (ed.), Evolution and Ecology of Zooplankton Communities. The University Press of New England, Hanover (N.H.); Lond.: 685–696.

Dumont, H. J., 1981. Cladocera and free-living Copepoda from the Fouta Djalon and adjacent mountain areas in West-Africa. Hydrobiologia 85: 97–116.

Dumont, H. J., 1982. Relict distribution patterns of aquatic animals: another tool in evaluating late Pleistocene climate changes in the Sahara and Sahel. Palaeoecology of Africa 14: 1–24.

Dumont, H. J., J. Pensaert & I. Van de Velde, 1981. The crustacean zooplankton of Mali. Hydrobiologia 80: 161–187.

Dumont, H. J. & I. Van de Velde, 1977. Cladocères et Conchostracés recoltes par le Professeur Th. Monod dans la moyenne vallee du Niger en decembre 1972 et janvier 1973. Bull. IFAN A 39: 76–93.

Dussart, B., 1977. Contribution a l'etude des copepodes des eaux douces du Ruanda. Bull. IFAN A 39: 821-840.

Dussart, B. & D. Defaye, 1983. Repertoire mondial des Crustacés Copépodes des eaux interieures. I. Calanoides. CNRS, Paris, 224 pp.

Dussart, B. & R. Gras, 1966. Faune planctonique du lac Tchad. 1. Crustacés Copépodes. Cah. ORSTOM Ser. Oceanogr. 4: 77-91.

Einsle, U., 1971. Copepodes libres. In: J. J. Symoens (ed.) Exploration Hydrobiologique du Bassin du Lac Bangweolo et du Luapala 13: 74 pp.

Elster, H.-J. & R. Vollenweider, 1961. Beitrage zur Limnologie Agyptens. Arch. Hydrobiol. 57: 241-343.

Frey, D. G., 1980. On the plurality of *Chydorus sphaericus* (O. F. Muller), and designation of a neotype from Sjaelso, Denmark. Hydrobiologia 69: 83-123.

Frey, D. G., 1982. Relocation of *Chydorus barroisi* and related species (Cladocera, Chydoridae) to a new genus and description of two new species. Hydrobiologia 86: 231-269.

Gasse, F., 1980. Late quaternary changes in lake-levels and diatom assemblages on the southeastern margin of the Sahara. Palaeoecology of Africa 12: 333-350.

Gauthier, H., 1939. Contribution à l'étude de la faune dulcaquicole de la region du Tchad et particulièrement des Branchiopodes et des Ostracodes. Bull. IFAN 1: 110-224.

Gauthier, H., 1951. Contribution à l'étude de la fauna des eaux douces du Senegal (Entomostracés). Minerva, Alger, 169 pp.

Goy, J., 1977. Sur les *Limnocnida* africaines (Cnidaires, Limnoméduses). Bull. IFAN A 39: 563-582.

Green, J., 1962. Zooplankton of the River Sokoto. The Crustacea. Proc. zool. Soc. Lond. 138: 415-453.

Greenwood, P. H., 1976. Fish fauna of the Nile. In: J. Rzoska (ed.), The Nile. Monographiae Biologicae 29. Junk, The Hague: 127-141.

Humes, A. G., 1960. On two new species of Tropodiaptomus from Upper Volta and the Ivory Coast. Bull. IFAN A 22: 153-163.

Jäckel, D., 1979. Run-off and fluvial processes in the Tibesti mountains as indicators of climatic history in the central Sahara during the late Pleistocene and Holocene. Palaeoecology of Africa 11: 13-44.

Kiefer, F., 1928. Crustacés. I. Copepoda aquae dulcis. In: Th. Monod (ed.), Contribution a l'étude de la faune du Cameroun. Faune Colonies francaises 1: 535-570.

Kiefer, F., 1933. Freilebende Binnengewassercopepoden. Diaptomiden und Cyclopiden Französisch Westafrika. Arch. Hydrobiol. 26: 121-142.

Kiefer, F., 1978. Zur Kenntnis der Copepodenfauna ägyptischer Binnengewassern. Arch. Hydrobiol. 84: 480-495.

Lindberg, K., 1958. Cyclopides du Soudan (A.O.F.). Bull. IFAN A 20: 115-116.

Livingstone, D. A., 1980. Environmental changes in the Nile headwaters. In: M. A. J. Williams & H. Faure (eds.), The Sahara and The Nile. Balkema, Rotterdam: 339-360.

Lowndes, A. G., 1936. Scientific results of the Cambridge expedition to the East African Lakes, 1930-1. No 16. The smaller crustacea. J. linn. Soc. Lond. (Zool.) 40: 1-31.

Maas, S., 1982. Morfologische studie van enkele Thermocyclopssoorten uit Afrika. M.Sc. Thesis, Universiteit Gent, 106 pp.

Michel, P., 1980. The southwestern Sahara margin: sediments and climatic changes during the recent quaternary. Palaeoecology of Africa 12: 297-306.

Monod, T., 1958. Majabat al Koubra. Contribution a l'étude de l'Empty Quarter Quest Saharien. Mem. IFAN 52, 406 pp.

Munson, P. J., 1981. A late Holocene (ca. 4 500-2 300 BP) climatic chronology for the southwestern Sahara. Palaeoecology of Africa 13: 53-60.

Pachur, H. J. & G. Brown, 1980. The paleoclimate of the central Sahara, Libya and the Libyan desert. Palaeoecology of Africa 12: 351-363.

Raynal-Roques, A., 1980. Les plantes aquatiques (plantes a fleurs et fougères). In: J.-R. Durand & C. Lévêque (eds.), Flore et Faune aquatiques de l'Afrique Sahelo-Soudanienne I. ORSTOM, Docmns Techqs 44: 65-152.

Rey, J. & L. St Jean, 1968. Les Cladocères (Crustacés Branchiopodes) du Tchad (première note). Cah. ORSTOM Ser. hydrobiol. 2: 79-118.

Rey, J. & L. St Jean, 1969. Les Cladocères (Crustacés Branchiopodes) du Tchad (deuxième note). Cah. ORSTOM. Ser. hydrobiol. 3: 21-42.

Rzoska, J. (ed.), 1976. The Nile. Biology of an Ancient river. Monographiae Biologicae 29. Junk, The Hague, 417 pp.

Said, R., 1981. The Geological Evolution of the River Nile. Springer, Heidelberg, New York, 151 pp.

Sars, G. O., 1909. Zoological results of the third Tanganyika expedition, conducted by Dr. W. A. Cunnington F.Z.S., 1904-1905. Report on the Copepoda. Proc. zool. Soc. Lond. 1909: 31-77.

Servant, M. & S. Servant, 1983. Paleolimnology of an upper quaternary endorheic lake in Chad basin. In: J.-P. Carmouze, J.-R. Durand & C. Leveque (eds.), Lake Chad. Monographiae Biologicae 53. Junk, The Hague: 11-26.

Thiel, H., 1973 (1972). Limnocnida indica in Africa. Publ. Seto mar. biol. Lab. 20: 73-79.

Van Damme, D., 1984. The Recent and Subfossil Mollusca of the Sahara, Sahel and Maghreb. Dev. Hydrobiol., in press.

Van de Velde, I., 1978. Cladocera and Copepoda from the valley of the River Senegal. Biol. Jaarb. 46: 192-201.

Van de Velde, I., 1982. Revisie van het genus Mesocyclops Sars, 1914 (Crustacea: Copepoda) in Afrika. Ph.D. Thesis, Universiteit Gent, 361 pp.

Van de Velde, I. & H. J. Dumont, 1982. Cladocères et Copépodes du parc national du Niokolo-Koba. Mem. IFAN 92: 123-132.

Verheye, H. & H. J. Dumont, 1984. The Calanoida of the Nile system. Hydrobiologia, in press.

Williams, M. A. J. & D. A. Adamson, 1980. Late quaternary depositional history of the Blue and White Nile rivers in Central Sudan. In: M. A. J. Williams & H. Faure (eds.), The Sahara and The Nile. Balkema, Rotterdam: 281-304.

Description of *Alona broaensis* sp. nov. (Crustacea: Cladocera)

T. Matsumura-Tundisi[1] & N. N. Smirnov[2]
[1] *Departamento de Ciencias Biologicas, Universidade Federal de Sao Carlos, Brazil*
[2] *Institute of Evolutionary Morphology and Ecology of Animals of the U.S.S.R. Academy of Sciences, Leninsky prospekt 33, Moscow, U.S.S.R. 117071*

Keywords: tropical zooplankton, *Alona broaensis*, Broa reservoir, Brazil, Cladocera

Abstract

A description and illustrations are given of *Alona broaensis*, a new species of Chydoridae. Its nearest relative is *A. brasiliensis* Bergamin.

Material. 2 females from Broa Reservoir, São Paulo State, Brazil.
Holotype. MGU ♀ 4000.
Slide. MGU ♀ 4000.

Alona broaensis sp. nov. (Figs. 15–21)
(The figures are numbered in: Continuation to the paper 'Some comments on tropical littoral Cladocera, with description of *Alona incredibiles* sp. nov.' in this volume).

Female. The body outline is oval, the length not much exceeding the height. The ventral margin of the shell is convex (much less convex than the dorsal margin). The inferior–posterior corner of the valve is approximately 90°, bearing a short denticle, which may be duplicated. There is no noticeable sculpturing on the shell. The dorso-posterior corner of the shell is hardly noticeable. The rostrum, when viewed laterally, seems to be rather acute (but not pointed); when viewed from a dorsal side it is seen as widely rounded. There are simple setae along the ventral margin of the valve which are placed on the edge of the valve. The setae are placed more densely in the middle part of the valve than in its anterior part, where they are longer, and in its posterior part, where they are diminishing in length. On the posterior margin of the valve just above the denticle there are very minute setae.

Hydrobiologia 113, 327–329 (1984).

The labrum is large and oval, its front outline is evenly convex. The antennule is rather thick, bearing on its distal end the sensory papillae which are of uniform length and ca. of the same length as the antennule. The sensory seta is placed distally of the middle of the antennule. The antenna is rather short, with setae 0-0-3/0-1-3. There is a spine on the basal segment (of a branch having 3 setae), the end of this spine reaching the middle of the distal segment.

The post-abdomen is rather wide, slightly nar-

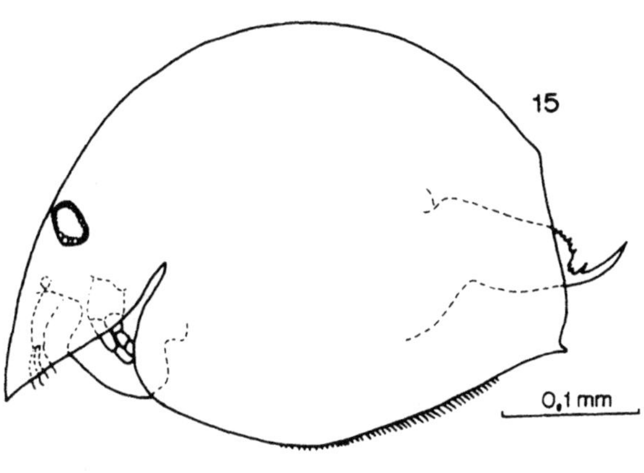

Fig. 15. ♀ *Alona broaensis* T. Matsumura-Tundisi et N. N. Smirnov sp. nov. Left-side lateral view. Broa Reservoir (São Paulo State, Brazil) 1981.3.30. The scale line corresponds to 0.1 mm. Holotype: 4000.

rowing distally. It is angular between the anal and post-anal margin. There are groups of small anal teeth along its post-anal margin, smaller teeth continue onto the anal margin. There are fascicles of long lateral setae on the postabdomen. The claw is long. There are minute setae along the concave side of the claw. The basal spine of the claw is thin and long, reaching the middle of the claw. There is a small proximal basal spine, its length approximately equals the widt of the claw in the place of its position.

Thoracic limb I has two setae on its external

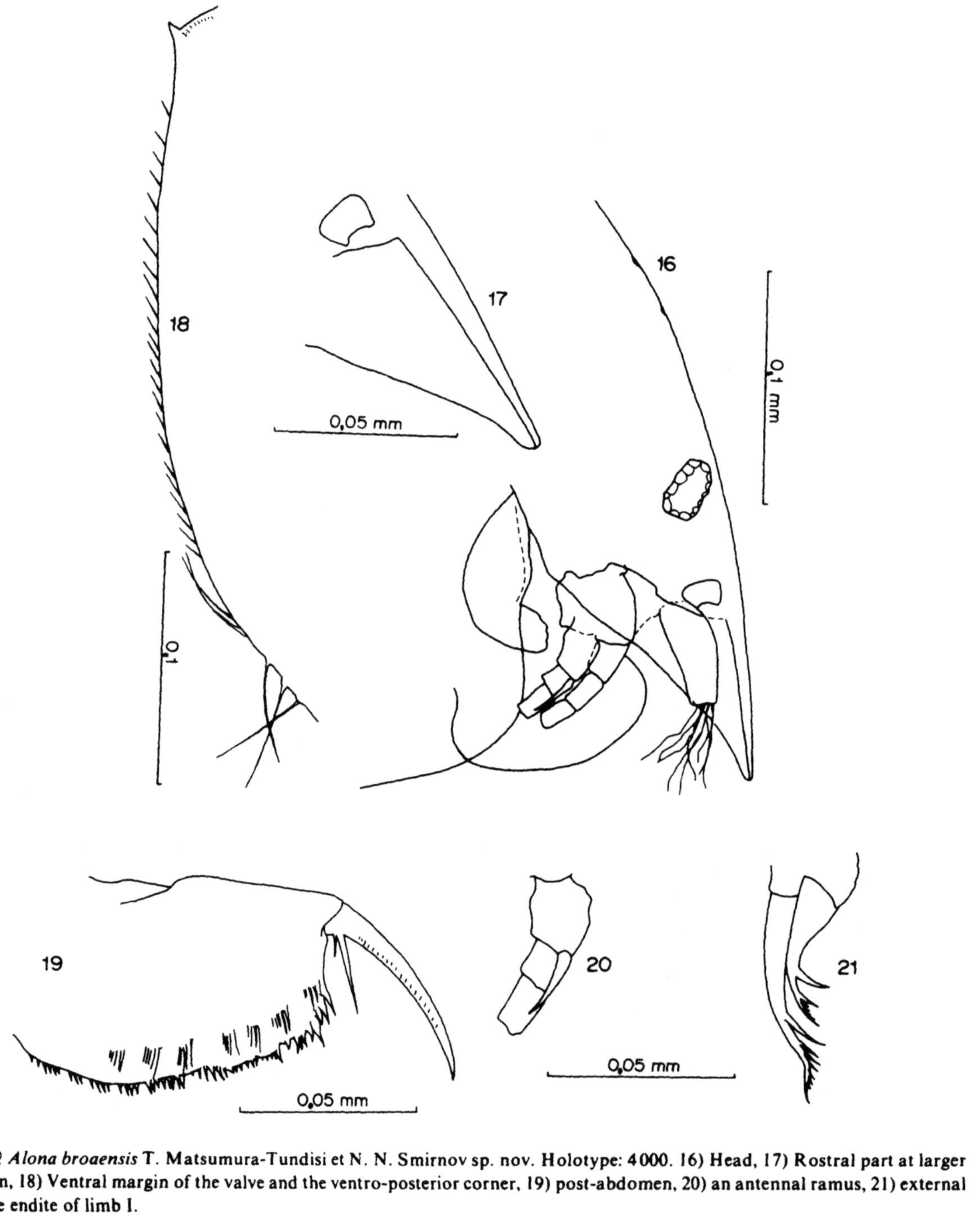

Fig. 16-21. ♀ ***Alona broaensis*** **T. Matsumura-Tundisi et N. N. Smirnov sp. nov. Holotype: 4000. 16) Head, 17) Rostral part at larger magnification, 18) Ventral margin of the valve and the ventro-posterior corner, 19) post-abdomen, 20) an antennal ramus, 21) external branch of the endite of limb I.**

branch of the endite. These setae are unilaterally setulated, the proximal setule being the largest and hook-like. The epipodites of limbs IV and V have long out-growths.

The exopodite of limb IV has 6 setae (on the left limb, while as a probable abnormal exception there are 5 setae on this exopodite in the right limb, in the female on slide 4001).

The ocellus is smaller than the eye, situated closer to the eye than to the tip of the rostrum. The intestine makes convolutions.

The head pores could not be discerned clearly in the female in slide 4000 (the holotype), and in the female in slide 4001 (both being in lateral position).

Length. 0.4 mm (holotype 4000).
Male. Unknown.

Differential diagnosis. This form is rather similar to *Alonella brasiliensis* Bergamin, 1935. The similar features are: short, high body, small size, setation of the ventral margin of the valve, large and oval labral plate, the presence of denticles on the inferior-posterior corner of the valve, a spine on a basal segment of an antennal ramus, the rounded tip of the rostrum.

Alona broaensis T. Matsumura-Tundisi et N. N. Smirnov sp. nov., differs, however, in the form of the anal teeth, the presence of fascicles of the lateral setae, the presence of a minute proximal basal spine, in the form of the denticle on the posterior-inferior corner of the valve, in larger spine on a basal segment of an antennal ramus.

It is attributed to *Alona* genus by reason of the exopodite of limb IV having 6 setae.

Etymology. The species name is given due to its occurrence in the Broa Reservoir (São Paulo State, Brazil).

Reference

Bergamin, F., 1935. Uma nova especie de Cladocera encontrade na Directoria de Industria Animal. Revta Ind. anim.: 284–285.

Epilogue: The future of tropical zooplankton studies

In order to assess the state-of-the-art of research on tropical zooplankton and to decide what directions they would like to see future developments take, the participants were divided into discussion groups of 8–12 persons. The deliberations took one day.

One group discussion topic was production. It was found that there is not enough data on growth rates, body weights and life cycle duration in tropical species of zooplankton. What is available comes mainly from Africa (e.g. 7 species of Cladocera and 5 species of Copepoda from Lake Chad) and much less from other tropical regions. Without information on instar body weights, it is impossible to obtain good estimates of biomass of populations; without data on instar or juvenile duration of life at various temperatures, it is impossible to obtain good estimates of secondary production of herbivorous or carnivorous species. Without these two (B and P), one cannot work towards a *quantitative* estimate of trophic zooplankton interactions.

The group recommended that more attention should be given to the measurement of instar body weights and duration of development in order to calculate growth rates, secondary production and biomass. The following species should be given priority: *Thermocyclos incognitus, Argyrodiaptomus furcatus, Ceriodaphnia cornuta, Moina micrura, Moina minuta, Daphnia gessneri, Daphnia laevis* and *Diaphanosoma brachyurum* from South America and Asia. Information on mean temperatures and min–max values of the waterbodies from which the species come should also be collected in order to produce an initial comparison of tropical species and a comparison of tropical species with temperate species (an unpublished manuscript of the Plankton Ecology Group of SIL deals with the same subject).

Another group addressed problems of succession and seasonality as determinants of community structure. It found that the degree of seasonality in tropical zooplankton is generally low. It is difficult to relate seasonality in phytoplankton with zooplankton, for lack of good data. There are no good analyses of clearance of phytoplankton by zooplankton. The nannoplankton assemblage is always more or less stable.

Problems of sampling were discussed next and the need for frequent sampling and for studying multi-annual fluctuations was stressed. Sampling should be adapted to the problem under study. Practical experience is very important in designing a sampling program.

Optimum use should be made of the samples, the 'raw material' such data as numbers, sizes, numbers of eggs, but also colour of eggs etc. should be obtained from one set of samples.

The group also discussed the subject of lake processes. Comparative work in tropical lakes indicates that processes are probably the same in many lakes, but their rates are different. Routine work may give some insight into those processes, but detailed and experimental work is needed if more is to be learned about them. Information on metabolic processes and rates for dominant organisms should be a priority. In particular, the effects of fish on the metabolism of lakes, and on the zooplankton specifically (by their feeding, excretion, etc.) should be given ample attention.

Finally, the group recommended that, in the

management of lake and reservoir ecosystems, and in the conservation of important waterbodies, the zooplankton should be part of an ecosystem approach, in which pure and applied limnologists should cooperate with experts from other disciplines.

Still another group discussed taxonomy, geographical distribution, and related problems. It was stressed that taxonomy should not be regarded as some sort of a hobby, but that species need to be described and named because they exist and have specific functions in their ecosystems. Therefore, incorrect naming of species will lead (and has led) to conflicting ecological results.

For the tropics, there is a chronic scarcity of good keys. This results in a literature that contains innumerable gross mistakes and often cannot be trusted. Editors of scientific journals should not accept papers that are very substandard from a taxonomical point of view. However, they do not always have control over this aspect of a paper.

Part of the problem is due to use of foreign keys (European and North American) in tropical regions. Students should not use such keys for other than exploratory identifications. Some local keys have been produced (e.g. by the University of Waterloo Biology series, usable in countries of South-East Asia and the Caribbean). Such efforts should be greatly stimulated in the future. Keys, while being simple, must still be *accurate*; if too general or too incomplete, they will not improve the situation. Another problem is that, in many plankton groups, major taxonomical developments have recently taken place, or are still taking place. These developments are often known only to a limited number of experts and centres and take a long time to become known and assimilated in the Third World. The representatives from Latin America remarked that, in their countries, problems related not only to taxonomy, but also to eutrophication and fisheries are not yet a governmental priority (unlike Africa and Asia, where freshwater fisheries are a priority now), and therefore they need help from foreign experts, and the possibility to travel to their laboratories, taking there own material to study. Great stress was laid on the necessity of increasing the number of training programs and centres now in operation, and of publicizing their existence. The experts present all agreed to host and train young taxonomists from Third World countries (alphabetically: Dumont, Ghent; Dussart, Les Eyzies; Fernando, Waterloo; Green, London; Smirnov, Moscow).

There is no adequate knowledge of the distribution of groups other than Cladocera, Copepoda and Rotifera. Ciliata may be important in tropical lakes, as indicated by a few documented cases, but the evidence is too scarce to be usable.

In Cladocera and Copepoda, it seems that instead of an increase in the number of limnetic species towards the equator, there is a decrease, with some genera (*Daphnia, Daphniopsis, Bosmina*) being lost or becoming extremely rare, especially in equatorial lowland areas. In altitudinal lakes, some of the diversity typical of temperate lakes is regained, but perhaps not all of it. This situation is locally complicated by different factors. In the African Sahel, for example, the transition of Saharan to Equatorial faunas is marked by a narrow zone with a high diversity.

Marked changes in species composition have been observed over historical time periods in some lakes, with several *Daphnia* species disappearing from Ugandan lakes where they used to be common (Green). Some tropical lakes have peculiar faunas, for example pelagic Ostracods, pelagic mites, or pelagic flatworms. In general, lakes which lose their *Daphnia* species tend to develop enriched rotifer faunas. Yet, rotifers as a group do not seem to increase in diversity towards the equator either. *Brachionus* and *Hexarthra* species tend to be more common in the tropics, and in South America *Trichocerca* appears to be remarkably common in the plankton of lakes. A typical sample from an average lake might contain about 6–8 rotifer species. If integrated over a year, a tropical lake might contain 30–50 species, but a temperate one 90–100. This scarcity needs to be investigated further. Possible causes are a lack of seasonality, as mentioned also by group 1, a higher and uninterrupted predation by a more diverse fish and insect fauna of the tropics (cf. recommendations of group 2), competition, intrazooplankton predation, lack of stimuli for going into diapause in some species, toxicity of algae, etc.

The conclusions of this working group at this point were that the causes of the discrepancy between the diversity of plankton and of fishes/insects in the tropics should be investigated in detail, and that there is a definite need for more data, especially in tropical South America (all altitudes), the Central African lowlands, and several areas in Asia (e.g. Burma and Cambodia).

With respect to the accelerating human impact and interference with tropical systems, it was stressed that there are probably many lakes that have not been investigated at all, but are losing species and may have lost their populations and species before anyone will have had a chance to look at them, unless surveying efforts are increased. As stressed by other discussion groups as well, there is thus a need for *simple* surveys, but covering as many lakes as possible, in order to enlarge the body of baseline data, in conjunction to a series of in-depth investigations.

All this wisdom, and the recommendations emanating therefrom, were subjected to a final plenary forum. There was general agreement on the numerous points brought forward. However, it was stressed that, before all other items, *training should be intensified* and scientific meetings multiplied. To achieve this priority, barriers of communication (by language, by distance) and of implementation (by lack of governmental interest, by bureaucratic inertia, or simply by lack of money) should be overcome. These last points, then, were unanimously adopted as the greatest priority for the immediate future.

Henri J. Dumont & Jose G. Tundisi

General index

Author index

GPSR Compliance

The European Union's (EU) General Product Safety Regulation (GPSR) is a set of rules that requires consumer products to be safe and our obligations to ensure this.

If you have any concerns about our products, you can contact us on ProductSafety@springernature.com

In case Publisher is established outside the EU, the EU authorized representative is:

Springer Nature Customer Service Center GmbH
Europaplatz 3
69115 Heidelberg, Germany

Batch number: 10800569

Printed by Printforce, the Netherlands